Carl Nordling Jonny Österman

Physics Handbook

for Science and Engineering

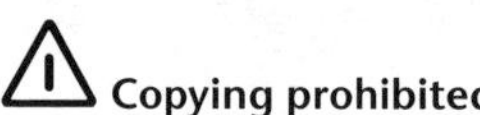

The papers and inks used in this product are eco-friendly.

Art. No 1657
ISBN 978-91-44-04453-8
Edition 8:13

www.studentlitteratur.se
Studentlitteratur AB, Lund, Sweden

Cover design: Tomas Dahlgren
Cover illustration: Erik Nordling

Printed by Dimograf, Poland 2014

‘Only by revaluation of the new era
and new extensions of the tensor calculus
was it possible to find a means
of discovering the split-up symmetry
which through the formula of five divided by three
was simplified and turned to real advantage
in every tour made by the Gopta chariot.’

*Excerpt from Harry Martinson’s epic ”Aniara”,
song 62, translated by H. Macdiarmid and
E. H. Schubert*

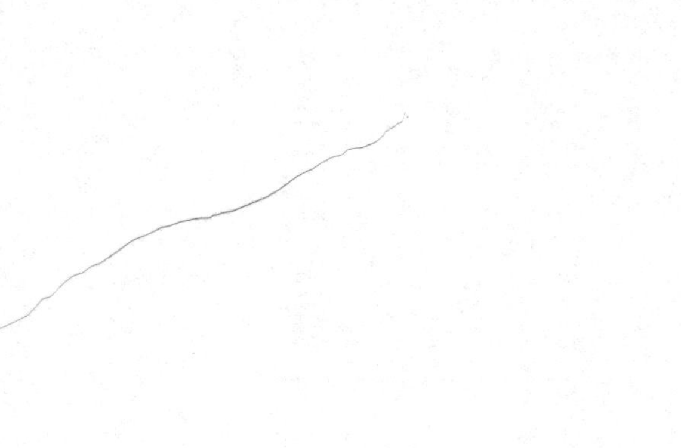

Contents

Foreword

For some time Swedish students and their teachers have had the benefit of the PHYSICS HANDBOOK. It has been used as a standard source of numerical data and formulae for exercise problems, and has been used as a companion in problem solving by both students and tutors. There are some who would decry reducing physics to a collection of data and formulae, but few would deny the need to have ready access to the basic relationships defined and expressed in these formulae. For many students on degree courses struggling with a deluge of new ideas, detailed arguments, and results of varying significance, a collection of important and useful formulae in each area of a subject can focus student awareness on those formulae and results which are of basic importance. It can help the student in the vital process of establishing pattern and order in the vast network of ideas in physics and reducing this to a manageable perspective in which he can find his own way around.

From experience in Sweden it can certainly be recommended as a valuable companion throughout a degree course and as a useful reference afterwards. Teachers may find it useful to recommend as a compact source to help students to learn to identify and select data and formulae for themselves instead of "spoonfeeding" them in each exercise problem.

For many years a close relationship has existed between Swedish and British universities, involving the exchange of students and faculty. I hope that this book will contribue to extend the benefit we have found from sharing our ideas in scientific education and resources.

Peter Unsworth
M.A. Ph D. (Cantab), Lecturer in Physics.
University of Sussex,
England

Preface

In physics and astrophysics one often deals with inconceivably large and incredibly small quantities. Thus, the proportion between the masses of the sun and the hydrogen atom is approximately 1 000 000 000 000 000 000 000 000 000 000 000 000 000 000 000 000 000 000 000 : 1 or 10^{57} : 1, but yet they follow the same laws of nature, for example the law of gravitation. The electric conductivity in different materials may differ by almost equally large amounts, but the physical laws which they have to abide are still the same.

In this PHYSICS HANDBOOK you will find tables and formulae which summarize the entire world of physics. The book consists of four major parts, namely CU, T, F, and M. CU for Constants and Units, T for Tables, F for Formulae, and M for Mathematics. In CU you will find the numerical values (in SI units as well as other suitable units) of the fundamental constants of physics, numerical values of non-SI units etc. Part T contains data on mechanical, thermal, electrical, atomic, nuclear, solid state and other properties of numerous materials, including the elements of the Periodic Table and astronomical objects. A wide selection of physical formulae is given in Part F, each chapter beginning with a list of notations used in that specific chapter. Other symbols are either fundamental constants, or they are carefully described in each section. The most important mathematical formulae are found in Part M.

Many formulae are described in some special case and in a more generalized form, so the book should suit the beginner as well as the more advanced student and professional physicist. Furthermore we have tried to include a sufficient amount of explanatory text to make this book useful as a reference book long after graduation, when the knowledge you once acquired is beginning to escape your active memory.

Throughout the book we have allowed ample space for your own notes. There are also a number of blank pages at the end for this purpose.

With the help of the extensive index you will have no difficulty in quickly locating the information you need in this PHYSICS HANDBOOK.

We want to express our gratitude to all those who have contributed valuable suggestions and observations during the course of our work. The support received from the Board of Pedagogical Development at the University of Uppsala is gratefully acknowledged.

Uppsala, Sweden, August 1980

Carl Nordling *Jonny Österman*

Minor additions and alterations have been made in the second edition, 1982, in the third edition, 1985, in the fourth edition, 1987, and in the fifth edition, 1996.

Two more appendices and several new or revised tables were included in the sixth edition, 1999. In the seventh edition, 2003, we updated several tables, added appendix F and a new chapter F–12 on solid mechanics. Professor Tore Dahlberg, university of Linköping, assumed the task of writing this chapter.

In the eighth edition, 2006, we have added a section on general relativity and made minor changes and additions throughout the book. Internet addresses have a tendency to change, so we have updated many of the references, and a number of new references has been added.

References to internet sites are tagged with the following symbol.

Information on Physics Handbook, including suggested changes and corrections can be found at

http://www.studentlitteratur.se/physicshandbook

If we find new interesting material we will notify you by upgrading this site.

C. N. *J. Ö.*

CU

Constants and Units

1 Fundamental Constants of Physics

1.1 Fundamental Constants – Numerical Values

The one-standard-deviation uncertainty is 0.13 % for the gravitational constant and at most nine units in the last digit for other constants. The speed of light is exact.

More information: http://physics.nist.gov/cuu/Constants

Quantity	Symbol	Relation to other constants	Value	Unit
Empty Space				
Speed of light	c_0		$2.997\,924\,58 \cdot 10^{8}$	m/s
Permeability	μ_0		$4\pi \cdot 10^{-7}$	Vs/Am
Permittivity	ε_0	$= 1/(\mu_0 c_0^2)$	$8.854\,187\,817 \cdot 10^{-12}$ ($\approx 10^{-9}/36\pi$)	As/Vm
Gravitation				
Gravitational constant	G		$6.6726 \cdot 10^{-11}$	Nm^2/kg^2
Acceleration of gravity at sea level (norm at 45° latitude)	g		9.806 65	m/s^2
Particle Masses				
Electron rest mass and rest energy	m_e		$9.109\,3819 \cdot 10^{-31}$	kg
			$5.485\,799\,11 \cdot 10^{-4}$	u
	$m_e c_0^2$		$5.109\,9890 \cdot 10^{5}$	eV
Muon rest mass and rest energy	m_μ		$1.883\,5311 \cdot 10^{-28}$	kg
			$1.134\,289\,17 \cdot 10^{-1}$	u
	$m_\mu c_0^2$		$1.056\,583\,57 \cdot 10^{8}$	eV
Proton rest mass and rest energy	m_p		$1.672\,6216 \cdot 10^{-27}$	kg
			1.007 276 467	u
	$m_p c_0^2$		$9.382\,7200 \cdot 10^{8}$	eV

Quantity	Symbol	Relation to other constants	Value	Unit
Ratio m_p/m_e	m_p/m_e		1836.152 667	
Neutron rest mass and rest energy	m_n		$1.674\,9272 \cdot 10^{-27}$	kg
			1.008 664 916	u
	$m_n c_0^2$		$9.395\,6533 \cdot 10^{8}$	eV
Ratio m_n/m_e	m_n/m_e		1838.683 655	
Atomic mass constant*	m_u		$1.660\,5387 \cdot 10^{-27}$	kg
			$9.314\,9413 \cdot 10^{8}$	eV
			1	u
Atomic Quantities				
Classical electron radius	r_e	$= e^2/4\pi\varepsilon_0 m_e c_0^2$	$2.817\,940\,28 \cdot 10^{-15}$	m
Bohr radius	a_0	$= h^2\varepsilon_0/\pi m_e e^2$	$5.291\,772\,08 \cdot 10^{-11}$	m
			0.529 177 208	Å
Nuclear radius constant	r_0		$(1.2–1.4) \cdot 10^{-15}$	m
Compton wavelength of electron and proton	λ_e	$= h/m_e c_0$	$2.426\,310\,22 \cdot 10^{-12}$	m
	λ_p	$= h/m_p c_0$	$1.321\,409\,85 \cdot 10^{-15}$	m
Hydrogen atom ground state energy (with negative sign)	E_H	$= e^2/8\pi\varepsilon_0 a_0$	13.605 692	eV
			$1.312\,750 \cdot 10^{6}$	J/mol
Electric Charge and Magnetic Moment				
Elementary charge	e		$1.602\,176\,46 \cdot 10^{-19}$	C
			$1.602\,176\,46 \cdot 10^{-20}$	emu
			$4.803\,2056 \cdot 10^{-10}$	esu
Electron charge-to-mass ratio	e/m_e		$1.758\,820\,17 \cdot 10^{11}$	C/kg
			$1.758\,820\,17 \cdot 10^{7}$	emu/g
			$5.272\,81 \cdot 10^{17}$	esu/g
Bohr magneton	μ_B	$= e\hbar/2m_e$	$9.274\,0090 \cdot 10^{-24}$	J/T (=Am^2)
			$9.274\,0090 \cdot 10^{-21}$	erg/G
			$5.788\,381\,75 \cdot 10^{-5}$	eV/T
			0.466 864 52	$cm^{-1}T^{-1}$
Nuclear magneton	μ_N	$= e\hbar/2m_p$	$5.050\,7832 \cdot 10^{-27}$	J/T (=Am^2)
			$5.050\,7832 \cdot 10^{-24}$	erg/G
			$3.152\,451\,24 \cdot 10^{-8}$	eV/T

* $1\ u = \frac{1}{12}$ of the atomic mass of ^{12}C.

Quantity	Symbol	Relation to other constants	Value	Unit
Electron magnetic moment	μ_e	$= g_e s \mu_B$	$-9.284\,7636 \cdot 10^{-24}$	J/T (=Am^2)
			$-1.001\,159\,652\,187$	μ_B
Proton magnetic moment	μ_p	$= g_p s \mu_N$	$1.410\,606\,63 \cdot 10^{-26}$	J/T (=Am^2)
			$2.792\,847\,337$	μ_N
Neutron magnetic moment	μ_n		$-1.913\,042\,72$	μ_N
Muon magnetic moment	μ_μ		$4.490\,4481 \cdot 10^{-26}$	J/T
Quantum Physics, Radiation				
Planck's constant	h		$6.626\,0688 \cdot 10^{-34}$	Js
			$6.626\,0688 \cdot 10^{-27}$	erg s
			$4.135\,6673 \cdot 10^{-15}$	eV s
	$\hbar$	$= h/2\pi$	$1.054\,571\,60 \cdot 10^{-34}$	Js
			$6.582\,1189 \cdot 10^{-16}$	eV s
Ratio	h/e		$4.135\,6673 \cdot 10^{-15}$	Vs
Product hc_0, conversion factor of quantum energy	$E \cdot \lambda$	$= hc_0$	$1.986\,4454 \cdot 10^{-25}$	Jm
			$1.239\,8419 \cdot 10^{-6}$	eV m
			$1.239\,8419 \cdot 10^{4}$	eV Å
			$1.2373 \cdot 10^{4}$	keV XU
Rydberg constant of infinite mass	R_∞	$= m_e e^4 / 8\,\varepsilon_0^2\, h^3 c_0$	$1.097\,373\,1569 \cdot 10^{7}$	m^{-1}
Fine-structure constant	α	$= e^2 / 4\pi\varepsilon_0 \hbar c_0$	$7.297\,352\,533 \cdot 10^{-3}$	
and inverted value	$1/\alpha$	$= 4\pi\varepsilon_0 \hbar c_0 / e^2$	$137.035\,999\,76$	
Stefan-Boltzmann constant	σ	$= 2\pi^5 k^4 / 15 h^3 c_0^2$	$5.670\,40 \cdot 10^{-8}$	W/m^2K^4
and constant of energy density	a	$= 4\sigma / c_0$	$7.565\,77 \cdot 10^{-16}$	J/m^3K^4
Constants in Wien's displacement law	b_ν		$5.8786 \cdot 10^{10}$	$s^{-1}K^{-1}$
	b_λ		$2.8978 \cdot 10^{-3}$	m K
Boltzmann constant	k (k_B)	$= R/N_A$	$1.380\,650 \cdot 10^{-23}$	J/K
			$8.617\,342 \cdot 10^{-5}$	eV/K
Lorenz constant	L	$= \pi^2 k^2 / 3e^2$	$2.443\,01 \cdot 10^{-8}$	V^2/K^2

Quantity	Symbol	Relation to other constants	Value	Unit
Planck length	ℓ_P	$= \sqrt{G\hbar / c_0^3}$	$1.616 \cdot 10^{-35}$	m
Planck mass	m_P	$= \sqrt{\hbar c_0 / G}$	$2.177 \cdot 10^{-8}$	kg
Planck time	t_P	$= \sqrt{G\hbar / c_0^5}$	$5.391 \cdot 10^{-44}$	s
Planck temperature	T_P	$= \sqrt{\hbar c_0^5 / G} / k$	$1.417 \cdot 10^{32}$	K
Planck density	ρ_P	$= c^5 / \hbar G^2$	$5.159 \cdot 10^{96}$	kg/m^3
Planck energy	E_P	$= \sqrt{\hbar c^5 / G}$	$1.956 \cdot 10^{9}$	J
			$1.221 \cdot 10^{19}$	GeV

Quantities Related to Amount of Substance

Quantity	Symbol	Relation to other constants	Value	Unit
Avogadro constant	N_A	$= 1 / m_u$	$6.022\,1420 \cdot 10^{23}$	molecules per mol
Molar volume of ideal gas at NTP (0° C, 1 atm)	V_0		22.414 00	dm^3/mol
Molar gas constant	R	$= kN_A$	$8.314\,472 \cdot 10^{3}$ 8.314 472 $8.206 \cdot 10^{-5}$	J/kmol K J/mol K m^3 atm/mol K
Faraday constant	F	$= eN_A$	$9.648\,5341 \cdot 10^{4}$	C/mol

Metrological Constants

Quantity	Symbol	Relation to other constants	Value	Unit
Josephson constant for electric voltage	K_j	$= 2e/h$	$4.835\,9790 \cdot 10^{14}$	Hz/V
von Klitzing constant for electric resistance (at plateau $i = 1$)	R_k	$= h/e^2$	$2.581\,280\,70 \cdot 10^{4}$	Ω

2 Units of Physics

2.1 SI Units

Symbol	Name of unit	Definition	Quantity measured
Basic Units			
m	metre	see CU – 2.2	length
kg	kilogram	see CU – 2.2	mass
s	second	see CU – 2.2	time
A	ampere	see CU – 2.2	electric current
K	kelvin	see CU – 2.2	thermodynamic temperature
cd	candela	see CU – 2.2	luminous intensity
mol	mole	see CU – 2.2	amount of substance
Supplementary Units			
rad	radian	see CU – 2.2	plane angle
sr	steradian	see CU – 2.2	solid angle
Derived Units			
Hz	hertz	$1\ \text{Hz} = 1\ \text{s}^{-1}$	frequency
N	newton	$1\ \text{N} = 1\ \text{kg m/s}^2$	force
J	joule	1 J = 1 Nm	energy
W	watt	1 W = 1 J/s	power
Pa	pascal	$1\ \text{Pa} = 1\ \text{N/m}^2$	pressure
V	volt	1 V = 1 W/A	voltage
C	coulomb	1 C = 1 As	electric charge
Ω	ohm	1 Ω = 1 V/A	resistance
F	farad	1 F = 1 C/V	capacitance
H	henry	1 H = 1 Ωs	inductance
S	siemens	1 S = 1 A/V	electric conductance
Wb	weber	1 Wb = 1 Vs	magnetic flux
T	tesla	$1\ \text{T} = 1\ \text{Wb/m}^2$	magnetic flux density
°C	degree Celsius	1 °C = 1 K*	temperature
Bq	becquerel	$1\ \text{Bq} = 1\ \text{s}^{-1}$	radioactivity
Gy	gray	1 Gy = 1 J/kg	absorbed dose of ionizing radiation

* 1 °C = 1 K is for temperature differences, 0 °C = 273.15 K.

Symbol	Name of unit	Definition	Quantity measured
Sv	sievert	$1\ \text{Sv} = 1\ \text{J/kg}$	dose equivalent
lm	lumen	$1\ \text{lm} = 1\ \text{cd} \cdot \text{sr}$	light flux
lx	lux	$1\ \text{lx} = 1\ \text{lm/m}^2$	illuminance

Additional Units

ℓ, L	litre	$1\ \ell = \frac{1}{1\,000}\ \text{m}^3 = 1\ \text{dm}^3$	volume
min	minute	$1\ \text{min} = 60\ \text{s}$	time
h	hour	$1\ \text{h} = 60\ \text{min}$	time
d	day	$1\ \text{d} = 24\ \text{h}$	time
t	ton(metric)	$1\ \text{ton} = 1\,000\ \text{kg}$	mass
°	degree	$1° = \frac{\pi}{180}\ \text{rad}$	plane angle
′	minute	$1' = \frac{1°}{60}$	plane angle
″	second	$1'' = \frac{1'}{60}$	plane angle
g	gon	$1\ \text{g} = \frac{\pi}{200}\ \text{rad}$	plane angle
bar	bar	$1\ \text{bar} = 10^5\ \text{Pa}$	pressure

2.2 Definition of SI Units

Definition of Basic Units

1 metre = the length of the path travelled by light in vacuum during a time interval of 1/299 792 458 of a second. (17th CGPM, 1983, Resolution 1)

1 kilogram = the mass of the international prototype of the kilogram. (3rd CGPM, 1901)

1 second = the duration of 9 192 631 770 periods of the radiation corresponding to the transition between the two hyperfine levels of the ground state of the cesium-133 atom. (13th CGPM, 1967, Resolution 1)

1 ampere = that constant current which, if maintained in two straight parallel conductors of infinite length, of negligible circular cross-section, and placed 1 metre apart in vacuum, would produce between these conductors a force equal to $2 \cdot 10^{-7}$ newton per metre of length. (9th CGPM, 1948)

1 kelvin = the fraction 1 / 273.16 of the thermodynamic temperature of the triple point of water. (13th CGPM, 1967)

1 candela = the luminous intensity, in a given direction, of a source that emits monochromatic radiation of frequency $540 \cdot 10^{12}$ hertz and that has a radiant intensity in that direction of 1 / 683 watt per steradian. (16th CGPM, 1948)

1 mole = the amount of substance of a system which contains as many elementary entities as there are atoms in 0.012 kilogram of carbon-12. (14th CGPM, 1971)

Definition of Supplementary Units

1 radian = the size of the plane angle between two radii in a circle which on the circumference cuts an arc with the length of the radius.

1 steradian = the size of a solid angle for a cone which has its apex at the centre of a sphere and cuts a surface of this sphere with an area which is equal to the area of a square with a side that has the same length as the radius of the sphere.

2.3 Conversion Factors of Non-SI Units

Many of the units below have been annexed to the SI system as multiple units, in which cases the conversion factors are exact by definition.

Length

1 Å (ångström) = 10^{-10} m
1 XU (x-unit) = 1.002 08 mÅ = $1.002\,08 \cdot 10^{-13}$ m
1 fermi = 10^{-15} m
1 au (atomic unit) = 1 bohr = 1 a_0 = 0.529 177 Å
1 AU (astronomical unit) = $1.495\,978\,70 \cdot 10^{11}$ m
(The astronomical unit is sometimes abbreviated ua.)
1 light-year* = $9.460\,55 \cdot 10^{15}$ m = $6.32 \cdot 10^{4}$ AU
1 pc (parsec) = $3.0857 \cdot 10^{16}$ m = $2.062\,65 \cdot 10^{5}$ AU = 3.262 light-year
1 nautical mile = 1852 m
1 English mile = 1609.344 m
1 yard = 0.9144 m
1 ft (foot) = 12 in (inches) = 0.3048 m
1 in (inch) = 2.54 cm
1 Swedish “fot” = 0.296 90 m

* A corresponding unit, 1 beard-second = 10^{-8} m, has been suggested for microscopic distances. One beard-second is the distance which a standard beard in a standard face grows in one second. Its chances of ever becoming an established unit in physics seem rather slim though.

Wave number	1 kayser = 1 cm^{-1}
Area	1 barn = 10^{-28} m^2 1 Swedish “tunnland” = 4 936 m^2 1 acre = 4 046.86 m^2 1 ha (hectare) = 10 000 m^2 1 in^2 (square inch) = 0.645 16 · 10^{-3} m^2 1 ft^2 (square foot) = 92.903 04 · 10^{-3} m^2 1 yd^2 (squqare yard) = 0.836 127 36 m^2 1 square mile = 2.589 988 1 · 10^6 m^2
Volume	1 barrel (British) = 163.66 ℓ 1 barrel (US petroleum) = 158.987 294 958 ℓ 1 barrel (US liq) = 119.24 ℓ 1 gallon (British) = 4.546 09 ℓ 1 gallon (US liq) = 3.785 411 784 ℓ 1 pint (British) = 20 Brit. fluid ounces = 0.568 25 ℓ 1 pint (US dry) = 0.550 60 ℓ 1 pint (US liq) = 16 US fluid ounces = 0.473 16 ℓ 1 cu in (cubic inch) = 0.016 387 064 ℓ 1 cu ft (cubic foot) = 28.316 847 ℓ 1 cu yd (cubic yard) = 0.764 554 86 m^3
Plane angle	1° = π/180 rad = 1/57.2958 rad = 0.017 453 3 rad 1′ (minute) = π/10 800 rad = 0.290 888 21 · 10^{-3} rad 1″ (second) = π/648 000 rad = 4.848 136 8 · 10^{-6} rad 1 gon = π/200 rad = 15.707 963 · 10^{-3} rad
Time	1 tropical year (solar year) = 31.556 925 974 · 10^6 s = = 365.242 198 78 d 1 sidereal year (stellar year) = 31.558 150 · 10^6 s = = 365.256 37 d 1 calender year = 365 d = 8 760 h = 31.536 · 10^6 s 1 leap-year = 366 d = 8784 h = 31.6224 · 10^6 s 1 au (atomic unit) = 2.418 884 · 10^{-17} s
Speed	1 mph (mile per hour) = 1.609 344 km/h = 0.447 04 m/s 1 knot = 1.852 km/h = 0.514 44 m/s 1 km/h = 1/3.6 m/s = 0.277 777 8 m/s 1 ft/s = 0.3048 m/s
Mass	1 u (atomic mass constant, atomic mass unit) = = 1.66054 · 10^{-27} kg 1 au (atomic unit) = 1 m_e = 9.109 38 · 10^{-31} kg 1 lb (pound) = 16 oz. (ounces) = 0.453 592 37 kg 1 stone = 14 lbs = 6.350 293 18 kg 1 slug = 0.453 592 37 · 9.806 65/0.3048 kg = 14.593 903 kg 1 metric ton = 1 000 kg

	1 ton (UK, in US long ton) = $1.016\ 046\ 908\ 8 \cdot 10^3$ kg 1 sh tn (short ton, US) = $0.907\ 184\ 74 \cdot 10^3$ kg 1 cwt (hundred-weight) = 50.802 345 44 kg 1 sh cwt (short hundred-weight, US) = 45.359 237 kg 1 carat (metric) = 200 mg
Density	1 amagat = 0.040 96 mole · dm^{-3} (ideal gas at STP) 1 lb/ft^3 = 16.0185 kg/m^3 1 lb/in^3 = $27.6799 \cdot 10^3$ kg/m^3
Temperature	x K = (x – 273.15) °C (degrees Celsius) x °C = ($x \cdot 9/5 + 32$) °F (degrees Fahrenheit) x °R (degrees Rankine) = (x – 459.67) °F (0 °R = 0 K) x °Réaumur = 1.25 · x °C
Energy	1 eV (electron volt) = $1.602\ 1765 \cdot 10^{-19}$ J 1 Ry (rydberg) = $2.1799 \cdot 10^{-18}$ J = 13.605 692 eV 1 au (atomic unit) = 1 H (hartree) = 2 Ry = $4.359 \cdot 10^{-18}$ J 1 erg = 10^{-7} J 1 kWh = $3.6 \cdot 10^6$ J 1 kcal = 1 000 cal = 4 186.8 J (= energy needed to raise the temperature of 1 kg of water from 14.5 °C to 15.5 °C when the pressure is 1 atm) 1 kcal/mol = $4.336 \cdot 10^{-2}$ eV/molecule 1 kiloton TNT = $4.18 \cdot 10^{12}$ J 1 Btu (British thermal unit) = 1 055.06 J 1 Q = 1 000 quad = 10^{18} Btu = $1.055\ 06 \cdot 10^{21}$ J
Power	1 hk (horsepower, metric) = 75 kpm/s = 735.50 W 1 hp (horsepower, UK and US) = 550 ft · lbf / s = = 745.70 W
Angular momentum	1 au (atomic unit) = 1 $\hbar$ = $1.054\ 571\ 6 \cdot 10^{-34}$ Js
Force	1 dyn = 10^{-5} N 1 kp (kilopond) = 9.806 65 N (conventional value) 1 lbf (pound force) = 4.4482 N = 0.453 59 kp
Pressure	1 torr (mm Hg) = $1.333\ 22 \cdot 10^2$ Pa 1 atm = 760 torr = $1.013\ 25 \cdot 10^5$ Pa 1 bar = 10^5 Pa 1 at = 1 kp/cm^2 = $9.806\ 65 \cdot 10^4$ Pa 1 psi (pound force per square inch, lbf/in^2) = $6.8948 \cdot 10^3$ Pa
Logarithmic units	1 B (bel) = (ln 10)/2 Np (neper) = 1.151 29 Np 1 dB (decibel) = (ln 10)/20 Np = 0.115 129 Np
Viscosity	1 P (poise) = 1 dyn s/cm^2 = 0.1 N s/m^2 (viscosity) 1 St (stokes) = 1 cm^2/s = 10^{-4} m^2/s (kinematic viscosity)

Electric dipole moment	1 au (atomic unit) = 1 $e\,a_0$ = 8.478 35 · 10^{-30} Cm 1 D (debye) = 10^{-18} esu = 3.335 64 · 10^{-30} Cm
Magnetic flux density	1 G (gauss) = 10^{-4} T 1 gamma = 10^{-5} G = 10^{-9} T
Magnetic flux	1 Mx (maxwell) = 10^{-8} Wb
Magnetizing field	1 oersted = $10^3/4\pi$ A / m
Magnetic dipole moment	1 au (atomic unit) = 1 $e\hbar/m_e = 2\,\mu_B$ = 1.8548 · 10^{-23} Am^2 (= J / T)
Activity	1 Ci (curie) = 3.7 · 10^{10} Bq (= 3.7 · 10^{10} s^{-1}) 1 Rd (rutherford) = 10^6 Bq
Exposure	1 R (röntgen) = 2.58 · 10^{-4} C/kg air (1 R ≈ 10^{-2} Sv)
Absorbed dose	1 rad = 10^{-2} Gy (= 10^{-2} J/kg)
Dose equivalent	1 rem = 10^{-2} Sv (= 10^{-2} J/kg)
Luminous intensity	1 hefner = 0.90 cd
Luminance	1 sb (stilb) = 10^4 cd/m^2 1 asb (apostilb) = $1/\pi$ cd/m^2 1 cd/ft^2 = 10.763 91 cd/m^2
Photometric brightness	1 lambert = $10^4/\pi$ cd / m^2
Gas exposure	1 L (langmuir) = 10^{-6} torr · s = 1.33 · 10^{-4} Pa · s
Gas fluence	1 ex = 10^{18} m^{-2} (suggested unit to replace the langmuir in surface physics)
Concentration of a solution	1 M (molarity) = 1 mole of a solute per litre solution 1 m (molality) = 1 mole of solute per kilogram solvent

2.4 Conversion Factors of Certain Units

Mass – Energy

1	2	3	4
kg	u	J	eV
1	6.022 142 · 10^{26}	8.987 552 · 10^{16}	5.609 589 · 10^{35}
1.660 539 · 10^{-27}	1	1.492 418 · 10^{-10}	9.314 940 · 10^{8}
1.112 650 · 10^{-17}	6.700 537 · 10^{9}	1	6.241 510 · 10^{18}
1.782 662 · 10^{-36}	1.073 544 · 10^{-9}	1.602 176 · 10^{-19}	1

Spectroscopical Units

1	2	3	4
Hz	cm^{-1}	Ry	eV
1	$3.335\,641 \cdot 10^{-11}$	$3.039\,660 \cdot 10^{-16}$	$4.135\,667 \cdot 10^{-15}$
$2.997\,924\,58 \cdot 10^{10}$	1	$9.112\,671 \cdot 10^{-6}$	$1.239\,842 \cdot 10^{-4}$
$3.289\,842 \cdot 10^{15}$	$1.097\,373 \cdot 10^{5}$	1	$1.360\,569 \cdot 10^{1}$
$2.417\,990 \cdot 10^{14}$	$8.065\,545 \cdot 10^{3}$	$7.349\,865 \cdot 10^{-2}$	1

Energy

Column 1 The absolute temperature corresponding to the energy kT is given.

1	2	3	4	5
K	kWh	kcal	J	eV
1	$3.835\,14 \cdot 10^{-30}$	$3.298 \cdot 10^{-27}$	$1.380\,65 \cdot 10^{-23}$	$8.617\,34 \cdot 10^{-5}$
$2.607\,47 \cdot 10^{29}$	1	$8.598 \cdot 10^{2}$	$3.600\,00 \cdot 10^{6}$	$2.246\,94 \cdot 10^{25}$
$3.0325 \cdot 10^{26}$	$1.1630 \cdot 10^{-3}$	1	$4.1868 \cdot 10^{3}$	$2.6132 \cdot 10^{22}$
$7.242\,97 \cdot 10^{22}$	$2.777\,78 \cdot 10^{-7}$	$2.388 \cdot 10^{-4}$	1	$6.241\,510 \cdot 10^{18}$
$1.160\,45 \cdot 10^{4}$	$4.450\,49 \cdot 10^{-26}$	$3.827 \cdot 10^{-23}$	$1.602\,176 \cdot 10^{-19}$	1

Pressure

1	2	3	4	5
Pa (N/m^2)	bar	kp/cm^2 (at)	torr (mm Hg, 0 °C)	atm (normal atmospheres)
1	10^{-5}	$1.020 \cdot 10^{-5}$	$7.5006 \cdot 10^{-3}$	$9.869 \cdot 10^{-6}$
10^{5}	1	1.020	$7.5006 \cdot 10^{2}$	$9.869 \cdot 10^{-1}$
$9.807 \cdot 10^{4}$	$9.807 \cdot 10^{-1}$	1	$7.3556 \cdot 10^{2}$	$9.678 \cdot 10^{-1}$
$1.333 \cdot 10^{2}$	$1.333 \cdot 10^{-3}$	$1.360 \cdot 10^{-3}$	1	$1.316 \cdot 10^{-3}$
$1.013 \cdot 10^{5}$	1.013	1.033	760	1

2.5 Prefixes Indicating Powers of Ten

10^{24}	yotta	Y	10^{-1}	deci	d
10^{21}	zetta	Z	10^{-2}	centi	c
10^{18}	exa	E	10^{-3}	milli	m
10^{15}	peta	P	10^{-6}	micro	µ
10^{12}	tera	T	10^{-9}	nano	n
10^{9}	giga	G	10^{-12}	pico	p
10^{6}	mega	M	10^{-15}	femto	f
10^{3}	kilo	k	10^{-18}	atto	a
10^{2}	hecto	h	10^{-21}	zepto	z
10^{1}	deca	da	10^{-24}	yocto	y

ppm means parts per million (10^6)
ppb means parts per billion (10^9, Am. English)
ppt means parts per trillion (10^{12})
ppq means parts per quadrillion (10^{15})

2.6 Prefixes for Binary Multiples

In December 1998 the Intrernational Electrotechnical Commission (IEC) approved an international standard for names and symbols for prefixes for binary multiples for use in the fields of data processing and transmission.

Factor	Name	Symbol	Example
2^{10}	kibi	Ki	one kibibit = 1 Kibit = 1024 bit
2^{20}	mebi	Mi	one mebibit = 1 Mibit = 1 048 576 bit
2^{30}	gibi	Gi	one gibibyte = 1 GiB = 1 073 741 284 B
2^{40}	tebi	Ti	1 TiB = 1 099 511 074 816 B
2^{50}	pebi	Pi	1 PiB = 1 125 899 340 611 584 B
2^{60}	exbi	Ei	1 EiB = 1 152 920 924 786 226 016 B

It is suggested that in English, the first syllable of the name should be pronounced in the same way as the first syllable of the corresponding SI prefix, and that the second syllable should be pronounced as "bee", for example pebi should be pronounced "pebee".

2.7 Atomic Units

When using atomic units $m_e = e = \hbar = a_0 = 1$, while $\alpha = e^2/4\pi\varepsilon_0\hbar c_0 = 1/137.036$ is dimensionless, and in particular $c_0 = \alpha^{-1}$ au = 137.036 au

$\varepsilon_0 = 1/4\pi$ au

$\mu_0 = 4\pi/c_0{}^2 = 4\pi\alpha^2$ in au

Quantity	**Atomic unit** (Numerical value = 1)	**Value and unit according to SI**	**Physical significance**
Length	$a_0 = 4\pi\,\varepsilon_0\hbar^2 / m_e\,e^2$	$5.291\,772\,1 \cdot 10^{-11}$ m	Bohr radius for atomic hydrogen.
Mass	m_e	$9.109\,382 \cdot 10^{-31}$ kg	Electron mass.
Time	$a_0/\alpha\,c_0$	$2.418\,884 \cdot 10^{-17}$ s	Time required for electron in first Bohr orbit to travel one Bohr radius.
Velocity	$\alpha\,c_0 = e^2/4\pi\,\varepsilon_0\hbar$	$2.187\,691 \cdot 10^{6}$ m/s	Electron velocity in first Bohr orbit.
Energy	$2\,h\,c_0\,R_\infty =$ 1 hartree $= \alpha^2\,m_e\,c_0{}^2$ $= e^2/4\pi\,\varepsilon_0\,a_0$	$4.359\,744 \cdot 10^{-18}$ J 27.2114 eV	Twice the ionisation energy of atomic hydrogen (with infinite nuclear mass).
Force	$2\,h\,c_0\,R_\infty/a_0$	$8.238\,72 \cdot 10^{-8}$ N	
Momentum and impulse	$m_e\,\alpha\,c_0$	$1.992\,85 \cdot 10^{-24}$ Ns	Electron momentum in first Bohr orbit.
Angular momentum	$\hbar$	$1.054\,572 \cdot 10^{-34}$ Js	
Electric charge	e	$1.602\,177 \cdot 10^{-19}$ C	Absolute value of electron charge.
Electric current	$e\,\alpha\,c_0/a_0$	$6.62362 \cdot 10^{-3}$ A	
Electric potential	$2\,h\,c_0\,R_\infty/e$	27.2114 V	
Electric field strength	$2\,h\,c_0\,R_\infty/e\,a_0$	$5.14220 \cdot 10^{11}$ V/m	
Electric dipole moment	$e\,a_0$	$8.478\,35 \cdot 10^{-30}$ Cm	
Dielectric constant	$4\pi\,\varepsilon_0$	$1.112\,65 \cdot 10^{-10}$ As/Vm	
Magnetic flux	$\hbar/e =$ $2\,h\,c_0\,R_\infty\,a_0/e\,\alpha\,c_0$	$6.582\,12 \cdot 10^{-16}$ Vs	
Magnetic flux density	$2\,h\,c_0\,R_\infty/e\,\alpha\,c_0\;a_0$	$2.350\,52 \cdot 10^{5}$ T	
Magnetic vector potential	$2\,h\,c_0\,R_\infty/e\,\alpha\,c_0$	$1.243\,84 \cdot 10^{-5}$ Vs/m	
Magnetic dipole moment	$2\,\mu_B = e\,\hbar/m_e$	$1.854\,80 \cdot 10^{-23}$ Am^2	
Permeability	$\mu_0/4\pi\,\alpha^2$	$1.877\,89 \cdot 10^{-3}$ Vs/Am	

T

Physical Tables

1 Mechanics and Thermal Physics

1.1 Solid Elements – Mechanical and Thermal Properties

Column 2	The density is given for 300 K and 0.1 MPa.
Columns 3, 4, 5	The figures refer to polycrystalline specimens.
Column 5	The linear thermal expansion as an average over the temperature range 0 °C – 100 °C is given.
Column 6	The specific heat capacity is given at constant pressure and the temperature 300 K. The molar heat capacitivity is obtained by multiplying with the relative atomic (molecular) mass.
Column 7	The thermal conductivity is given for 300 K.
Columns 8, 10	Melting and boiling points are given for the pressure of 0.1 MPa.
Column 9	Heat of fusion is also called enthalpy of fusion.
Column 11	Heat of vaporization is also called enthalpy of vaporization.

More information:

http://apamac.ch.adfa.oz.au/OzChemNet/web-elements/
http://www.allmeasures.com/Formulae

1	2	3	4	5	6	7	8	9	10	11
Element	**Density** 10^3 kg m^{-3}	**Young's modulus** 10^{10} Pa	**Shear modulus** 10^{10} Pa	**Thermal expansion** 10^{-6} K^{-1}	**Sp. heat capacity** J kg^{-1} K^{-1}	**Thermal conductivity** Wm^{-1} K^{-1}	**Melting point** K	**Heat of fusion** 10^3 J kg^{-1}	**Boiling point** K	**Heat of vaporization** 10^6 J kg^{-1}
aluminium	2.70	6.9	2.6	23.2	903	238	933	397	2 720	10.9
antimony	6.69	7.7	2.1	11.0	207	18	904	163	1 650	1.56
arsenic	5.73			6	330		883	433	1 090	1.86
barium	3.5	1.3	0.5	16.4	190		998	57	1 910	1.09
beryllium	1.85	30	14	11.5	1 825	230	1 551	1 384	3 240	32.6
bismuth	9.75	3.1	1.2	13.5	122	8.5	544	52	1 810	0.82
cadmium	8.65	5.0	2.0	31.5	232	92	594	57	1 037	0.95
calcium	1.55	2.0	0.8	22.3	658	98	1 115	228	1 765	3.75
carbon, diamond	3			1.1	509	1 000	3 900	17 000	4 600	50
graphite	2.25			8.8	711	150	3 800-4 000		4 500	50
chromium	7.2	2.5		8.5	448	87	2 160	280	2 915	6.15
cobalt	8.9	20	8.0	13.7	425		1 768	280	3 150	6.45
copper	8.96	12	4.6	16.8	385	400	1 356	205	2 855	4.75
dysprosium	8.54	0.64		9	173		1 680		2 900	
europium	5.26	0.15			176		1 099		1 712	
gadolinium	7.90			4	230		1 585		3 300	
gallium	5.91			19.2	375		303	80.1	2 676	3.64
germanium	5.32	8.1	3.1	5.7	322	60	1 211	480	3 100	4.6
gold	19.32	7.9	2.7	14.1	129	311	1 336	66	3 090	1.65
hafnium	13.29			6.0	144	22	2 420	140	5 700	3.72
indium	7.31	1.1	0.4	31.9	234	25	430	28	2 270	1.97
iodine	4.93			87	215		387	62	457.5	0.173
iridium	22.42	52	20	6.5	133	148	2 680	144	4 400	3.90
iron	7.87	21	8.4	11.7	449	82	1 808	276	3 160	6.80
lead	11.35	1.6	0.54	28.9	130	35	601	24.7	2 024	0.932

T – 1.1 Solid Elements – Mechanical and Thermal Properties

1	2	3	4	5	6	7	8	9	10	11
Element	**Density** 10^3 kg m^{-3}	**Young's modulus** 10^{10} Pa	**Shear modulus** 10^{10} Pa	**Thermal expansion** 10^{-6} K^{-1}	**Sp. heat capacity** J kg^{-1} K^{-1}	**Thermal conductivity** Wm^{-1} K^{-1}	**Melting point** K	**Heat of fusion** 10^3 J kg^{-1}	**Boiling point** K	**Heat of vaporization** 10^6 J kg^{-1}
lithium	0.534	0.49		47.0	3 570	71	452	420	1 590	24.5
magnesium	1.74	4.4	1.7	25.6	1 024	150	424	368	1 390	5.42
manganese	7.3	20	8	22.8	479		1 517	270	2 315	4.37
molybdenum	10.22	33	13	5.0	248	140	2 880	253	5 830	6.83
nickel	8.90	20	8	12.7	444	90	1 726	310	3 110	6.47
niobium	8.57	10	3.7	7.1	267	52	2 688	261	5 200	7.5
osmium	22.57	56	22	4.7	130		3 300	140	4 500	
palladium	12.02	11	4.4	11.6	244	70	1 825	162	3 200	4.0
phosphorus	1.82			127	750		317	21	553	0.40
platinum	21.45	16	6.1	8.9	138	69	2 042	113	4 100	2.67
plutonium	19.84			57		8	913	39	3 500	
potassium	0.862			83	757	99	337	59.7	1 030	2.15
rhenium	21.02	47	17.4		138	71	3 450	180	5 900	3.4
rhodium	12.41	37	15	8.3	242	150	2 239	210	4 000	5.2
rubidium	1.53			90	361		312	26	974	0.87
ruthenium	12.41			6.7	240		2 520	252	4 000	
selenium	4.79			26	322		490	66	958	0.76
silicon	2.33	1.9	8.0	2.5	707	170	1 680	165	2 628	10.6
silver	10.50	7.8	2.8	19.2	236	418	1 234	105	2 466	2.31
sodium	0.971			69.6	1 230	135	371	113	1 156	3.90
strontium	2.54				301		1 042	105	1 640	1.6
sulphur	2.07			61	736	0.20	392.2	38	717.8	0.28
tantalum	16.6	18	7	6.5	141	54	3 269	170	5 700	8.1
tellurium	6.24	4.1	1.6	18.2	202	1.7	723	140	1 260	0.89
thallium	11.85	0.8	0.3	29.2	129	41	577	21	1 740	0.80

1	2	3	4	5	6	7	8	9	10	11
Element	**Density** 10^3 kg m^{-3}	**Young's modulus** 10^{10} Pa	**Shear modulus** 10^{10} Pa	**Thermal expansion** 10^{-6} K^{-1}	**Sp. heat capacity** J kg^{-1} K^{-1}	**Thermal conductivity** Wm^{-1} K^{-1}	**Melting point** K	**Heat of fusion** 10^3 J kg^{-1}	**Boiling point** K	**Heat of vaporization** 10^6 J kg^{-1}
thorium	11.7	8.0	3.1	11.1	118	41	≈ 2 000	83	4 500	2.34
tin, white*	7.3	5.5	2.1	27	230	65	505	59	2 543	2.45
titanium	4.54	11	4	8.5	522	19	1 948	400	3 530	8.9
tungsten	19.3	38	15	4.5	133	170	3 653	192	5 800	8.8
uranium	18.9	18	7.2	13.5	116	25	1 405	53	4 600	1.73
vanadium	5.87	13	4.7	7.8	486	32	2 160	330	3 650	
yttrium	4.45	0.66			280	15	1 768	190	3 200	4.42
zinc	7.13	9.8	4	29.7	389	120	693	117	1 181	1.76
zirconium	6.53	7.0	2.5	5.4	275	21	2 125	220	4 650	6.4

* White tin (β tin) is a metal with tetragonal structure. It is stable above 13.2 °C.
Grey tin (α tin) is an intrinsic semiconductor with a diamond structure. It is stable below 13.2 °C.

1.2 Alloys – Mechanical and Thermal Properties

Column 3 The density at 290 K is given.

Column 4 The linear expansion coefficient is an average over the temperature range 20 °C–300 °C. For alloys 6, 7, 8, and 11, however, the range is 20 °C–100 °C and for alloy 9 it is 20 °C–1000 °C.

Column 7 The specific heat capacity at 293 K is given.

Column 8 The thermal conductivity at 293 K is given.

T – 1.2 Alloys – Mechanical and Thermal Properties

1		2	3	4	5	6	7	8	9
	Name	**Percentage composition by weight**	**Density** 10^3 kg m^{-3}	**Expansion coefficient** 10^{-6} K^{-1}	**Young's modulus** 10^{10} Pa	**Shear modulus** 10^{10} Pa	**Sp. heat capacity** 10^3 J kg^{-1} K^{-1}	**Thermal conductivity** W m^{-1} K^{-1}	**Melting point** K
1	aluminium bronze 5%	Cu 94.6, Al 5, Mn 0.4	8.1	18	12		0.42	84	1 333
2	argentan 18%	Cu 60, Zn 22, Ni 18	8.7	17	12–15		0.40	23	1 375
3	brass	Cu 62.7, Zn 37.3	8.4	21	10.5	3.5	0.38	79	1 188
4	cast iron	< 4% C	7.3	11	10		0.50	30–45	1 475
5	constantan	Cu 58, Ni 41, Mn 1	8.9	15	11		0.41	22	1 545
6	duralumi-nium	Cu 3–4, Mg 0.5, Mn 0.25–1, rest Al	2.8	24	7.2		0.93	160	925
7	electron	Mg 92, Al 5, Zn 3	1.8	25	4.4	1.7	1.00	115	900
8	invar	Fe 64, Ni 36	8.1	2.0	14.5		0.50	16	1 723
9	kanthal Al	Fe 67.5, Cr 25, Al 5.5, Co 2	7.1	17				17	
10	manganine	Cu 84.5, Mn 12.5, Fe 1, Ni 2	8.5	16	12.6		0.41	22	1 275
11	silumin	Al 87, Si 13	2.6	19	8.5	3	0.88	161	845
12	solder	Sn 60, Pb 40	8.5	24	3.0		0.15	50	456–461
13	steel	C 0.85	7.8	11.5	20	8.1	0.46	45	1 625
14	tin bronze 10%	Cu 90.75, Sn 9, P 0.25	8.9	19	10–12		0.38	46	1 285
15	wood's metal	Bi 44.5, Pb 35.5, Sn 10, Cd 10	9.7				0.15	13	344
16	wrought iron	C 0.04–0.4	7.6		22			60	

1.3 Other Solids – Mechanical and Thermal Properties

1		2	3	4	5
	Name	**Density** 10^3 kg m^{-3}	**Expansion coefficient** 10^{-6} K^{-1}	**Sp. heat capacity** 10^3 J kg^{-1} K^{-1}	**Thermal conductivity** W m^{-1} K^{-1}
1	acrylic	1.2	70–100	1.4–2.1	0.2
2	araldite (epoxy)	1.2	60	1.7	0.2
3	asbestos	0.58		0.84	0.2
4	brick	1.4–1.8	8–10	0.8	0.6–0.8
5	concrete (dry)	1.5–2.4	12	0.92	0.4–1.7
6	cork	0.20–0.35		1.7–2.1	0.045–0.06
7	ebonite	1.15	85	1.67	0.2
8	fibre board (porous)	0.3			0.06
9	glass (common)	2.5	8	0.84	0.9
10	granite	2.7	8	0.80	3.5
11	gypsum	0.97	25	1.1	1.3
12	ice (–4°C)	0.917	50	2.2	2.1
13	marble	2.5–2.8	5–16	0.9	3
14	mica	2.8	3	0.88	0.5
15	paper	0.7–1.2			0.2
16	paraffin	0.85	100–200	2.1–2.9	0.21–0.26
17	polyamide (nylon)	1.1	100–140	1.8	0.2
18	polyethene	0.92	100–200	2.1	0.23–0.29
19	polystyrene	1.05	60–80	1.3	0.07–0.08
20	polyvinyl chloride (PVC)	1.2–1.5	150–200	1.3–2.1	0.16
21	porcelain	2.3–2.5	2–5	0.8	1.0–1.7
22	quartz (fused)	2.2	0.4	0.8	0.2
23	rubber	0.92–0.96	150–200	2	0.13–0.16
24	teflon	2.1–2.3	60–100	1.0	0.2
25	wood (pine)	0.52	5–30	0.4	0.14

1.4 Approximate Coefficients of Friction

The values are strongly dependent on surface quality. (wet) means a wet surface instead of a lubricated surface.

Material	Static friction		Sliding friction	
	Clean surfaces	Lubricated surfaces	Clean surfaces	Lubricated surfaces
1 bronze/bronze	–	0.11	0.2	0.06
2 cast iron/cast iron	–	0.16	0.15–0.2	0.02–0.1
3 cast iron/bronze	0.15–0.2	–	0.15–0.2	0.07–0.01
4 glass/glass	0.9–1.0	0.3–0.6	–	–
5 ice/ice, 0 °C	0.05–0.15	–	0.02	–
6 ice/ice, –40 °C	0.4	–	0.075	–
7 leather/wood	0.27	–	0.4	–
8 metal/wood	0.5–0.6	0.1	0.2–0.5	0.02–0.07
9 rubber/concrete	1.00	0.3 (wet)	0.8	0.25 (wet)
10 steel/ice	0.027	–	0.18	0.01
11 steel/cast iron	0.18	0.1	0.18	0.1
12 steel/steel	0.15–0.3	0.1	0.15–0.2	0.01–0.1
13 steel/brake-shoe lining	–	–	0.4–0.6	0.3–0.5
14 wood/stone	0.7	0.4	0.3	–
15 wood/wood	0.3–0.6	0.16	0.25–0.5	0.04–0.16

1.5 Fluids – Mechanical and Thermal Properties

Column 3 The density is given at 20 °C and 0.1 MPa for liquids and at 0 °C and 0.1 MPa for gases. For some substances the density of the liquid state can be found in table T–7.5. The density of air and water at other temperatures and 0.1 MPa pressure is given by the following table.

air	1.2047 kg/m^3	at 20 °C
air	1.0600 kg/m^3	at 60 °C
water, common	$0.99987 \cdot 10^3$ kg/m^3	at 0 °C
water, common	$1.00000 \cdot 10^3$ kg/m^3	at 3.98 °C
water, common	$0.98234 \cdot 10^3$ kg/m^3	at 60 °C
water, common	$0.95838 \cdot 10^3$ kg/m^3	at 100 °C

$$\text{Density of air} \approx 1.2929 \, \frac{273.15 \, p}{0.1013 \, T}$$

T = temperature in K, p = pressure in MPa

Column 4 An average over the temperature range 20 °C–100 °C of the volume expansion coefficient is given. Gases in this interval are marked (g). Ideal gases have the volume expansion coefficient $1/T$ K^{-1}.

Column 5 The viscosity is strongly temperature dependent. It decreases with temperature for liquids, but increases for gases. The table gives the dynamic viscosity of liquids at 291 K, and the dynamic viscosity of gases at 273 K. The pressure dependence is negligible, even for gases.

Column 6 The surface tension is strongly temperature dependent. Its value at 291 K is given.

Column 7 The specific heat capacity at constant pressure is given. The values refer to the temperature 300 K for liquids and 273 K for gases.

Column 8 The thermal conductivity at 292 K is given. The pressure dependence is negligible, even for gases.

Column 9 The difference between c_p and c_v is important only for gases. Liquids at room temperature are marked (ℓ).

Columns 10 and 12 The values refer to the pressure 0.1 MPa.

Column 11 Heat of fusion is also called enthalpy of fusion.

Column 13 Heat of vaporization is also called enthalpy of vaporization.

T – 1.5 Fluids – Mechanical and Thermal Properties

	1	2	3	4	5	6	7
	Name	**Chemical formula**	**Density** $kg\ m^{-3}$	**Expansion coefficient** $10^{-3}\ K^{-1}$	**Viscosity** $10^{-6}\ Ns\ m^{-2}$	**Surface tension** $10^{-3}\ N\ m^{-1}$	**Sp. heat capacity** $10^{3}\ J\ kg^{-1}K^{-1}$
1	acetone	$(CH_3)_2CO$	$0.79 \cdot 10^3$	1.43	330	23	2.20
2	acetylene	C_2H_2	1.17	g	10		1.68
3	air		1.2929	g	16.7		1.01
4	ammonia	NH_3	0.77	g	9.1		2.05
5	aniline	$C_6H_5NH_2$	$1.02 \cdot 10^3$	0.85	4 700	42	2.05
6	argon	Ar	1.784	g	20.8		0.52
7	benzene	C_6H_6	$0.872 \cdot 10^3$	1.15	660	28	1.71
8	bromine	Br_2	$3.14 \cdot 10^3$	1.12	100	43	0.53
9	butane	C_4H_{10}	0.60	g	6.7		
10	carbon dioxide	CO_2	1.98	g	13.6		0.82
11	carbon disulphide	CS_2	$1.27 \cdot 10^3$	1.22	372	31	1.00
12	carbon monoxide	CO	1.25	g	15.9		1.04
13	carbon tetrachloride	CCl_4	$1.59 \cdot 10^3$	1.22	981	26	0.85
14	casto oil		$0.96 \cdot 10^3$	0.69	$1.25 \cdot 10^6$	36	1.80
15	chlorine	Cl_2	3.214	g	12		0.50
16	chloroform	$CHCl_3$	$1.48 \cdot 10^3$	1.27	569	26	
17	cyanogen	$(CN)_2$	2.32	g			1.71
18	diethyl ether	$(C_2H_5)_2O$	$0.705 \cdot 10^3$	1.62	248	17	0.67
19	ethane	C_2H_6	1.36	g	8.5		1.73
20	ethanol	C_2H_5OH	$0.789 \cdot 10^3$	1.10	1 230	22	2.43
21	ethylene	C_2H_4	1.26	g	9.4		1.51
22	fluorine	F_2	1.72	g			0.75
23	freon 12	CCl_2F_2	1.43	g	13.7		0.94
24	glycerol	$C_3H_5(OH)_3$	$1.26 \cdot 10^3$	0.505	$1.6 \cdot 10^6$	61	2.40
25	glycol	$(CH_2OH)_2$	$1.11 \cdot 10^3$			48	2.43
26	helium ^4He	^4He	0.178	g	18.2		5.2
27	^3He	^3He		g			
28	hydrogen ^1H	H_2	0.0899	g	8.4		14.2
29	^2H	D_2	0.180	g			
30	hydrogen chloride	HCl	1.64	g	13.5		0.81
31	krypton	Kr	3.74	g	22.9		
32	lubricating oil	Hydro-carbons	$0.88 \cdot 10^3$	0.96	$3.6 \cdot 10^5$	30	1.87
33	mercury	Hg	$13.54 \cdot 10^3$	0.1819	1 540	490	0.14
34	methane	CH_4	0.72	g	10.0		2.21
35	methanol	CH_3OH	$0.786 \cdot 10^3$	1.20	584	22	2.50
36	neon	Ne	0.900	g	29.7		1.03
37	nitrobenzene	$C_6H_5NO_2$	$1.20 \cdot 10^3$	0.83	2 030	22	1.48

8	9	10	11	12	13	14	15	
Thermal conductivity $W\,m^{-1}\,K^{-1}$	c_p/c_v	**Melting point** K	**Heat of fusion** $10^3\,J\,kg^{-1}$	**Boiling point** K	**Heat of vaporization** $10^3\,J\,kg^{-1}$	**Critical temperature** K	**Critical pressure** 10^6 Pa	
0.161	ℓ	178	98	329.7	509	508.7	4.72	1
0.019	1.23	191.4	150	189.6	670	309.5	6.24	2
0.026	1.40					130	3.8	3
0.022	1.31	195.5	332	239.73		405.50	11.28	4
0.172	ℓ	267.0	88	457.6	435	698.8	5.30	5
0.016	1.668	84.0	29.4	87.29	158	150.7	4.86	6
0.140	ℓ	278.66	127	353.24	393	562.7	4.92	7
	ℓ	266.0	68	332.36	188	584	10.3	8
	1.11	134.8		~274		425.16	3.80	9
0.015	1.30	216.6	189	194.7	573	304.19	7.38	10
0.143	ℓ	161	58	319.5	351	552	7.9	11
0.022	1.40	68	30	81	215	133.0	3.49	12
0.105	ℓ	249.4	18	349.8	193	556.4	4.56	13
0.181	ℓ							14
0.0076	1.35	172.18	90	239.10	282	417.2	7.71	15
0.121	ℓ	209.7	80	334.5	255	536.6	5.47	16
	1.26	238.8		252.5		401	6.2	17
0.138	ℓ	156.9	113	309.6	377	467.8	3.61	18
0.019	1.21	89.9		184.9		305.42	4.88	19
0.182	ℓ	155.9	102	351.7	841	516	6.38	20
0.016	1.18	104.0	120	169	483	283.05	5.12	21
		53.54	13.4	85.02	316	144	5.7	22
0.0077		118		244				23
0.285	ℓ	291.1	176	563				24
	ℓ	261.6	201	470.4	800			25
0.142	1.66		5.2	4.215	25	5.20	0.229	26
				3.191		3.37	0.124	27
0.19	1.41	13.8	58	20.26	446	33.23	1.30	28
	1.73	18		23.59	310	38.34	1.66	29
	1.41	159	55	188	443	324.6	8.26	30
	1.68	116.6	19.5	119.82	108	209.38	5.50	31
0.133	ℓ	210–270		650–900				32
10.3	ℓ	234.29	11.7	629.87	296	1900	360	33
0.025	1.30	89	59	111.7	510	191.1	4.64	34
0.212	ℓ	175.4	91.8	337.80	1100	513.2	7.95	35
0.046	1.64	24.19	16.7	27.24	86	44.3	2.72	36
0.16	ℓ	278.9	92	283.3	331			37

	1	2	3	4	5	6	7
	Name	**Chemical formula**	**Density**	**Expansion coefficient**	**Viscosity**	**Surface tension**	**Sp. heat capacity**
			$kg\ m^{-3}$	$10^{-3}\ K^{-1}$	$10^{-6}\ Ns\ m^{-2}$	$10^{-3}\ N\ m^{-1}$	$10^3\ J\ kg^{-1}K^{-1}$
38	nitrogen	N_2	1.250	g	16.5		1.04
39	nitrogen monoxide	NO	1.340	g	17.7		1.00
40	nitrogen oxide	N_2O	1.98	g	13.4		0.89
41	olive oil		$0.91 \cdot 10^3$	0.72	88 300	32	1,65
42	oxygen	O_2	1.429	g	19.2		0.92
43	ozone	O_3	2.22	g			
44	propane	C_3H_8	2.02	g	7.4		
45	pyridine	C_5H_5N	$0.98 \cdot 10^3$			38	1.65
46	sulphur dioxide	SO_2	2.93	g	11.5		0.61
47	sulphur acid	H_2SO_4	$1.84 \cdot 10^3$	0.56	27 500	55	1.38
48	terpentine	$C_{10}H_{16}$	$0.84 \cdot 10^3$	1.00	1 470		1.75
49	toluene	$C_6H_5CH_3$	$0.861 \cdot 10^3$	1.09	590	28	1.70
50	transformer oil	Hydro-carbons	$0.86 \cdot 10^3$	1.00	17 600	30	1.92
51	trichloroethylene	C_2HCl_3	$1.46 \cdot 10^3$	1.19			0.96
52	water, common	H_2O	$0.99820 \cdot 10^3$	0.51	1 040	73	4.19
53	heavy	D_2O	$1.1053 \cdot 10^3$	0.48			
54	xenon	Xe	5.89	g	22.2		

8	9	10	11	12	13	14	15	
Thermal conductivity W m^{-1} K^{-1}	c_p/c_v	**Melting point** K	**Heat of fusion** 10^3 J kg^{-1}	**Boiling point** K	**Heat of vaporization** 10^3 J kg^{-1}	**Critical temperature** K	**Critical pressure** 10^6 Pa	
0.027	1.404	63.30	25.7	77.36	200	126.3	3.40	38
0.023	1.40	109.5		121.38		366	7.26	39
0.015	1.28	170.8		184.68		312	10	40
0.169	ℓ							41
0.027	1.401	54.8	13.8	90.180	213	154.77	5.08	42
	1.29	21.7		165.65	410	285.3	5.53	43
	1.13	83		229	450	369.95	4.26	44
	ℓ	231		388.7		617.4	6.08	45
0.009	1.29	200		263.13	397	430.7	7.88	46
	ℓ		109	599	511			47
0.15	ℓ	263		353		594.0	4.21	48
0.151	ℓ	175	71	384	356			49
0.135	ℓ	220	190	500–650				50
0.15	ℓ	187		360	239			51
0.60	ℓ	273.150	333	373.125	2 260	647.4	22.11	52
	ℓ	277.0	318	374.6	2 070	644.1	21.8	53
0.005	1.66	161.3	17.5	165.1	102	289.74	5.88	54

1.6 Water Vapour

Column 2 The pressure of saturated vapour is given.

Column 3 The water mass per volume of vapour saturated air is given.

1	2	3	1	2	3
Temperature °C	**Pressure** mbar	**Density** g m^{-3}	**Temperature** °C	**Pressure** mbar	**Density** g m^{-3}
– 35	0.23	0.22	+ 52	136.1	90.8
– 30	0.37	0.35	+ 54	150.0	99.5
– 25	0.63	0.57	+ 56	165.0	108.8
– 20	1.03	0.91	+ 58	181.4	119.2
– 15	1.65	1.39	+ 60	199.2	130.1
– 10	2.60	2.15	+ 62	218.3	141.8
– 8	3.09	2.53	+ 64	239.0	153.5
– 6	3.68	2.99	+ 66	261.4	168.0
– 4	4.37	3.53	+ 68	285.5	182.5
– 2	5.17	4.14	+ 70	311.5	198.0
± 0	6.11	4.85	+ 72	339.4	214.6
+ 2	7.05	5.57	+ 74	369.5	232.4
+ 4	8.13	6.37	+ 76	401.8	251.4
+ 6	9.34	7.27	+ 78	436.3	271.7
+ 8	10.72	8.28	+ 80	473.3	293.3
+ 10	12.26	9.41	+ 82	513.1	316.2
+ 12	14.01	10.67	+ 84	555.6	340.7
+ 14	15.97	12.08	+ 86	601.1	366.7
+ 16	18.17	13.65	+ 88	649.3	394.3
+ 18	20.62	15.39	+ 90	700.9	423.5
+ 20	23.37	17.32	+ 92	755.8	454.2
+ 22	26.42	19.44	+ 94	814.3	486.9
+ 24	29.83	21.81	+ 96	876.6	521.7
+ 26	33.60	24.40	+ 98	942.8	558.7
+ 28	37.79	27.26	+ 100	1 013.2	598
+ 30	42.42	30.39	+ 105	1 208	703
+ 32	47.55	33.85	+ 110	1 433	824
+ 34	53.19	37.61	+ 115	1 690	962
+ 36	59.41	41.74	+ 120	1 985	1 120
+ 38	66.25	46.25	+ 130	2 701	1 494
+ 40	73.77	50.17	+ 140	3 612	1 970
+ 42	82.01	56.52	+ 160	6 180	3 260
+ 44	91.02	62.38	+ 200	15 540	7 850
+ 46	100.87	68.57	+ 300	85 800	46 300
+ 48	111.64	75.50	+ 374	220 000	315 000
+ 50	123.4	83.0			

1.7 Vapour Pressure of Various Materials

The vapour pressure is given in torr at the table head. The values in the table are the corresponding temperatures in kelvin. Vapour pressure over solid-state is given in boldface.

	Substance	Pressure (torr) 760	1	10^{-2}	10^{-4}	10^{-6}	10^{-8}
1	helium (^{4}He)	4.22	1.27	0.79	0.56		
2	nitrogen	77	**47**	**37**	**31**	**27**	**24**
3	oxygen	90	**54**	**43**	**37**	**32**	**28**
4	ammonia	240	**163**	**120**	**105**	**93**	
5	water	373	**256**	**214**	**185**	**162**	
6	mercury	630	400	320	267	229	201
7	silicone oil DC 704	683	520	430	380	340	300
8	sodium	1 180	716	564	468	400	**350**
9	iron	3 027	2 071	1 708	1 462	1 278	**1 130**
10	carbon	**4 627**	**3 404**	**2 816**	**2 490**	**2 195**	**1 963**
11	tungsten	6 200	4 330	3 565	3 032	2 648	2 346

1.8 Van der Waals Constants

	Name	Chemical formula	a 10^{-3} N m^4 mol^{-2}	b 10^{-6} m^3 mol^{-1}
1	acetone	$(CH_3)_2CO$	1405	99.4
2	acetylene	C_2H_2	443	51.36
3	ammonia	NH_3	421.2	37.07
4	carbon dioxide	CO_2	362.8	42.67
5	cabon monoxide	CO	150.0	39.85
6	ethane	C_2H_6	554.4	63.80
7	ethanol	C_2H_5OH	1214	84.07
8	helium	He	3.446	23.70
9	hydrogen	H_2	24.68	26.31
10	mercury	Hg	817	16.96
11	neon	Ne	21.28	17.09
12	nitrogen	N_2	140.4	39.13
13	oxygen	O_2	137.4	31.83
14	sulphur dioxide	SO_2	678	56.36
15	water	H_2O	551.9	30.49

1.9 Diffusion Coefficients

Entries 15–16: The diffusion coefficient at a total pressure of 0.1 MPa is given.

	Diffusive material	Medium	Temperature K	Diffusion coefficient $m^2\,s^{-1}$
1	Au	Pb	773	$3.7 \cdot 10^{-9}$
2	Au	Ag	1 134	$1.1 \cdot 10^{-15}$
3	Au	Si	1 680	$4.8 \cdot 10^{-11}$
4	Cu	Au	574	$1.5 \cdot 10^{-17}$
5	Cu	Au	1 013	$9.3 \cdot 10^{-14}$
6	Cu	Cu	923	$3.2 \cdot 10^{-16}$
7	Cu	Cu	1 356	$2.6 \cdot 10^{-13}$
8	Cu	Ge	$\geqslant$ 1 023	$2.7 \cdot 10^{-9}$
9	Cu	Ge	$\leqslant$ 1 023	$4.8 \cdot 10^{-11}$
10	Li	Ge	1 210	$1.5 \cdot 10^{-9}$
11	electrons	Ge	300	$9.5 \cdot 10^{-3}$
12	Si	Si	1 680	$5.5 \cdot 10^{-16}$
13	NaCl (0.1–1.0 molar)	H_2O	291	$1.2 \cdot 10^{-9}$
14	methanol (11%)	H_2O	291	$1.4 \cdot 10^{-9}$
15	CO	N_2	273	$1.9 \cdot 10^{-5}$
16	H_2	air	273	$6.1 \cdot 10^{-5}$
17	H_2	Pd	300	$3.6 \cdot 10^{-11}$
18	H_2	Pd	1 200	$4.1 \cdot 10^{-8}$

1.10 Acceleration of Gravity (g) at Various Latitudes

Latitude ϕ_g	0°	10°	20°	30°	40°
Acceleration (m/s^2)	9.780 49	9.782 05	9.786 52	9.793 37	9.801 80

Latitude ϕ_g	50°	60°	70°	80°	90°
Acceleration (m/s^2)	9.810 78	9.819 24	9.826 15	9.830 66	9.832 23

International gravitation formula:
$g = 9.780495\,(1 + 0.0052892 \sin^2 \phi_g - 0.0000073 \sin^2 2\phi_g)$

Free air correction for altitude:
$-3.086 \cdot 10^{-6}$ m/s^2 per metre.

2 Electricity

2.1 The Elements – Electric, Thermoelectric, and Magnetic Properties

Column 2 The resistivity at 300 K is given. For semiconductors the intrinsic value (no impurities) is given.

Column 4 The electromotive force for the material in thermocouple with platinum is given. One junction is taken to have the temperature 0 °C, and the other the temperature 100 °C. The sign gives the polarity of the material relative to platinum at the hot junction.

Column 5 The magnetic volume susceptibility $\chi = M/H$ at 300 K is given.
M = magnetization, i.e. magnetic moment per volume,
H = magnetizing field. χ (= $\mu_r - 1$) is a dimensionless quantity in SI.

1	2	3	4	5
Element	**Resistivity** $10^{-8}\,\Omega\,m$	**Temperature coefficient** $10^{-3}\,K^{-1}$	**Thermocouple emf** $\mu V\,K^{-1}$	**Magnetic susceptibility**
aluminium	2.65	4.29	+ 4	$+2.08 \cdot 10^{-5}$
antimony	39	5.1	+ 47	$-6.83 \cdot 10^{-5}$
argon				$-1.10 \cdot 10^{-8}$
arsenic	33.3			$-5.28 \cdot 10^{-6}$
barium	36	6.1		$+6.62 \cdot 10^{-6}$
beryllium	4.0	7.5		$-2.32 \cdot 10^{-5}$
bismuth	107	4.45	– 70	$-1.65 \cdot 10^{-4}$
boron	$1.8 \cdot 10^{12}$			$-1.82 \cdot 10^{-5}$
bromine Br_2				$-1.30 \cdot 10^{-5}$
cadmium	6.9	4.26	+ 9	$-1.91 \cdot 10^{-5}$
calcium	4.0	4.0		$+1.93 \cdot 10^{-5}$
carbon, diamond				$-2.17 \cdot 10^{-5}$
graphite	$1.3 \cdot 10^{3}$		+ 2	$-1.41 \cdot 10^{-5}$
cerium	75.0	0.87		$+1.49 \cdot 10^{-3}$
cesium	20	5.0	+ 5	$+5.15 \cdot 10^{-6}$

T – 2.1 The Elements – Electric, Thermoelectric, and Magnetic Properties

1	2	3	4	5
Element	**Resistivity** $10^{-8}\,\Omega\,\text{m}$	**Temperature coefficient** $10^{-3}\,\text{K}^{-1}$	**Thermocouple emf** $\mu\text{V}\,\text{K}^{-1}$	**Magnetic susceptibility**
chlorine Cl_2				$-2.3 \cdot 10^{-8}$
chromium	13	3		$+3.13 \cdot 10^{-4}$
cobalt	6.24	6.58	– 16	Ferromagn.
copper	1.67	4.33	+ 7	$-9.63 \cdot 10^{-6}$
dysprosium	57	1.2		$+6.85 \cdot 10^{-2}$
erbium	107	2.0		$+3.05 \cdot 10^{-2}$
europium	90			$+1.47 \cdot 10^{-2}$
gadolinium	141	1.76		$+4.79 \cdot 10^{-1}$
gallium	17.4	4.1		$-2.60 \cdot 10^{-5}$
germanium	$4.6 \cdot 10^{7}$			$-7.12 \cdot 10^{-5}$
gold	2.35	3.98	+ 7	$-3.45 \cdot 10^{-5}$
hafnium	35.1	3.8		$+7.03 \cdot 10^{-5}$
helium				$-1.05 \cdot 10^{-9}$
holmium	87	1.7		
hydrogen H_2				$-2.23 \cdot 10^{-9}$
indium	8.37	5.1		$-5.11 \cdot 10^{-5}$
iodine I_2	$1.3 \cdot 10^{15}$			$-2.16 \cdot 10^{-5}$
iridium	5.3	4.33	+ 6	$+3.75 \cdot 10^{-5}$
iron (α)	9.7	6.57	+ 18	Ferromagn.
krypton				$-1.61 \cdot 10^{-8}$
lanthanum	5.7	2.2		$+6.61 \cdot 10^{-5}$
lead	20.6	4.22	+ 4	$-1.58 \cdot 10^{-5}$
lithium	8.55	4.37		$+1.37 \cdot 10^{-5}$
magnesium	4.45	4.2	+ 4	$+1.18 \cdot 10^{-5}$
manganese (α)	7.10	0.17		$+8.71 \cdot 10^{-3}$
(β)	91	1.4		$+7.9 \cdot 10^{-3}$
(γ)	23	6.3		
mercury	98.4	0.99	0	$-2.85 \cdot 10^{-5}$
molybdenum	5.2	4.7	+ 12	$+1.19 \cdot 10^{-4}$
neodymium	64	1.6		$+3.43 \cdot 10^{-3}$
neon				$-3.78 \cdot 10^{-9}$
nickel	6.84	6.75	– 15	Ferromagn.
niobium	12.5	2.28		$+2.26 \cdot 10^{-4}$
nitrogen N_2				$-6.73 \cdot 10^{-9}$
osmium	95	4.2		$+1.47 \cdot 10^{-5}$
oxygen O_2				$+2.00 \cdot 10^{-6}$
palladium	10.8	3.8	– 3	$+8.02 \cdot 10^{-4}$
phosphorus	$1 \cdot 10^{17}$			$-1.97 \cdot 10^{-5}$
platinum	10.6	3.92	0	$+2.79 \cdot 10^{-4}$
plutonium	140	– 2.97		$+6.30 \cdot 10^{-4}$

1	2	3	4	5
Element	**Resistivity** $10^{-8}\,\Omega\,m$	**Temperature coefficient** $10^{-3}\,K^{-1}$	**Thermocouple emf** $\mu V\,K^{-1}$	**Magnetic susceptibility**
potassium	6.2	5.4	– 9	$+5.74 \cdot 10^{-6}$
praseodymium	68	1.65		$+3.03 \cdot 10^{-3}$
rhenium	19.3	3.1		$+9.37 \cdot 10^{-5}$
rhodium	4.51	4.57	+ 6	$+1.68 \cdot 10^{-4}$
rubidium	12.5	5.3		$+3.8 \cdot 10^{-6}$
ruthenium	7.6	4.5		$+6.61 \cdot 10^{-5}$
samarium	88	1.8		$+1.17 \cdot 10^{-3}$
scandium	61	2.8		$+2.63 \cdot 10^{-4}$
selenium	12			$-1.91 \cdot 10^{-5}$
silicon	$6.4 \cdot 10^{10}$			$-4.1 \cdot 10^{-6}$
silver	1.59	4.10	+ 7	$-2.38 \cdot 10^{-5}$
sodium	4.2	5.5	– 2	$+8.5 \cdot 10^{-6}$
strontium	23	~ 5		$+3.43 \cdot 10^{-5}$
sulphur, yellow	$2 \cdot 10^{23}$			$-1.26 \cdot 10^{-5}$
tantalum	12.5	3.6		$+1.78 \cdot 10^{-4}$
tellurium	$4.36 \cdot 10^{5}$		+ 500	$-2.33 \cdot 10^{-5}$
terbium				$+9.55 \cdot 10^{-2}$
thallium	18	5.2		$-3.71 \cdot 10^{-4}$
thorium	13.1	3.3		$+8.36 \cdot 10^{-5}$
thulium	79	2.0		$+1.77 \cdot 10^{-2}$
tin, grey	10.1	4.63	+ 5	$-1.93 \cdot 10^{-5}$
white	11	4.63		$+2.39 \cdot 10^{-6}$
titanium	42	5.5		$+1.81 \cdot 10^{-4}$
tungsten	5.65	4.83	+ 8	$+7.80 \cdot 10^{-5}$
uranium	25	2.1		$+4.11 \cdot 10^{-4}$
vanadium	25			$+3.75 \cdot 10^{-4}$
xenon				$-2.47 \cdot 10^{-8}$
ytterbium	29	1.3		$+1.26 \cdot 10^{-4}$
yttrium	57	2.7		$+1.32 \cdot 10^{-6}$
zinc	5.92	4.2	+ 7	$-1.56 \cdot 10^{-5}$
zirconium	40	4.0		$+1.09 \cdot 10^{-4}$

2.2 Electric Resistor and Conductor Materials

Column 1 Material composition: Nichrome 80 / 20: Ni 75–80 %, Cr 20 %, Mn 1–3 %. Silver alloy: Ag 91.22 %, Mn 8.78 %. Other materials are treated in Table 1.2.

Columns 2 and 3 Resistivity and temperature coefficient of the elements are given in Table 2.1.

Column 4 See comment on Column 4 in Table 2.1.

1	2	3	4
Name	**Resistivity** $10^{-8}\,\Omega$ m	**Temperature coefficient** $10^{-3}\,K^{-1}$	**Thermocouple emf** $\mu V\,K^{-1}$
Resistor materials			
1 constantan	50	± 0.03	– 34
2 kanthal A 1	145	0.03	
3 manganine	43	± 0.02	+ 6
4 nichrome 80/20	105	0.18	
5 silver alloy	28	– 0.001	+ 7
Other alloys			
6 argentan	35	0.3	
7 brass	6.5	1.5	
8 duraluminium	40	2.8	
9 invar	10	2	
10 steel	16	3.3	

2.3 Electric Insulators

Column 3 The low-frequency relative permittivity (dielectric constant) is given.

Column 4 The dielectric strength of a one-mm thickness is given. For a one-cm thickness, multiply by seven.

1		2	3	4
	Name	**Resistivity** Ω m	**Relative permittivity**	**Dielectric strength** kV/mm
1	acrylic	10^{19}	3.3	20
2	air, dry		1.0006	4.7
3	bakelite	10^{9}	5	10
4	ebonite	10^{8}	2.7	10
5	ethanol	$3 \cdot 10^{3}$	26	
6	glass, common	$5 \cdot 10^{11}$	7	15
7	quartz	10^{14}	4.2	25
8	mica	$5 \cdot 10^{14}$	6	80
9	neoprene	10^{10}	8.3	17
10	paper, hard	10^{10}	5	15
11	waxed	10^{14}	5	30
12	paraffin	10^{15}	2.1	40
13	polyethene	$3 \cdot 10^{15}$	2.3	18
14	polystyrene	$> 10^{14}$	2.6	20–28
15	polyvinyl	10^{12}	4–8.5	25
16	porcelain	10^{12}	5.5	35
17	rubber	$5 \cdot 10^{13}$	3.3	17
18	sulphur	10^{15}	4	
19	teflon	$> 10^{13}$	2.0	35
20	transformer oil	10^{11}	2.4	20
21	water, distilled	$5 \cdot 10^{3}$	81	30
22	wood	10^{12}	2.5–7	

2.4 Magnetic Materials

Column 3 μ_{rb} = relative permeability at initial magnetization

Column 4 μ_{rm} = maximum relative permeability

Column 5 B_m = saturation flux density

Column 6 B_r = residual induction

Column 7 H_c = coercive force

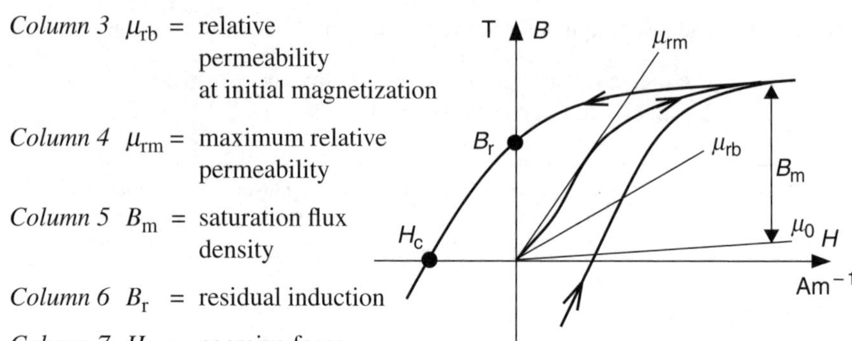

1	2	3	4	5	6	7
Trade name	**Per cent composition**	μ_{rb}	μ_{rm}	B_m T	B_r T	H_c $A\,m^{-1}$
Soft materials						
armco iron	Almost pure iron	250	700	2.15	0.6	40
hyperm 50 T	Ni 50, Fe 50	5 000	40 000	1.55	1.52	8
Mu metal	Ni 76, Fe 17, Cu 5, Cr 2	12 000	45 000	1.20		2.4
transformer steel	Fe 96, Si 4	500	7 000	2.0		40
alloy 1040	Ni 72, Cu 14, Fe 11, Mo 3	40 000	110 000	0.60	1.25	1.1
75-permalloy	Ni 75, Fe 25	10 000	105 000	1.10		4
Mn-Zn ferrite	$Mn_{1/2}\,Zn_{1/2}\,Fe_2\,O_4$	2 000	5 500	0.35		16
amorphous metal	glass	≈ 3 000	100 00	1.5	1.3	8

Hard materials		**Energy density** $(B \cdot H)_{max}$ $kJ\,m^{-3}$	B_r T	H_c $kA\,m^{-1}$
$Nd_2\,Fe_{14}\,B$		280	1.3	1 000
$Sm\,Co_5$		150	0.84	2 000
ticonal		19	0.63	96
alnico-mat	Fe 55, Ni 25, Al 12, Co 8	15	0.70	60
carbon steel	Fe 98.8, C 1.1, V 0.1	0.9	1.03	1.75

2.5 Electrochemical Series

Standard reduction potentials at 25 °C. All ions in a solution are combined with a number of water molecules. The ion concentration is 1 M and gases have the pressure 0.1 MPa. (g) = gas, (l) = liquid and (s) = solid. (1 M is explained in CU–2.3).

More reactions:
http://bilbo.chm.uri.edu/CHM112/tables/redpottable.htm

Reaction ox form red form	Standard potential V	Reaction ox form red form	Standard potential V
$Li^+ + e^- \rightleftharpoons Li$ (s)	–3.045	$Cu^{2+} + 2e^- \rightleftharpoons Cu$ (s)	0.3402
$K^+ + e^- \rightleftharpoons K$ (s)	–2.924	I_2 (s) $+ 2e^- \rightleftharpoons 2I^-$	0.535
$Ba^{2+} + 2e^- \rightleftharpoons Ba$ (s)	–2.90	$Fe^{3+} + e^- \rightleftharpoons Fe^{2+}$	0.7700
$Ca^{2+} + 2e^- \rightleftharpoons Ca$ (s)	–2.76	${Hg_2}^{2+} + 2e^- \rightleftharpoons 2Hg$ (l)	0.7961
$Na^+ + e^- \rightleftharpoons Na$ (s)	–2.7109	$Ag^+ + e^- \rightleftharpoons Ag$ (s)	0.7996
$Mg^{2+} + 2e^- \rightleftharpoons Mg$ (s)	–2.375	$Hg^{2+} + 2e^- \rightleftharpoons Hg$ (l)	0.851
$Al^{3+} + 3e^- \rightleftharpoons Al$ (s)	–1.706	${AuCl_4}^- + 3e^- \rightleftharpoons Au$ (s) $+ 4Cl^-$	0.994
$Mn^{2+} + 2e^- \rightleftharpoons Mn$ (s)	–1.029	Br_2 (l) $+ 2e^- \rightleftharpoons 2Br^-$	1.065
$2H_2O + 2e^- \rightleftharpoons 2OH^- + H_2$ (g)	–0.8277	O_2 (g) $+ 4H^+ + 4e^- \rightleftharpoons 2H_2O$ (l)	1.229
$Zn^{2+} + 2e^- \rightleftharpoons Zn$ (s)	–0.7628	MnO_2 (s) + 4H+ $+ 2e^- \rightleftharpoons$ $Mn^{2+} + 2H_2O$	1.208
$Fe^{2+} + 2e^- \rightleftharpoons Fe$ (s)	–0.409	${Cr_2O_7}^{2-} + 14H^+ + 6e^- \rightleftharpoons$ $2Cr^{3+} + 7H_2O$	1.33
AgI (s) $+ e^- \rightleftharpoons Ag$ (s) $+ I^-$	–0.1519	Cl_2 (g) $+ 2e^- \rightleftharpoons 2Cl^-$	1.3583
$Sn^{2+} + 2e^- \rightleftharpoons Sn$ (s)	–0.1364	$Au^{3+} + 3e^- \rightleftharpoons Au$ (s)	1.420
$Pb^{2+} + 2e^- \rightleftharpoons Pb$ (s)	–0.1263	${MnO_4}^- + 8H^+ + 5e^- \rightleftharpoons$ $Mn^{2+} + 4H_2O$	1.491
$2H^+ + 2e^- \rightleftharpoons H_2$ (g)	0	F_2 (g) $+ 2e^- \rightleftharpoons 2F^-$	2.870
AgBr (s) $+ e^- \rightleftharpoons Ag$ (s) $+ Br^-$	0.0713		
S (s) $+ 2H^+ + 2e^- \rightleftharpoons H_2S$ (g)	0.141		
AgCl (s) $+ e^- \rightleftharpoons Ag$ (s) $+ Cl^-$	0.2223		

3 Electronics

3.1 Resistor Colour Code

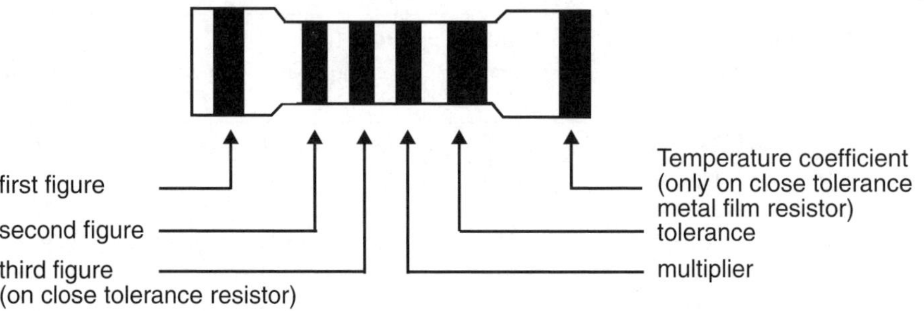

Figures		Multiplier		Tolerance (%)		Temp coeff ($10^{-6}\,K^{-1}$)	
black	0	silver	0.01	violet	0.1	black	200
brown	1	gold	0.1	blue	0.25	brown	100
red	2	black	1	green	0.5	red	50
orange	3	brown	10	brown	1	yellow	25
yellow	4	red	100	red	2	orange	15
green	5	orange	1k	gold	5		
blue	6	yellow	10k	silver	10		
violet	7	green	100k	colourless	20		
grey	8	blue	1M				
white	9	violet	10M				

3.2 E 12 and E 24 Series

The following values belong to the E 24 series for resistors (tolerance ±5 %). The E 12 series (tolerance ±10%) consists of the values in boldface.

10, 11, **12**, 13, **15**, 16, **18**, 20, **22**, 24, **27**, 30, **33**, 36, **39**, 43, **47**, 51, **56**, 62, **68**, 75, **82**, and 91.

3.3 Capacitor Colour Code

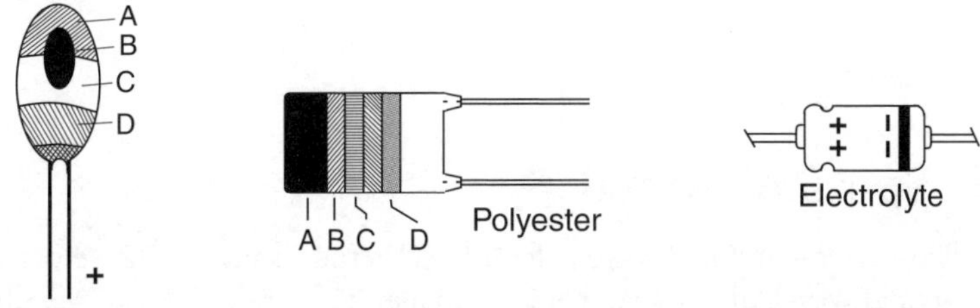

Colour	**A**	**B**	**C**		**D**	
			Multiply by		**Voltage** (V)	
	First figure	**Second figure**	**Tantalum**	**Polyester**	**Tantalum**	**Polyester**
black		0	1 μ	1 p	10	
brown	1	1	10 μ	10 p	1.6	100
red	2	2		100 p	4	250
orange	3	3		1 n	40	
yellow	4	4		10 n	6	400
green	5	5		100 n	16	
blue	6	6			20	630
violet	7	7	1 n			
grey	8	8	10 n	0.01 p	25	
white	9	9	100 n	0.1 p	3	

3.4 Gate Symbols

English name	**IEC/SEK symbol**	**DIN symbol**	**IEEE/ANSI symbol**	**Boolean algebra**
AND	&			$Z = A \cdot B$
OR	≥1			$Z = A + B$
NOT	1			$Z = \overline{A}$
NAND	&			$Z = \overline{A \cdot B}$
NOR	≥1			$Z = \overline{A + B}$
EXCLUSIVE OR	=1			$Z = A \cdot \overline{B} + \overline{A} \cdot B$ $Z = A \oplus B$

4 Waves

4.1 Speed of Sound

For solids the longitudinal wave velocity in thin rods is given. The velocity of plane longitudinal waves in massive media is greater by 15 %–45 % (for aluminium this velocity is 6420 m/s). The transverse wave velocity is generally about 60 % of the the given value. The temperature is 20 °C for solids and 0 °C (at the pressure 0.1 MPa) for gases. Other temperatures are given in parentheses.

More information: http://www.allmeasures.com/Formulae

Medium	**Sound velocity** m s^{-1}	**Medium**	**Sound velocity** m s^{-1}	**Medium**	**Sound velocity** m s^{-1}
Solid Elements		glass, flint	4 000	water (80)	1 555
aluminium	5 110	crown	5 300	water (100)	1 543
antimony	3 400	granite	4 000		
bismuth	1 800	ice (–4)	3 280	*Gases*	
cadmium	2 310	marble	5 260	air (–40)	307
cobalt	4 860	porcelain	4 880	air (0)	332
copper	3 800	wood	≈ 3 500	air (20)	343
germanium	3 910			air (+100)	387
gold	2 000	*Liquids*		ammonia	415
iron, steel	5 180	acetone	1 190	argon	308
lead	1 200	benzene	1 326	bromine	135
magnesium	5 070	carbon disulphide	1 150	carbon dioxide	260
molybdenum	6 250	carbon tetrachloride	930	carbon disulphide	189
nickel	4 970	ethanol	1 168	carbon monoxide	337
platinum	2 820	ethylether	990	chlorine	206
silver	2 790	glycerol	1 900	deuterium	890
tantalum	3 400	glycol	1 660	helium	971
tin (tetr)	2 690	mercury	1 451	hydrogen	1 286
tungsten	2 640	nitrobenzene	1 470	hydrogen sulphide	289
uranium	3 070	nitrogen (–197)	869	methane	430
zinc	3 600	oxygen (–183)	912	neon	433
		paraffin oil (34)	1 420	nitrogen	334
Other Solids		petroleum (15)	1 325	oxygen	315
acrylic	1 840	toluene	1 308	sulphur dioxide	213
brass	3 500	water (0)	1 403	water vapour (100)	405
cork	500	water (40)	1 529		

4.2 Spectrum of Electromagnetic Radiation

Frequency Hz	Photon energy J	Photon energy eV	Wavelength m
10^{25}	10^{-8}	10^{11}	10^{-17}
10^{24}	10^{-9}	10^{10}	10^{-16}
10^{23}	10^{-10}	10^{9}	10^{-15}
10^{22}	10^{-11}	10^{8}	10^{-14}
10^{21}	10^{-12}	10^{7}	10^{-13}
10^{20}	10^{-13}	10^{6}	10^{-12}
10^{19}	10^{-14}	10^{5}	10^{-11}
10^{18}	10^{-15}	10^{4}	10^{-10}
10^{17}	10^{-16}	10^{3}	10^{-9}
10^{16}	10^{-17}	10^{2}	10^{-8}
10^{15}	10^{-18}	10^{1}	10^{-7}
10^{14}	10^{-19}	10^{0}	10^{-6}
10^{13}	10^{-20}	10^{-1}	10^{-5}
10^{12}	10^{-21}	10^{-2}	10^{-4}
10^{11}	10^{-22}	10^{-3}	10^{-3}
10^{10}	10^{-23}	10^{-4}	10^{-2}
10^{9}	10^{-24}	10^{-5}	10^{-1}
10^{8}	10^{-25}	10^{-6}	10^{0}
10^{7}	10^{-26}	10^{-7}	10^{1}
10^{6}	10^{-27}	10^{-8}	10^{2}
		10^{-9}	10^{3}

Gamma rays

X-rays

Ultraviolet (UV)

Visible light

Infrared (IR)

Microwaves

UHF

SW

MW

Radio waves

LW

Visible light (approximate figures)

$4.3 - 7.5 \cdot 10^{14}$ Hz = $2.8 - 5.0 \cdot 10^{-19}$ J = 1.8 – 3.1 eV = 0.70 – 0.40 μm See also F–5.11.

Colours: 400 – 424 nm = violet, 424 – 491 nm = blue, 491 – 575 nm = green, 575 – 585 nm = yellow, 585 – 647 nm = orange, 647 – 700 nm = red.

4.3 Refractive Indices of Solids and Liquids. Optical Activity in Quartz

Refractive indices for five different Fraunhofer lines are given. The refractive index is taken relative to air at 20 °C (for ice 0 °C), 0.1 MPa, and normal moisture and CO_2 content. The Fraunhofer line wavelengths are measured in air. The C and F lines are also denoted H_α and H_β, respectively. The last entry of the table (in boldface) gives the rotation α_{20} in degrees along the polarization line in a quartz plate of thickness 1 mm at temperature 20 °C, cut at right angles to optical axis.

Fraunhofer line	**A**	**C**	**D**	**F**	**H**
Corresponding element	**oxygen**	**hydrogen**	**sodium**	**hydrogen**	**calcium**
Wavelength in nm	**760.8**	**656.3**	**589.3**	**486.1**	**396.8**
Colour	**red**	**orange**	**yellow**	**blue**	**violet**
Solids					
acrylic			1.491		
canada balsam			1.542		
diamond C			2.4173		
fluorite CaF_2	1.4310	1.4325	1.4338	1.4370	1.4421
glass, crown	1.5049	1.5076	1.5100	1.5157	1.5246
flint F3		1.6081	1.6128	1.6246	
flint SFS1		1.9104	1.9225	1.9545	
ice H_2O					
ordinary ray			1.3091		
extraordinary ray	1.3062	1.3091	1.3105	1.3147	
iceland spar $CaCO_3$					
ordinary ray	1.6500	1.6544	1.6584	1.6679	1.6832
extraordinary ray	1.4827	1.4846	1.4864	1.4908	1.4977
lithium fluoride LiF	1.3893	1.3905	1.3917	1.3949	1.3992
quartz SiO_2					
ordinary ray	1.5392	1.5419	1.5442	1.5497	1.5581
extraordinary ray	1.5481	1.5509	1.5533	1.5590	1.5677
rock salt Na Cl	1.5368	1.5406	1.5443	1.5533	1.5684
sylvine KCl	1.4838	1.4871	1.4903	1.4982	1.5115
Liquids					
aniline $C_6H_5NH_2$		1.577	1.5862	1.6042	
benzene C_6H_6	1.4910	1.4963	1.5013	1.5134	1.5340
carbon disulfide CS_2	1.6088	1.6182	1.6277	1.6523	1.6994
ethanol C_2H_5OH	1.3579	1.3599	1.3617	1.3662	1.3738
ethyl ether C_2H_5O	1.3488	1.3508	1.3526	1.3572	1.3643
glycerol $C_3H_5(OH)_3$	1.4646	1.4672	1.4695	1.4749	1.4836
pyridine C_5H_5N		1.5050	1.5094	1.5219	
silicon oil DC704			1.56		
water	1.3289	1.3312	1.3330	1.3371	1.3435
α_{20} for quartz		**17.313**	**21.724**	**32.764**	

4.4 Refractive Index of Water at Different Temperatures

Refractive index relative to air for Fraunhofer D line (589.3 nm, yellow Na light).

Temperature °C	n	Temperature °C	n	Temperature °C	n
16	1.3333	26	1.3324	60	1.3272
18	1.3332	28	1.3322	70	1.3251
20	1.3330	30	1.3319	80	1.3229
22	1.3328	40	1.3305	90	1.3205
24	1.3326	50	1.3289	100	1.3178

4.5 Refractive Index of Air

Refractive index n relative to vacuum at 15 °C and pressure 0.1 MPa. According to Cauchy $(n - 1) \cdot 10^7 = 2726.43 + 12.288/\lambda^2 + 0.3555/\lambda^4$ when the wavelength λ is given in µm. See F–5.5 for 0 °C.

Wavelength nm	n	Wavelength nm	n	Wavelength nm	n
200	1.000 325 6	500	1.000 278 1	800	1.000 274 6
250	301 4	550	277 1	850	274 4
300	290 7	600	276 3	900	274 2
350	285 0	650	275 8	950	274 0
400	281 7	700	275 3	1 000	273 9
450	1.000 279 6	750	1.000 274 9	2 000	1.000 273 0

4.6 Emission Lines of the Elements

A number of the brightest emission lines in the visible and UV regions, from a spark or discharge tube, are given. The wavelengths above 200 nm are for dry air at temperature 15 °C, pressure 0.1 MPa, and 0.03% CO_2 content. The wavelengths below 200 nm are vacuum wavelengths.

Element	Wavelength nm	Element	Wavelength nm	Element	Wavelength nm
aluminium	394.4932	barium	455.4042	cadmium	228.8018
	396.1527		553.5551		326.1057
argon	394.8979		577.7665		361.0510
	565.0703		614.1716		361.2875
	8103692		649.6901		467.8156
	811.5311		659.532		508.5824
					643.8470

T – 4.6 Emission Lines of the Elements

Element	Wavelength nm
carbon	247.8573
	283.6710
	426.702
	426.727
	657.803
cesium	455.5355
	459.3177
	672.3279
	807.8923
	852.110
	894.350
chromium	425.4346
	417.4803
	428.9721
	520.6039
copper	324.7540
	327.3962
	510.5541
	515.3235
	521.8202
	578.2132
gold	242.795
	267.595
	280.219
helium	30.3786
	58.4331
	388.6646
	447.1477
	471.3147
	492.1929
	501.5678
	587.5618
	667.8149
	706.5188
hydrogen	388.9055
	397.0074
	410.1735
	434.0465
	486.1327
	656.2725
iron	241.33087
	371.9935
	373.7133
	380.5345
	381.5842
	384.3259
	430.7906
	438.3547
	440.4752
iron	516.7491
	527.0360
	537.1493
krypton	427.39700
	431.95797
	450.23547
	557.02895
	587.09158
	642.1029
lead	220.3505
	368.3471
	405.7820
lithium	323.261
	460.2863
	610.3642
	670.7844
magnesium	279.553
	285.2129
	383.8285
manganese	257.6104
	403.0755
	403.3073
mercury	253.6519
	365.0146
	365.4833
	366.3276
	404.6561
	434.7496
	435.8343
	546.0740
	579.0654
neon	73.5886
	74.3709
	470.4395
	471.5344
	540.0562
	585.24878
	640.2246
nickel	225.386
	226.4457
	227.0213
	228.7084
	341.4765
	349.2956
nitrogen	409.994
	410.998
	566.664
	567.956
oxygen	686.72
	760.82
	777.1928
	777.4138
	777.5433
platinum	265.9454
	306.4712
potassium	404.4140
	404.7201
	691.130
	766.4907
	769.8979
rubidium	420.1851
	421.5566
	780.0227
	794.760
silicon	250.6896
	251.6111
	252.8513
	288.1595
silver	328.0683
	338.2891
	520.9067
	546.5487
sodium	330.2323
	330.2988
	588.9953
	589.5923
strontium	407.7714
	421.5524
	460.7331
	483.2075
thallium	351.924
	352.943
	377.572
	535.046
xenon	462.4276
	467.1226
zinc	213.856
	250.2001
	255.7958
	328.2333
	330.2588
	334.5020
	468.0138
	472.2159
	481.0534
	636.2347

4.7 Lasers

Pulsed character is denoted by p, continuous character by cw (continuous wave). Boldface indicates the lasing atom or molecule and most intense lines.

Type of laser	Wavelength (nm)	Output type	Comments
Excimer lasers			
Ar * F	193	p	Gas mixture containing
Kr * F	249	p	rare gas plus halogen and
Xe * Cl	308	p	buffer gas. Power ≦ 10 W.
Xe * F	351	p	
Dye lasers			
Organic dye in a solvent	300–1 000 (tunable)	p, cw	Output type depends on pump laser, tuning range depends on dye.
Gas lasers			
$\mathbf{N_2}$	337	p	Pumping dye lasers, flow tube.
He**Cd**	325, 442	cw	Sealed tube.
He**Ne**	543.5, **632.8**, 1 152, 3 391	cw	Sealed tube.
Ion lasers			
$\mathbf{Ar^+}$	**514.5**, 501.7, 496.5, **488.0**, 476.5	cw	Argon gas in sealed tube, can be mode locked.
$\mathbf{Kr^+}$	752.5, 676.4, **647.1**, 568.2, 530.9	cw	Krypton gas in sealed tube, can be mode locked.
Semiconductor lasers			
GaAs/GaAlAs	780–905	p, cw	Discrete line, wavelength depends on composition.
InGaAsP	670, 1 100–1 600	p, cw	Second wavelength depends on composition.
Solid-state lasers			
Ruby (**Cr**)	694	p	
Nd-doped glass	1 060	p	(YAG = yttrium-aluminium garnet)
Nd-doped YAG	**1 064**, 1 320	p, cw	Tunable.
Ti-sapphire	700–1 100	p, cw	Second wavelength depends on composition
Chemical lasers			
HF	2 600–3 000	p, cw	Many discrete lines.
DF	3 600–4 000	p, cw	
Carbon-dioxide lasers			
$\mathbf{CO_2}$: Flowing gas	9 000–11 000	p, cw	Power ≦ 10 kW Wavelength in pulsed mode is 10.6 μm.

4.8 Sound Intensity and Loudness

In the diagram are shown equal-loudness contours, connecting sound intensities of different frequencies perceived by the human ear to have the same loudness. The contour of 0 phon is named the threshold of hearing and that of 120 phon the threshold of feeling. Normal conversation is about 60 phon. The contours depend strongly on a person's age, those below refer to a thirty year old. At the extreme right is given the amplitude of the corresponding pressure wave in normal air.

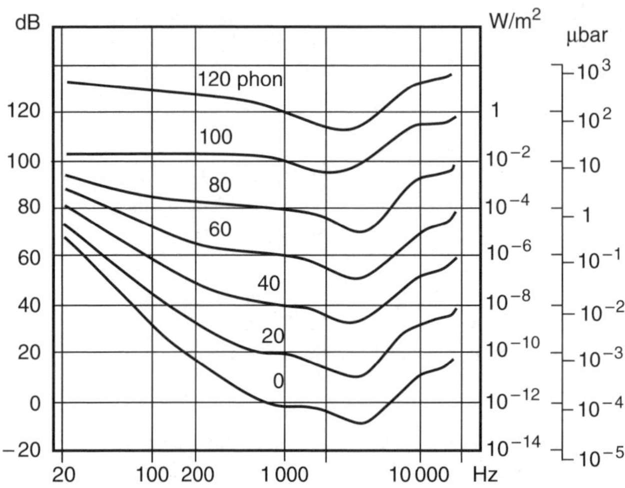

4.9 Luminous Efficiency

The following table shows the integral over all visible frequencies of the luminous flux (with respect to the sensitivity of the human eye to different frequencies) divided by the radiant flux for blackbody radiation according to Planck's radiation law.

Temperature K	**Efficiency** lm/W	**Temperature** K	**Efficiency** lm/W	**Temperature** K	**Efficiency** lm/W
1 500	0.085	2 300	4.67	4 000	54.9
1 600	0.182	2 400	6.20	5 000	81.6
1 700	0.350	2 500	8.00	6 000	93.9
1 800	0.623	2 600	10.1	6 600	95.7 (max)
1 900	1.03	2 700	12.4	7 000	95.2
2 000	1.61	2 800	15.0	8 000	90.2
2 100	2.40	2 900	17.8	9 000	82.3
2 200	3.41	3 000	20.7	10 000	73.7

4.10 Frequences and Intervals of Tones in the C Major Scale

The table gives the frequences and the intervals in relation to middle C for tones in the one-accented octave and for C in the two-accented octave. There are several tuning systems. According to the equally tempered tuning scale, which is often used by instruments with fixed frequences for tones, for exemple the piano, the octave is divided into 12 parts with intervals $\sqrt[12]{2^n}$. The note of A above middle C (a^1), which has a frequency of 440 Hz, is called the fundamental, or first harmonic frequency.

Tone	Name of interval	Interval			Frequency (Hz)	
		Just tuning		Equally tempered tuning	Just tuning	Equally tempered tuning
c^1	Prime	1	1.0000	1.00000	264.00	261.63
c^1 ♯	Augmented prime	25/24	1.0417	1.05946	275.00 }	277.19
d^1 ♭	Minor second	27/25	1.0800	1.05946	285.12 }	
d^1	Major second	9/8	1.1250	1.12246	297.00	293.66
d^1 ♯	Augmented second	75/64	1.1719	1.18921	309.38 }	311.13
e^1 ♭	Minor third	6/5	1.2000	1.18921	316.80 }	
e^1	Major third	5/4	1.2500	1.25992	330.00 }	329.63
f^1 ♭	Diminished fourth	32/25	1.2800	1.25992	337.92 }	
e^1 ♯	Augmented third	125/96	1.3021	1.33484	343.75 }	349.23
f^1	Perfect fourth	4/3	1.3333	1.33484	352.00 }	
f^1 ♯	Augmented fourth	25/18	1.3889	1.41421	366.67 }	370.00
g^1 ♭	Diminished fifth	36/25	1.4400	1.41421	380.16 }	
g^1	Perfect fifth	3/2	1.5000	1.49830	396.00	392.00
g^1 ♯	Augmented fifth	25/16	1.5625	1.58740	412.50 }	415.31
a^1 ♭	Minor sixth	8/5	1.6000	1.58740	422.40 }	
a^1	Major sixth	5/3	1.6667	1.68179	**440.00**	**440.00**
a^1 ♯	Augmented sixth	125/72	1.7361	1.78180	458.33 }	466.17
b^1	Minor seventh	9/5	1.8000	1.78180	475.20 }	
h^1	Major seventh	15/8	1.8750	1.88775	495.00 }	493.89
c^2 ♭	Diminished octave	48/25	1.9200	1.88775	506.88 }	
h^1 ♯	Augmented seventh	125/64	1.9531	2.00000	515.25 }	523.25
c^2	Perfect octave	2	2.0000	2.00000	528.00 }	

5 Atomic and Molecular Physics

5.1 The Elements – Properties of Free Atoms

Column 4 The relative atomic mass is an average over the natural mixture of isotopes. For elements with no stable isotope the mass number (number of protons plus number of neutrons) of the most long-lived isotope is given in parentheses. The relative atomic mass is a dimensionless quantity.

Number of atoms in 1 g = N_A/A_r where N_A = Avogadro number and A_r = relative atomic mass.

Column 5 The atomic radius has been calculated from the wavefunctions of the elements, and is equal to the expectation value (mean value) of the distance of the outermost electron to the nucleus.

Column 6 The level notation refers to the ground state of the atom.

Column 7 This column gives the energy required to split off the loosest bound electron of the neutral atom (I) and the singly ionized atom (II).

Column 8 The electron affinity is the energy released when a neutral atom accepts an electron, forming a monovalent negative ion. The electron affinity for most of the rare earths, which are marked with an asterisk, is 0.2 – 0.5 eV. The uncertainty is at most 9 units in the last digit.

Column 9 The electronegativity according to Pauling is given.

Internet sites with information on the elements:

http://pearl1.lanl.gov/periodic/
http://pol.spurious.biz/projects/chemglobe/ptoe/
http://www.chemicool.com/
http://www.gordonengland.co.uk/xelements/periodic.htm/
http://www.webelements.com/

T – 5.1 The Elements – Properties of Free Atoms

1	2	3	4	5	6	7		8	9
Element	**Symbol**	**Atomic number**	**Relative atomic mass**	**Atomic radius** pm	**Level**	**Ionization energy** eV		**Electron affinity** eV	**Electro-negativity**
						I	**II**		
actinium	Ac	89	(227)	245	$^2D_{3/2}$	5.17	12.1		1.1
aluminium	Al	13	26.98154	202	$^2P_{1/2}$	5.986	18.8	0.4411	1.5
americium	Am	95	(243)	242	$^8S_{7/2}$	5.993			1.3
antimony	Sb	51	121.75	168	$^4S_{3/2}$	8.641	16.5	1.075	1.9
argon	Ar	18	39.948	89	$^1S_0$	15.759	27.6	<0	
arsenic	As	33	74.9216	141	$^4S_{3/2}$	9.81	18.6	0.813	2.0
astatine	At	85	(210)	132	$^2P_{3/2}$	8.8		2.82	2.2
barium	Ba	56	137.327	248	$^1S_0$	5.212	10.0	0.144626	0.9
berkelium	Bk	97	(247)	226	$^8H_{17/2}$	6.23			
beryllium	Be	4	9.0122	149	$^1S_0$	9.322	18.2	<0	1.5
bismuth	Bi	83	208.980	188	$^4S_{3/2}$	7.289	16.7	0.9461	1.9
bohrium	Bh	107	(262)						
boron	B	5	10.811	134	$^2P_{1/2}$	8.298	25.2	0.2771	2.0
bromine	Br	35	79.904	114	$^2P_{3/2}$	11.814	21.6	3.3653	2.8
cadmium	Cd	48	112.412	152	$^1S_0$	8.993	16.9	<0	1.7
calcium	Ca	20	40.078	225	$^1S_0$	6.113	11.9	0.024551	1.0
californium	Cf	98	(251)	224		6.30			
carbon	C	6	12.01115	100	$^3P_0$	11.260	24.4	1.26293	2.5
cerium	Ce	58	140.12	241	$^1G_4$	5.466	12.3	*	1.1
cesium	Cs	55	132.9054	322	$^2S_{1/2}$	3.894	25.1	0.471630	0.7
chlorine	Cl	17	35.4527	90	$^1P_{3/2}$	12.967	23.8	3.6173	3.0
chromium	Cr	24	51.9961	197	$^7S_3$	6.766	16.5	0.6661	1.6
cobalt	Co	27	58.9332	162	$^4F_{9/2}$	7.864	17.1	0.6611	1.8
copper	Cu	29	63.546	163	$^2S_{1/2}$	7.478		1.2281	1.9
curium	Cm	96	(247)	228	$^8S_{7/2}$	6.02			
darmstadtium	Ds	110	(271)						
dubnium	Db	105	(262)						
dysprosium	Dy	66	162.50	236	$^5I_8$	5.927		*	
einsteinium	Es	99	(252)	222		6.42			1.2
erbium	Er	68	167.26	230	$^3H_6$	6.101		*	
europium	Eu	63	151.96	245	$^8S_{7/2}$	5.666		*	
fermium	Fm	100	(257)	221		6.50			
fluorine	F	9	18.998403	60	$^2P_{3/2}$	17.422		3.3993	4.0
francium	Fr	87	(223)	313	$^2S_{1/2}$	3.8			0.7
gadolinium	Gd	64	157.25	226	$^9D_2$	6.141		*	1.1
gallium	Ga	31	69.723	196	$^2P_{1/2}$	5.999		0.302	1.6
germanium	Ge	32	72.59	160	$^3P_0$	7.899		1.22	1.8
gold	Au	79	196.9665	162	$^2S_{1/2}$	9.225		2.308633	2.4
hafnium	Hf	72	178.49	191	$^3F_2$	6.65		≈ 0	1.3
hassium	Hs	108	(265)						

T – 5.1 The Elements – Properties of Free Atoms

1	2	3	4	5	6	7		8	9
Element	**Symbol**	**Atomic number**	**Relative atomic mass**	**Atomic radius** pm	**Level**	**Ionization energy** eV		**Electron affinity** eV	**Electro-negativity**
						I	**II**		
helium	He	2	4.002602	54	$^{1}S_{0}$	24.588	54.4	<0	
holmium	Ho	67	164.930	233	$^{4}I_{15/2}$	6.018		*	1.2
hydrogen	H	1	1.00798	79	$^{2}S_{1/2}$	13.598		0.754209	2.1
indium	In	49	114.82	216	$^{2}P_{1/2}$	5.786	18.9	0.32	1.7
iodine	I	53	126.9044	138	$^{2}P_{3/2}$	10.451	19.1	3.05914	2.5
iridium	Ir	77	192.22	165	$^{4}F_{9/2}$	9.1		1.5658	2.2
iron	Fe	26	55.844	168	$^{5}D_{4}$	7.870	16.2	0.1634	1.8
krypton	Kr	36	83.80	103	$^{1}S_{0}$	13.999	24.6	<0	
lanthanum	La	57	138.91	232	$^{2}D_{3/2}$	5.577	11.4	0.53	
lawrencium	Lr	103	(261)	216		8.6			1.1
lead	Pb	82	207.19	169	$^{3}P_{0}$	7.416	15.0	0.3648	
lithium	Li	3	6.941	205	$^{2}S_{1/2}$	5.392	75.6	0.61805	1.8
lutetium	Lu	71	174.97	200	$^{2}D_{3/2}$	5.426	14.7	*	1.0
magnesium	Mg	12	24.3051	178	$^{1}S_{0}$	7.646	15.0	<0	1.2
manganese	Mn	25	54.9380	173	$^{6}S_{5/2}$	7.437	15.6	<0	1.2
meitnerium	Mt	109	(266)						1.5
mendelevium	Md	101	(258)	219		6.58		1.5	
mercury	Hg	80	200.59	156	$^{1}S_{0}$	10.437	18.8	<0	1.9
molybdenum	Mo	42	95.93	194	$^{7}S_{3}$	7.099	16.2	0.7461	1.8
neodymium	Nd	60	144.24	255	$^{5}I_{4}$	5.489		*	
neon	Ne	10	20.180	53	$^{1}S_{0}$	21.564	41.1	<0	
neptunium	Np	93	(237)	234	$^{4}L_{11/2}$	6.19			1.3
nickel	Ni	28	58.69	157	$^{3}F_{4}$	7.638	18.2	1.1561	1.8
niobium	Nb	41	92.906	200	$^{6}D_{1/2}$	6.88	14.3	0.8933	1.6
nitrogen	N	7	14.00672	81	$^{4}S_{3/2}$	14.534	29.6	–0.072	3.0
nobelium	No	102	(259)	218		6.65			
osmium	Os	76	190.2	169	$^{5}D_{4}$	8.7	17	1.12	2.2
oxygen	O	8	15.9994	70	$^{3}P_{2}$	13.618	35.1	1.4611215	3.5
palladium	Pd	46	106.42	76	$^{1}S_{0}$	8.34	19.4	0.5578	2.2
phosphorus	P	15	30.97376	130	$^{4}S_{3/2}$	10.486	19.7	0.74653	2.1
platinum	Pt	78	195.08	175	$^{3}D_{3}$	9.0	18.6	2.1282	2.2
plutonium	Pu	94	(244)	244	$^{3}F_{0}$	6.06			1.3
polonium	Po	84	(209)	170	$^{3}P_{2}$	8.42	31.8	1.93	2.0
potassium	K	19	39.0983	280	$^{2}S_{1/2}$	4.341	31.8	0.501471	0.8
praseodymium	Pr	59	140.9076	258	$^{4}I_{9/2}$	5.422		*	1.1
prometium	Pm	61	(145)	251	$^{6}H_{5/2}$	5.554		*	
protactinium	Pa	91	(231)	239	$^{4}K_{11/2}$	5.89			1.5

1	2	3	4	5	6	7		8	9
Element	**Symbol**	**Atomic number**	**Relative atomic mass**	**Atomic radius** pm	**Level**	**Ionization energy** eV		**Electron affinity** eV	**Electro-negativity**
						I	**II**		
radium	Ra	88	(226)	262	$^1S_0$	5.279	10.1		0.9
radon	Rn	86	(222)	124	$^1S_0$	10.748		<0	
rhenium	Re	75	186.207	173	$^6S_{5/2}$	7.88	16.6	0.152	1.9
rhodium	Rh	45	102.9055	179	$^4F_{9/2}$	7.46	18.1	1.1378	2.2
roentgenium	Rg	111	(272)						
rubidium	Rb	37	85.4678	268	$^2S_{1/2}$	4.177	27.5	0.485922	0.8
ruthenium	Ru	44	101.07	183	$^5F_5$	7.37	16.8	1.052	2.2
rutherfordium	Rf	104	(261)						
samarium	Sm	62	150.36	248	$^7F_0$	5.631	11.2	*	1.2
scandium	Sc	21	44.95591	212	$^2D_{3/2}$	6.562	12.8	0.1882	1.3
seaborgium	Sg	106	(263)						
selenium	Se	34	78.96	126	$^3P_2$	9.752	21.5	2.020693	2.4
silicon	Si	14	28.0855	157	$^3P_0$	8.151	16.3	1.3855	1.8
silver	Ag	47	107.868	172	$^2S_{1/2}$	7.576	21.5	1.3027	1.9
sodium	Na	11	22.989767	221	$^2S_{1/2}$	5.139	47.3	0.5479303	0.9
strontium	Sr	38	87.62	220	$^1S_0$	5.695	11.0	0.052066	1.0
sulphur	S	16	32.064	101	$^3P_2$	10.360	23.4	2.077120	2.5
tantalum	Ta	73	180.948	184	$^4F_{3/2}$	7.89	16.2	0.3221	1.5
technetium	Tc	43	(98)	172	$^6S_{5/2}$	7.28	15.3	0.552	1.9
tellurium	Te	52	127.60	157	$^3P_2$	9.009	18.6	1.97083	2.1
terbium	Tb	65	158.9253	224	$^6H_{15/2}$	5.852		*	1.2
thallium	Tl	81	204.383	205	$^3F_2$	6.108	20.4	0.22	1.8
thorium	Th	90	232.0381	233	$^2P_{1/2}$	6.08	11.5		1.3
thulium	Tm	69	168.9342	227	$^2F_{7/2}$	6.184	12.1	*	1.2
tin	Sn	50	118.710	161	$^3P_0$	7.344	14.6	1.22	1.8
titanium	Ti	22	47.88	197	$^3F_2$	6.82	13.6	0.0791	1.5
tungsten	W	74	183.85	178	$^4D_0$	7.98	17.7	0.8158	1.7
uranium	U	92	238.0289	234	$^5L_4$	6.05			1.7
vanadium	V	23	50.9415	188	$^4F_{3/2}$	6.740	14.7	0.5251	1.6
xenon	Xe	54	131.29	127	$^1S_0$	12.130	21.2	<0	
ytterbium	Yb	70	173.03	215	$^1S_0$	6.254	12.1	*	1.1
yttrium	Y	39	88.9058	204	$^2D_{3/2}$	6.22	12.2	0.3071	1.3
zinc	Zn	30	65.40	148	$^1S_0$	9.394	18.0	<0	1.6
zirconium	Zr	40	91.224	193	$^3F_2$	6.84	13.1	0.4261	1.4

5.2 Periodic Table of the Elements

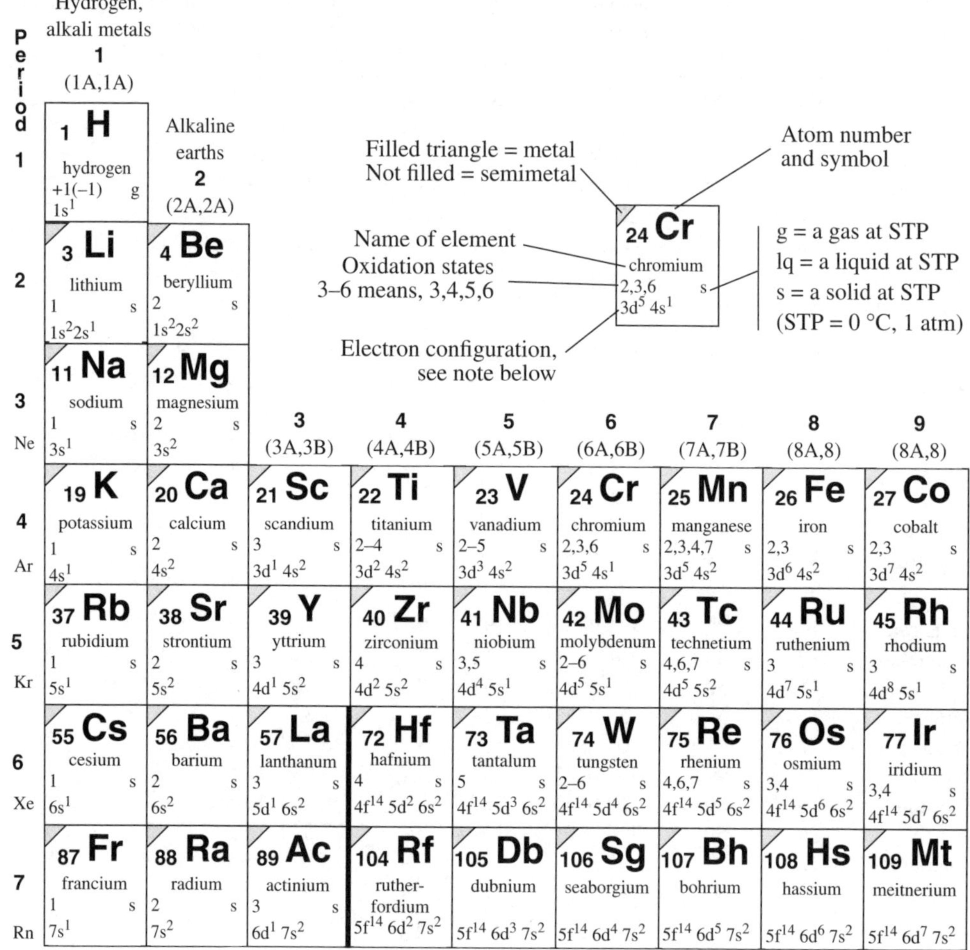

For periods 3–7 only the electron configuration of the outer shells is given.
Add the configuration for the inert gas at the extreme left.

Lanthanides →	Xe	58 Ce cerium 3,4 s $4f^1 5d^1 6s^2$	59 Pr praseody-mium 3 s $4f^3 6s^2$	60 Nd neodymium 3 s $4f^4 6s^2$	61 Pm promethium 3 s $4f^5 6s^2$	62 Sm samarium 2,3 s $4f^6 6s^2$	63 Eu europium 2,3 s $4f^7 6s^2$
Actinides →	Rn	90 Th thorium 4 s $6d^2 7s^2$	91 Pa protactinium 4,5 s $5f^2 6d^1 7s^2$	92 U uranium 3–6 s $5f^3 6d^1 7s^2$	93 Np neptunium 3–6 s $5f^4 6d^1 7s^2$	94 Pu plutonium 3–6 s $5f^6 7s^2$	95 Am americium 3–6 s $5f^7 7s^2$

Transuraniums →

The groups (columns) are now called 1–18. Earlier there were two systems: the IUPAC system and the CAS version. These old notations are found within parentheses with the IUPAC form first.

10 (8A,8)	Coin metals 11 (1B,1B)	12 (2B,2B)	Boron group 13 (3B,3A)	Carbon group 14 (4B,4A)	Nitrogen group 15 (5B,5A)	Oxygen group 16 (6B,6A)	Halo-gens 17 (7B,7A)	Inert gases 18 (0,8A)
								2 He helium 0 g $1s^2$
			5 B boron 3 s $1s^2 2s^2 2p^1$	**6 C** carbon ±4 (2) s $1s^2 2s^2 2p^2$	**7 N** nitrogen * g $1s^2 2s^2 2p^3$	**8 O** oxygen –2 (–1) g $1s^2 2s^2 2p^4$	**9 F** fluorine –1 g $1s^2 2s^2 2p^5$	**10 Ne** neon 0 g $1s^2 2s^2 2p^6$
			13 Al aluminium 3 s $3s^2 3p^1$	**14 Si** silicon 2±4 s $3s^2 3p^2$	**15 P** phosphorus ±3,5 s $3s^2 3p^3$	**16 S** sulphur –2,4,6 s $3s^2 3p^4$	**17 Cl** chlorine ±1,5,7 g $3s^2 3p^5$	**18 Ar** argon 0 g $3s^2 3p^6$
28 Ni nickel 2,3 s $3d^8 4s^2$	**29 Cu** copper 1,2 s $3d^{10} 4s^1$	**30 Zn** zinc 2 s $3d^{10} 4s^2$	**31 Ga** gallium 3 s $3d^{10} 4s^2 4p^1$	**32 Ge** germanium 2,4 s $3d^{10} 4s^2 4p^2$	**33 As** arsenic ±3,5 s $3d^{10} 4s^2 4p^3$	**34 Se** selenium –2,4,6 s $3d^{10} 4s^2 4p^4$	**35 Br** bromine ±1,5 1q $3d^{10} 4s^2 4p^5$	**36 Kr** krypton 0 g $3d^{10} 4s^2 4p^6$
46 Pd palladium 2,4 s $4d^{10}$	**47 Ag** silver 1 s $4d^{10} 5s^1$	**48 Cd** cadmium 2 s $4d^{10} 5s^2$	**49 In** indium 3 s $4d^{10} 5s^2 5p^1$	**50 Sn** tin 2,4 s $4d^{10} 5s^2 5p^2$	**51 Sb** antimony ±3,5 s $4d^{10} 5s^2 5p^3$	**52 Te** tellurium –2,4,6 s $4d^{10} 5s^2 5p^4$	**53 I** iodine ±1,5,7 s $4d^{10} 5s^2 5p^5$	**54 Xe** xenon 0 g $4d^{10} 5s^2 5p^6$
78 Pt platinum 2,4 s $4f^{14} 5d^9 6s^1$	**79 Au** gold 1,3 s $4f^{14} 5d^{10} 6s^1$	**80 Hg** mercury 1,2 1q $4f^{14} 5d^{10} 6s^2$	**81 Tl** thallium 1,3 s Hg + $6p^1$	**82 Pb** lead 2,4 s Hg + $6p^2$	**83 Bi** bismuth 3,5 s Hg + $6p^3$	**84 Po** polonium 2,4 s Hg + $6p^4$	**85 At** astatine ±1,5,7 s Hg + $6p^5$	**86 Rn** radon 0 g Hg + $6p^6$
110 Ds darmstadtium $5f^{14} 6d^9 7s^1$	**111 Rg** roentgenium $5f^{14} 6d^{10} 7s^1$							

* Nitrogen can have oxidation states –1,–2,–3,1,2,3,4,5

Light platinum metals: Ru, Rh, and Pd
Heavy platinum metals: Os, Ir, and Pt
Rare earths: Sc, Y, La, and Lanthanides

64 Gd gadolinium 3 s $4f^7 5d^1 6s^2$	**65 Tb** terbium 3 s $4f^9 6s^2$	**66 Dy** dysprosium 3 s $4f^{10} 6s^2$	**67 Ho** holmium 3 s $4f^{11} 6s^2$	**68 Er** erbium 3 s $4f^{12} 6s^2$	**69 Tm** thulium 3 s $4f^{13} 6s^2$	**70 Yb** ytterbium 2,3 s $4f^{14} 6s^2$	**71 Lu** lutetium 3 s $4f^{14} 5d^1 6s^2$
96 Cm curium 3 $5f^7 6d^1 7s^2$	**97 Bk** berkelium 3,4 $5f^9 7s^2$	**98 Cf** californium 3 $5f^{10} 7s^2$	**99 Es** einsteinium $5f^{11} 7s^2$	**100 Fm** fermium $5f^{12} 7s^2$	**101 Md** mendele-vium $5f^{13} 7s^2$	**102 No** nobelium 2,3 $5f^{14} 7s^2$	**103 Lr** lawrencium 3 ? $5f^{14} 6d^1 7s^2$

5.3 The Elements – Electron Binding Energies of the K, L, and M Shells. X-ray Energies of $K\alpha_1$ Lines

Listed are the energies in eV required to expel electrons from different subshells of the K, L, and M shells. The last column gives the energy of $K\alpha_1$ X-rays, i.e. the energy difference K–L_3. Reference level is the vacuum level for gases and the Fermi level for solids. In many cases the binding energies refer to the element in a chemical compound, usually an oxide (o) or a halide (h). Chemical shifts can be as large as 10 eV. The energies of the heaviest elements have been extrapolated, the uncertainties for these elements being of the order of 100 eV. Binding energies of atomic valence electrons are given in Table 5.1, Column 7.

Atomic number	Element	K $1s_{1/2}$	L_1 $2s_{1/2}$	L_2 $2p_{1/2}$	L_3 $2p_{3/2}$	M_1 $3s_{1/2}$	M_2 $3p_{1/2}$	M_3 $3p_{3/2}$	M_4 $3d_{3/2}$	M_5 $3d_{5/2}$	$K\alpha_1$
1	H	14									
2	He	25									
3	Li	55									54
4	Be	112									108
5	B	188									183
6	C	284									277
7	N	410	19								392
8	O	543	24								525
9	F	697	31								677
10	Ne	870	48		22						849
11	Na	1 072	63		30						1 042
12	Mg	1 303	89		49						1 254
13	Al	1 558	118		73						1 485
14	Si	1 839	149	100	99						1 740
15	P	2 143	187	131	130						2 013
16	S	2 472	229	165	164	16					2 308
17	Cl (h)	2 823	270	200	199	16					2 624

T – 5.3 The Elements – Electron Binding Energies of the K, L, and M Shells. X-ray Energies of $K\alpha_1$ Lines

Atomic number	Element		K $1s_{1/2}$	L_1 $2s_{1/2}$	L_2 $2p_{1/2}$	L_3 $2p_{3/2}$	M_1 $3s_{1/2}$	M_2 $3p_{1/2}$	M_3 $3p_{3/2}$	M_4 $3d_{3/2}$	M_5 $3d_{5/2}$	$K\alpha_1$
18	Ar		3 206	326	251	249	29	16				2 957
19	K		3 608	379	297	294	35	20				3 314
20	Ca		4 038	438	350	347	44	25				3 692
21	Sc		4 490	498	404	400	51	28				4 090
22	Ti		4 966	561	461	455	58	33				4 511
23	V		5 466	627	521	513	64	37				4 953
24	Cr		5 991	698	585	576	75	42				5 415
25	Mn		6 538	769	651	639	82	47				5 899
26	Fe		7 111	848	721	707	92	53				6 404
27	Co		7 711	927	781	769	101	59				6 930
28	Ni		8 332	1 010	871	854	111	67				7 478
29	Cu		8 981	1 099	953	933	122	77	75			8 048
30	Zn		9 661	1 196	1 045	1 022	140	91	89			8 639
31	Ga		10 367	1 298	1 143	1 117	158	107	103	18		9 250
32	Ge		11 104	1 413	1 249	1 217	181	129	122	29		9 886
33	As		11 867	1 527	1 359	1 323	204	147	141	42		10 544
34	Se		12 658	1 654	1 476	1 436	231	167	162	56		11 222
35	Br	(h)	13 474	1 782	1 596	1 550	257	189	182	70	69	11 924
36	Kr		14 326	1 925	1 731	1 678	293	222	214	95	94	12 648
37	Rb	(h)	15 200	2 065	1 865	1 805	321	248	239	112	111	13 395
38	Sr	(o)	16 105	2 216	2 007	1 940	358	280	269	135	133	14 165
39	Y	(o)	17 039	2 373	2 155	2 080	395	313	301	160	158	14 958
40	Zr	(o)	18 000	2 534	2 309	2 225	433	347	333	185	182	15 775
41	Nb		18 983	2 695	2 462	2 368	466	376	360	205	202	16 615
42	Mo		20 000	2 866	2 625	2 520	505	410	393	230	226	17 479
43	Tc		21 044	3 042	2 793	2 677	544	445	425	257	253	18 367

T – 5.3 The Elements – Electron Binding Energies of the K, L, and M Shells. X-ray Energies of $K\alpha_1$ Lines

Atomic number	Element	K $1s_{1/2}$	L_1 $2s_{1/2}$	L_2 $2p_{1/2}$	L_3 $2p_{3/2}$	M_1 $3s_{1/2}$	M_2 $3p_{1/2}$	M_3 $3p_{3/2}$	M_4 $3d_{3/2}$	M_5 $3d_{5/2}$	$K\alpha_1$
44	Ru	22 117	3 224	2 967	2 838	585	483	461	284	279	19 279
45	Rh	23 220	3 412	3 146	3 004	627	521	496	312	307	20 216
46	Pd	24 350	3 605	3 331	3 174	670	559	531	340	335	21 176
47	Ag	25 514	3 806	3 524	3 352	717	602	573	374	368	22 162
48	Cd	26 711	4 018	3 727	3 538	770	651	617	411	405	23 174
49	In	27 940	4 238	3 938	3 730	826	702	664	451	444	24 210
50	Sn	29 200	4 465	4 156	3 929	884	757	715	496	485	25 271
51	Sb	30 491	4 699	4 381	4 132	944	812	766	537	528	26 359
52	Te	31 814	4 939	4 612	4 341	1 006	870	819	583	573	27 472
53	I (h)	33 170	5 188	4 852	4 557	1 072	931	875	631	620	28 612
54	Xe	34 566	5 453	5 107	4 787	1 149	1 002	941	689	677	29 779
55	Cs (o)	35 985	5 713	5 360	5 012	1 217	1 065	998	740	724	30 973
56	Ba (o)	37 441	5 987	5 624	5 247	1 293	1 137	1 063	796	781	32 194
57	La (o)	38 925	6 267	5 891	5 483	1 362	1 205	1 124	852	835	33 442
58	Ce (o)	40 444	6 549	6 165	5 724	1 435	1 273	1 186	901	883	34 720
59	Pr (o)	41 991	6 835	6 441	5 965	1 511	1 338	1 243	951	931	36 026
60	Nd (o)	43 569	7 126	6 722	6 208	1 576	1 403	1 298	1 000	978	37 361
61	Pm (o)	45 185	7 430	7 015	6 465	1 656	1 478	1 364	1 060	1 034	38 720
62	Sm (o)	46 835	7 737	7 312	6 717	1 728	1 546	1 425	1 111	1 085	40 118
63	Eu (o)	48 519	8 052	7 618	6 977	1 805	1 619	1 486	1 166	1 136	41 542
64	Gd (o)	50 239	8 376	7 931	7 243	1 888	1 695	1 551	1 225	1 193	42 996
65	Tb (o)	51 996	8 708	8 252	7 515	1 970	1 770	1 614	1 278	1 244	44 482
66	Dy (o)	53 788	9 047	8 581	7 790	2 050	1 845	1 679	1 335	1 298	45 998
67	Ho (o)	55 618	9 395	8 918	8 071	2 123	1 918	1 736	1 386	1 346	47 547
68	Er (o)	57 486	9 752	9 265	8 358	2 211	2 010	1 816	1 457	1 413	49 128

T – 5.3 The Elements – Electron Binding Energies of the K, L, and M Shells. X-ray Energies of $K\alpha_1$ Lines

Atomic number	Element	K $1s_{1/2}$	L_1 $2s_{1/2}$	L_2 $2p_{1/2}$	L_3 $2p_{3/2}$	M_1 $3s_{1/2}$	M_2 $3p_{1/2}$	M_3 $3p_{3/2}$	M_4 $3d_{3/2}$	M_5 $3d_{5/2}$	$K\alpha_1$
69	Tm (o)	59 390	10 116	9 617	8 648	2 305	2 088	1 883	1 513	1 466	50 742
70	Yb (o)	61 332	10 487	9 978	8 943	2 397	2 172	1 949	1 576	1 527	52 389
71	Lu (o)	63 314	10 870	10 349	9 244	2 491	2 264	2 024	1 640	1 589	54 070
72	Hf (o)	65 351	11 272	10 739	9 561	2 601	2 365	2 108	1 716	1 662	55 790
73	Ta	67 413	11 680	11 136	9 881	2 705	2 466	2 191	1 790	1 732	57 532
74	W	69 523	12 099	11 542	10 205	2 817	2 572	2 278	1 869	1 807	59 318
75	Re	71 675	12 527	11 957	10 535	2 932	2 682	2 367	1 949	1 883	61 140
76	Os	73 871	12 968	12 385	10 871	3 049	2 792	2 458	2 031	1 960	63 000
77	Ir	76 111	13 419	12 824	11 215	3 174	2 909	2 551	2 116	2 041	64 896
78	Pt	78 395	13 880	13 273	11 564	3 298	3 027	2 646	2 202	2 121	66 832
79	Au	80 722	14 353	13 733	11 918	3 425	3 150	2 743	2 291	2 206	68 804
80	Hg	83 103	14 839	14 209	12 284	3 562	3 279	2 847	2 385	2 295	70 819
81	Tl	85 529	15 347	14 698	12 657	3 704	3 416	2 957	2 485	2 390	72 872
82	Pb	88 005	15 861	15 200	13 035	3 851	3 554	3 066	2 586	2 484	74 969
83	Bi	90 526	16 388	15 709	13 418	3 999	3 696	3 177	2 687	2 580	77 108
84	Po	93 105	16 939	16 244	13 814	4 149	3 854	3 302	2 798	2 683	79 290
85	At	95 730	17 493	16 785	14 214	4 317	4 008	3 426	2 909	2 787	81 520
86	Rn	98 400	18 049	17 337	14 619	4 482	4 159	3 538	3 022	2 892	83 780
87	Fr	101 140	18 639	17 906	15 031	4 652	4 327	3 663	3 136	3 000	86 100
88	Ra	103 920	19 237	18 484	15 444	4 822	4 490	3 792	3 248	3 105	88 470
89	Ac	106 760	19 840	19 083	15 871	5 002	4 656	3 909	3 370	3 219	90 884
90	Th	109 650	20 472	19 693	16 300	5 182	4 831	4 046	3 491	3 332	93 350
91	Pa	112 600	21 105	20 314	16 733	5 367	5 001	4 174	3 611	3 442	95 868
92	U (o)	115 610	21 758	20 948	17 168	5 548	5 181	4 304	3 728	3 552	98 439
93	Np	118 680	22 420	21 599	17 608	5 722	5 366	4 435	3 850	3 664	101 068

T – 5.3 The Elements – Electron Binding Energies of the K, L, and M Shells. X-ray Energies of $K\alpha_1$ Lines

Atomic number	Element	K $1s_{1/2}$	L_1 $2s_{1/2}$	L_2 $2p_{1/2}$	L_3 $2p_{3/2}$	M_1 $3s_{1/2}$	M_2 $3p_{1/2}$	M_3 $3p_{3/2}$	M_4 $3d_{3/2}$	M_5 $3d_{5/2}$	$K\alpha_1$
94	Pu	121 820	23 102	22 266	18 057	5 933	5 546	4 562	3 973	3 778	103 761
95	Am (o)	125 030	23 773	22 944	18 504	6 120	5 710	4 667	4 092	3 887	106 523
96	Cm	128 220	24 460	23 779	18 930	6 288	5 895	4 797	4 227	3 971	109 290
97	Bk	131 590	25 275	24 385	19 452	6 556	6 147	4 977	4 357	4 131	112 138
98	Cf	134 940	26 030	25 110	19 910	6 790	6 361	5 130	4 489	4 252	115 030
99	Es	138 390	26 800	25 860	20 390	7 011	6 574	5 273	4 622	4 373	118 000
100	Fm	141 920	27 590	26 640	20 870	7 240	6 793	5 419	4 757	4 497	121 050
101	Md	145 530	28 410	27 440	21 360	7 471	7 017	5 566	4 894	4 621	124 170
102	No	149 210	29 240	28 260	21 850	7 710	7 247	5 716	5 033	4 748	127 360
103	Lr	152 970	30 100	29 100	22 360	7 962	7 491	5 874	5 181	4 882	130 610
104	Rf	156 820	30 990	29 970	22 870	8 221	7 741	6 035	5 331	5 019	133 950

5.4 The Elements – Electron Binding Energies of the N and O Shells

The binding energies are in eV

Atomic number	Element	N_1 $4s_{1/2}$	N_2 $4p_{1/2}$	N_3 $4p_{3/2}$	N_4 $4d_{3/2}$	N_5 $4d_{5/2}$	N_6 $4f_{5/2}$	N_7 $4f_{7/2}$	O_1 $5s_{1/2}$	O_2 $5p_{1/2}$	O_3 $5p_{3/2}$	O_4 $5d_{3/2}$	O_5 $5d_{5/2}$
36	Kr	27	14										
37	Rb	29	14										
38	Sr	38	20										
39	Y	46	26										
40	Zr	54	31										
41	Nb	55	31										
42	Mo	62	35										
43	Tc	68	39										
44	Ru	75	43										
45	Rh	81	48										
46	Pd	86	51										
47	Ag	95	59										
48	Cd	108	67		9								
49	In	122	77		16								
50	Sn	137	89		25	24							
51	Sb	152	99		33	32			7				
52	Te	169	110		42	40			12				
53	I	186	123		50				14				
54	Xe	213	146		69	68			23	13	12		
55	Cs	231	162		77	75			23				
56	Ba	253	180		93	90			40				
57	La	271	196		103				36	18			

Atomic number	Element	N_1 $4s_{1/2}$	N_2 $4p_{1/2}$	N_3 $4p_{3/2}$	N_4 $4d_{3/2}$	N_5 $4d_{5/2}$	N_6 $4f_{5/2}$	N_7 $4f_{7/2}$	O_1 $5s_{1/2}$	O_2 $5p_{1/2}$	O_3 $5p_{3/2}$	O_4 $5d_{3/2}$	O_5 $5d_{5/2}$
58	Ce	289	207		112	108			38	18			
59	Pr	304	218		118	115			38	23			
60	Nd	321	230		124	121			38	22			
61	Pm	237	242		133	129			38	22			
62	Sm	351	251		137	132			33	26			
63	Eu	366	261		141	136			37	27			
64	Gd	383	311	272	147	142			43	28			
65	Tb	400	322	284	152	148			42	28			
66	Dy	419	339	297	158	155			66	29			
67	Ho	431	349	309	164	161			51	20			
68	Er	453	366	320	172	169			64	33			
69	Tm	470	382	333	180	176			51	30			
70	Yb	487	399	346	189	185			53	23			
71	Lu	507	412	359	206	196	7		57	28			
72	Hf	538	437	380	224	214	19	17	65	38	31		
73	Ta	563	462	402	239	227	25	23	68	42	34		
74	W	592	489	423	255	243	34	32	74	44	34		
75	Re	625	518	445	274	260	43	40	83	46	35		
76	Os	655	547	469	290	273	54	51	84	58	46		
77	Ir	690	577	495	312	296	64	61	96	63	51		
78	Pt	724	608	520	332	315	75	71	102	66	52		
79	Au	759	644	546	352	334	88	84	108	72	54		
80	Hg	800	677	571	379	360	104	100	120	81	58		
81	Tl	846	722	609	407	386	123	119	137	100	76	15	12
82	Pb	894	764	645	434	412	141	136	148	105	86	20	18

T – 5.4 The Elements – Electron Binding Energies of the N and O Shells

Atomic number	Element	N_1 $4s_{1/2}$	N_2 $4p_{1/2}$	N_3 $4p_{3/2}$	N_4 $4d_{3/2}$	N_5 $4d_{5/2}$	N_6 $4f_{5/2}$	N_7 $4f_{7/2}$	O_1 $5s_{1/2}$	O_2 $5p_{1/2}$	O_3 $5p_{3/2}$	O_4 $5d_{3/2}$	O_5 $5d_{5/2}$
83	Bi	939	805	679	464	441	162	157	160	117	93	27	25
84	Po	995	851	705	500	473	184		177	132	104	31	
85	At	1 042	886	740	533	507	210		195	148	115	40	
86	Rn	1 097	929	768	567	541	238		214	164	127	48	
87	Fr	1 153	980	810	603	577	268		234	182	140	58	
88	Ra	1 208	1 058	879	636	603	299		254	200	153	68	
89	Ac	1 269	1 080	890	675	639	319		272	215	167	80	
90	Th	1 330	1 168	968	714	677	344	335	290	229	182	95	88
91	Pa	1 387	1 224	1 007	743	708	371	360	310	223		94	
92	U	1 442	1 273	1 045	780	738	392	381	324	260	195	105	96
93	Np	1 501	1 328	1 087	817	773	415	404	338	283	206	109	101
94	Pu	1 558	1 377	1 120	849	801	422		352	279	212	116	105
95	Am	1 617	1 412	1 173	883	832	464	449	351	304	216	119	109
96	Cm	1 643	1 440	1 221	932	878	497	481	382	324	349	141	123
97	Bk	1 755	1 572	1 257	962	906	516	500	398	332	253	141	125
98	Cf	1 829	1 634	1 300	999	939	542	525	427	347	262	147	131
99	Es	1 898	1 698	1 344	1 037	974	568	551	444	362	272	153	137
100	Fm	1 968	1 764	1 388	1 075	1 008	595	576	462	378	281	159	142
101	Md	2 039	1 831	1 432	1 113	1 043	622	602	480	393	291	165	148
102	No	2 113	1 900	1 477	1 152	1 078	649	629	499	410	300	170	154
103	Lr	2 195	1 978	1 530	1 199	1 121	684	662	525	433	316	183	166
104	Rf	2 280	2 058	1 583	1 246	1 164	719	696	552	458	333	196	177

5.5 Excitation Energies and Resonance Lines of Atoms

The tables give energies in eV for the first levels of excitation, as well as wavelengths in nm of resonance lines (in boldface). For ionized atoms, the degree of ionization is given in Roman numerals, one unit greater than the number of electrons removed. For alkali metals and C IV the principal quantum number of the deepest d-term is $(n+1)$ for Li and C IV, n for Na, and $(n-1)$ for K and Cs.

Level	Atom H	D	He II
$n=2$	10.198	10.201	40.813
	121.567	**121.533**	**30.378**
$n=3$	12.087	12.090	
$n=4$	12.748	12.751	
$n=5$	13.054	13.058	
$n=6$	13.220	13.224	

Level	Atom He	C V
$2s^3S_1$	19.818	298.961
$2s^1S_0$	20.615	304.387
$2p^3P$	20.963	304.4
$2p^1P_1$	21.218	307.902
	58.433	**4.027**
$3s^3S_1$	22.717	352.064
$3s^1S_0$	22.919	353.505

Level	Atom Li	C IV	Na	K	Cs
	$n=2$	$n=2$	$n=3$	$n=4$	$n=6$
$np^2P_{1/2}$	1.848	7.995	2.102	1.610	1.39
	670.784	**155.077**	**589.592**	**769.898**	**894.350**
$np^2P_{3/2}$		8.008	2.104	1.617	1.45
		154.820	**588.995**	**766.491**	**852.110**
$(n+1)s^2S_{1/2}$	3.373	37.549	3.191	2.607	2.30
$(n+1)$ / n / $(n-1)$ d^2D	3.879	40.28	3.617	2.67	1.8
$(n+1)p^2P$	3.834	39.68	3.75	3.06	2.7

Level	Atom Ne	Ar	Kr	Xe
	$n=3$	$n=4$	$n=5$	$n=6$
ns^3P_2	16.62	11.55	9.91	8.31
ns^3P_1	16.671	11.62	10.03	8.43
	74.371	**106.666**	**123.582**	**149.102**
ns^3P_0	16.72	11.72	10.56	9.44
ns^3P_1	16.848	11.83	10.64	9.57
	73.589	**104.822**	**116.486**	**131.238**
np^3S_1	18.38	12.91	11.30	9.58

Level	Atom Cd $n = 5$	Hg $n = 6$
np^3P_0	3.73	4.667
np^3P_1	3.80	4.886
	326.106	**253.652**
np^3P_2	3.94	5.461
np^1P_1	5.29	6.704
	228.609	**184.957**
$(n+1)s^1S_0$	6.38	7.730

Level	Atom N
$2p^3\ ^2D$	2.38
$2p^3\ ^2P$	3.58
$3s^4P$	10.33
	119.955
	120.022
	120.071
$3s^2P$	10.68
$2p^4\ ^4P$	10.93
$3p^2S_{1/2}$	11.60

Level	Atom O
$2p^4\ ^1D_2$	1.97
$2p^4\ ^1S_0$	4.19
$3s^5S_2$	9.15
	135.560
	135.852
$3s^3S_1$	9.52
	130.217
	130.486
	130.602
$3p\ ^5P$	10.74

Level	Atom Al
$4s^2S_{1/2}$	3.143
	396.153
	394.403
$3p^2\ ^4P$	3.60
$3d^2D$	4.021
$4p^2P_{1/2}$	4.085
$4p^2P_{3/2}$	4.087
$5s^2S_{1/2}$	4.673
$4d\ ^2D$	4.827

5.6 Bond Angles

sp directed bonds		sp³ hybrids		sp³ hybrids	
Molecule	**Bond angle**	**Molecule**	**Bond angle**	**Molecule**	**Bond angle**
H_2O	104.5°	CH_4	109.5°	$GeHCl_3$	108.3°
H_2S	93.3°	CCl_4	109.5°	GeH_3Cl	110.9°
H_2Se	91.0°	C_2H_6	109.3°		
H_2Te	89.5°	C_2Cl_6	109.3°		
NH_3	107.3°	$CClF_3$	108.6°		
PH_3	93.3°	CH_3Cl	110.5°		
AsH_3	91.8°	$SiHF_3$	108.2°		
SbH_3	91.3°	SiH_3Cl	110.2°		

5.7 Diatomic Molecules

Column 2 The ground state equilibrium distance between the nuclei is given.

Column 3 The energy of the vibrational transition between $v = 1$ and $v = 2$ is given. (v = vibrational quantum number)

Columns 3 and 4 Vibrational constants according to $E_v = h\nu_e(v + 0.5) - h\nu_e x_e(v + 0.5)^2$ where v is the vibrational quantum number. Index e stands for 'equilibrium'.

Columns 5 and 6 Rotational constants according to $B_v = B_e - \alpha_e(v + 0.5)$ where v is the vibrational quantum number.

Column 7 The dissociation energy is equal to the difference between the atomic ground state and the lowest energy level in the molecule.

Column 9 The permanent ground state electrical dipole moment is given.

Column 11 s, sp and pp are covalent bonds.

1	2	3	4	5	6
Molecule	**Nuclear distance** r_e pm	**Vibrational energy** $h\nu_e$ meV	**Anharmonicity constant** $h\nu_e x_e$ meV	**Rotational constant** B_e meV	**Rovibrational constant** α_e meV
1 H_2	74.144	546.68	15.04	7.5448	0.3796
2 N_2	109.768	292.43	1.76	0.2478	0.00215
3 O_2	120.752	196.92	1.49	0.1792	0.00198
4 Na_2	307.88	19.73	0.090	0.0192	0.00011
5 Cl_2	198.7	69.39	0.331	0.0302	0.00017
6 NaCl	236.079	45.38	0.254	0.0270	0.00020
7 KCl	266.66	26.04			
8 RbCl	278.673	28.27	0.114	0.01087	$5.62 \cdot 10^{-5}$
9 HF	91.68	513.09	11.14	2.5982	0.09894
10 HCl	127.455	370.83	6.55	1.3134	0.03809
11 HBr	141.443	328.43	5.61	1.0495	0.02892
12 HI	160.916	286.28	4.92	0.7968	0.02093
13 CO	112.832	269.02	1.65	0.2394	0.00217
14 NO	115.077	236.09	1.75	0.2073	0.00212

	7	8	9	10	11
Molecule	**Dissociation energy** D_0 eV	**Ionization energy** E_0 eV	**Electric dipole moment** 10^{-30} Cm	**Characteristic temperature of rotation** K	**Type of bond**
1 H_2	4.4781	15.4258	0.000	85.5	s
2 N_2	9.759	15.580	0.000		pp
3 O_2	5.115	12.074	0.000	2.09	pp
4 Na_2	0.720	4.90	0.000	0.224	s
5 Cl_2	2.4794	11.480	0.000	0.347	pp
6 NaCl	4.228	8.9	30.02		ionic, sp
7 KCl	4.393	8.4	9.106		ionic, sp
8 RbCl	4.367	8.3	10.340		ionic, sp
9 HF	5.869	16.0409	6.091		sp
10 HCl	4.433	12.7447	3.699	15	sp
11 HBr	3.758	11.668	2.759		sp
12 HI	3.054	10.384	1.494		sp
13 CO	11.09	14.0139	0.367	2.77	pp
14 NO	6.496	9.26436	0.530	2.4	pp

5.8 Molecular Symmetry

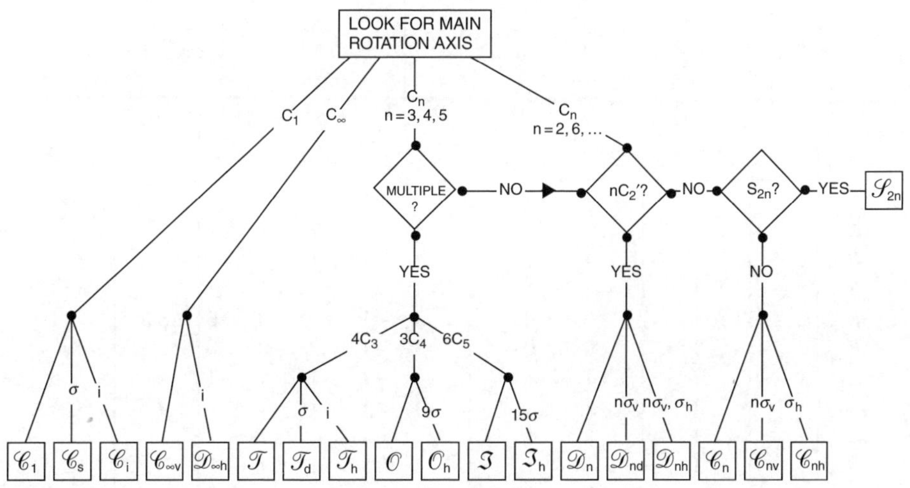

Character tables of point groups

IR = Irreducible representation
$\mathscr{C}_{2v}$ = Group notation
* = IRs of transl. and rot. coordinates
** = IRs generated by p and d atomic orbitals

IR	Symmetry classes				*	**
$\mathscr{C}_{2v}$	E	C_2	$\sigma_v(xz)$	$\sigma_v(yz)$		
A_1	1	1	1	1	T_z	$x^2; y^2; z^2$
A_2	1	1	−1	−1	R_z	xy
B_1	1	−1	1	−1	$T_x; R_y$	xz
B_2	1	−1	−1	1	$T_y; R_x$	yz

$\mathscr{C}_{3v}$	E	$2C_3$	$3\sigma_3$		
A_1	1	1	1	T_z	$x^2+y^2; z^2$
A_2	1	1	−1	R_z	
E	2	−1	0	$(T_x, T_y); (R_x, R_y)$	$(x^2-y^2, xy); (xz, yz)$

$\mathscr{D}_{2h} = \mathscr{V}_h$	E	$C_2(z)$	$C_2(y)$	$C_2(x)$	i	$\sigma(xy)$	$\sigma(xz)$	$\sigma(yz)$		
A_g	1	1	1	1	1	1	1	1		$x^2; y^2; z^2$
B_{1g}	1	1	−1	−1	1	1	−1	−1	R_z	xy
B_{2g}	1	−1	1	−1	1	−1	1	−1	R_y	xz
B_{3g}	1	−1	−1	1	1	−1	−1	1	R_x	yz
A_u	1	1	1	1	−1	−1	−1	−1		
B_{1u}	1	1	−1	−1	−1	−1	1	1	T_z	
B_{2u}	1	−1	1	−1	−1	1	−1	1	T_y	
B_{3u}	1	−1	−1	1	−1	1	1	−1	T_x	

$\mathscr{D}_{2d} = \mathscr{V}_d$	E	$2S_4$	C_2	$2C_2'$	$2\sigma_d$		
A_1	1	1	1	1	1		$x^2+y^2; z^2$
A_2	1	1	1	−1	−1	R_z	
B_1	1	−1	1	1	−1		x^2-y^2
B_2	1	−1	1	−1	1	T_z	xy
E	2	0	−2	0	0	$(T_x, T_y); (R_x, R_y)$	(xz, yz)

$\mathscr{D}_{6h}$	E	$2C_6$	$2C_3$	C_2	$3C_2'$	$3C_2''$	i	$2S_3$	$2S_6$	σ_h	$3\sigma_d$	$3\sigma_v$		
A_{1g}	1	1	1	1	1	1	1	1	1	1	1	1		$x^2+y^2; z^2$
A_{2g}	1	1	1	1	−1	−1	1	1	1	1	−1	−1	R_z	
B_{1g}	1	−1	1	−1	1	−1	1	−1	1	−1	1	−1		
B_{2g}	1	−1	1	−1	−1	1	1	−1	1	−1	−1	1		
E_{1g}	2	1	−1	−2	0	0	2	1	−1	−2	0	0	(R_x, R_y)	(xz, yz)
E_{2g}	2	−1	−1	2	0	0	2	−1	−1	2	0	0		(x^2-y^2, xy)
A_{1u}	1	1	1	1	1	1	−1	−1	−1	−1	−1	−1		
A_{2u}	1	1	1	1	−1	−1	−1	−1	−1	−1	1	1	T_z	
B_{1u}	1	−1	1	−1	1	−1	−1	1	−1	1	−1	1		
B_{2u}	1	−1	1	−1	−1	1	−1	1	−1	1	1	−1		
E_{1u}	2	1	−1	−2	0	0	−2	−1	1	2	0	0	(T_x, T_y)	
E_{2u}	2	−1	−1	2	0	0	−2	1	1	−2	0	0		

$\mathcal{T}_d$	E	$8C_3$	$3C_2$	$6S_4$	$6\sigma_d$		
A_1	1	1	1	1	1		$x^2+y^2+z^2$
A_2	1	1	1	-1	-1		
E	2	-1	2	0	0		$(2z^2-x^2-y^2, x^2-y^2)$
T_1, F_1	3	0	-1	1	-1	(R_x, R_y, R_z)	
T_2, F_2	3	0	-1	-1	1	(T_x, T_y, T_z)	(xy, xz, yz)

$\mathcal{O}_h$	E	$8C_3$	$3C_2$	$6C_4$	$6C_2'$	i	$8S_6$	$3\sigma_h$	$6S_4$	$6\sigma_d$		
A_{1g}	1	1	1	1	1	1	1	1	1	1		$x^2+y^2+z^2$
A_{2g}	1	1	1	-1	-1	1	1	1	-1	-1		
E_g	2	-1	2	0	0	2	-1	2	0	0		$(2z^2-x^2-y^2, x^2-$
T_{1g}, F_{1g}	3	0	-1	1	-1	3	0	-1	1	-1	(R_x, R_y, R_z)	$y^2)$
T_{2g}, F_{2g}	3	0	-1	-1	1	3	0	-1	-1	1		
A_{1u}	1	1	1	1	1	-1	-1	-1	-1	-1		(xy, xz, yz)
A_{2u}	1	1	1	-1	-1	-1	-1	-1	1	1		
E_u	2	-1	2	0	0	-2	1	-2	0	0		
T_{1u}, F_{1u}	3	0	-1	1	-1	-3	0	1	-1	1	(T_x, T_y, T_z)	
T_{2u}, F_{2u}	3	0	-1	-1	1	-3	0	1	1	-1		

$\mathcal{C}_{\infty v}$	E	$2C(\phi)$	…	$\infty\sigma_v$		
$A_1 = \Sigma^+$	1	1	…	1	T_z	$x^2+y^2;\ z^2$
$A_2 = \Sigma^-$	1	1	…	-1	R_z	
$E_1 = \Pi$	2	$2\cos\phi$	…	0	$(T_x, T_y); (R_x, R_y)$	(xy, xz)
$E_2 = \Delta$	2	$2\cos 2\phi$	…	0		(x^2-y^2, xy)
$E_3 = \Phi$	2	$2\cos 3\phi$	…	0		
…	…	…	…	…		

$\mathcal{D}_{\infty v}$	E	$2C(\phi)$	…	$\infty\sigma_v$	i	$2S(\phi)$	…	$\infty C_2'$		
Σ_g^+	1	1	…	1	1	1	…	1		$x^2+y^2;\ z^2$
Σ_g^-	1	1	…	-1	1	1	…	-1	R_z	
Π_g	2	$2\cos\phi$	…	0	2	$-2\cos\phi$	…	0	(R_x, R_y)	(xz, yz)
Δ_g	2	$2\cos 2\phi$	…	0	2	$2\cos 2\phi$	…	0		(x^2-y^2, xy)
…	…	…	…	…	…	…	…	…		
Σ_u^+	1	1	…	1	-1	-1	…	-1	T_z	
Σ_u^-	1	1	…	-1	-1	-1	…	1		
Π_u	2	$2\cos\phi$	…	0	-2	$2\cos\phi$	…	0	(T_x, T_y)	
Δ_u	2	$2\cos 2\phi$	…	0	-2	$-2\cos 2\phi$	…	0		
…	…	…	…	…	…	…	…	…		

More information:

http://www.chemsoc.org/exemplarchem/entries/2004/hull_booth/CharacterMaps/whatsa_character_table.htm

http://spider.chemphys.lu.se/~per/symmetry.pdf

6 Nuclear Physics

6.1 Nuclear Binding Energies

The uncertainty is at most five units in the last digit. *A* = the mass number, i.e. proton number plus neutron number. The energy corresponding to the mass difference between the most common elementary particles is also given.

A	Element	Binding energy MeV
	$m_n - m_p$	1.29331
	$m_n - m_p - m_e$	0.78232
2	D	2.2245
3	T	8.4819
	He	7.7181
4	H	3
	He	28.2961
5	H	10
	He	27.34
	Li	26.33
6	He	29.266
	Li	31.993
	Be	26.93
7	Li	39.245
	Be	37.601
8	He	28
	Li	41.278
	Be	56.498
	B	37.736
9	Li	45.33
	Be	58.163
	B	56.312
10	Be	64.978
	B	64.750
	C	60.36

A	Element	Binding energy MeV
11	Be	65.48
	B	76.206
	C	73.443
12	B	79.575
	C	92.163
	N	74.02
13	B	84.455
	C	97.109
	N	94.106
14	C	105.286
	N	104.659
	O	98.733
15	C	106.504
	N	115.494
	O	111.952
16	C	110.76
	N	117.981
	O	127.620
	F	111.20
17	N	123.87
	O	131.763
	F	128.221
18	O	139.809
	F	137.371
	Ne	132.142

T – 6.1 Nuclear Binding Energies

A	**Element**	**Binding energy** MeV	*A*	**Element**	**Binding energy** MeV	*A*	**Element**	**Binding energy** MeV
19	O	143.765	30	Al	249.1	40	Cl	337.1
	F	147.801		Si	255.628		Ar	343.812
	Ne	143.781		P	250.60		K	341.524
20	O	151.37		S	243.71		Ca	342.056
	F	154.399	31	Si	262.22		Sc	327.3
	Ne	160.646		P	262.916	41	Ar	349.91
	Na	144.6		S	256.69		K	351.615
21	F	162.50	32	Si	271.53		Ca	350.42
	Ne	167.406		P	270.852		Sc	343.14
	Na	163.08		S	271.780	42	Ar	359.34
22	Ne	177.772		Cl	257.8		K	359.15
	Na	174.147	33	P	280.955		Ca	361.891
	Mg	168.3		S	280.421		Sc	354.71
23	Ne	182.967		Cl	274.07		Ti	345.16
	Na	186.565	34	P	287.5	43	K	368.78
	Mg	181.726		S	291.843		Ca	369.819
24	Ne	191.84		Cl	285.58		Sc	366.82
	Na	193.526		Ar	278		Ti	359.2
	Mg	198.258	35	S	298.828	44	K	375.6
	Al	183.5		Cl	298.213		Ca	380.954
25	Na	202.54		Ar	291.47		Sc	376.53
	Mg	295.587	36	S	308.71		Ti	375.59
	Al	200.55		Cl	306.790	45	K	385.0
26	Na	208.9		Ar	306.719		Ca	388.374
	Mg	216.682		K	292		Sc	387.844
	Al	211.896	37	S	313.1		Ti	385.00
	Si	206.04		Cl	317.106	46	K	392
27	Mg	223.122		Ar	315.510		Ca	398.78
	Al	224.953		K	308.59		Sc	396.611
	Si	219.361	38	S	321.0		Ti	398.195
28	Mg	231.63		Cl	323.22		V	390.35
	Al	232.684		Ar	327.349	47	K	400.7
	Si	236.536		K	320.63		Ca	406.06
	P	221.9		Ca	313		Sc	407.253
29	Al	242.12	39	Cl	331.28		Ti	407.070
	Si	245.011		Ar	333.94		V	403.37
	P	239.28		K	333.723	48	Ca	416.00
				Ca	326.44		Sc	415.50
							Ti	418.698
							V	413.903
							Cr	411.7

A	Element	Binding energy MeV
49	Ca	421.14
	Sc	425.62
	Ti	426.844
	V	425.45
	Cr	422.11
50	Sc	432.1
	Ti	437.789
	V	434.791
	Cr	435.042
	Mn	426.66
51	Ti	444.17
	V	445.846
	Cr	444.312
	Mn	440.34
52	Ti	452
	V	453.16
	Cr	456.347
	Mn	450.86
	Fe	447.70
53	V	462
	Cr	464.288
	Mn	462.91
	Fe	458.14
54	V	467
	Cr	474.01
	Mn	471.85
	Fe	471.76
	Co	462.73
55	Cr	480.26
	Mn	482.073
	Fe	481.059
	Co	476.82
56	Cr	488.5
	Mn	489.34
	Fe	492.26
	Co	486.91
	Ni	484.01
57	Mn	498.0
	Fe	499.90
	Co	498.29
	Ni	494.27
58	Mn	504
	Fe	509.95
	Co	506.86
	Ni	506.46
	Cu	497.11
59	Fe	516.53
	Co	517.32
	Ni	515.47
	Cu	509.88
60	Fe	525.45
	Co	524.81
	Ni	526.85
	Cu	519.94
61	Fe	531
	Co	534.16
	Ni	534.67
	Cu	531.65
	Zn	525.5
62	Co	540.83
	Ni	545.27
	Cu	540.55
	Zn	538.08
63	Co	549.3
	Ni	552.11
	Cu	551.39
	Zn	547.24
	Ga	541
64	Ni	561.77
	Cu	559.31
	Zn	559.10
	Ga	551.24
65	Ni	567.87
	Cu	569.22
	Zn	567.09
	Ga	563.05
	Ge	556
66	Ni	576.86
	Cu	576.28
	Zn	578.12
	Ga	572.17
	Ge	568.4
67	Cu	585.39
	Zn	585.18
	Ga	583.40
	Ge	578.2
68	Cu	591.6
	Zn	595.38
	Ga	591.68
	Ge	590
69	Zn	601.88
	Ga	602.00
	Ge	598.99
	As	594.3
70	Zn	611.08
	Ga	609.64
	Ge	610.52
	As	603.50
71	Zn	617.12
	Ga	618.95
	Ge	617.94
	As	615.14
	Se	610.0
72	Zn	625.81
	Ga	625.47
	Ge	628.68
	As	623.54
	Se	622
73	Ga	634.70
	Ge	635.47
	As	634.32
	Se	630.78
	Br	625

T – 6.1 Nuclear Binding Energies

A	Element	Binding energy MeV
74	Ga	640.85
	Ge	645.67
	As	642.32
	Se	642.90
	Br	635
	Kr	631
75	Ge	652.15
	As	652.57
	Se	650.92
	Br	647.42
	Kr	642
76	Ge	661.60
	As	659.89
	Se	662.08
	Br	656.7
	Kr	655
77	Ge	667.63
	As	669.60
	Se	669.50
	Br	667.35
	Kr	663.7
78	As	676.5
	Se	679.99
	Br	675.63
	Kr	675.55
79	As	685.5
	Se	686.96
	Br	686.33
	Kr	683.93
80	As	691.7
	Se	696.87
	Br	694.21
	Kr	695.44
	Rb	690
81	Se	703.58
	Br	704.37
	Kr	703.3
	Rb	700.3
82	Se	712.84
	Br	711.97
	Kr	714.28
	Rb	709.33
	Sr	708
83	Br	721.56
	Kr	721.75
	Rb	720
	Sr	717
84	Br	728.35
	Kr	732.27
	Rb	728.80
	Sr	728.91
	Y	721.8
85	Br	737.4
	Kr	739.39
	Rb	739.28
	Sr	737.39
	Y	733.35
86	Br	742.9
	Kr	749.24
	Rb	747.92
	Sr	748.91
	Y	742.85
	Zr	741
87	Kr	754.75
	Rb	757.86
	Sr	757.35
	Y	754.9
	Zr	750.6
88	Kr	762.0
	Rb	764.0
	Sr	768.45
	Y	764.04
	Zr	763
	Nb	755
89	Kr	767.9
	Rb	771.70
	Sr	774.84
	Y	775.52
	Zr	771.91
	Nb	767.2
90	Kr	773.0
	Rb	776.8
	Sr	782.63
	Y	782.39
	Zr	783.90
	Nb	777.01
	Mo	773.7
91	Rb	784
	Sr	788.45
	Y	790.34
	Zr	791.10
	Nb	789.2
	Mo	783.93
92	Rb	789
	Sr	795.8
	Y	796.89
	Zr	799.74
	Nb	796.92
	Mo	796.51
	Tc	787.7
93	Sr	800.4
	Y	804.38
	Zr	806.49
	Nb	805.77
	Mo	804.57
	Tc	800.60
94	Sr	807.8
	Y	810.5
	Zr	814.68
	Nb	812.98
	Mo	814.26
	Tc	809.22

T – 6.1 Nuclear Binding Energies

A	Element	Binding energy MeV
95	Y	818
	Zr	821.15
	Nb	821.49
	Mo	821.63
	Tc	819.19
	Ru	816.38
96	Y	823
	Zr	828.99
	Nb	828.42
	Mo	830.79
	Tc	827.07
	Ru	826.50
97	Zr	834.57
	Nb	836.46
	Mo	837.61
	Tc	837
	Ru	835
	Rh	830
98	Zr	842
	Nb	842
	Mo	846.25
	Tc	843.9
	Ru	844.80
	Rh	839.8
99	Nb	850
	Mo	852.17
	Tc	852.75
	Ru	852.26
	Rh	849.38
	Pd	844.8
100	Nb	855
	Mo	860.47
	Tc	859.4
	Ru	861.94
	Rh	857.51
	Pd	856
101	Mo	865.86
	Tc	867.89
	Ru	868.74
	Rh	867.40
	Pd	864.86
102	Mo	874
	Tc	874
	Ru	877.96
	Rh	874.85
	Pd	875.22
	Ag	869
103	Tc	882.6
	Ru	884.20
	Rh	884.16
	Pd	882.83
	Ag	879.5
104	Tc	888.0
	Ru	893.09
	Rh	891.16
	Pd	892.85
	Ag	887.80
	Cd	886
105	Tc	896.5
	Ru	899.07
	Rh	900.16
	Pd	899.94
	Ag	898
	Cd	894
106	Ru	907.47
	Rh	906.73
	Pd	906.49
	Ag	905.74
	Cd	905.14
	In	897.9
107	Ru	912.9
	Rh	915.29
	Pd	916.02
	Ag	915.27
	Cd	913.07
	In	908.8
108	Ru	921
	Rh	922
	Pd	925.25
	Ag	922.55
	Cd	923.41
	In	917.5
109	Rh	930
	Pd	931.40
	Ag	931.73
	Cd	930.79
	In	927.99
110	Rh	935.5
	Pd	940.20
	Ag	938.55
	Cd	940.64
	In	935.93
111	Pd	945.94
	Ag	947.35
	Cd	947.62
	In	945.8
	Sn	942.4
112	Pd	954.28
	Ag	953.79
	Cd	957.02
	In	953.65
	Sn	953.52
113	Ag	962.34
	Cd	963.56
	In	963.07
	Sn	961.27
	Sb	956.01
114	Ag	968.8
	Cd	972.60
	In	970.38
	Sn	971.59
	Sb	964.5
115	Ag	976.3
	Cd	978.75
	In	979.42
	Sn	979.12
	Sb	975.31
116	Ag	982
	Cd	987.44
	In	986.14
	Sn	988.69
	Sb	983.35
	Te	981.0

A	Element	Binding energy MeV
117	Cd	993.21
	In	994.94
	Sn	995.63
	Sb	993.03
	Te	988.7
118	Cd	1 002
	In	1 001.5
	Sn	1 004.96
	Sb	1 000.48
	Te	999
119	Cd	1 007.0
	In	1 009.7
	Sn	1 011.44
	Sb	1 010.08
	Te	1 007.00
120	In	1 016
	Sn	1 020.55
	Sb	1 017.08
	Te	1 017.29
	I	1 011
121	In	1 024
	Sn	1 026.73
	Sb	1 026.33
	Te	1 024.26
	I	1 021.1
	Xe	1 016.6
122	In	1 030
	Sn	1 035.54
	Sb	1 033.13
	Te	1 034.32
	I	1 029.40
123	In	1 038
	Sn	1 041.47
	Sb	1 042.11
	Te	1 041.26
	I	1 039
	Xe	1 036
124	In	1 043.4
	Sn	1 049.97
	Sb	1 048.54
	Te	1 050.67
	I	1 046.72
	Xe	1 046.1
125	Sn	1 055.74
	Sb	1 057.30
	Te	1 057.28
	I	1 056.34
	Xe	1 054
	Cs	1 050
126	Sn	1 064
	Sb	1 063.4
	Te	1 066.37
	I	1 063.43
	Xe	1 063.90
	Cs	1 058.3
127	Sn	1 070
	Sb	1 071.86
	Te	1 072.68
	I	1 072.59
	Xe	1 071.1
	Cs	1 068.2
	Ba	1 064
128	Sn	1 077.4
	Sb	1 077.9
	Te	1 081.44
	I	1 079.38
	Xe	1 080.74
	Cs	1 076.03
	Ba	1 075
129	Sb	1 086
	Te	1 087.55
	I	1 088.25
	Xe	1 087.66
	Cs	1 086
	Ba	1 083
	La	1 078
130	Sb	1 091
	Te	1 095.94
	I	1 094.75
	Xe	1 096.91
	Cs	1 093.14
	Ba	1 092.80
	La	1 086
131	Te	1 101.83
	I	1 103.33
	Xe	1 103.52
	Cs	1 102.38
	Ba	1 100.43
	La	1 096.69
	Ce	1 090.7
132	Te	1 109.95
	I	1 109.67
	Xe	1 112.45
	Cs	1 109.59
	Ba	1 110.0
	La	1 104.4
	Ce	1 102
133	I	1 118.0
	Xe	1 118.98
	Cs	1 118.63
	Ba	1 117.36
	La	1 114.4
	Ce	1 111
134	I	1 124.1
	Xe	1 127.44
	Cs	1 125.33
	Ba	1 126.61
	La	1 122.1
	Ce	1 121.1
135	I	1 132
	Xe	1 134.0
	Cs	1 134.4
	Ba	1 133.8
	La	1 132
	Ce	1 129

T – 6.1 Nuclear Binding Energies

A	Element	Binding energy MeV
136	I	1 135.7
	Xe	1 141.89
	Cs	1 141.0
	Ba	1 143.0
	La	1 139.4
	Ce	1 138.9
137	Xe	1 146.3
	Cs	1 149.6
	Ba	1 150.0
	La	1 149
	Ce	1 147
	Pr	1 143
138	Xe	1 152
	Cs	1 154
	Ba	1 158.5
	La	1 155.97
	Ce	1 156.2
	Pr	1 151.1
139	Xe	1 156.2
	Cs	1 160.0
	Ba	1 163.3
	La	1 164.76
	Ce	1 163.71
	Pr	1 160.9
	Nd	1 157
140	Cs	1 164
	Ba	1 169.49
	La	1 169.76
	Ce	1 172.75
	Pr	1 168.60
	Nd	1 168
141	Ba	1 174.3
	La	1 176.54
	Ce	1 178.18
	Pr	1 177.98
	Nd	1 175.40
	Pm	1 171.0
142	Ba	1 180.3
	La	1 181.7
	Ce	1 185.39
	Pr	1 183.83
	Nd	1 185.21
	Pm	1 179.6
143	La	1 188.0
	Ce	1 190.50
	Pr	1 191.16
	Nd	1 191.31
	Pm	1 189.4
	Sm	1 185.3
144	La	1 193
	Ce	1 197.39
	Pr	1 196.93
	Nd	1 199.14
	Pm	1 196
	Sm	1 195.76
145	Ce	1 202
	Pr	1 203.86
	Nd	1 204.88
	Pm	1 203.96
	Sm	1 202.52
	Eu	1 198.95
146	Ce	1 208.8
	Pr	1 209.0
	Nd	1 212.44
	Pm	1 210.22
	Sm	1 210.96
	Eu	1 206.32
	Gd	1 204
147	Pr	1 216
	Nd	1 217.73
	Pm	1 217.85
	Sm	1 217.29
	Eu	1 214.7
	Gd	1 212
148	Pr	1 221
	Nd	1 225.06
	Pm	1 223.76
	Sm	1 225.43
	Eu	1 221.56
	Gd	1 220.78
	Tb	1 214.4
149	Nd	1 230.10
	Pm	1 230.99
	Sm	1 231.28
	Eu	1 230
	Gd	1 227.7
	Tb	1 223.18
150	Nd	1 237.44
	Pm	1 236.6
	Sm	1 239.26
	Eu	1 236.23
	Gd	1 236.46
	Tb	1 230.88
	Dy	1 228
151	Nd	1 242.8
	Pm	1 244.46
	Sm	1 244.87
	Eu	1 244.16
	Gd	1 243
	Tb	1 239.5
	Dy	1 236
152	Pm	1 250
	Sm	1 253.09
	Eu	1 250.45
	Gd	1 251.49
	Tb	1 246.5
	Dy	1 245.33
	Ho	1 238.2
153	Pm	1 258.0
	Sm	1 258.98
	Eu	1 259.00
	Gd	1 257.97
	Tb	1 255
	Dy	1 252.5
	Ho	1 247.5

A	Element	Binding energy MeV
154	Sm	1 266.88
	Eu	1 265.38
	Gd	1 265.58
	Tb	1 262
	Dy	1 261.8
	Ho	1 256
	Er	1 252
155	Sm	1 272.70
	Eu	1 273.57
	Gd	1 273.03
	Tb	1 271
	Dy	1 268
156	Sm	1 279.96
	Eu	1 279.90
	Gd	1 281.56
	Tb	1 278
	Dy	1 278.4
157	Eu	1 287.42
	Gd	1 287.91
	Tb	1 287.07
	Dy	1 285
158	Eu	1 293.1
	Gd	1 295.84
	Tb	1 293.86
	Dy	1 294.02
	Ho	1 289.12
159	Eu	1 300.4
	Gd	1 301.87
	Tb	1 302.03
	Dy	1 300.87
	Ho	1 298
160	Eu	1 306.4
	Gd	1 309.24
	Tb	1 308.43
	Dy	1 309.46
	Ho	1 305.38
161	Gd	1 314.9
	Tb	1 310.11
	Dy	1 315.91
	Ho	1 314
	Er	1 312
	Tm	1 307
162	Gd	1 322
	Tb	1 322
	Dy	1 324.11
	Ho	1 321.17
	Er	1 320.7
	Tm	1 315.1
163	Tb	1 329.47
	Dy	1 330.37
	Ho	1 329.57
	Er	1 327.58
	Tm	1 324.53
164	Tb	1 335
	Dy	1 338.02
	Ho	1 336.13
	Er	1 336.38
	Tm	1 331.63
165	Dy	1 343.66
	Ho	1 344.18
	Er	1 343.02
	Tm	1 341
	Yb	1 337
166	Dy	1 350.81
	Ho	1 350.51
	Er	1 351.57
	Tm	1 347.8
	Yb	1 346.7
167	Ho	1 357.8
	Er	1 358.01
	Tm	1 356
	Yb	1 354
	Lu	1 350
168	Ho	1 363.3
	Er	1 365.78
	Tm	1 363.28
	Yb	1 362.6
	Lu	1 357
169	Ho	1 370.5
	Er	1 371.78
	Tm	1 371.33
	Yb	1 369
	Lu	1 366
170	Ho	1 375.5
	Er	1 379.0
	Tm	1 377.7
	Yb	1 377.9
	Lu	1 373.6
171	Er	1 384.6
	Tm	1 385.4
	Yb	1 384.7
	Lu	1 382
172	Er	1 391.6
	Tm	1 391.7
	Yb	1 392.8
	Lu	1 389
173	Tm	1 398.7
	Yb	1 399.3
	Lu	1 397.8
174	Tm	1 404.5
	Yb	1 406.7
	Lu	1 404.4
	Hf	1 403.6
175	Tm	1 411
	Yb	1 412.55
	Lu	1 412.24
	Hf	1 411
176	Tm	1 415.8
	Yb	1 419.2
	Lu	1 418.4
	Hf	1 418.66

T – 6.1 Nuclear Binding Energies

A	Element	Binding energy MeV
177	Yb	1 424.7
	Lu	1 425.3
	Hf	1 425.0
	Ta	1 423.1
178	Yb	1 431
	Lu	1 431.2
	Hf	1 432.7
	Ta	1 430.0
179	Lu	1 438.2
	Hf	1 438.7
	Ta	1 437.8
180	Lu	1 443.5
	Hf	1 446.1
	Ta	1 444.60
	W	1 444.32
181	Hf	1 452.00
	Ta	1 452.24
	W	1 451.27
182	Hf	1 458.6
	Ta	1 458.30
	W	1 459.26
	Re	1 455.61
183	Hf	1 463.7
	Ta	1 465.16
	W	1 465.44
	Re	1 464
184	Ta	1 470.90
	W	1 472.86
	Re	1 470
	Os	1 469.7
185	Ta	1 477.5
	W	1 478.61
	Re	1 478.26
	Os	1 476.49
186	Ta	1 482.9
	W	1 485.82
	Re	1 484.5
	Os	1 484.8
	Ir	1 480.2
187	W	1 491.28
	Re	1 491.82
	Os	1 491.03
	Ir	1 489
188	W	1 497.89
	Re	1 497.54
	Os	1 498.87
	Ir	1 495.26
	Pt	1 494.0
189	Re	1 504.7
	Os	1 504.9
	Ir	1 504
	Pt	1 501
190	Re	1 510.3
	Os	1 512.6
	Ir	1 509.8
	Pt	1 509.8
	Au	1 505
191	Os	1 518.5
	Ir	1 518.1
	Pt	1 517
	Au	1 514
192	Os	1 526.2
	Ir	1 524.2
	Pt	1 524.9
	Au	1 520.9
	Hg	1 519
193	Os	1 531.64
	Ir	1 531.99
	Pt	1 531.17
	Au	1 529
	Hg	1 526
194	Os	1 538.78
	Ir	1 538.10
	Pt	1 539.55
	Au	1 536.26
	Hg	1 535
	Tl	1 529
195	Os	1 544.2
	Ir	1 545.5
	Pt	1 545.68
	Au	1 544.67
	Hg	1 542
	Tl	1 539
196	Ir	1 551
	Pt	1 553.60
	Au	1 551.34
	Hg	1 551.24
	Tl	1 545.9
	Pb	1 542
197	Ir	1 558.2
	Pt	1 559.46
	Au	1 559.43
	Hg	1 557.88
	Tl	1 554.9
	Pb	1 550
198	Ir	1 563.4
	Pt	1 567.02
	Au	1 565.92
	Hg	1 566.52
	Tl	1 562.3
	Pb	1 560
	Bi	1 551
199	Pt	1 572.59
	Au	1 573.49
	Hg	1 573.17
	Tl	1 571.3
	Pb	1 567
	Bi	1 561
200	Pt	1 580
	Au	1 579.8
	Hg	1 581.19
	Tl	1 577.96
	Pb	1 576
	Bi	1 569
	Po	1 565

A	**Element**	**Binding energy** MeV
201	Pt	1 584.8
	Au	1 586.7
	Hg	1 587.42
	Tl	1 586.2
	Pb	1 583
	Bi	1 578
	Po	1 572
202	Au	1 593
	Hg	1 595.18
	Tl	1 593.18
	Pb	1 592.35
	Bi	1 586
	Po	1 582
	At	1 573
203	Au	1 600
	Hg	1 601.17
	Tl	1 600.88
	Pb	1 599.28
	Bi	1 595.31
	Po	1 590
	At	1 583
204	Hg	1 608.67
	Tl	1 607.54
	Pb	1 607.52
	Bi	1 602
	Po	1 599
	At	1 591
	Rn	1 586
205	Hg	1 614.2
	Tl	1 615.07
	Pb	1 614.26
	Bi	1 610.77
	Po	1 606
	At	1 601
	Rn	1 594
206	Hg	1 612.07
	Tl	1 621.60
	Pb	1 622.34
	Bi	1 617.91
	Po	1 615.32
	At	1 689
	Rn	1 604
	Fr	1 595
207	Tl	1 628.41
	Pb	1 629.07
	Bi	1 625.93
	Po	1 622.24
	At	1 617.73
	Rn	1 612
	Fr	1 605
208	Tl	1 634.24
	Pb	1 636.45
	Bi	1 632.80
	Po	1 630.61
	At	1 625
	Rn	1 621
	Fr	1 613
209	Tl	1 637.25
	Pb	1 640.39
	Bi	1 640.25
	Po	1 637.58
	At	1 633.31
	Rn	1 629
	Fr	1 622
210	Tl	1 640.89
	Pb	1 645.57
	Bi	1 644.85
	Po	1 645.23
	At	1 640.57
	Rn	1 637.45
	Fr	1 630
211	Pb	1 649.40
	Bi	1 649.96
	Po	1 649.78
	At	1 648.24
	Rn	1 644.57
	Fr	1 639.4
	Ra	1 633
212	Pb	1 654.53
	Bi	1 654.33
	Po	1 655.79
	At	1 653.28
	Rn	1 652.51
	Fr	1 647
	Ra	1 642
213	Pb	1 659
	Bi	1 659.57
	Po	1 660.17
	At	1 659.2
	Rn	1 657.58
	Fr	1 654.70
	Ra	1 650
	Ac	1 643
214	Pb	1 663.35
	Bi	1 663.57
	Po	1 666.03
	At	1 664.19
	Rn	1 664
	Fr	1 660.15
	Ra	1 658.44
	Ac	1 651
215	Bi	1 668.7
	Po	1 670.17
	At	1 670.10
	Rn	1 669.3
	Fr	1 666.95
	Ra	1 663.97

A	**Element**	**Binding energy** MeV	*A*	**Element**	**Binding energy** MeV	*A*	**Element**	**Binding energy** MeV
216	Bi	1 673	224	Fr	1 718	233	Th	1 771.60
	Po	1 675.92		Ra	1 720.31		Pa	1 772.06
	At	1 674.68		Ac	1 718.2		U	1 771.85
	Rn	1 675.89		Th	1 717.58		Np	1 770.0
	Fr	1 672	225	Ra	1 725.30		Pu	1 767.07
	Ra	1 671.32		Ac	1 724.87	234	Th	1 777.70
217	Po	1 680		Th	1 723.37		Pa	1 777.18
	At	1 680.66		Pa	1 720		U	1 778.63
	Rn	1 680.58	226	Ra	1 731.67		Np	1 776.0
	Fr	1 679.0		Ac	1 730.1		Pu	1 774.82
	Ra	1 676.70		Th	1 730.53	235	Pa	1 783.3
218	Po	1 685.53		Pa	1 727.0		U	1 783.90
	At	1 685.05	227	Ra	1 736.20		Np	1 782.99
	Rn	1 687.06		Ac	1 736.73		Pu	1 781.1
	Fr	1 684.5		Th	1 735.99	236	Pa	1 787.8
	Ra	1 684		Pa	1 734.18		U	1 790.36
219	At	1 690.6	228	Ra	1 742.43		Np	1 788.66
	Rn	1 691.52		Ac	1 741.70		Pu	1 788.40
	Fr	1 690.95		Th	1 743.09		Am	1 785
	Ra	1 689.4		Pa	1 740.2	237	Pa	1 794.2
220	At	1 695		U	1 739.07		U	1 795.67
	Rn	1 697.81	229	Ac	1 748		Np	1 795.40
	Fr	1 696.15		Th	1 748.46		Pu	1 794.39
	Ra	1 696.59		Pa	1 747.33		Am	1 792.2
221	Rn	1 702		U	1 745.19	238	U	1 801.73
	Fr	1 702.50	230	Ac	1 753		Np	1 800.83
	Ra	1 701.99		Th	1 755.19		Pu	1 801.33
	Ac	1 699.6		Pa	1 753.15		Am	1 798
222	Rn	1 708.24		U	1 752.83		Cm	1 796.49
	Fr	1 707		Np	1 749	239	U	1 806.51
	Ra	1 708.68	231	Ac	1 759.0		Np	1 807.01
	Ac	1 705.7		Th	1 760.28		Pu	1 806.95
223	Fr	1 713.47		Pa	1 759.88		Am	1 805.36
	Ra	1 713.84		U	1 758.7		Cm	1 803
	Ac	1 712.46		Np	1 756.1	240	U	1 812.46
	Th	1 710.0	232	Th	1 766.64		Np	1 812.2
				Pa	1 765.41		Pu	1 813.42
				U	1 765.97		Am	1 811
				Np	1 763		Cm	1 810.30
				Pu	1 760.7			

T – 6.1 Nuclear Binding Energies

A	Element	Binding energy MeV	*A*	Element	Binding energy MeV	*A*	Element	Binding energy MeV
241	Np	1 818.3	246	Pu	1 846.8	251	Bk	1 875
	Pu	1 818.83		Am	1 846.4		Cf	1 875
	Am	1 818.07		Cm	1 847.87		Es	1 874.0
	Cm	1 816.51		Bk	1 846		Fm	1 872
	Bk	1 813		Cf	1 844.85		Md	1 868
242	Np	1 823		Es	1 841	252	Cf	1 881
	Pu	1 825.05	247	Am	1 852		Es	1 879.4
	Am	1 823.54		Cm	1 853		Fm	1 878.86
	Cm	1 823.42		Bk	1 852.3		Md	1 875
	Bk	1 820		Cf	1 851	253	Cf	1 886.2
243	Pu	1 830.10		Es	1 848		Es	1 885.70
	Am	1 829.88	248	Cm	1 859		Fm	1 884.73
	Cm	1 829.09		Bk	1 857.9		Md	1 882
	Bk	1 826.82		Cf	1 857.74		No	1 877.2
	Cf	1 824		Es	1 854	254	Es	1 890.8
244	Pu	1 836		Fm	1 851.67		Fm	1 891.02
	Am	1 835.16	249	Cm	1 864.1		Md	1 888
	Cm	1 835.81		Bk	1 864.15		No	1 885.5
	Bk	1 833		Cf	1 863.49	255	Fm	1 896
	Cf	1 831.31		Es	1 861.30		Md	1 894.9
245	Pu	1 841		Fm	1 857.8		No	1 892
	Am	1 841.39	250	Bk	1 869.1	257	Lr	1 902
	Cm	1 841.51		Cf	1 870.04			
	Bk	1 839.89		Es	1 867			
	Cf	1 837.58		Fm	1 865.6			
	Es	1 834						

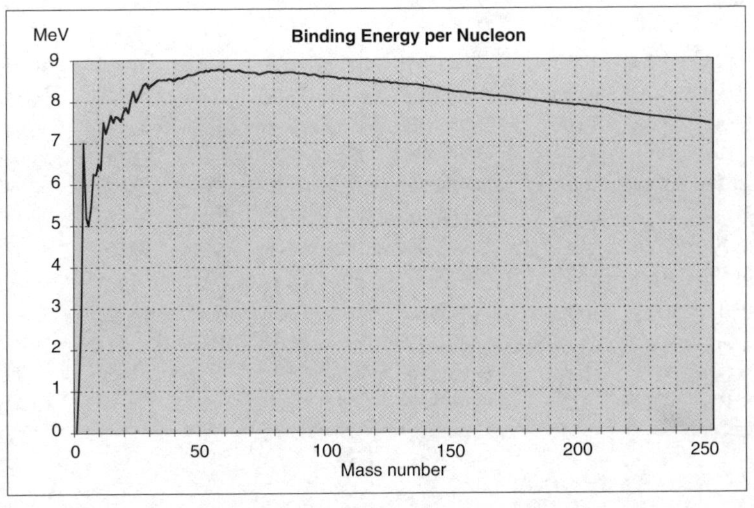

6.2 Properties of Naturally Occurring Nuclides

Column 1 Z = Atomic number = number of protons.

Column 2 Symbol and mass number = number of protons + number of neutrons.

Column 4 For isotopes with a trailing asterisk (*) significant differences in the abundance values have been found in some samples.

Column 5 Total atomic mass, including electrons.

1 Z	2 Nuclide	3 Spin and parity	4 Abundance %	5 Mass u
1	H 1	1/2 +	99.985	1.00782503
	H 2	1 +	0.015	2.01410178
2	He 3	1/2 +	0.00014	3.01602930
	He 4	0 +	99.99986	4.00260323
3	Li 6	1 +	7.5*	6.015122
	Li 7	3|2 –	92.5*	7.016003
4	Be 9	3/2 –	100	9.0121822
5	B 10	3 +	19.9*	10.0129372
	B 11	3|2 –	80.1*	11.0093056
6	C 12	0 +	98.90	12.00000000
	C 13	1/2 –	1.10	13.00335483
7	N 14	1 +	99.63	14.00307400
	N 15	1/2 –	0.37	15.00010896
8	O 16	0 +	99.762	15.99491462
	O 17	5/2 +	0.038	16.9991314
	O 18	0 +	0.200	17.999160
9	F 19	1/2 +	100	18.9984032
10	Ne 20	0 +	90.51	19.992434
	Ne 21	3/2 +	0.27	20.993841
	Ne 22	0 +	9.22	21.991382
11	Na 23	3/2 +	100	22.989768
12	Mg 24	0 +	78.99	23.985042
	Mg 25	5/2 +	10.00	24.985838
	Mg 26	0 +	11.01	25.982594
13	Al 27	5/2 +	100	26.981539
14	Si 28	0 +	92.23	27.976927
	Si 29	1/2 +	4.67	28.976495
	Si 30	0 +	3.10	29.973770
15	P 31	1/2 +	100	30.973762
16	S 32	0 +	95.02	31.972071
	S 33	3|2 +	0.75	32.971459
	S 34	0 +	4.21	33.967867
17	Cl 35	3/2 +	75.77	34.968853
	Cl 37	3/2 +	24.23	36.965903
18	Ar 36	0 +	0.337	35.967545
	Ar 38	0 +	0.063	37.962732
	Ar 40	0 +	99.600	39.962383
19	K 39	3/2 +	93.2581	38.963707
	K 40	4 –	0.0117	
	K 41	3/2 +	6.7302	40.961825
20	Ca 40	0 +	96.941	39.962591
	Ca 42	0 +	0.647	41.958618
	Ca 43	7/2 –	0.135	42.958767
	Ca 44	0 +	2.086	43.955481
	Ca 46	0 +	0.004	45.953690
	Ca 48	0 +	0.187	47.952534
21	Sc 45	7/2 –	100	44.955910
22	Ti 46	0 +	8.0	45.952630
	Ti 47	5/2 –	7.3	44.951764
	Ti 48	0 +	73.8	47.947948
	Ti 49	7/2 –	5.5	48.947871
	Ti 50	0 +	5.4	49.944792
23	V 50	6 +	0.250	49.947161
	V 51	7/2 –	99.750	50.943962
24	Cr 50	0 +	4.35	49.946047
	Cr 52	0 +	83.79	51.940510
	Cr 53	3/2 –	9.50	52.940652
	Cr 54	0 +	2.36	53.938883
25	Mn 55	5/2 –	100	54.938047
26	Fe 54	0 +	5.8	53.939613
	Fe 56	0 +	91.72	55.934940
	Fe 57	1/2 –	2.2	56.935396
	Fe 58	0 +	0.28	57.933278
27	Co 59	7/2 –	100	58.933198

T – 6.2 Properties of Naturally Occurring Nuclides

1	2	3	4	5
Z	**Nuclide**	**Spin and parity**	**Abundance** %	**Mass** u
28	Ni 58	0 +	68.27	57.935347
	Ni 60	0 +	26.10	59.930789
	Ni 61	3/2 –	1.13	60.931058
	Ni 62	0 +	3.59	61.928346
	Ni 64	0 +	0.91	63.927968
29	Cu 63	3/2 –	69.17	62.929599
	Cu 65	3/2 –	30.83	64.927793
30	Zn 64	0 +	48.6	63.929146
	Zn 66	0 +	27.9	65.926035
	Zn 67	5/2 –	4.1	66.927129
	Zn 68	0 +	18.8	67.924846
	Zn 70	0 +	0.6	69.925324
31	Ga 69	3/2 –	60.1	68.925580
	Ga 71	3|2 –	39.9	70.924701
32	Ge 70	0 +	20.5	69.924250
	Ge 72	0 +	27.4	71.922080
	Ge 73	9/2 +	7.8	72.923462
	Ge 74	0 +	36.5	73.921177
	Ge 76	0 +	7.8	75.921402
33	As 75	3/2 –	100	74.921593
34	Se 74	0 +	0.9	73.922474
	Se 76	0 +	9.0	75.919211
	Se 77	1/2 –	7.6	76.919911
	Se 78	0 +	23.5	77.917306
	Se 80	0 +	49.6	79.916521
	Se 82	0 +	9.4	81.91670
35	Br 79	3/2 –	50.69	78.918336
	Br 81	3/2 –	49.31	80.91629
36	Kr 78	0 +	0.35	77.92040
	Kr 80	0 +	2.25	79.91638
	Kr 82	0 +	11.6	81.91348
	Kr 83	9/2 +	11.5	82.914136
	Kr 84	0 +	57.0	83.911508
	Kr 86	0 +	17.3	85.910615
37	Rb 85	5/2 –	72.17	84.911793
	Rb 87	3|2 –	27.83	86.909188
38	Sr 84	0 +	0.56	83.913429
	Sr 86	0 +	9.86	85.909267
	Sr 87	9/2 +	7.00	86.908884
	Sr 88	0 +	82.58	87.905619
39	Y 89	1/2 –	100	88.905850
40	Zr 90	0 +	51.45	89.904703
	Zr 91	5/2 +	11.27	90.905642
	Zr 92	0 +	17.17	91.905037
	Zr 94	0 +	17.33	93.906314
	Zr 96	0 +	2.78	95.908274
41	Nb 93	9/2 +	100	92.906376
42	Mo 92	0 +	14.84	91.906807
	Mo 94	0 +	9.25	93.905084
	Mo 95	5/2 +	15.92	94.905840
	Mo 96	0 +	16.68	95.904678
	Mo 97	5/2 +	9.55	96.906020
	Mo 98	0 +	24.13	97.905406
	Mo 100	0 +	9.63	99.90747
44	Ru 96	0 +	5.52	95.90760
	Ru 98	0 +	1.88	97.90529
	Ru 99	5/2 +	12.7	98.905938
	Ru 100	0 +	12.6	99.904218
	Ru 101	5/2 +	17.0	100.905581
	Ru 102	0 +	31.6	101.904348
	Ru 104	0 +	18.7	103.90542
45	Rh 103	1/2 –	100	102.905499
46	Pd 102	0 +	1.02	101.90563
	Pd 104	0 +	11.14	103.90403
	Pd 105	5/2 +	22.33	104.90508
	Pd 106	0 +	27.33	105.90348
	Pd 108	0 +	26.46	107.903896
	Pd 110	0 +	11.72	109.90517
47	Ag 107	1/2 –	51.84	106.90509
	Ag 109	1/2 –	48.16	108.904757
48	Cd 106	0 +	1.25	105.90646
	Cd 108	0 +	0.89	107.90418
	Cd 110	0 +	12.49	109.903006
	Cd 111	1/2 +	12.80	110.904182
	Cd 112	0 +	24.13	111.902758
	Cd 113	1/2 +	12.22	112.904400
	Cd 114	0 +	28.73	113.903357
	Cd 116	0 +	7.49	115.904754
49	In 113	9/2 +	4.3	112.904061
	In 115	9/2 +	95.7	114.903880

T – 6.2 Properties of Naturally Occurring Nuclides

1	2	3	4	5	1	2	3	4	5
Z	**Nuclide**	**Spin and parity**	**Abun-dance %**	**Mass u**	**Z**	**Nuclide**	**Spin and parity**	**Abun-dance %**	**Mass u**
50	Sn 112	0 +	1.0	111.90483	58	Ce 136	0 +	0.19	135.9071
	Sn 114	0 +	0.7	113.902784		Ce 138	0 +	0.25	137.90598
	Sn 115	1/2 +	0.4	114.903348		Ce 140	0 +	88.48	139.905433
	Sn 116	0 +	14.7	115.901747		Ce 142	0 +	11.08	141.909241
	Sn 117	1/2 +	7.7	116.902956	59	Pr 141	5/2 +	100	140.90765
	Sn 118	0 +	24.3	117.901609					
	Sn 119	1/2 +	8.6	118.903310	60	Nd 142	0 +	27.13	141.90772
	Sn 120	0 +	32.4	119.902200		Nd 143	7/2 –	12.18	142.90981
	Sn 122	0 +	4.6	121.903440		Nd 144	0 +	23.80	143.910084
	Sn 124	0 +	5.6	123.90527		Nd 145	7/2 –	8.30	144.912570
						Nd 146	0 +	17.19	145.913114
51	Sb 121	5/2 +	57.3	120.903823		Nd 148	0 +	5.76	147.916890
	Sb 123	7/2 +	42.7	122.904220		Nd 150	0 +	5.64	149.920888
					62	Sm 144	0 +	3.1	143.91200
52	Te 120	0 +	0.096	119.90405		Sm 147	7/2 –	15.0	146.914895
	Te 122	0 +	2.60	121.903054		Sm 148	0 +	11.3	147.914820
	Te 123	1/2 +	0.908	122.904276		Sm 149	7/2 –	13.8	148.917181
	Te 124	0 +	4.816	123.902823		Sm 150	0 +	7.4	149.917273
	Te 125	1/2 +	7.14	124.904433		Sm 152	0 +	26.7	151.919729
	Te 126	0 +	18.95	125.903314		Sm 154	0 +	22.7	153.922206
	Te 128	0 +	31.69	127.904467	63	Eu 151	5/2 +	47.8	150.919847
	Te 130	0 +	33.80	129.906232		Eu 153	5/2 +	52.2	152.921226
53	I 127	5/2 +	100	126.904478	64	Gd 152	0 +	0.20	151.919787
						Gd 154	0 +	2.18	153.920862
54	Xe 124	0 +	0.10	123.9061		Gd 155	3/2 –	14.80	154.922618
	Xe 126	0 +	0.09	125.90429		Gd 156	0 +	20.47	155.922119
	Xe 128	0 +	1.91	127.90353		Gd 157	3/2 –	15.65	156.923956
	Xe 129	1/2 +	26.4	128.904780		Gd 158	0 +	24.84	157.924100
	Xe 130	0 +	4.1	129.903510		Gd 160	0 +	21.86	159.927051
	Xe 131	3/2 +	21.2	130.905075	65	Tb 159	3/2 +	100	158.925341
	Xe 132	0 +	26.9	131.904147					
	Xe 134	0 +	10.4	133.90540	66	Dy 156	0 +	0.06	155.92428
	Xe 136	0 +	8.9	135.90721		Dy 158	0 +	0.10	157.92440
						Dy 160	0 +	2.34	159.925193
55	Cs 133	7/2 +	100	132.90543		Dy 161	5/2 +	18.9	160.926929
						Dy 162	0 +	25.5	161.926795
56	Ba 130	0 +	0.106	129.90628		Dy 163	5/2 –	24.9	162.928726
	Ba 132	0 +	0.101	131.90504		Dy 164	0 +	28.2	163.929172
	Ba 134	0 +	2.417	133.90448	67	Ho 165	7/2 –	100	164.930319
	Ba 135	3/2 +	6.592	134.90566					
	Ba 136	0 +	7.854	135.90455	68	Er 162	0 +	0.14	161.92878
	Ba 137	3|2 +	11.23	136.90581		Er 164	0 +	1.61	163.929199
	Ba 138	0 +	71.70	137.90523		Er 166	0 +	33.6	165.930292
						Er 167	7/2 +	22.95	166.932047
57	La 138	5/2 +	0.09	137.90711		Er 168	0 +	26.8	167.932369
	La 139	7/2 +	99.91	138.906346		Er 170	0 +	14.9	169.935461

T – 6.2 Properties of Naturally Occurring Nuclides

1	2	3	4	5
Z	**Nuclide**	**Spin and parity**	**Abundance** %	**Mass** u
69	Tm 169	1/2 +	100	168.934212
70	Yb 168	0 +	0.13	167.933897
	Yb 170	0 +	3.05	169.934760
	Yb 171	1/2 –	14.3	170.936324
	Yb 172	0 +	21.9	171.936379
	Yb 173	5/2 –	16.12	172.938208
	Yb 174	0 +	31.8	173.938860
	Yb 176	0 +	12.7	175.942563
71	Lu 175	7/2 +	97.40	174.940771
	Lu 176	7 –	2.60	175.942680
72	Hf 174	0 +	0.16	173.940042
	Hf 176	0 +	5.2	175.941404
	Hf 177	7/2 –	18.6	176.943219
	Hf 178	0 +	27.1	177.943697
	Hf 179	9/2 +	13.74	178.945814
	Hf 180	0 +	35.2	179.946547
73	Ta 180	9/2 –	0.012	179.947464
	Ta 181	7/2 +	99.988	180.947995
74	W 180	0 +	0.13	179.946702
	W 182	0 +	26.3	181.948202
	W 183	1/2 –	14.3	182.950221
	W 184	0 +	30.67	183.950929
	W 186	0 +	28.6	185.954358
75	Re 185	5/2 +	37.40	184.952952
	Re 187	5/2 +	62.60	186.955747
76	Os 184	0 +	0.020	183.952487
	Os 186	0 +	1.58	185.953835
	Os 187	1|2 –	1.6	186.955744
	Os 188	0 +	13.3	187.955832
	Os 189	3/2 –	16.1	188.958140
	Os 190	0 +	26.4	189.958439
	Os 192	0 +	41.0	191.961469
77	Ir 191	3/2 +	37.3	190.960585
	Ir 193	3/2 +	62.7	192.962916
78	Pt 190	0 +	0.01	189.95992
	Pt 192	0 +	0.79	191.961027
	Pt 194	0 +	32.9	193.962655
	Pt 195	1/2 –	33.8	194.964766
	Pt 196	0 +	25.3	195.964926
	Pt 198	0 +	7.2	197.967867
79	Au 197	3/2 +	100	196.966543
80	Hg 196	0 +	0.15	195.965806
	Hg 198	0 +	10.1	197.966743
	Hg 199	1/2 –	17.0	198.968254
	Hg 200	0 +	23.1	199.968300
	Hg 201	3/2 –	13.2	200.970276
	Hg 202	0 +	29.65	201.970617
	Hg 204	0 +	6.8	203.973467
81	Tl 203	1/2 +	29.524	202.972320
	Tl 205	1/2 +	70.476	204.974401
82	Pb 204	0 +	1.4	203.973020
	Pb 206	0 +	24.1	205.974440
	Pb 207	1/2 –	22.1	206.975871
	Pb 208	0 +	52.4	207.976627
83	Bi 209	9/2 –	100	208.980373
90	Th 232	0 +	100	232.038051
92	U 234	0 +	0.0055	234.040947
	U 235	7/2 –	0.7200	235.043924
	U 238	0 +	99.2745	238.050785

6.3 Properties of Radioactive Nuclides

Column 1 Z = atomic number = number of protons.

Column 2 Symbol and mass number = number of protons + number of neutrons. A bold font indicates a nuclide naturally occurring or otherwise available.

Column 4 Total atomic mass, including electrons. For some nuclides data is also given for a meta state. For these rows, the nucleus' excitation energy is given within parentheses.

Column 5 s = seconds, m= minutes, h = hours, d = days, a = years. Boldface indicates an important product in nuclear power plants.

Column 6 Types of decay: α = alpha particle, β^- = negative electron, β^+ = positron, ε = electron capture, γ = gamma ray, n = neutron, p = proton, d = deuteron, IT = isomeric transition, D = delayed radiation, SF = spontaneous fission, e^- = conversion electron, $\beta^-\beta^-$ = double beta-decay, and C14 and Ne24 = particle emission. A value within parenthesis means weak decay mode intensity (abundance < 1 %).

For nuclides not mentioned in this table, information can be found on the Internet:
http://atom.kaeri.re.kr/
http://www.nndc.bnl.gov/mird/

1	2	3	4	5	6
Z	**Nu-clide**	**Spin and parity**	**Mass** u	**Half life**	**Modes of decay and energy of radiation** MeV for particles, keV for γ (and IT)
1	**H 3**	1/2 +	3.016049	**12.33 a**	β^- 0.0186 (no γ)
2	He 6	0 +	6.018888	807 ms	β^- 3.510 (no γ)
	He 8	0 +	8.033922	119 ms	β^- 9.7, γ 980.7
3	Li 8	2 +	8.022486	0.84 s	β^- 12.5, (2α 1.57)
	Li 9	3/2 –	9.026789	178 ms	β^- 13.5, 11.0
4	**Be 7**	3/2 –	7.016929	53.3 d	ε, γ 477.8
	Be 8	0 +	8.005305	0.07 fs	2α 0.0461
	Be 10	0 +	10.013534	1.6 Ma	β^- 0.556 (no γ)
	Be 11	1/2 +	11.021658	13.8 s	β^- 11.5,... γ 2124.5, 6791,...
5	B 8	2 +	8.024607	0.77 s	β^+ 14.1,...
	B 9	3/2 –	9.013329	0.85 as	p + 2 α
	B 12	1 +	12.014352	20.2 ms	β^- 13.37,..., γ 4439,...
	B 13	3/2 –	13.017780	17.4 ms	β^- 13.4,... γ 3680, n 3.61, 2.40,...
	B 14	2 –	14,025404	13.8 ms	β^- 14,... γ 6094, 6730
6	C 10	0 +	10.016853	19.3 s	β^+ 1.87,... γ 718.3, 1022
	C 11	3/2 –	11.011433	20.3 m	β^+ 0.960 (no γ)
	C 14	0 +	14.003242	5730 a	β^- 0.157 (no γ)
	C 15	1/2 +	15.010599	2.45 s	β^- 4.51, 9.82,... γ 5297.8,...
	C 16	0 +	16,014701	0.75 s	β^- 4.7, 7.9, n 0.81, 1.71
	C 17		17.022584	193 ms	β^-, n 1.62, γ 1375, 1849, 1906

T – 6.3 Properties of Radioactive Nuclides

1	2	3	4	5	6
Z	**Nu-clide**	**Spin and parity**	**Mass** u	**Half life**	**Modes of decay and energy of radiation** MeV for particles, keV for γ(and IT)
7	N 12	1 +	12,018613	11.00 ms	β^+ 16.3,... γ 4439,...
	N 13	1/2 –	13.005739	9.97 m	β^+ 1.190
	N 16	2 –	16.006100	7.13 s	β^- 4.27, 10.44,... γ 6129, 7115,... (α 1.85,...)
	N 17	1/2 –	17.008450	4.17 s	β^- 3.77,... n 1.17, 0.38,... γ 870.7, 2184
	N 18	1 –	18,014081	0.62 s	β^- 9.4, γ 1981.9, 1651.5, 821.7,...
	N 19		19.017027	0.329 s	β^-, γ 96, 3138, 709,...
8	O 14	0 +	14.008595	70.60 s	β^+ 1,81,... γ 2312.7,...
	O 15	1/2 –	15.003066	122.2 s	β^+ 1.72 (no γ)
	O 19	5/2 +	19.003577	26.9 s	β^- 3.3, 4.60,... γ 197.1, 1356.8,...
	O 20	0 +	20.004076	13.5 s	β^- 2.75,... γ 1056.8,...
	O 21		21.008655	3.4 s	β^- 6.4, γ 1730.3, 3517.4, 280.1, 1787.2,...
9	F 17	5/2 +	17.002095	64.5 s	β^+ 1.74 (no γ)
	F 18	1 +	18.000938	109.8 m	β^+ 0.635, ε (no γ)
	F 20	2 +	19.999981	11.00 s	β^- 5.40,... γ 1636,...
	F 21	5/2 +	20.999949	4.16 s	β^- 5.4,... γ 350.7, 1395.1,...
	F 22	4 +	22,002999	4.23 s	β^- 5.5,... γ 1274.5, 2082.5, 2166.0,...
10	Ne 17	1/2 –	17.017698	109 ms	β^+, p 4.59, 3.77, 5.12,... γ 495
	Ne 18	0 +	18.005710	1.67 s	β^+ 3.42,..., γ 1041,...
	Ne 19	1/2 +	19.001880	17.22 s	β^+ 3.24, ε, γ 109.9, 1356.8
	Ne 23	5/2 +	22.994467	37.2 s	β^- 4.38, 3.95,... γ 439.8,...
	Ne 24	0 +	23.993615	3.38 s	β^- 1.98,... γ 472.3D,...
	Ne 25	1/2 +	24.997790	0.61 s	β^- 7.2, 6.3,... γ 89.5, 979.8,...
11	Na 20	2 +	20,007348	447 ms	β^+ 11.25, γ 1633.6, (α 2.15, 4.44)
	Na 21	3/2 +	20.997665	22.48 s	β^+ 2.51,..., γ 350.7
	Na 22	3 +	21.994437	2.605 a	β^+ 0.546, ε, γ 1274.5
	Na 24	4 +	23.990963	14.96 h	β^- 1.391, γ 1368.6, 2754.0,...
	Na 25	5/2 +	24.989954	60 s	β^- 3.8,... γ 947.7, 585.0, 387.7, 1611.7,...
	Na 26	3 +	25.992590	1.07 s	β^- 7.4,... γ 1808.6,...
	Na 27	5/2 +	26.994009	0.290 s	β^- 8.0,... γ 984.7, 1698.0,... (n 0.46)
12	Mg 22	0 +	21.999574	3.86 s	β^+ 3.1,... γ 582, 72.9,...
	Mg 23	3/2 +	22.994125	11.32 s	β^+ 3.09,... γ 439.8,...
	Mg 27	1/2 +	26.984341	9.45 m	β^- 1.75, 1.59,... γ 843.8, 1014.4,...
	Mg 28	0 +	27.983877	21.0 h	β^- 0.459,... γ 30.6, 1342.3,...
	Mg 29	3/2 +	28.988555	1.30 s	β^- 5.4,... γ 2224.0, 1398.0, 960.4,...
13	Al 25	5/2 +	24.990429	7.17 s	β^+ 3.26,... γ 1611.7,...
	Al 26	5 +	25.986892	0.73 Ma	β^+ 1.17, ε, γ 1808.6,...
	Al 28	3 +	27.981910	2.25 m	β^- 2.86,... γ 1779.0
	Al 29	5/2 +	28.980445	6.56 m	β^- 2.5,... γ 1273.4, 2426, 2028,...
14	Si 26	0 +	25.992330	2.23 s	β^+ 3.83,... γ 829, 1622...
	Si 27	5/2 +	26.986704	4.15 s	β^+ 3.85,... (γ 2210,...)
	Si 31	3/2 +	30.975363	2.62 h	β^- 1.48,... γ 1266.2
	Si 32	0 +	31.974148	172 a	β^- 0.221 (no γ)
15	P 29	1/2 +	28.981801	4.14 s	β^+ 3.94, γ 1273.4,...
	P 30	1 +	29.978314	2.50 m	β^+ 3.24, ε, (γ 2235.2,...)
	P 32	1 +	31.973907	14.28 d	β^- 1.709 (no γ)
	P 33	1/2 +	32.971725	25.3 d	β^- 0.249 (no γ)

T – 6.3 Properties of Radioactive Nuclides

1	2	3	4	5	6
Z	**Nu-clide**	**Spin and parity**	**Mass** u	**Half life**	**Modes of decay and energy of radiation** MeV for particles, keV for γ(and IT)
16	S 30	0 +	29.984903	1.18 s	β^+ 4.42, 5.09... γ 677.2,...
	S 31	1/2 +	30.979555	2.57 s	β^+ 4.39,... γ 1266.2,...
	S 35	3/2 +	34.969032	87.3 d	β^- 0.1674 (no γ)
	S 37	7/2 –	36.971126	5.05 m	β^- 1.76,... γ 3104.0,...
17	Cl 33	3/2 +	32.977452	2.511 s	β^+ 4.5,... γ 1966.2, 2866.3,...
	Cl 34	3 +	(146 keV)	32.2 m	β^+ 2.5, 1.3... γ 2127.7, 1176.0,... IT 146.4
	Cl 34	0 +	33,973762	1.528 s	β^+ 4.47 (no γ)
	Cl 36	2 +	35.968307	0.301 Ma	β^-, 0.709, ε, (β^+ 0.12) (no γ)
	Cl 38	2 –	37.968011	37.2 m	β^- 4.91, 1.11,... γ 2167.7, 1642.4,...
	Cl 39	3/2 +	38.968009	55.6 m	β^- 1.91,... γ 1267.2, 250.3, 1517.5,...
18	Ar 34	0 +	33.980270	844 ms	β^+ 5.037,... γ 666.5, 3129, 461.0, 2580
	Ar 35	3/2 +	34.975257	1.77 s	β^+ 4.943,... γ 1219.2, (1763.0), 2693.6,...
	Ar 37	3/2 +	36.966776	35.0 d	ε (no γ)
	Ar 39	7/2 –	38.964313	269 a	β^- 0.565 (no γ)
	Ar 41	3/2 +	40.964501	1.82 h	β^- 1.198, 2.5,... γ 1293.6,...
	Ar 42	2 –	41.963035	32.9 a	β^- 0.6, (no γ)
19	K 37	3/2 +	36.973377	1.23 s	β^+ 5.13,... γ 2796,...
	K 38	3 +	37.969080	7.63 m	β^+ 2.68,... γ 2167.7,...
	K 40	4 –	39.963999	1.28 Ga	β^- 1.33, ε, γ 1460.8, (β^+)
	K 42	2 –	41.962403	12.36 h	β^- 3.52,... γ 1524.6,...
	K 43	3/2 +	42.960716	22.3 h	β^- 0.83, 1.8,... γ 372.8, 617.5,...
20	Ca 38	0 +	37.976319	0.44 s	β^+ 5.6, γ 1568,...
	Ca 39	3/2 +	38.970718	861 ms	β^+ 5.49,... (γ 2522)
	Ca 41	7/2 –	40.962278	0.103 Ma	ε (no γ)
	Ca 45	7/2 –	44.956186	162.7 d	β^- 0.258,... (γ 12.4D)
	Ca 47	7/2 –	46.954546	4.536 d	β^- 0.694, 1.990,... γ 1297.1, 808, 489...
	Ca 49	3/2 –	48.955673	8.72 m	β^- 2.18, 2.9,... γ 3084.4, 4072...
21	Sc 43	7/2 –	42.961151	3.89 h	β^+ 1.20, 0.82,... ε, γ 372.8,...
	Sc 44	6 +	(271 keV)	2.44 d	IT 271.2, ε, γ 1001.8, 1226.1, 1157.0
	Sc 44	2 +	43.959403	3.93 h	β^+ 1.47, ε, γ 1157.0,...
	Sc 46	1 –	(143 keV)	18.7 s	IT 142.5
	Sc 46	4 +	45.955170	83.8 d	β^- 0.357,... γ 1120.5, 889.3,...
	Sc 47	7/2 –	46.952408	3.349 d	β^- 0.439, 0.600,... γ 159.4
	Sc 48	6 +	47.952235	43.7 h	β^- 0.66,... γ 983.5, 1312.1, 1037.5,...
22	Ti 44	0 +	43.959690	49 a	ε, γ 78.4D, 67.8D,...
	Ti 45	7/2 –	44.958124	3.078 h	β^+ 1.04,... ε, γ (719.4), 1407.8,...
	Ti 51	3/2 –	50.946616	5.76 m	β^- 2.14,... γ 320.1, 928,...
	Ti 52	0 +	51.946898	1.7 m	β^- 1.8,... γ 124.5, 17.0, e^-
23	V 47	3/2 –	46.954907	32.6 m	β^+ 1.89,... ε, γ 1794.0,...
	V 48	4 +	47.952254	15.98 d	β^+ 0.694,... ε, γ 983.5, 1312.1, 944,...
	V 49	7/2 –	48.948517	337 d	ε (no γ)
	V 50	6 +	49.947163	144 Pa	ε, γ 1553,8, (β^-, γ 783)
	V 52	3 +	51.944780	3.76 m	β^- 2.47,... γ 1434.1,...
	V 53	7/2 –	52.944342	1.61 m	β^- 2.5,... γ 1006.2, 1289,...

T – 6.3 Properties of Radioactive Nuclides

1	2	3	4	5	6
Z	**Nu-clide**	**Spin and parity**	**Mass** u	**Half life**	**Modes of decay and energy of radiation** MeV for particles, keV for γ(and IT)
24	Cr 48	0 +	47.954036	21.6 h	ε , β^+, γ 308.3, 112.4,...
	Cr 49	5/2 –	48.951341	42.3 m	β^+ 1.39, 1.45,... γ 90.6, 152.9. 62.3,...
	Cr 51	7/2 –	50.944772	27.70 d	ε , γ 320.1
	Cr 55	3/2 –	54.940844	3.497 m	β^- 2.49,... γ 1528.1,...
	Cr 56	0 +	55.940645	5.9 m	β^- 1.5,... γ 83.9, 26.6,...
25	Mn 52	2 +	(378 keV)	21.1 m	β^+ 2.63,... ε , γ 1434.1,... IT 377.7
	Mn 52	6 +	51.945570	5.591 d	ε , β^+ 0.575,... γ 1434.1, 935.5, 744.2,...
	Mn 53	7/2 –	52.941294	3.7 Ma	ε (no γ)
	Mn 54	3 +	53.940363	312 d	ε , γ 834.8
	Mn 56	3 +	55.938909	2.578 h	β^- 2.84, 1.04,... γ 846.8, 1810.8, 2113.1,...
	Mn 57	5/2 –	56.938287	1.45 m	β^- 2.55,... γ 122.1, 14.4, 692.0,...
26	Fe 52	12 +	(6820 keV)	46 s	β^+ 4.4,... ε , γ 622, 870, 929, 1460,...
	Fe 52	0 +	51.948116	8.28 h	β^+ 0.80, ε , γ 168.7,...
	Fe 53	19/2 –	(3040 keV)	2.6 m	IT 701.2,... γ 1328.2, 1011.6, 2340,...
	Fe 53	7/2 –	52.945312	8.51 m	β^+ 2.8, 2.4,... ε , γ 377.9,...
	Fe 55	3/2 –	54.938298	2.73 a	ε (no γ)
	Fe 59	3/2 –	58.934880	44.51 d	β^- 0.466, 0.271,... γ 1099.2, 1291.6,...
	Fe 60	0 +	59.934077	1.5 Ma	β^- 0.147, γ 58.6D, e^-
	Fe 61		60.936749	6.0 m	β^- 2.8, 2.6,... γ 1205.1, 1027.5, 297.9,...
27	Co 56	4 +	55.939844	77.26 d	ε , β^+ 1.459,... γ 846.8, 1238.3,...
	Co 57	7/2 –	56.936296	271.8 d	ε , γ 122.1, 136.5, 14.4,...
	Co 58	5 +	(25 keV)	9.0 h	IT 24.9, e^-
	Co 58	2 +	57.935757	70.88 d	ε , β^+ 0.474,... γ 810.8,...
	Co 60	2 +	(59 keV)	10.47 d	IT 58.6, e^- (β^- 1.6,...), γ (1332.5),...
	Co 60	5 +	59.933822	5.271 a	β^- 0.318, 1.5,... γ 1332.5, 1173.2,...
	Co 61	7/2 –	60.932479	1.650 h	β^- 1.22,... γ 67.4,...
28	Ni 56	0 +	55.942136	6.10 d	ε , γ 158.4, 811.8, 750, 480,...
	Ni 57	3/2 –	56.939800	35.6 h	ε , β^+ 0.85,... γ 1377.8, 1920,...
	Ni 59	3/2 –	58.934351	0.076 Ma	ε (no γ)
	Ni 63	1/2 –	62.929673	100 a	β^- 0.0669 (no γ)
	Ni 65	5/2 –	64.930088	2.517 h	β^- 2.14, 6.5,... γ 1481.9, 1115.5,...
29	Cu 61	3/2 –	60.933462	3.35 h	β^+ 1.21,... ε , γ 283.0, 656.0,...
	Cu 62	1 +	61.932587	9.74 m	β^+ 2.93,... ε , γ 1173.0, 875.7,...
	Cu 64	1 +	63.292768	12.701 h	ε , β^- 0.578, β^+ 0.651, γ 1345.8
	Cu 66	1 +	65.928873	5.10 m	β^- 2.63,... γ 1039.3,...
	Cu 67	3/2 –	66.927750	2.580 d	β^- 0.39, 0.48, 0.58,... γ 184.6, 93.3D,...
30	Zn 62	0 +	61.934334	9.22 h	ε , β^+ 0.86,... γ 596.7, 40.8, 548.4, 507.6,...
	Zn 63	3/2 –	62.933215	38.5 m	β^+ 2.32,... ε , γ 669.7, 962.1, 1412,...
	Zn 65	5/2 –	64.929245	243.8 d	ε , γ 1115.5,... (β^+ 0.325)
	Zn 69	9/2 +	(439 keV)	13.76 h	IT 438.6, β^-, γ 574.1
	Zn 69	1/2 –	68.926553	56 m	β^- 0.90,... γ 318.5,...
	Zn 72	0 +	71.926861	46.5 h	β^- 0.30,... γ 144.7,...

T – 6.3 Properties of Radioactive Nuclides

1	2	3	4	5	6
Z	**Nu-clide**	**Spin and parity**	**Mass** u	**Half life**	**Modes of decay and energy of radiation** MeV for particles, keV for γ(and IT)
31	Ga 67	3/2 –	66.928205	3.260 d	ε , γ 93.3D, 184.6, 300.2...
	Ga 68	1 +	67.927983	67.7 m	β^+ 1.899,... ε , γ 1077.3,...
	Ga 70	1 +	69.926027	21.1 m	β^- 1.65,... γ 1039.3, 176.2,... (ε)
	Ga 72	3 –	71.926372	14.10 h	β^- 0.96, 0.64,... γ 834.1, 2201.7, 630.0,...
	Ga 73	3/2 –	72.925170	4.87 h	β^- 1.2,... γ 297.3, 325.7, 53.4D, 13.3D,... e^-
32	Ge 68	0 +	67.928097	270.8 d	ε (no γ)
	Ge 69	5/2 –	68.927972	39.1 h	ε , β^+ 1.21,... γ 1106.8, 574.1, 872.0,...
	Ge 71	1 –	70.924954	11.4 d	ε (no γ)
	Ge 75	7/2 +	(140 keV)	48 s	IT 139.6,... e^-, β^- (γ)
	Ge 75	1/2 –	74.922860	82.80 m	β^- 1.19,... γ 264.7,...
	Ge 76	0 +	75.921401	~ 1.5 Za	$\beta^-\beta^-$
	Ge 77	1/2 –	(160 keV)	53 s	β^- 2.9,... γ 215.5,... IT 159.7
	Ge 77	7/2 +	76.923549	11.30 h	β^- 2.20, 1.38,... γ 264.4, 211.0, 215.5,...
33	As 71	5/2 –	70.927115	2.72 d	ε , β^+ 0.81,... γ 175.0,...
	As 72	2 –	71.926753	26.0 h	β^+ 2.48, 3.3... ε , γ 834.1, 630,...
	As 73	3/2 –	72.923825	80.3 d	ε , γ 53.4D, 13.3D, e^-
	As 74	2 –	73.923929	17.78 d	ε , β^+ 0.941,... γ 595.9,... β^- 1.350, 0.717, γ 634.8,...
	As 76	2 –	75.922394	26.3 h	β^- 2.97, 2.41,... γ 559.1,...
	As 77	3/2 –	76.920648	38.8 h	β^- 0.68,... γ 239.0,...
34	Se 72	0 +	71.927112	8.5 d	ε , γ 46.0
	Se 73	3/2 –	(26 keV)	40 m	IT 25.7, e^-, β^+ 1.7,... ε , γ 67.1, 253.9, 84.3,...
	Se 73	9/2 +	72.926767	7.1 h	β^+ 1.32,... ε , γ 361.0, 67.1,...
	Se 75	5/2 +	74.922524	119.78 d	ε , γ 264.7, 136.0, 279.5,...
	Se 79	7/2 +	78.918500	≤ 65 ka	β^- 0.16 (no γ)
	Se 81	7/2 +	(103 keV)	57.3 m	IT 103.0, (β^-), γ (260), 276,...
	Se 81	1/2 –	80.917993	18.5 m	β^- 1.58,... γ 276.0, 290.1,...
	Se 82	0 +	81.916700	~ 0.11 Za	$\beta^-\beta^-$
35	Br 77	3/2 –	76.921380	2.376 d	ε , γ 239.0, 520.6,... (β^+ 0.34)
	Br 78	1 +	77.921146	6.45 m	β^+ 2.5,... ε , γ 613.7,... (β^-, γ)
	Br 80	5 –	(86 keV)	4.42 h	IT 48.8, e^-, γ 37.1,...
	Br 80	1 +	79.918530	17.66 m	β^- 2.00, γ 616.6, ε , β^+ 0.85, γ 665.9,...
	Br 82	2 –	(46 keV)	6.1 m	IT 45.9, e^-, β^- (γ 776.5,...)
	Br 82	5 –	81.916805	35.31 h	β^- 0.444,... γ 776.5, 554.3, 619.1,...
	Br 83	3/2 –	82.915181	2.40 h	β^- 0.93,... γ 9.4D(e^-), 32.2D(e^-), 526.6,...
36	Kr 76	0 +	75.925950	14.8 h	ε , γ 315.7, 270.2, 45.5,...
	Kr 77	5/2 +	76.924669	74.4 m	β^+ 1.88, 1.70,... ε , γ 129.7, 146.4,...
	Kr 79	1/2 –	78.920083	34.92 h	ε , β^+ 0.60,... γ 261.3, 397.5, 606.1,...
	Kr 81	7/2 +	80.916593	0.213 Ma	ε , γ 276.0
	Kr 85	1/2 –	(305 keV)	4.48 h	β^- 0.839, γ 151.2, IT 304.9
	Kr 85	9/2 +	84.912530	**10.73 a**	β^- 0.687,... (γ 514.0D)
	Kr 87	5/2 +	86.913359	76.2 m	β^- 3.5, 3.9,... γ 402.6, 2555,...

T – 6.3 Properties of Radioactive Nuclides

1	2	3	4	5	6
Z	**Nu-clide**	**Spin and parity**	**Mass** u	**Half life**	**Modes of decay and energy of radiation** MeV for particles, keV for γ(and IT)
37	Rb 83	5/2 –	82.915114	82.2 d	ε, γ 520.4, 529.6, 552.6,...
	Rb 84	6 –	(464 keV)	20.3 m	IT 216.1, 464.3, γ 248.2
	Rb 84	2 –	83.914385	32.9 d	ε, β^+ 1.66, 0.78,... γ 881.7,... β^- 0.892
	Rb 86	2 –	85.911167	18.65 d	β^- 1.775,... γ 1076.7, (ε)
	Rb 87	3/2 –	86.909184	47.5 Ga	β^- 0.273 (no γ)
	Rb 88	2 –	87.911319	17.7 m	β^- 5.31,... γ 1836.1, 898.1,...
38	Sr 82	0 +	81.918401	25.36 d	ε (no γ)
	Sr 83	7/2 +	82.917555	1.350 d	ε, β^+ 1.23,... γ 762.7, 381.5,...
	Sr 85	1/2 –	(239 keV)	1.127 h	IT 238.7,... γ 231.7, ε, γ 151.2
	Sr 85	9/2 +	84.912933	64.84 d	ε, γ 514.0D,...
	Sr 89	5/2 +	88.907453	50.52 d	β^- 1.49,... γ 909.2D,...
	Sr 90	0 +	89.907738	**29.1 a**	β^- 0.546 (no γ)
39	Y 87	1/2 –	86.910878	3.35 d	ε, β^+ (0.8),... γ 484.8, 388.5D,...
	Y 88	4 –	87.909504	106.6 d	ε, β^+ (0.76), γ 1836.1, 898.1,...
	Y 90	7 +	(682 keV)	3.19 h	IT 479.5, γ 202.5, (β^-), γ 2318.9D,...
	Y 90	2 –	89.907151	2.67 d	β^- 2.281,... γ (2186.2)
	Y 91	1/2 –	90,907303	58.5 d	β^- 1.545,... γ 1205
40	Zr 88	0 +	87,910226	83.4 d	ε, γ 392.9D
	Zr 89	9/2 +	88.908889	3.27 d	ε, β^+ 0.90, γ 909.2D,...
	Zr 93	5/2 +	92.906475	1.53 Ma	β^- 0.060, γ 30.4D
	Zr 95	5/2 +	94.908043	64.02 d	β^- 0.366, 0.400,... γ 756.7, 724.2,...
	Zr 97	1/2 +	96.910951	16.8 h	β^- 1.92,... γ 743.3D,...
41	Nb 91	1/2 –	(104 keV)	62 d	IT 104.5, e^-, ε, γ 1205
	Nb 91	9/2 +	90.906990	680 a	ε, (β^+)
	Nb 92	2 +	(135 keV)	10.13 d	ε, (β^+), γ 934.5
	Nb 92	7 +	91,907193	35 Ma	ε, γ 561.1, 934.5
	Nb 93	1/2 –	(031 keV)	16.1 a	IT 30.4, e^-
	Nb 94	6 +	93.907283	0.020 Ma	β^- 0.473, γ 871.1, 702.6
	Nb 95	1/2 –	(236 keV)	3.61 d	IT 235.7, β^- 1.16,... γ 204.1,...
	Nb 95	9/2 +	94,906835	34.97 d	β^- 0.160,... γ 765.8,...
42	Mo 90	0 +	89.913936	5.7 h	ε, β^+ 1.085, γ 257.9, 122.9D,...
	Mo 91	9/2 +	90.911751	15.5 m	β^+ 3.44,... ε, γ 1637.0, 1581.2,...
	Mo 93	5/2 +	92.906811	3.50 ka	ε, (γ 30.4D)
	Mo 99	1/2 +	98.907712	2.7476 d	β^- 1.214,... γ 140.5D, 739.5...
	Mo 101	1/2 +	100.910347	14.6 m	β^- 0.7, 2.23,... γ 191.9, 590.9, 1012.5, 506.0,...
43	Tc 95	1/2 –	(39 keV)	61 d	ε, γ 204.1, 582.1, 835.1,... IT 38.9, e^-, β^+ (0.71) 0.51
	Tc 95	9/2 +	94.907657	20.0 h	ε, γ 765.8
	Tc 96	7 +	95.907871	4.3 d	ε, γ 778.2, 849.9, 812.5,...
	Tc 97	1/2 –	(97 keV)	90 d	IT 96.5, e^-
	Tc 97	9/2 +	96.906365	2.6 Ma	ε (no γ)
	Tc 98	6 +	97.907215	4.2 Ma	β^- 0.40, γ 745.4, 652.4
	Tc 99	9/2 +	98.906254	**0.213 Ma**	β^- 0.292, (γ 89.7)
	Tc 100	1 +	99.907658	15.8 s	β^- 3.4, 2.9,..., γ 539.5, 590.8,...

T – 6.3 Properties of Radioactive Nuclides

1	2	3	4	5	6
Z	**Nu-clide**	**Spin and parity**	**Mass** u	**Half life**	**Modes of decay and energy of radiation** MeV for particles, keV for γ(and IT)
44	Ru 94	0 +	93.911360	52 m	ε , γ 367, 892,...
	Ru 95	5/2 +	94.910413	1.64 h	ε , β^+ 1.20, 0.91,... γ 336.4, 1096.9, 626.9,...
	Ru 97	5/2 +	96.907555	2.89 d	ε , γ 215.7, 324.5,...
	Ru 103	3/2 +	102.906324	39.27 d	β^- 0.223,..., γ 497.1,...
	Ru 105	3/2 +	104.907750	4.44 h	β^- 1.187, 1.11, 1.8,..., γ 724.3, 469.4, 676.3,...
	Ru 106	0 +	105.907327	**372.6 d**	β^- 0.0394 (no γ)
45	Rh 101	9/2 +	(157 keV)	4.35 d	ε , γ 306.9, 545,... IT 157.3, e$^-$
	Rh 101	1/2 –	100.906163	3.3 a	ε , γ 127.2, 198.0,...
	Rh 102	6 +	(141 keV)	~ 2.9 a	ε , γ 475.1, 631.3, 697.5,... IT 42, e$^-$
	Rh 102	2 –	101.906843	207 d	ε , β^- 1.15,... β^+ 1.30, 0.82,... γ 475.1,...
	Rh 104	5 +	(129 keV)	4.36 m	IT 77.5(e$^-$), 31.8(e$^-$), γ 51.4, 91.7, (β^- 1.3), γ 555.8
	Rh 104	1 +	103.906655	42.3 s	β^- 2.44,... (ε), γ 555.8,...
	Rh 105	1/2 –	(130 keV)	40 s	IT 129.6
	Rh 105	7/2 +	104.905692	35.4 h	β^- 0.566, 0.248,..., γ 319.2,...
46	Pd 100	0 +	99.908505	3.7d	ε , γ 84.0, 74.7,...
	Pd 101	5/2 +	100.908289	8.4 h	ε , β^+ 0.776,... γ 296.3, 590.5,...
	Pd 103	5/2 +	102.906087	16.99 d	ε , γ 38.8D (e$^-$) 357.5, ,...
	Pd 107	5/2 +	106.905129	6.5 Ma	β^- 0.040 (no γ)
	Pd 109	5/2 +	108.905954	13.5 h	β^- 1.028,..., γ 88.0D,...
	Pd 111	11/2 –	(172 keV)	5.5 h	IT 172.2, β^- 0.35, 0.77,... γ 70.4, 391.2,...
	Pd 111	5/2 +	110.907644	23.4 m	β^- 2.2,..., γ (580.0), 70.4,...
47	Ag 105	1/2 –	104.906528	41.3 d	ε , γ 344.5, 280.5,... (β^+)
	Ag 106	6 +	(90 keV)	8.4 d	ε , γ 511.9, 1046,...
	Ag 106	1 +	105.906666	24.0 m	β^+ 1.96,... ε , γ 51.9,... β^-
	Ag 108	6 +	(109 keV)	130 a	ε , β^+, γ 722.9, 433.9, 614.3, IT 30.4, e$^-$
	Ag 108	1 +	107.905954	2.39 m	β^- 1.65, γ 633.0, ε , β^+ 0.88, γ (433.9), 618.8,...
	Ag 110	6 +	(117 keV)	249.8 d	β^- 0.087, 0.530,..., γ 657.8, 884.7,... IT 116.5 e$^-$
	Ag 110	1 +	109.906111	24.6 s	β^- 2.981,... γ 657.8,... ,
48	Cd 104	0 +	103.909848	58 m	ε , (β^+ 0.29), γ 83.5, 709.3,...
	Cd 105	5/2 +	104.909468	55.5 m	ε , β^+ 1.69,... γ 961.8, 346.6, 1302.5,...
	Cd 107	5/2 +	106.906614	6.52 h	ε , γ 91.3D, 828.9... (β^+ 0.302)
	Cd 109	5/2 +	108.904985	462.6 d	ε , γ 88.0D, e$^-$
	Cd 113	11/2 –	(264 keV)	14.1 a	β^- 0.59, (IT 263.7)
	Cd 113	1/2 +	112.904402	9 Pa	β^- 0.3
	Cd 115	11/2 –	(181 keV)	44.6 d	β^- 1.62,... γ 933.8, 1290.6,...
	Cd 115	1/2 +	114.905431	2.228 d	β^- 1.31, 0.593,... γ 336.3D, 527.9,...
	Cd 117	1/2 +	116.907218	2.49 h	β^- 0.67, 2.2,... γ 315.3D, 273.3, 1303.3,...
49	In 111	9/2 +	110.905111	2.8049 d	ε , γ 245.4, 171.3,...
	In 114	5 +	(190 keV)	49.51 d	IT 190.3, ε , γ 558.4, 725.2
	In 114	1 +	113.904917	1.198 m	β^- 1.984,... γ 1299.9, ε , (β^+ 0.40), γ 558.4, 576,...
	In 115	9 +	114.903878	0.44 Pa	β^- 0.49 (no γ)
50	Sn 110	0 +	109.907853	4.1 h	ε , γ 283
	Sn 111	7/2 +	110.907735	35 m	ε , β^+ 1.5,... γ 1153.0, 1915.0, 762.0, 1610.5,...
	Sn 113	1/2 +	112.905174	115.1 d	ε , γ 391.7D,...

1	2	3	4	5	6
Z	**Nu-clide**	**Spin and parity**	**Mass** u	**Half life**	**Modes of decay and energy of radiation** MeV for particles, keV for γ(and IT)
	Sn 121	11/2 –	(6 keV)	55 a	IT 6.3, e^-, β^- 0.35, γ 37.1, e^-
	Sn 121	3/2 +	120.904237	1.128 d	β^- 0.383 (no γ)
	Sn 123	11/2 –	122.905722	129.2 d	β^- 1.42,... γ 1088.6,...
	Sn 125	11/2 –	124.907785	9.63 d	β^- 2.35,... γ 1067.0, 1089.2, 822.4, 915.5,...
	Sn 126	0 +	125.907654	**0.1 Ma**	β^- 0.25, γ 87.6,...
51	Sb 119	5/2 +	118.903947	38.1 h	ε, γ 23.9, e^-
	Sb 120	8 –	(0 keV)	5.76 d	ε, γ 1171.4, 1023.1, 197.3D, 89.8,...
	Sb 120	1 +	119.902197	15.89 m	ε, β^+ 1.72,... γ 1171.4,...
	Sb 122	2 –	121.905175	2.70 d	β^- 1.414, 1.980,... γ 564.1, ε, β^+ 0.57
	Sb 124	3 –	123.905938	60.20 d	β^- 0.61, 2.301,... γ 602.7, 1691.0,...
	Sb 125	7/2 +	124.905248	**2.758 a**	β^- 0.302, 0.13,... γ 427.9, 600.5, 635.9, 463.4,...
52	Te 118	0 +	117.905825	6.00 d	ε (no γ)
	Te 119	11/2 –	(261 keV)	4.69 d	ε, (β^+), γ 153.6, 1212.7, 270.5,...
	Te 119	1/2 +	118.906408	16.0 h	ε, β^+ 0.627,... γ 644.0, 700,...
	Te 121	11/2 –	(294 keV)	154 d	IT 818.8, e^-, γ 212.2, ε, γ 1102.2, 37.1 (e^-),...
	Te 121	1/2 +	120.904930	16.8 h	ε, γ 573.1, 507.6,...
	Te 123	11/2 –	(248 keV)	119.7 d	IT 88.5, e^-, γ 159.0
	Te 123	1/2 +	122.904273	12 Ta	ε (no γ)
	Te 127	11/2 –	(88 keV)	109 d	IT 88.3, e^-, β^- 0.7, γ 57.6,...
	Te 127	3/2 +	126.905217	9.4 h	β^- 0.69,... γ 417.9, (360.3),...
	Te 132	0 +	131.908524	**3.27 d**	β^- 0.215,... γ 228.3, 49.7,...
53	I 124	2 –	123.906211	4.18 d	ε, β^+ 2.14, 1.53,... γ 602.7,...
	I 125	5/2 +	124.904624	60.1 d	ε, γ 35.5, e^-
	I 126	2 –	125.905619	13.0 d	ε, β^- 0.87,... γ 388.6,... β^+ (1.13),...
	I 128	1 +	127.905805	25.0 m	β^- 2.13,... γ 442.9,... ε, (β^+), (γ 743.4)
	I 129	7/2 +	128.904988	15.7 Ma	β^- 0.15, γ 39.6, e^-
	I 130	5 +	129.906674	12.36 h	β^- 1.04, 0.62,... γ 536.1, 668.6, 739.5
	I 131	7/2 +	130.906124	**8.040 d**	β^- 0.606,... γ 264.5,...
	I 132	4 +	131.907995	**2.28 h**	β^- 1.22, 2.16,... γ 667.7, 772.7,...
54	Xe 127	1/2 +	126.905180	36.4 d	ε, γ 202.9, 172.1,...
	Xe 133	3/2 +	132.905906	5.243 d	β^- 0.346,... γ 81.0,...
	Xe 135	3/2 +	134.907208	9.10 h	β^- 0.91,... γ 249.8,...
55	Cs 131	5/2 +	130.905460	9.69 d	ε (no γ)
	Cs 132	2 –	131.906430	6.48 d	ε, β^+ 0.40,... γ 667.7,... β^- 0.8, γ 464.5,...
	Cs 134	4 +	133.906714	**2.065 a**	β^- 0.658, 0.089,... γ 604.7, 795.8,... ε, (β^+)
	Cs 135	7/2 –	134.905972	2.3 Ma	β^- 0.21 (no γ)
	Cs 136	5 +	135.907306	13.16 d	β^- 0.341,... γ 818.5, 1048.1, 340.6,...
	Cs 137	7/2 +	136.907084	**30.17 a**	β^- 0.514,... γ 661.65D
56	Ba 128	0 +	127.908309	2.43 d	ε, γ 273.4,...
	Ba 129	1/2 +	128.908674	2.2 h	β^+ 1.42,... γ 214.3, 220.9, 129.1,...
	Ba 131	1/2 +	130.906931	11.7 d	ε, (β^+),... γ 496.3, 123.8, 216.1,...
	Ba 133	1/2 +	132.906002	10.53 a	ε, γ 356.0, 81.0, 302.9,...
	Ba 139	7/2 –	138.908835	1.40 h	β^- 2.27, 2.14,... γ 165.9,...
	Ba 140	0 +	139.910599	**12.75 d**	β^- 1.0, 0.48, 1.02,... γ 537.3, 30.0,...

T – 6.3 Properties of Radioactive Nuclides

1	2	3	4	5	6
Z	**Nu-clide**	**Spin and parity**	**Mass** u	**Half life**	**Modes of decay and energy of radiation** MeV for particles, keV for γ(and IT)
57	La 136	1 +	135.907651	9.87 m	ε , β^+ 1.8,.... γ 818.5,...
	La 137	7/2 +	136.906470	0.06 Ma	ε (no γ)
	La 138	5 +	137.907107	105 Ga	ε , β^- 0.25 γ 1435.8, 788.7,...
	La 140	3 –	139.909473	1.678 d	β^- 1.35, 1.24, 1.67,... γ 1596.5, 487.0, 815.8,...
	La 141	7/2 +	140.910957	3.90 h	β^- 2.43,... γ 1354.5,...
58	Ce 134	0 +	133.909026	75.9 h	ε , γ 162.3, 130.4,...
	Ce 135	1/2 +	134.909146	17.7 h	ε , β^+ 0.8,.... γ 265.6, 300.1, 606.8,...
	Ce 137	11/2 –	(254 keV)	34.3 h	IT 254.3, ε , γ 824.7, 169.2, 762.2,...
	Ce 137	3/2 +	136.907778	9 h	ε , β^+, γ 447.2,...
	Ce 139	3/2 +	138.906647	137.6 d	ε , γ 165.9,...
	Ce 141	7/2 –	140.908271	32.50 d	β^- 0.436, 0581,... γ 145.4,...
	Ce 143	3/2 –	142.912381	1.38 d	β^- 1.110, 1.404,... γ 293.3, 57.4,...
	Ce 144	0 +	143.913643	**284.6 d**	β^- 0.318, 0.185,... γ 133.5, 80.1,...
59	Pr 140	1 +	139.909071	3.39 m	ε , β^+ 2.37,... (γ 1596.5, 306.9,...)
	Pr 142	2 –	141.910040	19.12 h	β^- 2.162,... γ 1575.5,... ε , γ 641.2
	Pr 143	7/2 +	142.910812	13.57 d	β^- 0.933,... γ (742.0)
60	Nd 140	0 +	139.909310	3.37 d	ε (no γ)
	Nd 141	3/2 +	140.909605	2.49 h	ε , β^+ 0.802,... (γ 1127.0, 1292.7, 1147.3,...)
	Nd 144	0 +	143.910082	2.3 Pa	α 1.83
	Nd 147	5/2 –	146.916096	10.98 d	β^- 0.805,... γ 91.1, 531.0,...
	Nd 149	5/2 –	148.920144	1.72 h	β^- 1.42, 1.13, 1.03,... γ 211.3, 114.3, 270.2,...
61	Pm 143	5/2 +	142.910928	265 d	ε , γ 742.0,...
	Pm 144	5 –	143.912586	360 d	ε , γ 696.5, 618.0, 476.8,...
	Pm 145	5/2 +	144.912744	17.7 a	ε , γ 72.5, 67.2, (e^-), (α 2.24)
	Pm 146	3 –	145.914692	5.53 a	ε , γ 453.9, 735.8,... β^-, 0.795,... γ 747.2
	Pm 147	7/2 +	146.915134	**2.6234 a**	β^- 0.224,... γ (121.3),...
	Pm 148	1 –	147.917468	5.37 d	β^- 2.47, 1.02,... γ 1465.1, 550.3,...
	Pm 149	7/2 +	148.918329	53.1 h	β^- 1.072,... γ 286.0,...
62	Sm 145	7/2 –	144.913406	340 d	ε , γ 61.2, (492),...
	Sm 146	0 +	145.913038	103 Ma	α 2.455
	Sm 147	7/2 –	146.914893	0.11 Ta	α 2.235
	Sm 148	0 +	147.914818	7 Pa	α 1.96
	Sm 149	7/2 –	148.917179	2 Pa	α 1.07
	Sm 151	5/2 –	150.919929	90 a	β^- 0.076,... γ (21.5), e^-
	Sm 153	3/2 +	152.922094	46.3 h	β^- 0.69, 0.64,... γ 103.2, 69.7,...
	Sm 155	3/2 –	154.924636	22.2 m	β^- 1.52,... γ 104.3, 246, 141,...
63	Eu 147	5/2 +	146.916741	24.4 d	ε , β^+ 0.701, 0.58, 0.505... γ 197.4, 121.3,... (α 2.91)
	Eu 148	5 –	147.918154	54.5 d	ε , β^+ (0.92),... γ 550.3, 630.0,... (α 2.63)
	Eu 149	5/2 +	148.917926	93.1 d	ε , γ 327.5, 277.1,...
	Eu 150	5 –	149.919699	36.9 a	ε , γ 333.9, 439.4, 584.3,...
	Eu 152	3 –	151.921741	13.54 a	ε , β^+ 0.727,... γ 121.8, 1408.0,... β^- 0.696,... γ 344.3,...
	Eu 154	3 –	153.922976	8.59 a	β^- 0.58, 0.27,... γ 123.1, 1274.5,... ε , (γ)
	Eu 155	5/2 +	154.922890	4.71 a	β^- 0.15,... γ 86.5, 105.3,...
	Eu 156	0 +	155.924751	15.2 a	β^- 2.45, 0.49,... γ 811.8, 89.0, 1230.7,...

T – 6.3 Properties of Radioactive Nuclides

1	2	3	4	5	6
Z	**Nu-clide**	**Spin and parity**	**Mass** u	**Half life**	**Modes of decay and energy of radiation** MeV for particles, keV for γ (and IT)
64	Gd 148	0 +	147.918110	75 a	α 3.1828
	Gd 149	7/2 –	148.919336	9.3 d	ε, γ 149.7, 298.6, 346.7D, (α 3.016)
	Gd 150	0 +	149.918656	1.79 Ma	α 2.726
	Gd 151	7/2 –	150.920344	124 d	ε, γ 153.6, 243.2,... (α 2.60)
	Gd 152	0 +	151.919788	0.11 Pa	α 2.14
	Gd 153	3/2 –	152.921746	241.6 d	ε, γ 97.4, 103.2,...
	Gd 159	3/2 –	158.926385	18.6 h	β^- 0.96,... γ 363.6, 58.0(e^-),...
	Gd 161	5/2 –	160.929666	3.66 m	β^- 1.56,... γ 360.9, 314.9, 102.3,...
65	Tb 157	3/2 +	156.924021	99 a	ε, γ 54.5, e^-
	Tb 158	3 –	157.925410	180 a	ε, γ 944.2, 962.2,... β^- 0.85,... γ 99.0,...
	Tb 160	3 –	159.927164	72.3 d	β^- 0.57, 0.86,... γ 879.4, 298.6, 966.2,...
	Tb 161	3/2 +	160.927566	6.90 d	β^- 0.52, 0.46,... γ 25.7, 48.9, 74.6,...
66	Dy 154	0 +	153.924423	3.0 Ma	α 2.870
	Dy 155	3/2 –	154.925749	9.9 h	ε, β^+ 0.845,... γ 226.9,...
	Dy 157	3/2 –	156.925461	8.1 h	ε, γ 326.2,...
	Dy 159	3/2 –	158.925736	144.4 d	ε, γ 58.0, e^-,...
	Dy 165	7/2 +	164.931700	2.33 h	β^- 1.29,... γ 94.7, 361.7D,...
	Dy 166	7/2 +	165.932803	3.400 d	β^- 0.40,... γ 82.5, (426),...
67	Ho 163	7/2 –	162.928730	4570 a	ε (no γ)
	Ho 164	1 +	163.930231	29 m	ε, γ 73.4,... β^- 0.96, 0.88,... γ 91.4
	Ho 166	7 –	(6 keV)	1.2 ka	β^- < 0.065,... γ 184.4, 810.3, 711.7,...
	Ho 166	0 –	165.932281	26.80 h	β^- 1.855, 1.773,... γ 80.6, 1379.4,...
	Ho 167	7/2 –	166.933126	3.1 h	β^- 0.32, 0.97, 0.61,... γ 346.5, 321.3...
68	Er 160	0 +	159.929079	28.58 h	ε, γ (60.0 (e^-)),...
	Er 161	3/2 –	160.930001	3.21 h	ε, (β^+ 0.82),... γ 826.5, 211.2D,...
	Er 163	5/2 –	162.930029	1.25 h	ε, (β^+ 0.19),... γ (1113.5), 436.1, 439.9,...
	Er 165	5/2 –	164.930723	10.36 h	ε (no γ)
	Er 169	1/2 –	168.934588	9.40 d	β^- 0.344, 0.34,... γ (8.4 (e^-)),...
	Er 171	5/2 –	170.938026	7.52 h	β^- 1.065,... γ 308.3D, 295.9, 111.6,...
69	Tm 167	1/2 +	166.932849	9.24 d	ε, γ 207.8D,...
	Tm 168	3 +	167.934170	93.1 d	ε, (β^+), (β^-), γ 198.2, 816.0, 447.5, 184.3,....
	Tm 170	1 –	169.935798	128.6 d	β^- 0.968, 0.883, γ 84.3, (ε), (γ 78.7)
	Tm 171	1/2 +	170.936426	1.92 a	β^- 0.097,... γ 66.7, e^-
	Tm 172	2 –	171.938396	2.65 d	β^- 1.79, 1.87,... γ 78.8, 1093.6, 1387.1, 1529,...
70	Yb 166	0 +	165.933880	56.7 h	ε, γ 82.3
	Yb 167	5/2 –	166.934947	17.5 m	ε, (β^+ 0.64),... γ 113.3D, 106.2, 176.2D,...
	Yb 169	7/2 +	168.935187	32.03 d	ε, γ 63.1, 198.0, 177,...
	Yb 175	7/2 –	174.941273	4.19 d	β^- 0.466,... γ 396.3, 282.5, 113.8,...
71	Lu 173	7/2 +	172.938927	1.37 a	ε, γ 272.0, 78.7, 100.7, 171.4,...
	Lu 174	1 –	173.940334	3.3 a	ε, γ 1241.8, 76.5,... β^+ 0.38,...
	Lu 176	7 –	175.942682	37 Ga	β^- 0.57,... γ 306.9, 201.8,...
	Lu 177	23/2 –	(970 keV)	160 d	β^- 0.152, γ 208.4, 228.5, (IT 115.8), e^-, γ 413.7,...
	Lu 177	7/2 +	176.943755	6.71 d	β^- 0.497,... γ 208.4, 112.9,...

1	2	3	4	5	6
Z	**Nu- clide**	**Spin and parity**	**Mass** u	**Half life**	**Modes of decay and energy of radiation** MeV for particles, keV for γ(and IT)
72	Hf 172	0 +	171.939458	1.87 a	ε, γ 24.0, 125.8, 67.4, 81.8,...
	Hf 173	1/2 –	172.940650	23.6 h	ε, γ 123.7D, 297.0,...
	Hf 174	0 +	173.940040	2.0 Pa	α 2.50
	Hf 175	5/2 –	174.941503	70 d	ε, γ 343.4,...
	Hf 178	16 +	(2446 keV)	31 a	IT 12.7, e^-, 426.4, 325.6,...
	Hf 181	1/2 –	180.949099	42.4 d	β^- 0.405,... γ 482.1, 133.0, 345.9,...
	Hf 182	0 +	181.950553	9 Ma	β^-, γ 270.4,...
73	Ta 179	7/2 +	178.945934	1.8 a	ε (no γ)
	Ta 180	9 –	(75 keV)	> 1.2 Pa	ε, β^+, γ 350, 332,...
	Ta 180	1 +	179.947466	8.15 h	ε, γ 93.3, β^- 0.71, 0.61, γ 103.4
	Ta 182	3 –	181.950152	114.43 d	β^- 0.522, 0.25,... γ 67.8, 1121.3, 1221.4,...
74	W 178	0 +	177.945848	21.6 d	ε (no γ)
	W 181	9/2 +	180.948198	121.2 d	ε, γ 6.2D (e^-),...
	W 185	3/2 –	184.953421	74.8 d	β^- 0.433, ... (γ 125.4)
	W 187	3/2 –	186.957158	23.9 h	β^- 0.622, 1.312,... γ 685.7, 479.6,...
	W 188	0 +	187.958487	69.4 d	β^- 0.349,... (γ 290.7, 227.1, 63.6,...)
75	Re 183	5/2 +	182.950821	70 d	ε, γ 162.3, 46.5,...
	Re 184	8 +	(188 keV)	165 d	IT 83.3, e^-, 104.7, ε, γ 252.8, 216.6, 920.9,...
	Re 184	3 –	183.952524	38 d	ε, γ 902.3, 792.1,...
	Re 186	8 +	(149 keV)	0.20 Ma	IT (~ 50), e^-, γ 59.0, 40.4, 99.4,...
	Re 186	1 –	185.954987	3.777 d	β^- 1.071, 0.933,... γ 137.1,... ε, γ 122.4
	Re 187	5/2 +	186.955751	41 Ga	β^- 0. 00264 (no γ)
	Re 188	1 –	187.958112	16.94 h	β^- 2.118, 1.962, ... γ 155.0,...
	Re 189	5/2 +	188.959228	24 h	β^- 1.01, ... γ 216.7, 219.4, 245.1,...
76	Os 185	1/2 –	184.954043	93.6 d	ε, γ 646, 874.8, 880.4, 717.4,...
	Os 186	0 +	185.953838	2 Pa	α 2.757
	Os 191	9/2 –	190.960928	15.4 d	β^- 0.143, ... γ 129.4D,...
	Os 193	3/2 –	192.964148	30.5 h	β^- 1.13,... γ 138.9, 460.5, 73.0,...
	Os 194	0 +	193.965179	6.0 a	β^- 0.096, 0.054, ... γ 43 e^-,...
77	Ir 188	2 –	187.958852	41.3 h	ε, β^+ 1.65, 1.13,... γ 155.1, 2214.7, 633.1, 478.0,...
	Ir 189	3/2 +	188.958717	13.2 d	ε, γ 245.0, 69.5, 59.1,...
	Ir 190	4 +	189.960592	11.8 d	ε, γ 186.7, 605.2, 518.5,...
	Ir 192	9 +	(155 keV)	241 a	IT (155.2), e^-
	Ir 192	4 –	191.962602	73.83 d	β^- 0.672, 0.54,... γ 316.5, 468.1,... ε, γ 205.8, 484.6,...
	Ir 194	11 ?	(190 keV)	170 d	β^-, γ 482.9, 328.5,...
	Ir 194	1 –	193.965076	19.3 h	β^- 2.24,... γ 328.5,...
78	Pt 188	0 +	187.959396	10.2 d	ε, γ 187.5, 195.0,... (α 3.92)
	Pt 189	3/2 –	188.960832	10.9 h	ε, β^+ 0.89,... γ 721.4, 607.6, 94.3, 568.8, 243.5,...
	Pt 190	0 +	189.959930	0.65 Ta	α 3.18
	Pt 191	3/2 –	190.961685	2.96 d	ε, γ 538.9, 409.5, 359.9,...
	Pt 193	1/2 –	192.962984	50 a	ε (no γ)
	Pt 197	1/2 –	196.967323	19.8 h	β^- 0.642, 0.719,... γ 77.3, 191.4,...

T – 6.3 Properties of Radioactive Nuclides

1	2	3	4	5	6
Z	**Nu-clide**	**Spin and parity**	**Mass** u	**Half life**	**Modes of decay and energy of radiation** MeV for particles, keV for γ(and IT)
79	Au 194	1 –	193.965339	39.4 h	ε , β^+ 1.49,... γ 328.5, 293.6,...
	Au 195	3/2 +	194.965018	186.12 d	ε , γ 98.9,...
	Au 196	2 –	195.966551	6.18 d	ε , γ 355.6, 332.9,... β^- 0.259, γ 425.6, (β^+)
	Au 198	2 –	197.968225	64.66 h	β^- 0.962,... γ 411.8,...
	Au 199	3/2 +	198.968748	3.14 d	β^- 0.292, 0.25, 0.453,... γ 158.4, 208.2,...
80	Hg 194	0 +	193.965382	520 a	ε (no γ)
	Hg 195	13/2 +	(176 keV)	40.1 h	IT 122.8, e⁻, γ 37.1, e⁻, ε , γ 261.8D, 560.3,...
	Hg 195	1/2 –	194.966639	9.5 h	ε , γ 779.8, 61.4,...
	Hg 197	1/2 –	196.967195	64.13 h	ε , γ 77.3,...
	Hg 203	5/2 –	202.972857	46.61 d	β^- 0.213 γ 279.2
	Hg 205	1/2 –	204.976056	5.2 m	β^- 1.54,... γ 203.7,...
81	Tl 200	2 –	199.970945	26.1 h	ε , β^+ (1.07), 1.44,... γ 368.0, 1205.7,...
	Tl 201	1/2 +	200.970804	72.9 h	ε , γ 167.4, 135.3,...
	Tl 202	2 –	201.972091	12.23 d	ε , (β^+), γ 439.6,...
	Tl 204	2 –	203.973849	3.78 a	β^- 0.7634, , (no γ)
	Tl 206	0 –	205.976095	4.20 m	β^- 1.528,... (γ 803.1)
	Tl 207	1/2 +	206.977408	4.77 m	β^- 1.44,... γ (897.2), ...
	Tl 208	5 +	207.982005	3.053 m	β^- 1.796, 1.28, 1.52,... γ 2614.5, 583.2, 510.7,...
	Tl 209	1/2 +	208.985349	2.2 m	β^- 1.83, γ 1566, 117, 467
	Tl 210	5 +	209.990066	1.30 m	β^- 1.9, 1.3, 2.3,... γ 799.7, 298,... (n)
82	Pb 202	0 +	201.972144	53 ka	ε (no γ)
	Pb 203	5/2 –	202.973376	51.88 h	ε , γ 279.2,...
	Pb 205	5/2 –	204.974467	15.2 Ma	ε (no γ)
	Pb 209	9/2 +	208.981075	3.25 h	β^- 0.645 (no γ)
	Pb 210	0 +	209.984173	22.3 a	β^- 0.017, 0.061, γ 46.5, e⁻, (α 3.72)
	Pb 211	9/2 +	210.988731	36.1 m	β^- 1.38,... γ 404.9, 831.9, 427.0,...
	Pb 212	0 +	211.991887	10.64 h	β^- 0.331, 0.569,...γ 238.6, 300.0,...
	Pb 214	0 +	213.999798	27 m	β^- 0.67, 0.73,...γ 351.9, 295.2, 242.0,...
83	Bi 205	9/2 –	204.977375	15.31 d	ε , (β^+ 0.985), γ 1764.3, 703.5, 987.6D,...
	Bi 206	6 +	205.978483	6.243 d	ε , (β^+ 0.977), γ 803.1, 881.0, 516.2,...
	Bi 207	9/2 –	206.978455	32.2 a	ε , (β^+ 0.808), γ 569.7D, 1063.7D, 1770.2,...
	Bi 208	5 +	207.979727	0.368 Ma	ε , γ 2614.4
	Bi 210	9 –	(271 keV)	3 Ma	α 4.946, 4.908,...γ 266.2, 305.2,...
	Bi 210	1 –	209.984105	5.01 d	β^- 1.162, (α 4.648, 4.687), (γ 305, 266)
	Bi 211	9/2 –	210.987258	2.14 m	α 6.623, 6.279, γ 350, (β^-)
	Bi 212	1 –	211.991271	60.5 m	β^- 0.2.251,...γ 727.2, α 6.051,...γ 39.8,...
	Bi 213	9/2 –	212.994375	45.6 m	β^- 1.42, 1.02,...γ 440.4,...α 5.869, (5.549), γ (323.8)
	Bi 214	1 –	213.998699	19.9 m	β^- 3.27, 1.54, 1.51,...γ 609.3, 1764.5, 1120.3,...α 5.450, 5.513, γ 63,...

1	2	3	4	5	6
Z	**Nu-clide**	**Spin and parity**	**Mass** u	**Half life**	**Modes of decay and energy of radiation** MeV for particles, keV for γ (and IT)
84	Po 206	0 +	205.980465	8.8 d	ε , γ 1032.3, 511.3, 286.4, 807.4,... α 5.223
	Po 208	0 +	207.981231	2.90 a	α 5.115,... γ 899, (ε), (γ 292, 571, 603,...)
	Po 209	1/2 –	208.982416	102 a	α 4.880,... γ (260.5), 262.8, ε , γ (896.4)
	Po 210	0 +	209.982857	138.4 d	α 5.3044, γ (803.1)
	Po 211	9/2 +	210.986637	0.516 s	α 7.451,... γ (569.2D), 897.2
	Po 212	0 +	211.988852	0.298 μs	α 8.7844
	Po 214	0 +	213.995186	163.7 μs	α 7.6869,... (γ 799,...)
	Po 215	9/2 +	214.999415	1.78 ms	α 7.386,... (γ 439,...), (β^-)
	Po 216	0 +	216.001905	0.145 s	α 6.7785,... γ (805)
	Po 218	0 +	218.008966	3.10 m	α 6.0024,... (γ 510,...), (β^-)
85	At 210	5 +	209.98713	8.1 h	ε , γ 1181, 245.3, 1483.3,... α 5.524, 5.442, 5.361,... (γ 83, 106,...)
	At 211	9/2 –	210.987481	7.21 h	ε , γ (687), α 5.868,... γ (669.6),...
	At 215	9/2 –	214.998641	0.10 ms	α 8.026, γ (404.9)
	At 217	9/2 –	217.004710	32 ms	α 7.067,... γ 260, 440, 594,... (β^-)
86	Rn 211	1/2 –	210.990585	14.6 h	ε , β^+, γ 674.1, 1363.0, 678.4,... α 5.784, 5.851,... (γ 68.6), e^-, ...
	Rn 219	5/2 +	219.009475	3.96 s	α 6.8193, 6.553, 6.4254 ,... γ 271.1, 401.7,...
	Rn 220	0 +	220.011384	55.6 s	α 6.2882,... γ 549.7
	Rn 221	7/2 +	221.01546	25 m	β^- 0.83,... γ 186.4,... α 6.037, 5.788,... γ 254, 265
	Rn 222	0 +	222.017571	3.8235 d	α 5.4895,... γ 510
87	Fr 212	5 +	211.99620	20 m	ε , β^+, γ 1275, 227.7,... α 6.261, 6.383, 6.406,... γ 124.84, (84), 72,...
	Fr 221	5/2 –	221.014246	4.8 m	α 6.341, 6.127,... γ 218.0,...
	Fr 223	3/2 –	223.019731	21.8 m	β^- 1.17,... γ 50, 79.8, 235,... (α 5.340)
88	Ra 223	1/2 +	223.018497	11.435 d	α 5.7164, 5.607,... γ 269.4, 154.2, 323.9,... (C14)
	Ra 224	0 +	224.020202	3.66 d	α 5.685, 5.449,... γ 241.0,... (C14)
	Ra 225	1/2 +	225.023605	14.9 d	β^- 0.32,... γ 40.3
	Ra 226	0 +	226.025403	1.60 ka	α 4.7844, 4.602,... γ 186.1,... (C14)
	Ra 227	3/2 +	227.029171	42 m	β^- 1.31, 1.03,... γ 27.4, 300.1, 302.7, 283.7,...
	Ra 228	0 +	228.031064	5.76 a	β^- 0.039, 0.015, (γ 14, 16, 13,...)
89	Ac 225	3/2 –	225.023221	10.0 d	α 5.829, 5.793, 5.731,... γ 100, 150, 63, e^-, (C14)
	Ac 226	1 –	226.026090	29.4 h	β^- 0.89, 1.11,... γ 230.3, 158.1,... ε , γ 253.7, 186.0, (α 5.399)
	Ac 227	3/2 –	227.027747	21.77 a	β^- 0.045,... γ (15.2 (e^-)),... α 4.9534, (4.941), γ (100),...
	Ac 228	3 +	228.031015	6.15 h	β^- 1.2, 2.1,... γ 911.2, 969.0, 338.3,... (α 4.27)

T – 6.3 Properties of Radioactive Nuclides

1	2	3	4	5	6
Z	**Nuclide**	**Spin and parity**	**Mass** u	**Half life**	**Modes of decay and energy of radiation** MeV for particles, keV for γ(and IT)
90	Th 227	3/2 +	227.027699	18.72 d	α 6.038, 5.978, 5.757,... γ 236.0, 50.2,...
	Th 228	0 +	228.028731	1.913 a	α 5.423, 5.340,... γ 84.4 e^-, 216.0, 131.6, 166.4,...
	Th 229	5/2 +	229.031755	7880 a	α 4.845, 4.901, 4.815,... γ 193.6, 86.4, 210.9, 31.5,... (Ne24)
	Th 230	0 +	230.033127	75.4 ka	α 4.688, 4.621,... γ 67.7 e^-, (SF), (Ne24)
	Th 231	5/2 +	231.036297	25.5 h	β^- 0.305, 0.138,... γ 25.6, 84.2,...
	Th 232	0 +	232.038050	14.0 Ga	α 4.013, 3.950,... (γ 64), (e^-)
	Th 233	1/2 +	233.041577	22.3 m	β^- 1.245,... γ 86.5, 29.4, 459.3,...
	Th 234	0 +	234.043596	24.10 d	β^- 0.198,... γ 63.3, 92.4, 92.8,...
91	Pa 230	2 –	230.034533	17.4 d	ε, γ 952,... β^- 0.51,... (γ 314.8,...), (α 5.345)
	Pa 231	3/2 –	231.035879	32.8 ka	α 5.013, 4.950, 5.028,... γ 27.4, 300.0,... (Ne24)
	Pa 232	2 –	232.038582	1.31 d	β^- 0.314, 0.294,... γ 969.3, 894.3,... ε
	Pa 233	3/2 –	233.040240	27.0 d	β^- 0.256, 0.15,... γ 312.0,...
	Pa 234	0 –	(74 keV)	1.17 m	β^- 2.29,... γ 1001.0, 766.4,... (IT < 73.9 e^-)
	Pa 234	4 +	234.043302	6.69 h	β^- 0.48, 0.65,... γ 131.3, 881, 883,...
92	U 232	0 +	232.037146	70 a	α 5.3203, 5.2635,... (γ 57.8, (e^-)), (SF), (Ne24)
	U 233	5/2 +	233.039628	159.2 ka	α 4.824, 4.783,... γ (42.5, 97.1, 54.7,...), (SF), (Ne24)
	U 234	0 +	234.040946	246 ka	α 4.776, 4.725,... γ 53.2 (e^-), 120.9,... (SF)
	U235	1/2 +	(0.0 keV)	26 m	IT ~ 76.8 eV, e^-
	U 235	7/2 –	235.043923	704 Ma	α 4.400, 4.365,... γ 185.7, 143.8,... (SF)
	U 236	0 +	236.045562	23.42 Ma	α 4.494, 4.445,... (γ 49.4 e^-, 112.8,...), (SF)
	U 237	1/2 +	237.048724	6.75 d	β^- 0.24, 0.25,... γ 59.5, 208.0,...
	U 238	0 +	238.050783	4.47 Ga	α 4.197, 4.147,... (γ 49.6 e^-,...), (SF)
	U 239	5/2 +	239.054288	23.5 m	β^- 1.21, 1.28,... γ 74.7, 43.5,...
93	Np 235	5/2 +	235.044056	1.085 a	α 5.021, 5.004,... (γ 25.6–188.8)
	Np 236	6 –	236.04656	0.115 Ma	ε, γ 160.3, β^- 0.2, γ 44.6 e^-, 104,...
	Np 237	5/2 +	237.048167	**2.14 Ma**	α 4.788, 4.771,... γ 29.4, 86.5,...
	Np 238	2 +	238.050941	2.117 d	β^- 0.263, 1.248,... γ 984.5, 1028.5,...
	Np 239	5/2 +	239.052931	**2.355 d**	β^- 0.438, 0.341,... γ 106.1, 277.6, 228.2,...
94	Pu 236	0 +	236.046048	2.87 a	α 5.7677, 5.7210,... (γ 47.6–643.7), (SF)
	Pu 237	7/2 –	237.048404	45.2 d	ε, γ 59.5,... α 5.344, (γ 280.4, 298.9,...)
	Pu 238	0 +	238.049553	87.74 a	α 5.4992, 5.4565,... (γ 43.5 e^-, 99.9e^-,...), (SF)
	Pu 239	1/2 +	239.052156	**24.10 ka**	α 5.156, 5.143, 5.105,... γ 51.6 e^-, (30.1–1057.3), (SF)
	Pu 240	0 +	240.053807	6563 a	α 5.1683, 5.1237,... (γ 45.2 e^-, 104.2 e^-), (SF)
	Pu 241	5/2 +	241.056845	14.4 a	β^- 0.0208, (α 4.897, 4.853,...), (γ 148.6, 103.7,...)
	Pu 242	0 +	242.058736	0.373 Ma	α 4.901, 4.856,... (γ 44.9 e^-,...), (SF)
	Pu 243	7/2 +	243.061997	4.9656 h	β^- 0.578, 0.485,... γ 84.0,...
	Pu 244	0 +	244.064198	80 Ma	α 4.589, 4.546, (SF)

1	2	3	4	5	6
Z	**Nu-clide**	**Spin and parity**	**Mass** u	**Half life**	**Modes of decay and energy of radiation** MeV for particles, keV for γ(and IT)
95	Am 241	5/2 –	241.056823	432.7 a	α 5.4857, 5.4430,... γ 26.3–955, (SF)
	Am 242	5 –	(48 keV)	141 a	IT 48.6, e^-, (α 5.207,...), (γ 49.2), (SF)
	Am 242	1 –	242.059543	16.02 h	β^- 0.63, 0.67, γ 42.2 e^-,... ε, γ 44.5 e^-
	Am 243	5/2 –	243.061372	7380 a	α 5.276, 5.234,... γ 74.7, 31.1–662.2, (SF)
96	Cm 242	0 +	242.058829	162.8 d	α 6.1127, 6.0694,... (γ 44.1 e^-,...), (SF)
	Cm 243	5/2 +	243.061382	29.1 a	α 5.785, 5.742,... γ 277.6, 228.2,... ε, (SF)
	Cm 244	0 +	244.062747	18.1 a	α 5.8048, 5.7627,... (γ 42.8 e^-,...), (SF)
	Cm 245	7/2 +	245.065486	8.5 ka	α 5.362, 5.304,... γ 174.9, 133.0,... (SF)
	Cm 246	0 +	246.067218	4.76 ka	α 5.386, 5.343, (γ 44.5, e^-,...), (SF)
	Cm 247	9/2 –	247.070347	15.6 Ma	α 4.869, 5.266,... γ 403, 279, 289,...
	Cm 248	0 +	248.072342	348 ka	α 5.078, 5.035,... SF
97	Bk 247	3/2 –	247.070299	1.38 ka	α 5.532, 5.711, 5.687,... γ 84.0, 268,...
98	Cf 249	9/2 –	249.074847	351 a	α 5.812,5.945,... γ 388.3, 333.4,... (SF)
	Cf 250	0 +	250.076400	13.1 a	α 6.0304, 5.989,... (γ 42.9, e^-,...), (SF)
	Cf 251	1/2 +	251.079580	898 a	α 5.677, 5.852, 6.014,... γ 176.7, 226.8, 285,...
	Cf 252	0 +	252.081620	2.64 a	α 6.118, 6.076,... (γ 43.4, e^-, 100, e^-,...), SF
99	Es 252	5 –	252.082972	472 d	α 6.632 6.562,... (γ 52.3, 64.4, 418,...)
100	Fm 257	9/2 +	257.095099	101 d	α 6.519,... γ 241.0, 179.4,... (SF)
101	Md 258		258.098425	51.5 d	α 6.716, 6.763,... γ 369, 448,...
102	No 259		259.10102	58 m	α 7.520, 7.551, 7.581, ,
103	Lr 260		260.10557	3.0 m	α 8.03, ε, (SF)
	Lr 262		262.10969	3.6 h	ε, (SF)
104	Rf 261		261.10875	65 s	α 8.28, ε, SF ?
105	Db 262		262.11415	34 s	α 8.45, 8.63, 8.53, ε, SF
106	Sg 261		261.1162	0.23 s	α 9.56, 9.52, SF?
	Sg 263		263.11831	0.9 s	α 9.06, 9.25, SF
107	Bh 262		262.1230	0.10 s	α 10.06, 9.91, 9.74, SF?
	Bh 264		264.1247	~ 440 ms	α 9.48, 9.62
108	Hs 267		267.13177	19 ms	α 9.83, SF?
109	Mt 268		268.1388	70 ms	α 10.10, 10.24, SF?
110	Ds 271		271.1461	1.1 ms	α 10.74, 10.68
111	Rg 272		272.1535	1.5 ms	α 10.82

6.4 Thermal Neutron Absorption Cross-Section of the Elements

Z = atomic number = number protons. σ_a = thermal neutron absorption cross-section for the naturally occurring abundances of different isotopes. Elements with no naturally occurring isotope are marked –. σ_a is in barn unless otherwise indicated. 1 b = 10^{-28} m^2.

Z	Element	σ_a barn	Z	Element	σ_a barn	Z	Element	σ_a barn
1	H	0.3326	32	Ge	2.3	63	Eu	4.6 kb
2	He	6.9 mb	33	As	4.48	64	Gd	49.0 kb
3	Li	70.5	34	Se	11.7	65	Tb	23.2
4	Be	7.6 mb	35	Br	6.9	66	Dy	0.94 kb
5	B	767	36	Kr	24.5	67	Ho	64.7
6	C	3.50 mb	37	Rb	0.35	68	Er	162
7	N	1.90	38	Sr	1.28	69	Tm	103
8	O	0.190 mb	39	Y	1.28	70	Yb	36.6
9	F	9.6 mb	40	Zr	0.185	71	Lu	77
10	Ne	0.039	41	Nb	1.15	72	Hf	102
11	Na	0.530	42	Mo	2.55	73	Ta	21.6
12	Mg	0.063	43	Tc	–	74	W	18.5
13	Al	0.232	44	Ru	2.57	75	Re	88.7
14	Si	0.171	45	Rh	145	76	Os	15.3
15	P	0.172	46	Pd	6.9	77	Ir	426
16	S	0.53	47	Ag	63.3	78	Pt	10.0
17	Cl	33.5	48	Cd	2.52 kb	79	Au	98.8
18	Ar	0.675	49	In	193.8	80	Hg	375
19	K	2.1	50	Sn	0.626	81	Tl	3.4
20	Ca	0.43	51	Sb	5.1	82	Pb	0.171
21	Sc	27.2	52	Te	4.7	83	Bi	0.033
22	Ti	6.09	53	I	6.2	84	Po	–
23	V	5.07	54	Xe	23.9	85	At	–
24	Cr	3.07	55	Cs	29.0	86	Rn	–
25	Mn	13.3	56	Ba	1.2	87	Fr	–
26	Fe	2.56	57	La	8.97	88	Ra	–
27	Co	37.45	58	Ce	0.63	89	Ac	–
28	Ni	4.49	59	Pr	11.5	90	Th	7.37
29	Cu	3.78	60	Nd	50.5	91	Pa	–
30	Zn	1.11	61	Pm	–	92	U*	3.35
31	Ga	2.9	62	Sm	5.8 kb	93	Pu	–

* Cross-section for fission σ_f = 4.19 barn.

6.5 Thermal Neutron Absorption Cross-Sections of Nuclides

Column 1 Z = atomic number = number of protons.

Column 2 Symbol and mass number = number of protons + number of neutrons. A bold font indicates a naturally occurring nuclide. A following asterisk indicates a metastable state of the nuclide.

Column 3 No symbol means $\sigma_\gamma = (n_{th}, \gamma)$ cross-section (n_{th} = thermal neutron). $\sigma_p = (n_{th}, p)$ cross-section, $\sigma_\alpha = (n_{th}, \alpha)$ cross-section, and σ_f = thermal neutron fission cross-section. $\sigma = 9.8 + 17.4$ (for ^{45}Sc) = cross-section for the formation of the metastable and the ground state of the product nuclide (^{46}Sc).

All values are in barn unless otherwise indicated. 1 b = 10^{-28} m^2.

1	2	3
Z	**Nuclide**	σ barn
1	**H 1**	0.3326
	H 2	0.519 mb
	H 3	< 6 μb
2	**He 3**	0.031 mb, σ_p 5333
	He 4	0.0
3	**Li 6**	38.5 mb, σ_α 940
	Li 7	45.4 mb
4	**Be 9**	7.6 mb
	Be 10	< 1 mb
5	**B 10**	0.5, σ_α 3837, σ_p < 0.178
	B 11	5.5 mb
6	**C 12**	3.53 mb
	C 13	1.37 mb
	C 14	< 1 μb
7	**N 14**	75.0 mb
	N 15	0.024 mb
8	**O 16**	0.190 mb
	O 17	0.523, σ_α 0.235
	O 18	0.16 mb
9	**F 19**	9.6 mb
10	**Ne 20**	0.037
	Ne 21	0.666, σ_α < 1.5
	Ne 22	45.5 mb
11	Na 22	29 kb
	Na 23	0.40+0.13
12	**Mg 24**	0.051
	Mg 25	0.190
	Mg 26	0.035
	Mg 27	0.07
13	**Al 27**	0.232
14	**Si 28**	0.177
	Si 29	0.101
	Si 30	0.107
15	**P 31**	0.172
16	**S 32**	0.53, σ_α 7 mb
	S 33	0.35, σ_α 0.190, σ_p < 0.2 mb
	S 34	0.240

1	2	3
Z	**Nu-clide**	σ barn
17	**Cl 35**	43.6, σ_α 0.08, σ_p 0.049
	Cl 36	< 10
	Cl 37	0.047+0.376
18	**Ar 36**	5.2, σ_α 5.5 mb
	Ar 37	37, σ_α 1970, σ_p 69
	Ar 38	0.8
	Ar 39	600
	Ar 40	0.660
	Ar 41	0.5
19	**K 39**	2.1, σ_α 4.3 mb
	K 40	30, σ_p 4.4, σ_α 0.39
	K 41	1.41
20	**Ca 40**	0.41, σ_α 2.5 mb
	Ca 41	4
	Ca 42	0.680
	Ca 43	6.2
	Ca 44	0.88
	Ca 45	15
	Ca 46	0.74
	Ca 48	1.09
21	**Sc 45**	9.8+17.4
	Sc 46	8.0
22	**Ti 46**	0.59
	Ti 47	1.7
	Ti 48	7.84
	Ti 49	2.2
	Ti 50	0.179
23	**V 50**	60
	V 51	4.93
24	**Cr 50**	15.9
	Cr 52	0.76
	Cr 53	18.2
	Cr 54	0.39

1	2	3
Z	**Nu-clide**	σ barn
25	Mn 53	~ 70
	Mn 54	38
	Mn 55	13.3
26	**Fe 54**	2.25
	Fe 55	13
	Fe 56	2.59
	Fe 57	2.48
	Fe 58	1.28
	Fe 59	<10
27	Co 58*	0.14 Mb
	Co 58	1.9 kb
	Co 59	18.80+18.65
	Co 60*	58
	Co 60	2.0
28	**Ni 58**	4.6, σ_α < 0.03 mb
	Ni 59	77.7, σ_α 12.3, σ_p 2.0
	Ni 60	2.9
	Ni 61	2.5, σ_α <0.03 mb
	Ni 62	14.5
	Ni 63	24.4
	Ni 64	1.58
	Ni 65	22.4
29	**Cu 63**	4.50
	Cu 64	~270
	Cu 65	2.17
	Cu 66	135
30	**Zn 64**	0.76
	Zn 65	66, σ_α < 250
	Zn 66	0.85, σ_α < 0.02 mb
	Zn 67	6.8
	Zn 68	0.072+1.0
	Zn 70	8.7 mb+83 mb
31	**Ga 69**	1.68
	Ga 71	4.56
	Ga 72	4.25

1	2	3
Z	**Nu-clide**	σ barn
32	**Ge 70**	0.28+3.15
	Ge 72	0.98
	Ge 73	15
	Ge 74	0.17+0.34
	Ge 76	0.09
33	**As 75**	4.48
	As 76	60.8
	As 77	12.69
34	**Se 74**	51.8
	Se 75	0.33 kb
	Se 76	22+63
	Se 77	42
	Se 78	0.38+0.05
	Se 80	0.08+0.53
	Se 82	39 mb+5.2 mb
35	**Br 79**	2.4+8.6
	Br 81	2.43+0.26
36	**Kr 78**	0.17+6.03
	Kr 80	4.55+6.95
	Kr 82	14.0+16
	Kr 83	180
	Kr 84	90 mb+42 mb
	Kr 85	1.66
	Kr 86	3 mb
37	**Rb 85**	53 mb+427 mb
	Rb 87	0.120
	Rb 88	1.2
38	**Sr 84**	0.60+0.35
	Sr 86	0.84+0.20
	Sr 87	16
	Sr 88	5.8 mb
	Sr 89	0.42
	Sr 90	14 mb
	Sr 91	0.148

1	2	3
Z	**Nu-clide**	σ barn
39	**Y 89**	1 mb+1.279
	Y 90	<6.5
	Y 91	1.4
	Y 93	0.078
40	**Zr 90**	11 mb
	Zr 91	1.24
	Zr 92	0.220
	Zr 93	2.6
	Zr 94	49.9 mb
	Zr 95	0.49
	Zr 96	22.9
	Zr 97	0.202
41	**Nb 93**	0.15+1.0
	Nb 94	0.6+14.9
	Nb 95	<7
42	**Mo 92**	6 mb+45 mb
	Mo 94	15 mb
	Mo 95	14.0, σ_{α} 0.032 mb
	Mo 96	0.5
	Mo 97	2.1
	Mo 98	0.130
	Mo 99	1.733
	Mo 100	0.199
43	Tc 98	0.93+1.67
	Tc 99	20
44	**Ru 96**	0.29
	Ru 98	<8
	Ru 99	7.1
	Ru 100	5.0
	Ru 101	3.4
	Ru 102	1.21
	Ru 103	7.71
	Ru 104	0.32
	Ru 105	0.39
	Ru 106	0.146

1	2	3	1	2	3
Z	**Nu-clide**	σ barn	*Z*	**Nu-clide**	σ barn
45	**Rh 103**	10+135	50	**Sn 112**	0.30+0.71
	Rh 104*	~800		Sn 113	~9
	Rh 104	39.53+0.47		**Sn 114**	0.115
	Rh 105	5 kb+11 kb		**Sn 115**	30
				Sn 116	6 mb+134 mb
46	**Pd 102**	3.4		**Sn 117**	2.3
	Pd 104	0.6		**Sn 118**	10 mb+0.21
	Pd 105	20		**Sn 119**	2.2
	Pd 106	0.013+0.292		**Sn 120**	1 mb+140
	Pd 107	1.8		**Sn 122**	180 mb+1 mb
	Pd 108	0.183+8.3		**Sn 124**	130 mb+4 mb
	Pd 109	5.24			
	Pd 110	0.190	51	**Sb 121**	0.06+5.84
	Pd 112	0.29		**Sb 123**	19 mb+37 mb
				Sb 124	17.4
47	**Ag 107**	0.33+37.27			
	Ag 109	4.7+86.3	52	**Te 120**	0.34+2.0
	Ag 110*	82		**Te 122**	1.1+2.3
	Ag 111	3		Te 123*	42.89
				Te 123	418
48	**Cd 106**	1		**Te 124**	0.040+6.76
	Cd 108	1.1		**Te 125**	155
	Cd 109	0.7 kb, σ_α 0.05		**Te 126**	0.135+0.90
	Cd 110	0.14+10.9		Te 127*	3.4 kb
	Cd 111	24		**Te 128**	15 mb+199.7 mb
	Cd 112	0.04+2.2		**Te 130**	0.02+0.27
	Cd 113	20.6 kb			
	Cd 114	0.036+0.30	53	I 125	894
	Cd 115*	31.2		I 126	5.96 kb
	Cd 115	5.43		**I 127**	6.2
	Cd 116	25 mb+3.9		I 128	22
				I 129	20.7+10.3
49	**In 113**	5.0+3.9		I 130	18
	In 115	81+81.3+40		I 131	0.7

1	2	3	1	2	3
Z	**Nu-clide**	σ barn	*Z*	**Nu-clide**	σ barn
54	**Xe 124**	28+137	58	**Ce 136**	0.95+6.3
	Xe 125	$\sigma_\alpha < 0.03$		**Ce 138**	15 mb+1.1
	Xe 126	0.45+3.05		Ce 139	0.50 kb
	Xe 127	$\sigma_\alpha < 0.01$		**Ce 140**	0.57
	Xe 128	0.48+6.02		Ce 141	29
	Xe 129	21		**Ce 142**	0.95
	Xe 130	0.46+6		Ce 143	6.1
	Xe 131	85		Ce 144	1.0
	Xe 132	0.05+0.40			
	Xe 133	190	59	**Pr 141**	3.9+7.6
	Xe 134	3 mb+262 mb		Pr 142	20
	Xe 135	2.65 Mb		Pr 143	~90
	Xe 136	0.26		Pr 145	18.44
55	**Cs 133**	2.5+26.5	60	**Nd 142**	18.7
	Cs 134	~140		**Nd 143**	3.25, σ_α 174 mb
	Cs 135	8.7		**Nd 144**	3.6
	Cs 136	1.3		**Nd 145**	0.012 mb
	Cs 137	0.11		**Nd 146**	1.4
				Nd 147	0.44 kb
56	**Ba 130**	2.5+8.8		**Nd 148**	2.5
	Ba 132	0.5+6.5		**Nd 150**	1.2
	Ba 134	0.158+1.84			
	Ba 135	13.9 mb+5.78	61	Pm 146	8.4 kb
	Ba 136	10 mb+0.39		Pm 147	85+97
	Ba 137	5.1		Pm 148*	22 kb
	Ba 138	0.360		Pm 148	~2 kb
	Ba 139	6.2		Pm 149	1.4 kb
	Ba 140	1.6			
57	**La 138**	57.2			
	La 139	8.93			
	La 140	2.7			

1	2	3
Z	**Nu-clide**	σ barn
62	**Sm 144**	0.7
	Sm 145	~280
	Sm 147	64, σ_α 0.7 mb
	Sm 148	2.4
	Sm 149	41 kb, σ_α 0.043
	Sm 150	102
	Sm 151	15.2 kb
	Sm 152	206
	Sm 153	334.5
	Sm 154	5.5
63	**Eu 153**	603, σ_α<1 μb
	Eu 154	1.5 kb
	Eu 155	4.0 kb
	Eu 156	480
	Eu 157	190
64	**Gd 152**	1.0 kb
	Gd 153	~20 kb
	Gd 154	854
	Gd 155	61 kb, σ_α<0.03 mb
	Gd 156	1.5, σ_α <0.08 mb
	Gd 157	254 kb, σ_α <0.05 mb
	Gd 158	2.5
	Gd 160	0.77
	Gd 161	~20 kb
65	**Tb 159**	23.2
	Tb 160	525
66	**Dy 156**	33, σ_α <9 mb
	Dy 158	43, σ_α <6 mb
	Dy 159	8 kb
	Dy 160	95, σ_α <0.3 mb
	Dy 161	510, σ_α <0.03 mb
	Dy 162	245
	Dy 163	305, σ_α <0.02 mb
	Dy 164	1.7 kb+1.0 kb
	Dy 165	3.6 kb

1	2	3
Z	**Nu-clide**	σ barn
67	**Ho 165**	3.5+61.2, σ_α < 0.02 mb
68	**Er 162**	19, σ_α < 11 mb
	Er 164	13, σ_α < 1.2 mb
	Er 166	15+20, σ_α < 0.07 mb
	Er 167	670
	Er 168	1.95, σ_α < 0.09 mb
	Er 170	5.7
	Er 171	0.28 kb
69	**Tm 169**	103
	Tm 170	92
70	Yb 167	2.3 kb
	Yb 168	3.47 kb, σ_α < 0.1 mb
	Yb 169	3.6 kb
	Yb 170	10, σ_α < 0.01 mb
	Yb 171	50, σ_α < 1.5 μb
	Yb 172	1.3, σ_α < 1 μb
	Yb 173	19, σ_α < 1 μb
	Yb 174	65, σ_α < 0.02 mb
	Yb 176	2.4, σ_α < 1 μb
71	**Lu 175**	15.1+7, σ_α < 0.06 mb
	Lu 176	2.1+1.78 kb
72	**Hf 174**	390
	Hf 176	38
	Hf 177	1.1+363
	Hf 178	53+33
	Hf 179	0.34+44.66
	Hf 180	12.6, σ_α 0.5
	Hf 181	~30
73	**Ta 180***	0.7 kb
	Ta 181	21.5, σ_α 0.15 mb
	Ta 182	8.5 kb

1	2	3
Z	**Nu-clide**	σ barn
74	**W 180**	3.5
	W 182	20.7
	W 183	10.2
	W 184	2 mb + 1.8
	W 185	3.3
	W 186	37.0
	W 187	64
75	Re 184	9 kb
	Re 185	0.3 + 110
	Re 187	1.6 + 75
76	**Os 184**	3005, $\sigma_\alpha < 10$ mb
	Os 186	80, $\sigma_\alpha < 0.1$ mb
	Os 187	336, $\sigma_\alpha < 0.1$ mb
	Os 188	4.3, $\sigma_\alpha < 0.03$ mb
	Os 190	13.2+3.9, $\sigma_\alpha < 0.02$ mb
	Os 191	0.38 kb
	Os 192	1.97, $\sigma_\alpha < 0.01$ mb
	Os 193	~40
77	Ir 192	1.4 kb
	Ir 193	5.8+110
	Ir 194	1.5 kb
78	**Pt 190**	150, $\sigma_\alpha < 8$ mb
	Pt 192	22 + < 14, $\sigma_\alpha < 0.2$ mb
	Pt 194	90 mb+1.11, $\sigma_\alpha < 5$ μb
	Pt 195	27, $\sigma_\alpha < 5$ μb
	Pt 196	50 mb+0.74
	Pt 198	27 mb+3.673
79	**Au 197**	98.8
	Au 198	26.74 kb
	Au 199	~30

1	2	3
Z	**Nu-clide**	σ barn
80	**Hg 196**	120+3.08 kb
	Hg 198	18 mb+1.882
	Hg 199	2.0 kb
	Hg 200	< 60
	Hg 201	< 60
	Hg 202	4.9
	Hg 204	0.43
81	**Tl 203**	11.0, σ_α 0.025
	Tl 204	21.6
	Tl 205	0.10
82	**Pb 204**	661 mb
	Pb 205	5
	Pb 206	30.5 mb
	Pb 207	709 mb
	Pb 208	0.487 mb, σ_α 8 μb
	Pb 210	< 0.5
83	Bi 209	19 mb+14 mb, σ_α 0.15 mb
	Bi 210*	54 mb
84	Po 210	< 0.5 mb
86	Rn 220	< 0.2
	Rn 222	0.74
88	Ra 223	130, σ_f 0.7
	Ra 224	12.0
	Ra 226	11.5, $\sigma_f < 7$ μb
	Ra 228	36, $\sigma_f < 2$
89	Ac 227	762, σ_f 0.35 mb
90	Th 227	σ_f 0.2 kb
	Th 228	123, $\sigma_f < 0.3$
	Th 229	54, σ_f 30.5
	Th 230	23.2, $\sigma_f < 1.2$ mb
	Th 231	160.1, σ_f 26.68
	Th 232	7.37, σ_f 39 μb
	Th 233	1.5 kb, σ_f 15
	Th 234	1.8, $\sigma_f < 0.01$

1	2	3	1	2	3
Z	**Nuclide**	σ barn	*Z*	**Nuclide**	σ barn
92	U 230	σ_f 25	94	Pu 237	σ_f 2.4 kb
	U 231	σ_f 400		Pu 238	547, σ_f 16.5
	U 232	73.1, σ_f 74		Pu 239	268.8, σ_f 744.4
	U 233	47.7, σ_f 522.6		Pu 240	289.5, σ_f 0.030
	U 234	100.2, $\sigma_f < 0.65$		Pu 241	368, σ_f 1009
	U 235	93.6, σ_f 582.2		Pu 242	18.5, $\sigma_f < 0.2$
	U 236	5.2, σ_f 0.04		Pu 243	87.4, σ_f 180
	U 237	411, σ_f 2	95	Am 241	83.8+748, σ_f 3.15
	U 238	2.7, $\sigma_f < 0.5$ mb		Am 242*	1.4 kb, σ_f 6.6 kb
	U 239	22, σ_f 14		Am 243	75.2 + 4.1, σ_f 0.20
	U 240	1.53	96	Cm 243	138, σ_f 672
93	Np 234	σ_f 900		Cm 244	13.9, σ_f 1.2
	Np 235	1.6 kb+184		Cm 245	345, σ_f 2.02 kb
	Np 236	σ_f 2.6 kb		Cm 246	1.3, σ_f 0.17
	Np 237	169, σ_f 0.019		Cm 247	60, σ_f 80
	Np 238	43, σ_f 2.2 kb			
	Np 239	31+14, $\sigma_f < 1$			

More information: http://ie.lbl.gov/ngdata/sig.txt

6.6 Naturally Radioactive Decay Chains

For the half life the following abbreviations are used: s = seconds, m = minutes, h = hours, d = days and a = years. In some cases there are several decays leading to the same isotope. In these cases the upper decay is more frequent than the one below, as indicated with dashed arrows. Nowadays the neptunium chain ($A = 4n + 1$) does not exist in nature. A = mass number, n = an integer.

Thorium Chain ($A = 4n$)

$$ {}^{232}_{90}\mathrm{Th} \xrightarrow[14.0\ \mathrm{Ga}]{\alpha} {}^{228}_{88}\mathrm{Ra} \xrightarrow[5.76\ \mathrm{a}]{\beta^-} {}^{228}_{89}\mathrm{Ac} \xrightarrow[6.15\ \mathrm{h}]{\beta^-} {}^{228}_{90}\mathrm{Th} \xrightarrow[1.913\ \mathrm{a}]{\alpha} $$

$$ \longrightarrow {}^{224}_{88}\mathrm{Ra} \xrightarrow[3.66\ \mathrm{d}]{\alpha} {}^{220}_{86}\mathrm{Rn} \xrightarrow[55.6\ \mathrm{s}]{\alpha} {}^{216}_{84}\mathrm{Po} \xrightarrow[0.145\ \mathrm{s}]{\alpha} {}^{212}_{82}\mathrm{Pb} \xrightarrow[10.64\ \mathrm{h}]{\beta^-} $$

$$ \longrightarrow {}^{212}_{83}\mathrm{Bi} \left\{ \begin{array}{l} \xrightarrow[1.009\ \mathrm{h}]{\beta^-} {}^{212}_{84}\mathrm{Po} \xrightarrow[0.298\ \mu\mathrm{s}]{\alpha} \\ \underset{25\ \mathrm{m}}{\overset{\alpha}{\dashrightarrow}} {}^{208}_{81}\mathrm{Tl} \underset{3.053\ \mathrm{m}}{\overset{\beta^-}{\dashrightarrow}} \end{array} \right\} {}^{208}_{82}\mathrm{Pb}\ \text{(stable)} $$

Neptunium Chain ($A = 4n + 1$)

$$ {}^{241}_{94}\mathrm{Pu} \xrightarrow[13.2\ \mathrm{a}]{\beta^-} {}^{241}_{95}\mathrm{Am} \xrightarrow[458\ \mathrm{a}]{\alpha} {}^{237}_{93}\mathrm{Np} \xrightarrow[2.14\ \mathrm{Ma}]{\alpha} {}^{233}_{91}\mathrm{Pa} \xrightarrow[27.4\ \mathrm{d}]{\beta^-} $$

$$ \longrightarrow {}^{233}_{92}\mathrm{U} \xrightarrow[162\ \mathrm{ka}]{\alpha} {}^{229}_{90}\mathrm{Th} \xrightarrow[7340\ \mathrm{a}]{\alpha} {}^{225}_{88}\mathrm{Ra} \xrightarrow[14.8\ \mathrm{d}]{\beta^-} {}^{225}_{89}\mathrm{Ac} \xrightarrow[10.0\ \mathrm{d}]{\alpha} $$

$$ \longrightarrow {}^{221}_{87}\mathrm{Fr} \xrightarrow[4.8\ \mathrm{m}]{\beta^-} {}^{217}_{85}\mathrm{At} \xrightarrow[0.0323\ \mathrm{s}]{\alpha} {}^{213}_{83}\mathrm{Bi} \left\{ \begin{array}{l} \xrightarrow[47\ \mathrm{m}]{\beta^-} {}^{213}_{84}\mathrm{Po} \xrightarrow[4.2\ \mu\mathrm{s}]{\alpha} \\ \overset{\alpha}{\dashrightarrow} {}^{209}_{81}\mathrm{Tl} \underset{2.2\ \mathrm{m}}{\overset{\beta^-}{\dashrightarrow}} \end{array} \right\} $$

$$ \longrightarrow {}^{209}_{82}\mathrm{Pb} \xrightarrow[5.01\ \mathrm{d}]{\beta^-} {}^{211}_{83}\mathrm{Bi}\ \text{(stable)} $$

Uranium Chain ($A = 4n + 2$)

$$
{}^{238}_{92}\mathrm{U} \xrightarrow[4.47\ \mathrm{Ga}]{\alpha} {}^{234}_{90}\mathrm{Th} \xrightarrow[24.10\ \mathrm{d}]{\beta^-} {}^{234}_{91}\mathrm{Pa} \xrightarrow[6.69\ \mathrm{h}]{\beta^-} {}^{234}_{92}\mathrm{U} \xrightarrow[0.246\ \mathrm{Ma}]{\alpha}
$$

$$
\longrightarrow {}^{230}_{90}\mathrm{Th} \xrightarrow[75.4\ \mathrm{ka}]{\alpha} {}^{226}_{88}\mathrm{Ra} \xrightarrow[1600\ \mathrm{a}]{\alpha} {}^{222}_{86}\mathrm{Rn} \xrightarrow[3.8235\ \mathrm{d}]{\alpha} {}^{218}_{84}\mathrm{Po} \xrightarrow[3.10\ \mathrm{m}]{\alpha}
$$

$$
\longrightarrow {}^{214}_{82}\mathrm{Pb} \xrightarrow[27\ \mathrm{m}]{\beta^-} {}^{214}_{83}\mathrm{Bi} \left\{ \begin{array}{l} \xrightarrow[19.9\ \mathrm{m}]{\beta^-} {}^{214}_{84}\mathrm{Po} \xrightarrow[164\ \mu\mathrm{s}]{\alpha} \\ \overset{\alpha}{\dashrightarrow} {}^{210}_{81}\mathrm{Tl} \underset{1.30\ \mathrm{m}}{\overset{\beta^-}{\dashrightarrow}} \end{array} \right\} {}^{210}_{82}\mathrm{Pb} \xrightarrow[22.3\ \mathrm{a}]{\beta^-}
$$

$$
\longrightarrow {}^{210}_{83}\mathrm{Bi} \xrightarrow[5.01\ \mathrm{d}]{\beta^-} {}^{210}_{84}\mathrm{Po} \xrightarrow[138.4\ \mathrm{d}]{\alpha} {}^{206}_{82}\mathrm{Pb}\ \text{(stable)}
$$

Actinium Chain ($A = 4n + 3$)

$$
{}^{235}_{92}\mathrm{U} \xrightarrow[704\ \mathrm{Ma}]{\alpha} {}^{231}_{90}\mathrm{Th} \xrightarrow[1.603\ \mathrm{d}]{\beta^-} {}^{231}_{91}\mathrm{Pa} \xrightarrow[32.8\ \mathrm{ka}]{\alpha} {}^{227}_{89}\mathrm{Ac} \xrightarrow[21.77\ \mathrm{a}]{\beta^-}
$$

$$
\longrightarrow {}^{227}_{90}\mathrm{Th} \xrightarrow[18.72\ \mathrm{d}]{\alpha} {}^{223}_{88}\mathrm{Ra} \xrightarrow[11.435\ \mathrm{d}]{\alpha} {}^{219}_{86}\mathrm{Rn} \xrightarrow[3.96\ \mathrm{s}]{\alpha} {}^{215}_{84}\mathrm{Po} \left\{ \begin{array}{l} \xrightarrow[1.78\ \mathrm{ms}]{\alpha} \\ \overset{\beta^-}{\dashrightarrow} \end{array} \right.
$$

$$
\begin{array}{l} \longrightarrow {}^{211}_{82}\mathrm{Pb} \xrightarrow[36.1\ \mathrm{m}]{\beta^-} \\ \dashrightarrow {}^{215}_{85}\mathrm{At} \underset{0.10\ \mathrm{ms}}{\overset{\alpha}{\dashrightarrow}} \end{array} \left\} {}^{211}_{83}\mathrm{Bi} \left\{ \begin{array}{l} \xrightarrow[2.14\ \mathrm{m}]{\alpha} {}^{207}_{81}\mathrm{Tl} \xrightarrow[4.77\ \mathrm{m}]{\beta^-} \\ \overset{\beta^-}{\dashrightarrow} {}^{211}_{84}\mathrm{Po} \underset{0.516\ \mathrm{s}}{\overset{\alpha}{\dashrightarrow}} \end{array} \right\} {}^{207}_{82}\mathrm{Pb}\ \text{(stable)} \right.
$$

6.7 Fission Product Yields

A = atomic mass number = number of protons + number of neutrons. *Y*(*A*) = isomeric fission yield for all nuclides with atomic mass number = *A*. *Y*(X) = fission yield for the most probable nuclide.

An asterisk (*) in front of the half-life means that the most probable nuclide can be in several states, and only the half life of the most long-lived state is given.

At the temperature of 293 K the energy of thermal neutrons ≈ 0.038 eV.

For detailed information on fission yields:
http://ie.lbl.gov/fission.html

Uranium 235 thermal neutron fission yields

A	*Y*(*A*) %	Most probable nuclide	*Y*(X) %	Half life of X	*A*	*Y*(*A*) %	Most probable nuclide	*Y*(X) %	Half life of X
76	0.003	Zn 76	0.0018	5.7 s	101	5.18	Zr 101	2.79	2.1 s
77	0.008	Ga 77	0.0041	13 s	102	4.29	Zr 102	1.78	2.9 s
78	0.021	Ga 78	0.0103	5.1 s	103	3.03	Nb 103	1.41	1.5 s
79	0.045	Ge 79	0.0233	19 s	104	1.88	Mo 104	1.13	60 s
80	0.129	Ge 80	0.102	29.5 s	105	0.96	Mo 105	0.668	36 s
81	0.204	Ge 81	0.126	7.6 s	106	0.402	Mo 106	0.359	8.4 s
82	0.325	As 82	0.129	*19 s	107	0.146	Mo 107	0.121	3.5 s
83	0.535	As 83	0.291	13.4 s	108	0.054	Mo 108	0.030	1.5 s
84	1.00	Se 84	0.631	3.3 s	109	0.031	Mo 109	0.0155	1.4 s
85	1.32	Se 85	0.894	39 s	110	0.025	Tc 110	0.012	0.83 s
86	1.95	Ge 86	0.629	0.25 s	111	0.018	Ru 111	0.012	1.5 s
87	2.56	Br 87	1.27	56 s	112	0.013	Ru 112	0.0099	4.5 s
88	3.58	Kr 88	1.73	2.84 h	113	0.014	Rh 113	0.0064	2.7 s
89	4.74	Kr 89	3.44	3.15 m	114	0.012	Rh 114	0.0050	1.8 s
90	5.78	Kr 90	4.40	32.3 s	115	0.012	Rh 115	0.0036	0.99 s
91	5.83	Kr 91	3.16	8.6 s	116	0.013	Pd 116	0.0068	12.7 s
92	6.01	Rb 92	3.13	4.5 s	117	0.008	Ag 117	0.0030	*1.2 m
93	6.36	Rb 93	3.07	5.85 s	118	0.011	Ag 118	0.0064	*4.0 s
94	6.47	Sr 94	4.51	1.25 m	119	0.013	Ag 119	0.0073	2.1 s
95	6.50	Sr 95	4.54	25.1 s	120	0.013	Cd 120	0.0084	51 s
96	6.27	Sr 96	3.57	1.06 s	121	0.013	Cd 121	0.0072	13.5 s
97	6.00	Y 97	3.14	3.76 s	122	0.016	Cd 122	0.012	5.3 s
98	5.76	Zr 98	2.57	30.7 s	123	0.016	Cd 123	0.010	2.1 s
99	6.11	Zr 99	3.58	2.2 s	124	0.027	Cd 124	0.012	1.24 s
100	6.29	Zr 100	4.98	7.1 s	125	0.034	Sn 125	0.011	*9.63 d

A	*Y*(*A*) %	**Most probable nuclide**	*Y*(X) %	**Half life of X**	*A*	*Y*(*A*) %	**Most probable nuclide**	*Y*(X) %	**Half life of X**
126	0.059	Sn 126	0.045	0.1 Ma	143	5.95	Ba 143	4.10	14.3 s
127	0.157	Sn 127	0.095	*2.1 h	144	5.50	Ba 144	3.98	11.4 s
128	0.35	Sn 128	0.301	59 m	145	3.93	La 145	1.92	24 s
129	0.76	Sn 129	0.43	*6.9 m	146	3.00	La 146	1.49	6.3 s
130	1.81	Sn 130	1.08	3.7 m	147	2.25	Ce 147	1.00	56 s
131	2.89	Sb 131	1.65	39 s	148	1.67	Ce 148	1.24	56 s
132	4.31	Sb 132	2.16	*4.2 m	149	1.08	Ce 149	0.70	5.2 s
133	6.70	Te 133	4.14	*55.4 m	150	0.653	Ce 150	0.39	4.4 s
134	7.84	Te 134	6.22	42 m	151	0.419	Pr 151	0.24	22 s
135	6.54	Te 135	3.22	19 s	152	0.267	Nd 152	0.141	11.4 m
136	6.32	I 136	2.57	*1.39 m	153	0.158	Nd 153	0.111	29 s
137	6.19	Xe 137	3.19	3.82 m	154	0.074	Nd 154	0.058	26 s
138	6.71	Xe 138	4.81	14.1 m	155	0.032	Nd 155	0.018	8.9 s
139	6.41	Xe 139	4.32	39.7 s	156	0.0149	Pm 156	0.0071	27 s
140	6.22	Xe 140	3.51	13.6 s	157	0.0062	Pm 157	0.0029	11 s
141	5.85	Cs 141	2.92	25 s	158	0.0033	Sm 158	0.0024	5.5 s
142	5.84	Ba 142	3.01	10.7 m	159	0.0010	Sm 159	0.0007	11 s

Plutonium 239 thermal neutron fission yields

A	*Y*(*A*) %	**Most probable nuclide**	*Y*(X) %	**Half life of X**	*A*	*Y*(*A*) %	**Most probable nuclide**	*Y*(X) %	**Half life of X**
76	0.0029	Ga 76	0.0015	29 s	90	2.16	Kr 90	1.10	32 s
77	0.007	Ga 77	0.0038	13 s	91	2.49	Rb 91	1.38	58 s
78	0.019	Ge 78	0.012	19 s	92	2.95	Rb 92	1.61	4.5 s
79	0.044	Ge 79	0.030	19 s	93	3.75	Sr 93	2.14	7.4 m
80	0.093	Ge 80	0.055	29.5 s	94	4.35	Sr 94	2.92	1.25 m
81	0.184	As 81	0.103	33 s	95	4.85	Sr 95	2.61	25 s
82	0.228	As 82	0.114	*19 s	96	4.9	Y 96	2.54	*9.6 s
83	0.297	Se 83	0.156	*22 m	97	5.40	Y 97	2.96	3.8 s
84	0.470	Se 84	0.327	3.3 s	98	5.76	Zr 98	2.92	31 s
85	0.576	Se 85	0.406	*39 s	99	6.23	Zr 99	3.76	2.2 s
86	0.678	Br 86	0.378	*55 s	100	6.77	Zr 100	4.78	7.1 s
87	0.989	Br 87	0.55	56 s	101	6.02	Nb 101	3.46	7.1 s
88	1.32	Kr 88	0.75	2.8 h	102	6.13	Nb 102	3.08	1.3 s
89	1.72	Kr 89	1.10	3.15 m	103	6.99	Mo 103	3.81	1.13 s

A	*Y*(*A*) %	**Most probable nuclide**	*Y*(X) %	**Half life of X**	*A*	*Y*(*A*) %	**Most probable nuclide**	*Y*(X) %	**Half life of X**
104	6.06	Mo 104	4.29	60 s	134	7.59	Te 134	4.40	42 m
105	5.65	Mo 105	3.51	36 s	135	7.61	I 135	4.29	6.6 h
106	4.36	Mo 106	2.17	8.4 s	136	7.1	I 136	2.89	1.4 m
107	3.33	Tc 107	1.82	21 s	137	6.71	Xe 137	3.68	3.8 m
108	2.14	Ru 108	1.28	4.5 m	138	6.11	Xe 138	3.93	14 m
109	1.6	Ru 109	1.00	34.5 s	139	5.66	Xe 139	2.79	40 s
110	0.64	Ru 110	0.57	15 s	140	5.37	Cs 140	2.28	64 s
111	0.30	Ru 111	0.25	1.5 s	141	5.25	Cs 141	2.87	25 s
112	0.129	Ru 112	0.092	4.5 s	142	4.93	Ba 142	3.08	10.7 m
113	0.082	Rh 113	0.0435	0.9 s	143	4.42	Ba 143	2.89	14 s
114	0.060	Rh 114	0.032	1.8 s	144	3.74	Ba 144	2.16	11.4 s
115	0.043	Pd 115	0.022	47 s	145	2.99	La 145	1.70	24 s
116	0.051	Pd 116	0.036	13 s	146	2.46	La 146	1.16	6.3 s
117	0.045	Ag 117	0.019	*1.22 m	147	2.01	Ce 147	1.22	56 s
118	0.032	Ag 118	0.020	*4.0 s	148	1.64	Ce 148	0.89	56 s
119	0.032	Cd 119	0.015	*2.69 s	149	1.22	Pr 149	0.57	2.3 m
120	0.030	Cd 120	0.021	51 s	150	0.97	Pr 150	0.51	6.2 s
121	0.037	Cd 121	0.023	13.5 s	151	0.74	Pr 151	0.37	22 s
122	0.044	In 122	0.022	*10 s	152	0.58	Nd 152	0.37	11.4 s
123	0.044	In 123	0.024	*47 s	153	0.361	Nd 153	0.24	29 s
124	0.078	Sn 124	0.044	stable	154	0.262	Nd 154	0.15	26 s
125	0.112	Sn 125	0.077	9.6 d	155	0.166	Pm 155	0.092	48 s
126	0.202	Sn 126	0.169	0.1 Ma	156	0.124	Pm 156	0.062	27 s
127	0.51	Sn 127	0.43	*2.1 h	157	0.074	Sm 157	0.041	8.0 m
128	0.73	Sn 128	0.55	59 m	158	0.041	Sm 158	0.029	5.5 m
129	1.37	Sn 129	0.98	*2.4 m	159	0.021	Sm 159	0.012	11.3 s
130	2.36	Sn 130	0.94	3.7 m	160	0.0097	Eu 160	0.0046	38 s
131	3.86	Sb 131	1.90	23 m	161	0.0048	Eu 161	0.0028	27 s
132	5.41	Te 132	2.25	3.3 d	162	0.0023	Gd 162	0.0012	8.4 m
133	7.02	Te 133	4.66	*55 m	163	0.009	Gd 163	0.0006	68 s

7 Particle Physics

Extensive information on particle properties can be found at
http://pdg.web.cern.ch/pdg/pdg.html

7.1 Intermediate Bosons (Field Particles)

The graviton has not been observed experimentally and is therefore still a hypothetical particle.

Bosons are particles with integer spin. They do not obey the Pauli exclusion principle. (Fermions have half-integer spin and obey Pauli's exclusion principle.)

Inter-mediate boson	**Mass** GeV/c^2	**Spin** J	**Parity** π	**Charge** q	**Decay modes and fraction** (%)	**Associated force**
Gluon	0	1	−1	0	–	strong
Photon	0	1	−1	0	stable	electromagnetic
$W^{\pm}$	80.42 ± 0.06	1	–	±1	$W^+ \rightarrow e^+ + \nu_e$ (10.7) $\rightarrow \mu^+ + \nu_\mu$ (10.7) $\rightarrow \tau^+ + \nu_\tau$ (10.7) → Hadrons (68)	weak
Z^0	91.18 ± 0.03	1	–	0	$\rightarrow e^+ + e^-$ (3.4) $\rightarrow \mu^+ + \mu^-$ (3.4) $\rightarrow \tau^+ + \tau^-$ (3.4) $\rightarrow \nu + \bar{\nu}$ (20) → Hadrons (70)	weak
Graviton	0	2	+1	0	stable	gravitational

7.2 Hadrons

Baryons have baryon number = 1 (antibaryons –1) and mesons have baryon number = 0. The strangeness can be found by adding the strangeness for included quarks, see T–7.4. The parity values assume angular momentum $L = 0$. Protons and neutrons are called nucleons. Baryons with none-zero strangeness are called hyperons.

Particle	**Sym-bol**	**Anti-par-ticle**	**Mass** MeV/c^2	**Quark struc-ture**	**Charge** q	**Spin** J	**Isospin** τ	τ_z	**Hyper-charge** Y	**Parity** π
Baryons										
proton	p	$\bar{p}$	938.2723	uud	+1	1/2	1/2	1/2	+1	+1
neutron	n	$\bar{n}$	939.5656	udd	0	1/2	1/2	–1/2	+1	+1
Λ hyperon	Λ^0	$\bar{\Lambda}^0$	1115.7	uds	0	1/2	0	0	0	+1
Σ hyperon	Σ^+	$\bar{\Sigma}^+$	1189.4	uus	+1	1/2	1	+1	0	+1
	Σ^0	$\bar{\Sigma}^0$	1192.6	uds	0	1/2	1	0	0	+1
	Σ^-	$\bar{\Sigma}^-$	1197.4	dds	–1	1/2	1	–1	0	+1
Ξ hyperon	Ξ^0	$\bar{\Xi}^0$	1314.9	uss	0	1/2	1/2	1/2	–1	+1
	Ξ^-	$\bar{\Xi}^-$	1321.3	dss	–1	1/2	1/2	–1/2	–1	+1
Ω hyperon	Ω^-	$\bar{\Omega}^-$	1672.5	sss	–1	3/2	0	0	–2	+1
Δ	Δ^{++}	$\bar{\Delta}^{++}$	1232	uuu	+2	3/2	3/2	3/2	+1	+1
	Δ^+	$\bar{\Delta}^+$	1232	uud	+1	3/2	3/2	1/2	+1	+1
	Δ^0	$\bar{\Delta}^0$	1232	udd	0	3/2	3/2	–1/2	+1	+1
	Δ^-	$\bar{\Delta}^-$	1232	ddd	–1	3/2	3/2	–3/2	+1	+1
$\Lambda_C{}^+$	$\Lambda_C{}^+$	$\bar{\Lambda}_C{}^+$	2285	udc	+1	1/2	0	0	+2	+1
Mesons										
pion	π^+	π^-	139.567	$u\bar{d}$	+1	0	1	+1	0	–1
	π^0	π^0	134.974	$u\bar{u},d\bar{d}$*	0	0	1	0	0	–1
	π^-	π^+	139.567	$d\bar{u}$	–1	0	1	–1	0	–1
η meson	η	η	547.5	$u\bar{u},d\bar{d},s\bar{s}$	0	0	0	0	0	–1
kaon	K^+	K^-	493.65	$u\bar{s}$	+1	0	1/2	1/2	+1	–1
	K^0	$\bar{K}^0$	497.67	$d\bar{s}$	0	0	1/2	–1/2	+1	–1
D meson	D^+	D^-	1869.3	$c\bar{d}$	+1	0	1/2	1/2	+1	–1
	D^0	$\bar{D}^0$	1864.5	$c\bar{u}$	0	0	1/2	–1/2	+1	–1
	$D_S{}^+$	$D_S{}^-$	1968.5	$c\bar{s}$	+1	0	0	0	+2	–1
B meson	B^+	B^-	5278.9	$u\bar{b}$	+1	0	1/2	1/2	+1	–1
	B^0	B^0	5279.2	$d\bar{b}$	0	0	1/2	–1/2	+1	–1
	$B_S{}^0$	$\bar{B}_S{}^0$	5369.3	$s\bar{b}$	0	0	0	0	0	–1
J/Ψ meson	J/Ψ		3090.6	$c\bar{c}$	0	1	0	0	0	–1
upsilon	ϒ		9460.4	$b\bar{b}$	0	1	0	0	0	–1

$* \; \pi^0 = \frac{1}{\sqrt{2}}(u\bar{u} - d\bar{d})$

Decay of hadrons

K^0_S and K^0_L means "short lifetime" and "long lifetime" neutral kaons. Mean life of the proton is mode dependent.

Particle	Mean life s	Decay Mode	Decay Fraction %
Baryons			
p	$\geqslant 10^{31}$ years		
n	887	p, e^-, $\overline{\nu}_e$	100
Λ^0	$2.6 \cdot 10^{-10}$	p, π^-	64.1
		n, π^0	35.7
Σ^+	$0.8 \cdot 10^{-10}$	p, π^0	51.6
		n, π^+	48.3
Σ^0	$7.4 \cdot 10^{-20}$	Λ^0, γ	100
Σ^-	$1.5 \cdot 10^{-10}$	n, π^-	100
Ξ^0	$2.9 \cdot 10^{-10}$	Λ^0, π^0	100
Ξ^-	$1.6 \cdot 10^{-10}$	Λ^0, π^-	100
Ω^-	$0.82 \cdot 10^{-10}$	Λ^0, K^-	67.8
		Ξ^0, π^-	23.6
		Ξ^-, π^0	8.6
Δ^{++}	$\sim 10^{-23}$		
Δ^+	$\sim 10^{-23}$		
Δ^0	$\sim 10^{-23}$		
Δ^-	$\sim 10^{-23}$		
$\Lambda_C{}^+$	$2.06 \cdot 10^{-13}$	Λ^0, …	27 ± 9
		Σ^+, π^+, π^+, π^-	10 ± 5
		other Σ^+, …	10 ± 5
		p, $\overline{K}^0$, π^+, π^-	8 ± 4
		e^+, …	5 ± 2
Mesons			
π^+	$2.60 \cdot 10^{-8}$	μ^+, ν_μ	100
π^0	$8.4 \cdot 10^{-17}$	γ, γ	98.8
		e^+, e^-, γ	1.2
π^-	$2.60 \cdot 10^{-8}$	μ^-, $\overline{\nu}_\mu$	100

Particle	Mean life (Width) s	Decay Mode	Decay Fraction %
***Mesons** (continued)*			
η	$5.6 \cdot 10^{-19}$	γ, γ	38.9
	(1.18 keV)	π^0, π^0, π^0	31.9
		π^+, π^-, π^0	23.6
		π^+, π^-, γ	4.6
K^+	$1.24 \cdot 10^{-8}$	μ^+, ν_μ	63.5
		π^+, π^0	21.2
		π^+, π^+, π^-	5.6
		π^0, e^+, ν_e	4.8
		π^0, μ^+, ν_μ	3.2
K^0_S	$0.89 \cdot 10^{-10}$	π^+, π^-	68.6
		π^0, π^0	31.4
K^0_L	$5.2 \cdot 10^{-8}$	$\pi^\pm$, $e^\mp$, ν_e ($\overline{\nu}_e$)	38.7
		$\pi^\pm$, $\mu^\mp$, ν_μ ($\overline{\nu}_\mu$)	27.0
		π^0, π^0, π^0	21.6
		π^+, π^-, π^0	12.4
D^+	$1.06 \cdot 10^{-12}$	e^+, …	19 ± 2
		K^-, …	16 ± 4
		K^+, …	7 ± 3
		K^0 ($\overline{K}^0$), …	48 ± 15
D^0	$4.2 \cdot 10^{-13}$	K^-, …	43 ± 5
		e^+, …	8 ± 1
		K^+, …	6 ± 2
		K^0 ($\overline{K}^0$), …	33 ± 10
$D_S{}^+$	$4.7 \cdot 10^{-13}$	K^0, …	
B^+	$1.65 \cdot 10^{-12}$		
B^0	$1.55 \cdot 10^{-12}$		
$B_S{}^0$	$1.49 \cdot 10^{-12}$	$D_S{}^-$, …	
J/Ψ	$7.6 \cdot 10^{-21}$	hadrons	
	(87 keV)		
Υ	$1.3 \cdot 10^{-20}$		
	(52.5 keV)		

7.3 Leptons

All leptons have spin (J) = 1/2 and baryon number = 0. The antiparticle of the electron is called positron.

Particle	Symbol	Mass MeV/c^2	Charge q	Anti particle	Mean life s
e neutrino	ν_e	$< 3 \cdot 10^{-6}$	0	$\overline{\nu}_e$	stable
μ neutrino	ν_μ	< 0.19	0	$\overline{\nu}_\mu$	stable
τ neutrino	ν_τ	< 19	0	$\overline{\nu}_\tau$	stable (?)
electron	e^-	0.511 00	–1	e^+	stable
muon	μ^-	105.658 37	–1	μ^+	$2.1970 \cdot 10^{-6}$
tauon	τ^-	1777	–1	τ^+	$2.91 \cdot 10^{-13}$

Decay modes of the muon and the tauon

$\mu^- \to e^- + \overline{\nu}_e + \nu_\mu$ (100%)

$\tau^- \to \mu^- + \overline{\nu}_\mu + \nu_\tau$ (18%)

$\to e^- + \overline{\nu}_e + \nu_\tau$ (18%)

$\to \overline{\nu}_\tau$ + Hadrons (~ 64%)

7.4 Quarks

Quark and lepton families

	1st family		2nd family		3rd family		charge
quarks	u	= up	c	= charm	t	= top (truth)	2/3
	d	= down	s	= strange	b	= bottom (beauty)	– 1/3
leptons	ν_e		ν_μ		ν_τ		0
	e		μ		τ		– 1

u, d, c, s, t, b are the quark "flavours".

According to Quantum Chromo Dynamics (QCD) each quark (antiquark) also carries "colour" ("anticolour"), either "blue", "green", or "red". In this way the table generates 36 different quarks. Only "colourless" ("white") particles can be observed.

Quark quantum numbers

Columns 3–9 Q = charge, S = strangeness, C = charm, $\tilde{B}$ = beauty, T = truth, B = baryon number, I = isospin. For antiquarks, the signs of all these quantum numbers are reversed. Quantum numbers derived from these are: hypercharge $Y \equiv S + C + \tilde{B} + T + B$
third component of isospin $I_3 \equiv Q - Y/2$

1	2	3	4	5	6	7	8	9
Name	**Symbol**	Q	S	C	$\tilde{B}$	T	B	I
down	d	−1/3	0	0	0	0	1/3	1/2
up	u	2/3	0	0	0	0	1/3	1/2
strange	s	−1/3	−1	0	0	0	1/3	0
charmed	c	2/3	0	1	0	0	1/3	0
bottom	b	−1/3	0	0	−1	0	1/3	0
top	t	2/3	0	0	0	1	1/3	0

Quark masses

The *constituent* mass is the effective mass when the quark is confined in a hadron and includes effects of the binding energy. Constituent masses are only defined in the context of a particular hadronic model, and the values below are only approximate.

The *current* mass is the mass of the quark in absence of confinement. No isolated quark has ever been found, and the mass is determined indirectly through their influence on hadronic properties.

More information: http://pdg.lbl.gov/2002/qxxx.html

WWW

Name	**Symbol**	**Constituent mass** GeV/c^2	**Current mass** GeV/c^2
down	d	0.311	0.005 – 0.0085
up	u	0.315	0.0015 – 0.0045
strange	s	0.46	0.080 – 0.155
charmed	c	1.65	1.0 – 1.4
bottom	b	5.1	4.0 – 4.5
top	t	*	169 – 173

* The top quark has such a short half-life that it has never been observed in a hadron, and thus a constituent mass cannot be determined.

7.5 Particle Interaction with Matter

(g) stands for gaseous state. All other values are for solid and liquid states.

Column 3 n_a = number density of atoms.

Column 4 n_e = number density of electrons.

Column 5 I = mean ionization energy averaged over all electrons.

Column 6 L_R = radiation length.

Column 7 $X_R = \rho L_R$ where ρ = density according to column 8.

1	2	3	4	5	6	7	8
Material	Z	n_a $10^{23}/cm^3$	n_e $10^{23}/cm^3$	I eV	L_R cm	X_R g/cm^2	**Density** g/cm^3
H_2	1	0.423	0.423	21.8	891	63.05	0.0708
He	2	0.188	0.376	41.8	755	94.32	0.125
Li	3	0.463	1.39	40.0	155	82.76	0.534
Be	4	1.23	4.94	63.7	35.3	65.19	1.85
B	5	1.32	6.60	76	22.2	52.69	2.37
C	6	1.146	6.82	78	18.8	42.70	2.27
N_2	7	0.347	2.43	85.1	47.0	37.99	0.808
O_2	8	0.429	3.43	98.3	30.0	34.24	1.14
Ne	10	0.358	3.58	137 (g)	24.0	28.94	1.20
Al	13	0.603	7.84	166	8.89	24.01	2.70
Si	14	0.500	6.99	173	9.36	21.82	2.33
Ar	18	0.211	3.80	188 (g)	14.0	19.55	1.40
Fe	26	0.849	22.1	286	1.76	13.84	7.87
Cu	29	0.845	24.6	322	1.43	12.82	8.96
Zn	30	0.658	19.6	330	1.75	12.48	7.13
Kr	36	0.155	5.59	352 (g)	5.26	11.37	2.16
Ag	47	0.586	27.6	470	0.85	8.97	10.5
Sn	50	0.371	18.5	488	1.21	8.82	7.31
W	74	0.632	46.8	727	0.35	6.76	19.3
Pt	78	0.662	51.5	790	0.31	6.54	21.45
Au	79	0.591	46.7	790	0.33	6.46	19.32
Pb	82	0.330	27.0	823	0.56	6.37	11.34
U	92	0.476	43.8	890	0.32	6.05	18.9

Ranges for 1 MeV particles

See also diagram in section F–8.6.

Material	Density g/cm^3	Protons mg/cm^2	Deuterons mg/cm^2	Tritons mg/cm^2	^{3}He mg/cm^2	^{4}He mg/cm^2
C	2.25	2.8	2.0	1.7	0.60	0.60
Al	2.70	4.0	3.0	2.7	0.93	0.96
Si	2.33	3.7	2.7	2.3	0.81	0.81
Fe	7.87	5.3	4.1	3.8	1.4	1.5
Ge	5.32	6.5	5.3	5.0	1.8	1.9
Pb	11.35	11.6	9.5	9.0	3.3	3.5
Na I	3.67	6.5	4.9	4.4	1.6	1.7

Ranges for 10 MeV particles

Even if three digits are sometimes given, at most two of these are significant.

Material	Density g/cm^3	Protons mg/cm^2	Deuterons mg/cm^2	Tritons mg/cm^2	^{3}He mg/cm^2	^{4}He mg/cm^2
C	2.25	135	80	60	15	12
Al	2.70	171	103	78	20	17
Si	2.33	162	99	76	19	16
Fe	7.87	207	130	100	25	22
Ge	5.32	223	143	113	29	25
Pb	11.35	353	239	194	49	43
Na I	3.67	244	156	122	31	26

A photon cross-section database:
http://physics.nist.gov/PhysRefData/Xcom/Text/XCOM.html

8 Solid State Physics

8.1 Metals – Quantum Physical Properties

Column 3 The photoelectric work function is given.

Column 5 Different authors often give widely differing values of the Hall coefficient. Disputed values are set within parentheses.

Column 7 Concentrations have been computed from Column 2 and room temperature densities.

Column 8 The Fermi temperature in free-electron model at room temperature has been computed from Column 7. The Fermi energy in electron volts is obtained by multiplying by the constant $8.617 \cdot 10^{-5}$.

Columns 9 and 10 Starred values indicate that the element is superconductive only in thin films or under high pressure in a crystal structure that is normally unstable.

Column 10 The critical magnetic flux density at zero kelvin is given.

	1	2	3	4	5	6	7	8	9	10
	Metal	**Number of valences**	**Work function**	**Mobility**	**Hall coefficient**	**Debye temperature**	**Free electron concentration**	**Fermi temperature**	**Transition temperature of super-conductivity**	**Critical magnetic field for super-conductivity**
			eV	$10^{-3}\,m^2\,V^{-1}\,s^{-1}$	$10^{-10}\,m^3\,A^{-1}\,s^{-1}$	K	$10^{28}\,m^{-3}$	10^4 K	K	10^{-4} T
1	Ag	1	4.54	5.6	– 0.84	225	5.85	6.36		
2	Al	3	4.20	1.2	– 0.30	428	18.06	13.49	1.18	105
3	Au	1	4.83	3.0	– 0.72	165	5.90	6.39		
4	Ba	2	2.52	0.5		110	3.20	4.24	*	*
5	Be	2	3.92	4.4	+ 2.44	1 460	24.2	16.41		
6	Bi		4.25		(– 5 400)	120			*	*
7	Ca	2	3.20	3.8	– 1.8	230	4.60	5.43		
8	Cd	2	4.11	0.8	+ 0.6	210	9.28	8.66	0.56	30
9	Co		4.97		(– 1.3)	445				
10	Cr		4.45		(+ 6.5)	630				
11	Cs	1	1.94	3.4	– 7.8	38	0.85	1.76	*	*
12	Cu	1	4.84	3.2	– 0.55	345	8.45	8.12		
13	Fe (α)		4.63		(+ 0.2)	462				
14	Ga	3	4.45	0.3	– 0.63	320	15.30	12.01	1.09	51
15	Hf		3.53		+ 0.43	254				
16	Hg		4.53		– 0.73	69			4.15	412
17	In	3	4.08	0.6	– 0.24	109	11.49	9.98	3.40	293
18	Ir		4.57		+ 0.32	420			0.14	
19	K	1	2.25	6.2	– 4.2	91	1.33	2.37		
20	La		3.3		– 0.35	142			6.00	1 100
21	Li	1	2.46	1.8	– 1.70	343	4.66	5.46		
22	Mg	2	3.70	1.7	– 0.94	400	8.60	8.27		

	1	2	3	4	5	6	7	8	9	10
	Metal	**Number of valences**	**Work function**	**Mobility**	**Hall coefficient**	**Debye temperature**	**Free electron concentration**	**Fermi temperature**	**Transition temperature of super-conductivity**	**Critical magnetic field for super-conductivity**
			eV	10^{-3} m^2 V^{-1} s^{-1}	10^{-10} m^3 A^{-1} s^{-1}	K	10^{28} m^{-3}	10^4 K	K	10^{-4} T
23	Mn		4.10		(– 0.93)	410				
24	Mo		4.19		+ 1.91	450			0.92	95
25	Na	1	2.28	5.3	– 2.50	160	2.56	3.66		
26	Nb		3.99		+ 0.72	275			9.20	1 980
27	Ni		5.09		– 0.61	453				
28	Os		4.55			500			0.66	65
29	Pb	4	4.02	0.23	+ 0.09	105	13.20	10.87	7.19	803
30	Pd		5.40		(– 0.68)	275				
31	Pt		5.66		– 0.24	240				
32	Rb	1	2.13	4.3	– 5.9	56	1.08	2.06		
33	Re		4.97		+ 3.15	430			1.70	198
34	Rh		5.03		+ 0.48	480				
35	Ru		4.52		+ 22	600			0.51	70
36	Sb		4.60		(+ 210)	210			*	*
37	Sn	4	4.31	0.39	– 0.048	200	14.48	11.64	3.72	309
38	Sr	2	2.74	0.8		148	3.56	4.58		
39	Ta		4.13		+ 1.01	247			4.48	830
40	Th		3.47		– 0.88	163			1.37	1.6
41	Ti		3.87		– 2.4	420			0.39	100
42	Tl		4.05		+ 0.12	78			2.39	171
43	U		3.45		+ 0.34	207			0.68	

	1	2	3	4	5	6	7	8	9	10
	Metal	**Number of valences**	**Work function**	**Mobility**	**Hall coefficient**	**Debye temperature**	**Free electron concentration**	**Fermi temperature**	**Transition temperature of super-conductivity**	**Critical magnetic field for super-conductivity**
			eV	10^{-3} m^2 V^{-1} s^{-1}	10^{-10} m^3 A^{-1} s^{-1}	K	10^{28} m^{-3}	10^4 K	K	10^{-4} T
44	V		4.11		+ 0.76	380			5.38	1 420
45	W		4.57		+ 0.86	400			0.01	1.1
46	Zn	2	4.34	0.80	+ 0.33	327	13.10	10.90	0.88	53
47	Zr		3.69		+ 0.21	290			0.55	47

8.2 Brillouin Zones

bcc lattice

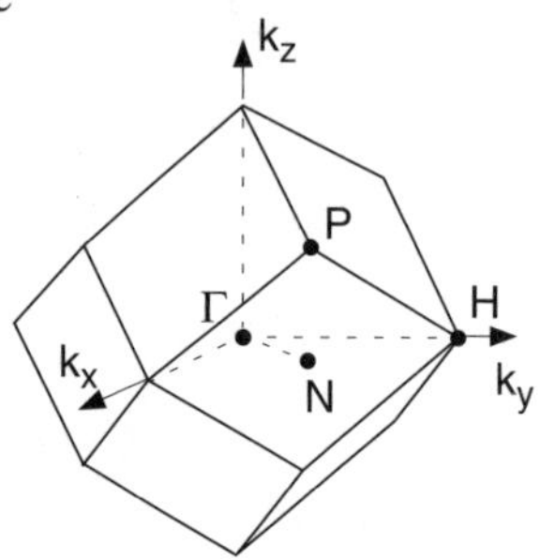

Γ: $\frac{\pi}{a}$ (0, 0, 0)

H: $\frac{2\pi}{a}$ (0, 1, 0)

P: $\frac{\pi}{a}$ (1, 1, 1)

N: $\frac{\pi}{a}$ (1, 1, 0)

fcc lattice

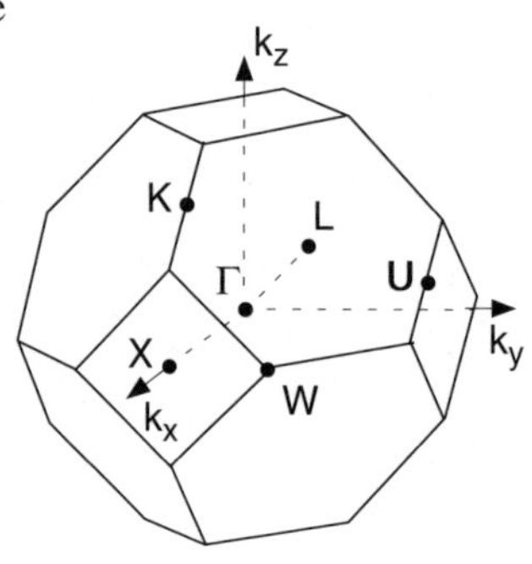

Γ: $\frac{\pi}{a}$ (0, 0, 0)

L: $\frac{\pi}{a}$ (1, 1, 1)

X: $\frac{2\pi}{a}$ (1, 0, 0)

W: $\frac{\pi}{a}$ (2, 1, 0)

hexagonal lattice

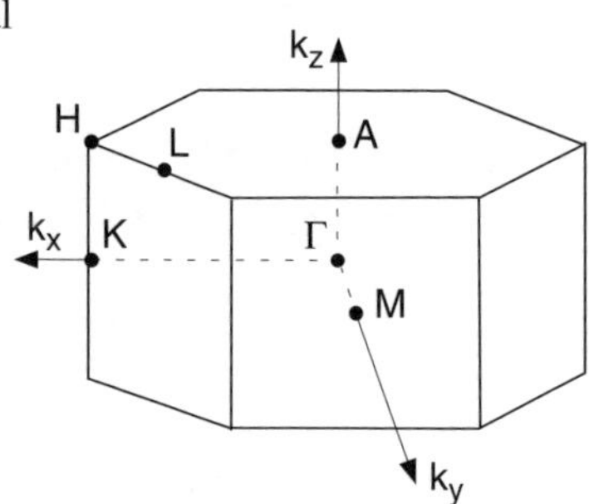

Γ: (0, 0, 0)

A: $\frac{\pi}{c}$ (0, 0, 1)

K: $\frac{4\pi}{3a}$ (1, 0, 0)

M: $\frac{2\pi}{a\sqrt{3}}$ (0, 1, 0)

8.3 Crystal Structure of Chemical Elements and Compounds

Unless otherwise indicated, the values are taken at room temperature (18–25 °C). The uncertainty is at most one unit in the last digit.

Lattice type [Number of atoms in cell]	Coordinates of lattice points [Basis atom coordinates]	Material	Lattice constants a pm	 c pm
Face-centred cubic, fcc (A_1)	000, $\frac{1}{2}\frac{1}{2}0$, $\frac{1}{2}0\frac{1}{2}$, $0\frac{1}{2}\frac{1}{2}$	Ag	408.6	
[4]	[000]	Al	405	
		Ar	531	(at 4 K)
		Au	407.8	
		Ca	557	
		Ce(α)	515	
		Cu	361	
		Ir	383.9	
		Kr	564	(at 4 K)
		Ne	453	(at 4 K)
		Ni	352.4	
		Pb	495	
		Pd	389	
		Pt	392.4	
		Rh	380.4	
		Sr	606	
		Th	508	
		Yb	547.9	
Body-centred cubic, bcc (A_2)	000, $\frac{1}{2}\frac{1}{2}\frac{1}{2}$	Ba	502	
[2]	[000]	Cr	288	
		Cs	614	
		Eu	457	
		Fe(α)	286.6	
		K	534	
		Li	351	
		Mo	315	
		Na	429	
		Nb	330	
		Rb	562	
		Ta	330	
		V	304	
		W	316.5	
Hexagonal close-packed, hcp (A_3)	000, $\frac{1}{3}\frac{2}{3}\frac{1}{2}$	Be	229	358
[2] $c/a \approx 1.633$	[000]	Cd	298	562
		Co(α)	251	407

Lattice type / Number of atoms in cell	Coordinates of lattice points / Basis atom coordinates	Material	Lattice constants a (pm)	c (pm)
Hexagonal close-packed, hcp (A_3)	000, $\frac{1}{3}\frac{2}{3}\frac{1}{2}$	Dy	358	566
		H_2	376	613
[2] $c/a \approx 1.633$	[000]		(at 4 K)	
		He	358	581
			(at 1.5 K, 37 atm)	
		Hf	320	507
		La(α)	377	606
		Mg	321	521
		Os	273	431
		Pr(α)	367	593
		Re	276	446
		Ru	270	428
		Sc	331	526
		Tc	275	440
		Tl	346	552
		Ti	296	474
		Y	364	574
		Zn	266	494
		Zr	324	515
Diamond structure (A_4)	000, $\frac{1}{2}\frac{1}{2}0$,	C	356.683	
a = cube edge	$\frac{1}{2}0\frac{1}{2}$, $0\frac{1}{2}\frac{1}{2}$	Ge	565.791	
		Si	543.010	
[8]	[000, $\frac{1}{4}\frac{1}{4}\frac{1}{4}$]	Sn(α, gray)	648	
Graphite structure (A_9)	000	C	247	672
[4] a = shortest lattice vector (distance to nearest neighbour = $a/\sqrt{3}$)	[000, $\frac{2}{3}\frac{1}{3}0$, $\frac{2}{3}\frac{1}{3}\frac{1}{2}$, $\frac{1}{3}\frac{2}{3}\frac{1}{2}$]			
Sodium chloride structure (B_1)	000, $\frac{1}{2}\frac{1}{2}0$,	NaCl	563.9	
a = cube edge	$\frac{1}{2}0\frac{1}{2}$, $0\frac{1}{2}\frac{1}{2}$	AgBr	577	
[8]	[000, $\frac{1}{2}\frac{1}{2}\frac{1}{2}$]	AgCl	554	
		BaO	552	
		CaO	480	
		CdO	469	
		FeO	428–436	
		KBr	659	
		KCl	628	
		KF	533	
		KI	705	
		LiF	402	

Lattice type Number of atoms in cell	Coordinates of lattice points Basis atom coordinates	Material	Lattice constants *a* pm	 *c* pm
		MgO	420	
		PbS	591	
		PbSe	614	
		TiC	431	
Cesium chloride structure (B_2)	000	CsCl	412	
2	000, $\frac{1}{2}\frac{1}{2}\frac{1}{2}$	CsBr	429	
Zinc sulphide structure (B_3)	000, $\frac{1}{2}\frac{1}{2}0$,	ZnS	541	
		ZnSe	566	
8	$\frac{1}{2}0\frac{1}{2}$, $0\frac{1}{2}\frac{1}{2}$	CdS	581	
		CdSe	604	
	000, $\frac{1}{4}\frac{1}{4}\frac{1}{4}$	GaAs	563	
		GaP	544	
		InSb	646	

8.4 Ferromagnetic Elements

Magnetic moment is given with the Bohr magneton as unit. The temperature is zero kelvin.

1	2	3
Element	**Curie temperature** K	**Magnetic moment per atom** μ_B
Co	1 390	1.715
Dy	85	10.0
Er	20	7.2
Fe	1 043	2.219
Gd	289	7.12
Ho	20	8.54
Ni	631	0.604
Tb	230	4.95

8.5 Semiconductors

Column 3 The thermal band gap at 300 K is given.

Column 4 D = direct, I = indirect.

Column 5 The mobilities have been obtained from measurements of the electrical conductivity and Hall effect, the so-called Hall mobility. Given values apply at 300 K and doping concentration ≈ 10^{14} per cm^3.

Column 7 The static relative permittivity (dielectric constant) is given.

1	2	3	4	5		6			7
Material	**Lattice constant** 10^{-10}m	**Band gap** eV	**Band type**	**Mobility** m^2/Vs **electrons**	**holes**	**Effective mass,** m^*/m **electrons**	**heavy holes**	**light holes**	**Relative permittivity**
Ge	5.657906	0.664	I	0.39	0.19	0.12	0.28	0.044	16.2
Si	5.430102	1.124	I	0.14	0.048	0.26	0.49	0.16	11.9
AlAs	5.6605	2.153	I	0.10	0.01	0.19	0.48	0.020	10.1
AlP	5.4635	2.410	I	–	–	0.21	0.51	0.21	9.8
AlSb	6.1355	1.615	I	0.01	0.04	0.33	0.47	0.16	12.0
GaAs	5.6533	1.424	D	0.85	0.04	0.067	0.45	0.082	13.2
GaP	5.4505	2.272	I	0.017	0.01	0.254	0.67	0.17	11.1
GaSb	6.0959	0.68	D	0.50	0.06	0.047	0.27	0.06	15.7
InAs	6.0584	0.354	D	3.2	0.045	0.023	0.41	0.025	15.1
InP	5.8688	1.351	D	0.45	0.015	0.027	0.34	0.027	12.5
InSb	6.4794	0.175	D	7.8	0.4	0.014	0.40	0.016	16.8
CdTe	6.482	1.475	D	0.11	0.01	0.096	0.4		10.2
ZnSe	5.6676	2.822	D	0.045	0.0028	0.21	0.6		9.1

8.6 Impurity Levels

For trivalent impurity atoms (a) the distance in eV to the edge of the valence band is given, for pentavalent impurities (d) the distance to the edge of the conduction band is given.

Crystal	**Impurity** Al (a)	**As** (d)	**B** (a)	**Ga** (a)	**In** (a)	**P** (d)	**Sb** (d)
Si	0.057	0.049	0.045	0.065	0.16	0.045	0.039
Ge	0.0102	0.0127	0.0104	0.0108	0.0112	0.0120	0.0096

9 Astrophysics and Geophysics

9.1 The Sun and the Planets

Column 4 The density of the Sun varies greatly with the distance from its centre. The density of the innermost region is up to $160 \cdot 10^3$ kg/m^3.

Columns 5, 16, and 17 The gas planets (Jupiter, Saturn, Uranus, and Neptune) do not have a solid surface. The "surface" is considered to be where the atmospheric pressure is 1 atmosphere.

Column 6 The rotation period of the sun at its poles is 36 days. Venus and Pluto rotate in a retrograde or "backward" direction compared to the Earth. The rotation period of Uranus and Neptune is based on the rotation of the magnetic field, measured by Voyager 2, and this is presumed to trace the rotation of the core. The rotation axis of Uranus is inclined 97.9°, so the poles lie nearly in the plane of its orbit around the sun.

Jupiter, which mainly consists of fluid hydrogen, has three rotation systems. System I, within 9° from the equator, has a rotation period of 9 h 50 min 30.003 s. System II, more than 9° from the equator, has a rotation period of 9 h 55 min 40.632 s. System III has a rotation period of 9 h 55 min 29.700 s, which is based on observations of radio emission from layers far below the surface.

Column 7 The sidereal year is determined with respect to the stars. On the Earth it is longer than the tropical year because it is not subject to the shortening effects of precession. A tropical year is the time interval between successive occurrences of the spring equinox. In the table 1 a = 365.26 days.

Column 16 The surface temperature of Mercury varies greatly between day and night. Thus the average day temperature is about 630 kelvin and the average night temperature about 100 kelvin.

Column 17 Counting the number of atoms Jupiter and Saturn consist of about 90% hydrogen and 10% helium. By mass the ratio is 75/25%. The ratio of hydrogen decreases with the depth of the atmosphere. (In addition to this Saturn may have a small core of silicate material.)

Pluto's atmosphere may exist as a gas only when Pluto is near its perihelion, that is when the planet is closest to the sun. The atmospheric pressure at the surface is 93 atm at Venus and 0.007 atm at Mars.

T – 9.1 The Sun and the Planets

	1	2	3	4	5	6
	Mean distance from sun		**Relative mass**	**Density**	**Equatorial radius**	**Period of rotation**
	AU	Gm		10^3 kg m^{-3}	Mm	
Sun	–	–	$333 \cdot 10^3$	1.41	696	25 d (at equator)
Mercury	0.3871	57.9	0.054	5.43	2.48	58.6 d
Venus	0.7233	108.2	0.815	5.24	6.10	243 d
Earth	1.0000	149.6	1	5.51	6.378	23 h 56 min 4 s
Mars	1.5237	227.9	0.107	3.94	3.397	24 h 50 min 23 s
Jupiter	5.2028	778.3	317.9	1.32	71.35	9 h 50 min 30 s
Saturn	9.540	1 427.0	95.15	0.69	60.27	10 h 45 min 45 s
Uranus	19.22	2 875.0	14.58	1.21	25.6	17 h 14 min
Neptune	30.07	4 496.6	17.22	1.67	24.75	16 h 3 min
Pluto*	39.44	5 900	0.0022	2.1	1.14	6 d 9 h

	7	8	9	10
	Sidereal period of revolution	**Excentricity of orbit**	**Known satellites**	**Largest satellites** (diameter in km)
Mercury	87.95 d	0.2056	0	
Venus	224.7 d	0.0068	0	
Earth	365.26 d	0.0167	1	Moon (3476)
Mars	687.0 d	0.0934	2	Phobos (27×21.5×19 km), Deimos
Jupiter	11.862 a	0.0485	63	Ganymede (5262), Callisto (4808), Io (3630), Europa (3138), Amalthea (270×160 km)
Saturn	29.458 a	0.0556	47	Titan (5140), Rhea (1530), Iapetus (1460), Dione, Thetys, Enceladus, Mimas, Hyperion
Uranus	84.013 a	0.0472	21	Titania (1578), Oberon (1523), Umbriel (1170), Ariel (1158), Miranda (472)
Neptune	164.79 a	0.0086	13	Triton (2700), Proteus (≈400), Nereid (340)
Pluto*	248.4 a	0.249	3	Charon (≈1270)

	11	12	13	14	15	16	17
	Gravity	**Escape velocity**	**Mean orbital velocity**	**Axial inclination**	**Incl of orbit**	**Surface temp**	**Atmospheric composition**
	g	km/s	km/s	°	°	K	(% of volume)
Mercury	0.378	4.25	47.87	0	7.004	90–700	–
Venus	0.905	10.36	35.02	177.36	3.394	730	CO_2 (96), N_2 (3)
Earth	1.000	11.18	29.79	23.45	0.000	287	N_2 (78.1), O_2 (20.9), Ar (0.9)
Mars	0.379	5.02	24.13	25.19	1.850	218	CO_2 (95.3), N_2 (2.7), Ar(1.6)
Jupiter	2.529	59.56	13.06	3.13	1.308	120	H_2, He
Saturn	1.066	35.49	9.66	26.73	2.488	88	H_2, He
Uranus	0.903	21.30	6.80	97.86	0.774	59	H_2 (83), He (15), CH_4 (2)
Neptune	1.096	23.50	5.44	29.60	1.774	48	H_2 (80), He (18), CH_4 (2)
Pluto*	0.069	1.22	4.74	122.52	17.148	37	N_2, CH_4, CO

* Since August 2006 Pluto is considered a "dwarf planet" instead of a planet.

More information: http://www.solarviews.com/eng/toc.htm
and http://www.seds.org/nineplanets/nineplanets/nineplanets.html

9.2 Classification of Stars

Each spectral class has ten subdivisions designated by the numbers 0 through 9, for example A0, A1, A2, … A9. Level 0 is hotter than level 9. Classes R and N are carbon stars (also called class C) and class S stars are zirconium stars. To remember the letters, Henry Norris Russell devised the mnemonic "O Be A Fine Girl, Kiss Me Right Now, Sweetheart."

The luminosity L of stars is approximately proportional to the mass M to the power of 3.5, i.e. $L \sim M^{3.5}$. The classification below is the generally accepted classification developed by Yerkes observers Morgan and Keenan. The sun has spectral type G2 V.

Spectral Classes

Class	Temperature (K)
O	>25 000
B	11 000–25 000
A	7 500–11 000
F	6 000– 7 500
G	5 000– 6 000
K	3 500– 5 000
M	<3 500
R,N	2 000– 4 500
S	<2 000

Luminosity Classes

Class	Characteristics
Ia	Supergiants with very high luminosity
Ib	Supergiants with high luminosity
II	Giants with high luminosity
III	Normal giants
IV	Sub-giants
V	Main sequence dwarfs (e.g. the sun)
(VI)	sub-dwarfs
(VII)	white dwarfs

9.3 Well-known Stars

Column 1 If the star is a visual double, the letter A indicates the data for the brighter component. Proxima is also called Alpha Centauri C.

Column 3 The distance to distant stars varies greatly between different references, so the values on the next page are approximate.

Column 4 Variable stars are indicated with a trailing v.

Column 6 The spectral types are explained in table 9.2. WD = White dwarf.

Column 7 RA = right ascension measured clockwise around Polaris, starting with the sun's position at vernal equinox. One rotation = 24 hours and one hour = 60 minutes. The pointer stars of the big dipper are 11 hours, Cassiopeia (the big W) is about 0 hours.

Column 8 The declination (latitude) is measured in degrees and minutes, with positive values north of the equator and negative south of the equator. One minute of right ascension is fifteen times as great an angle as one minute of declination.

T – 9.3 Well-known Stars

1	2	3	4	5	6	7	8
Star Proper name	**Constellation**	**Distance** ly	**Magnitude**		**Spectral type**	**RA** (h:min) α	**Declination** δ
			Apparent	**Absolute**			
The nearest stars							
Sun		–	–26.75	4.8	G2 V	–	–
Proxima	Centaurus	4.2	11.1	15.5	M5 V	14:40	–60° 41′
α Centaruri A	Centaurus	4.3	–0.01	4.4	G2 V	14:40	–60° 49′
α Centaruri B	Centaurus	4.3	1.33	5.7	K1 V	14:40	–60° 50′
Barnard's star	Ophiuchus	5.9	9.5	13.5	M5 V	17:58	+4° 34′
Wolf 359	Leo	7.6	13.5	16.7	M6 V	10:56	+7° 01′
HD 95735	Ursa Major	8.1	7.5	10.5	M2 V	11:01	+36° 18′
Sirius A	Canis Major	8.7	–1.47	1.4	A1 V	06:45	–16° 43′
Sirius B	Canis Major	8.7	8.3	11.2	WD VII	06:45	–16° 49′
UV Ceti A	Cetus	8.9	12.5	15.3	M5 V	01:39	–17° 57′
UV Ceti B	Cetus	8.9	13.5	15.8	M6 V	01:39	–17° 57′
Ross 154	Sagittarius	9.4	10.6	13.3	M5	18:50	–23° 50′
Ross 248	Andromeda	10.3	12.3	14.8	M6	23:42	+44° 10′
ε Eridani	Eridanus	10.8	3.7	6.1	K2 V	03:33	–9° 28′
Luyten 789-6	Aquarius	10.8	12.2	14.6	M7	22:38	–15° 19′
Ross 128	Virgo	10.8	11.1	13.5	M5	11:48	+0° 48′
61 Cygni A	Cygnus	11.1	5.2	7.6	K5 V	21:07	+38° 45′
61 Cygni B	Cygnus	11.1	6.0	8.4	K7 V	21:07	+38° 45′
ε Indi	Indus	11.3	4.7	7.0	K5 V	22:03	–56° 47′
Other well-known stars (sorted by the apparent magnitude)							
Canopus	Carina	310	–0.62	–2.5	F0 Ib	06:24	–52° 41′
Arcturus	Bootes	37	–0.04	0.2	K1 III	14:16	+19° 12′
Vega	Lyra	25	0.03	0.6	A0 V	18:37	+38° 46′
Capella	Auriga	43	0.08	–0.4	G6 III	05:16	+45° 59′
Rigel	Orion	770	0.12	–8.1	B8 Ia	05:15	–8° 13′
Procyon	Canis Minor	11	0.38	2.7	F5 IV	07:39	+5° 14′
Achernar	Eridanus	69	0.46	–2.2	B3 V	01:37	–57° 15′
Betelgeuse	Orion	650	0.5 v	–7.2	M2 Iab	05:55	+7° 24′
Hadar	Centaurus	320	0.6 v	–4.4	B1 III	14:04	–60° 21′
Altair	Aquila	16	0.77	2.3	A7 V	19:51	+8° 51′
Aldebaran	Taurus	60	0.85 v	–0.3	K5 III	04:36	+16° 30′
Acrux	Crux	500	0.87	–3.5	B0 IV	12:26	–63° 04′
Spica	Virgo	220	0.96 v	–3.2	B1 V	13:25	–11° 08′
Antares	Scorpius	520	1.0 v	–5.2	M1 Ib	16:29	–26° 26′
Pollux	Gemini	35	1.14	0.7	K0 III	07:45	+28° 02′
Fomalhaut	Piscis Austr.	22	1.16	2.0	A3 V	22:58	–29° 39′
Deneb	Cygnus	1500	1.25	–7.2	A2 Ia	20:41	+45° 16′
Polaris	Ursa Minor	320	2.0 v	–2.9	F6 Ib	02:28	+89° 15′

More information: http://en.wikipedia.org

9.4 Astronomical Quantities

Time

1 ephemeris second = 1 / 31556925.975 of the tropical year 1900
1 day = 24 hours = 1 440 minutes = 86 400 seconds
Mean solar day = 24 hours of mean solar time = 24 h 03 min 56.55536 s mean sidereal time
Mean sidereal day ("stellar day") = 24 hours of mean sidereal time = = 23 h 56 min 4.09054 s mean solar time
Synodic month (the time between two successive new moon) = 29.53059 mean solar days (29 d 12 h 44 min 03 s)
Tropical month = 27.32158 mean solar days (27 d 07 h 43 min 05 s)
Sidereal month = 27.32166 mean solar days (27 d 07 h 43 min 12 s)
Anomalistic month = 27.55455 mean solar days (27 d 13 h 18 min 33 s)
Draconitic month = 27.21222 mean solar days (27 d 05 h 05 min 36 s)
Tropical year = 365.24220 mean solar days (365 d 05 h 48 min 46 s)
Sidereal year ("stellar year") = 365.25636 mean solar days (365 d 06 h 09 min 10 s)
Anomalistic year = 365.25964 mean solar days (365 d 06 h 13 min 53 s)
Eclipse year = 346.62003 mean solar days (346 d 14 h 52 min 51 s)
Julian year = 365.25 mean solar days (365 d 06 h 00 min 00 s)
The Saros period = 223 synodic months = 18 years 11 days = $6\,585\frac{1}{3}$ mean solar days ($\approx$ 19 eclipse years)
Sun's sidereal period of rotation (mean value) = 25.38 mean solar days
Time for one rotation of the Sun about the galactic centre = $225 \cdot 10^6$ years = 225 Ma

Length

Earth's radius, equatorial = 6 378.16 km = 6.37816 Mm
Earth's radius, polar = 6 356.78 km = 6.35678 Mm
Moon's radius = 1 738 km = 1.738 Mm
Sun's radius = 696 000 km = 696 Mm
Semi-major axis of earth's orbit = 1 astronomical unit (AU) = 149 597 870 km
Mean distance sun-to-earth = 149.600 Gm
Mean distance moon to centre of earth = 384 403 km = 384.403 Mm
1 parcec (1 pc) = 3.262 light-years (l.y.) = 206 265 AU = $3.0857 \cdot 10^{13}$ km = 30.857 Pm
Distance to nearest known star = 1.3 pc = 40 Pm
Distance from the Sun to the galactic plane = 14 pc = 0.43 Em
Distance from the Sun to the centre of the Milky Way = 8.5 kpc = 0.26 Zm
Thickness of the galactic disc = 2 kpc = 60 Em
Diameter of the Milky Way $\approx$ 30 kpc = 1 Zm
Diameter of the halo around the Milky Way $\approx$ 60 kpc = 2 Zm

Distance to nearest known galaxy (the Large Magellanic Cloud) = 46 kpc = 1.4 Zm
Distance to the Andromeda galaxy = 690 kpc = 20 Zm
Distance to the Virgo galaxy cluster = 20 Mpc = 0.6 Ym
Distance to quasars (redshift $z = 2$) ≈ 5 Gpc = $1.5 \cdot 10^{26}$ m

Mass

Earth's mass = $5.977 \cdot 10^{24}$ kg
Moon's mass = $7.349 \cdot 10^{22}$ kg
Sun's mass = $1.989 \cdot 10^{30}$ kg
Mass of lightest observed stars = 0.01–0.04 solar masses
Mass of heaviest observed stars = approx 100 solar masses
Mass of the galactic disc of the Milky Way ≈ $2 \cdot 10^{11}$ solar masses
Mass of the Milky Way, including the halo ≈ $1 \cdot 10^{12}$ solar masses

Measures of Arc

Constant of nutation (1900) = 9.120″
Solar mean semi-diameter = 15′59.63″ = 0.266564°
Solar parallax = 8.798″ = 0.00244°
Lunar mean semi-diameter = 15′32.6″ = 0.25906°
Mean lunar parallax = 57′02.62″ = 0.950728°
Mean obliquity of Ecliptic = 23°27′08.26″–46.84″ T =
23.452294°–0.013011° T (T measured in centuries from 1900)
Mean inclination to the ecliptic of lunar orbit = 5°09′ = 5.15°
Annual lunisolar precession = 50.3508″ = 0.0139919°
Horizontal refraction = 34′ = 0.57°

Speed

Earth's rotational speed at equator = 0.465 km s^{-1}
Moon's orbital velocity = 1.023 km s^{-1}
Earth's orbital velocity = 29.8 km s^{-1}
Sun's movement relative to the local system of reference = 19.4 km s^{-1} =
4.09 AU $year^{-1}$
Rotation velocity of the Milky Way at position of the Sun = 220 km s^{-1}

Density

Mean density of Earth = 5.52 g cm^{-3}
Mean density of Moon = 3.34 g cm^{-3}
Mean density of Sun = 1.41 g cm^{-3}
Density of Sirius B (a white dwarf) = 10^5 g cm^{-3}
Density of a neutron star = 10^{14} g cm^{-3}
Density of VV Cephei (a red supergiant) = $5 \cdot 10^{-9}$ g cm^{-3}

Assumed value of mean density of interstellar matter in the Milky Way = 10^{-26} g cm^{-3}
Density needed for a closed universe $(3H^2/8\pi G) = 0.9 \cdot 10^{-29}$ g cm^{-3} (assuming a value of 70 km s^{-1}/Mpc for Hubble's constant H)
Mean density of matter contributed by known galaxies $\approx 10^{-30}$ g cm^{-3}

Further information on the Sun

Volume = $1.412 \cdot 10^{27}$ m^3
Escape velocity = 618 km/s
Surface temperature = 5 805 K
Temperature at centre = 15.6 MK
Solar constant (flux of electromagnetic radiation reaching the Earth above the atmosphere) = 1.367 kW m^{-2}
Total radiation = $3.92 \cdot 10^{26}$ W
Mass turned into energy = $4 \cdot 10^9$ kg s^{-1}
Absolute visual magnitude = 4.79
Apparent visual magnitude = – 26.78
Composition (by number of atoms): 92.1% hydrogen, 7.8% helium, and 0.1% heavier elements. (By mass about 74% hydrogen, 25% helium, 0.8% oxygen, 0.3% cabon, and 0.4% other elements.)
(See also T – 9.1)

Further information on the Moon

Acceleration of gravity = 1.62 m/s^2
Escape velocity = 2.38 km/s

Further information on the Earth

Volume = $1.083 \cdot 10^{21}$ m^3
Escape velocity = 11.19 km/s
Surface area of land = $1.48 \cdot 10^{14}$ m^2 (29%)
Surface area of sea = $3.62 \cdot 10^{14}$ m^2 (71%)
Mass of atmosphere = $5.1 \cdot 10^{18}$ kg
Mass of oceans = $1.4 \cdot 10^{21}$ kg
Mean geothermal flux = $6.2 \cdot 10^{-2}$ Wm^{-2}
Magnetic dipole moment (1975) = $7.94 \cdot 10^{22}$ Am^2
Magnetic **B**-field at north magnetic pole = $6 \cdot 10^{-5}$ T
Magnetic **B**-field at south magnetic pole = $7 \cdot 10^{-5}$ T
Magnetic **B**-field at magnetic equator = $3 \cdot 10^{-5}$ T

Cosmological quantities

Age of the universe = (13.7 ± 0.2) · 10^9 years = (13.7 ± 0.2) Ga (according to calculations in 2003 based on observation of the background radiation by the WMAP satellite).

Age of solar system = (4.56 ± 0.01) · 10^9 years = (4.56 ± 0.01) Ga

Radius of observable universe = (10 – 40) · 10^9 light-years = (10 – 40) · 10^{25} m = (100 – 400) Ym

Number of protons in observable universe = 5 · 10^{77} – 1 · 10^{79}

Hubble's constant = (70.8 ± 1.6) km s^{-1}/Mpc (according to calculations in 2004 based on data from the WMAP satellite)

Blackbody temperature corresponding to the cosmic background radiation = 2.725 K

9.5 Beaufort Wind Scale

Column 3 Amount of kinetic energy contained in the mass of air that passes through a 1 m^2 area per second is given, energy associated with thermal motion excluded.

1	2	3	4	5
Beaufort number	**Wind speed (10 m above the ground)** m/s	**Energy flow** W m^{-2}	**Name of wind**	**Description of sea surface**
0	0 – 0.2	0–0.005	calm	like a mirror
1	0.3– 1.5	0.02–2.0	light air	ripples
2	1.6– 3.3	2.5–22	light breeze	small wavelets
3	3.4– 5.4	24 –94	gentle breeze	large wavelets; beginning to break
4	5.5– 7.9	100 –300	moderate breeze	small waves, some white horses
5	8.0–10.7	310 –740	fresh breeze	moderate waves, many white horses
6	10.8–13.8	760 –1 600	strong breeze	large waves, extensive foam crests
7	13.9–17.1	1 600 –3 000	moderate gale	white foam and spray
8	17.2–20.7	3 100 –5 300	fresh gale	foam in well-marked streaks
9	20.8–24.4	5 400 –8 700	strong gale	high waves
10	24.5–28.4	8 800 –14 000	whole gale	very high waves, visibility affected
11	28.5–32.6	14 000 –21 000	storm	small ships sometimes lost to view
12	32.7–36.9	21 000 –30 000	hurricane	the air is filled with foam and spray; very poor visibility

9.6 Richter Scale

The Richter scale gives a measure of the magnitude of earthquakes.

Column 1 The magnitude is always given with one decimal.

Column 2 $\lg E \approx 4.8 + 1.5M$, where E = the released seismic energy in joule, and M = magnitude. Notice that this is only the energy radiated from the earthquake as seismic waves, which is a small fraction of the total energy transferred during an earthquake process.

Column 3 Approximate equivalent TNT explosives, exploded below ground, to achieve the same seismic energy, assuming 1.5 % of the released energy becomes seismic energy. As a comparison, the largest man-made nuclear explosion released 58 megaton, and the energy from the Hiroshima nuclear bomb was equivalent to 13 kiloton TNT.

Column 4 Approximate frequency of earthquakes with magnitude from the magnitude in column 1 to this magnitude + 0.9 inclusive.

1	2	3	4	5
Magnitude	**Released energy** J	**TNT for the same seismic energy yield**	**Frequency of occurence up to magnitude + 0.9**	**Example, equivalent energy (approximate)**
1.0	$2 \cdot 10^6$	32 kg	8 000 per day	Large blast at a construction site.
2.0	$6 \cdot 10^7$	1 ton	1 000 per day	Large mine blast.
3.0	$2 \cdot 10^9$	32 ton	50 000 per year	Generally not felt.
4.0	$6 \cdot 10^{10}$	1 kiloton	6 000 per year	Felt 10–50 km away, but rarely causes damage.
5.0	$2 \cdot 10^{12}$	32 kiloton	800 per year	At most slight damage to well-designed buildings.
6.0	$6 \cdot 10^{13}$	1 megaton	120 per year	Destructive in areas up to 100 km.
7.0	$2 \cdot 10^{15}$	32 magaton	18 per year	Can cause serious damage over large areas
8.0	$6 \cdot 10^{16}$	1 gigaton	1 per year	San Francisco, 1906. ($M = 7.8$)
9.0	$2 \cdot 10^{18}$	32 gigaton	1 per 20 years	Indian Ocean quake/tsunami, 2004 ($M = 9.2$) and Great Chilean quake, 1960 ($M = 9.5$).
10.0	$6 \cdot 10^{19}$	1 teraton	–	An earthquake fault break that nearly circles the entire Earth.
12.0	$6 \cdot 10^{22}$	1 petaton	–	The energy of an asteroid with diameter 3 km that enters with 30 km/s, or the Earth's daily receipt of solar energy.

9.7 Energy Content in Different Fuels

Type of fuel	**Energy content** kWh / kg	MJ / kg
Urban wastes	2–5	7–18
Wood	5	18
Peat (50 % water content)	2.3	8.3
Coal and coke	4.5–9.0	16–32
Oil for domestic heating (no 6)	11.6	42
Petrol (gasoline), kerosene	12	44
Natural gas (95 % methane)	14.3	50
Hydrogen	34.5	124
Uranium (average fission reaction)	$2.3 \cdot 10^7$	$8.3 \cdot 10^7$
Uranium (natural, electrical energy produced in conventional reactors, with reprocessing)	$5.4 \cdot 10^4$	$19.4 \cdot 10^4$
Uranium dioxide (practical value for electrical energy produced in breeders)	$3.6 \cdot 10^6$	$13.0 \cdot 10^6$
Deuterium (fusion reaction $6D \to 2\,^4He + 2n + 2p$)	$9.6 \cdot 10^7$	$35 \cdot 10^7$
Sea water (assuming total deuterium content to be available for fusion reaction above)	$5 \cdot 10^3$	$1.8 \cdot 10^4$
Arbitrary material, total annihilation energy $E = mc^2$	$2.5 \cdot 10^{10}$	$9 \cdot 10^{10}$

F

Physical Formulae and Diagrams

2 Thermal Physics

2.1 Kinetic Theory of Gases
2.2 Equations of State
2.3 Thermal Capacity
2.4 Thermodynamic Relations
2.5 Thermodynamic Potentials
2.6 Thermal Transport
2.7 Diffusion
2.8 Special Relations

3 Electromagnetic Theory

3.1 The Coulomb Field
3.2 Motion of Charged Particles
3.3 The Quasi-Stationary B-field
3.4 Electric and Magnetic Dipoles
3.5 Dielectric and Magnetic Media
3.6 The Electromagnetic Field
3.7 Relativity and Electromagnetism
3.8 The Magnetic Circuit
3.9 Induction and Inductance
3.10 Capacitance
3.11 The Electric Circuit
3.12 Alternating (AC) Currents
3.13 Series and Parallel Circuits
3.14 Three-Phase Current
3.15 Rotating Electric Machines

4 Electronics

4.1 The PN Diode
4.2 The Bipolar Transistor
4.3 The Field Effect Transistor (FET)
4.4 The Transistor as Amplifier
4.5 Filters and Cut-Off Frequencies
4.6 Feedback
4.7 The Differential Amplifier
4.8 Transmission Lines
4.9 Digital Circuits and Boolean Algebra

5 Waves

5.1 Wave Motion
5.2 Superposition of Waves
5.3 Energy Transport
5.4 Doppler Effect

1 Classical Mechanics and Relativity

Quantity	Symbol	SI Unit
Position vector	$\boldsymbol{r}$	m
Velocity	$\dot{\boldsymbol{r}}$, $\boldsymbol{v}$ (v)	m/s
Mass	m	kg
Angular velocity	$\boldsymbol{\omega}$	rad/s
Angular frequency	ω	rad/s
Period	T	s
Area	A	m^2
Volume	V	m^3
Density	ρ	kg/m^3
Force	$\boldsymbol{F}$ (F)	$N = kg\ m/s^2$
Linear momentum	$\boldsymbol{p}$ (p)	Ns
Work	W	J = Nm
Kinetic energy	E_k	J
Potential energy	E_p	J
Total energy	E	J
Angular momentum relative to a point P	$\boldsymbol{L}^P$ (L)	Js
Torque relative to a point P	$\boldsymbol{M}^P$ (M)	Nm
Power	P	W = J/s
Position vector of centre of mass	$\boldsymbol{R}$	m
Moment of inertia	I	$kg\ m^2$
Static pressure	p	$Pa = N/m^2$
Coefficient of viscosity	η	$Ns/m^2 = kg/m\,s$
Acceleration of gravity at sea level	$\boldsymbol{g}$ (g)	m/s^2

1.1 Particle Mechanics

Newton's Laws (valid in inertial frame of reference)

1. Law of Inertia

If a body is not subject to any net external force, it either remains at rest or continues in uniform motion.

2. Law of Acceleration

$$\frac{\mathrm{d}}{\mathrm{d}t}\,(m\boldsymbol{v}) = \boldsymbol{F}$$

3. Law of Action and Reaction

When two particles interact, the force on one particle is equal and opposite to the force on the other.

Instantaneous velocity and instantaneous acceleration

$$v = \frac{\mathrm{d}d}{\mathrm{d}t} \qquad a = \frac{\mathrm{d}v}{\mathrm{d}t} \qquad d = \text{displacement}$$

Constant acceleration

$$v = v_0 + at \qquad a = \text{acceleration} = \text{constant}$$

$$d = v_0 t + \tfrac{1}{2} a t^2 = \frac{v + v_0}{2}\, t \qquad v_0 = \text{initial speed}$$

$$v^2 - v_0^2 = 2\,ad \qquad d = \text{distance}$$

Projectile ballistics

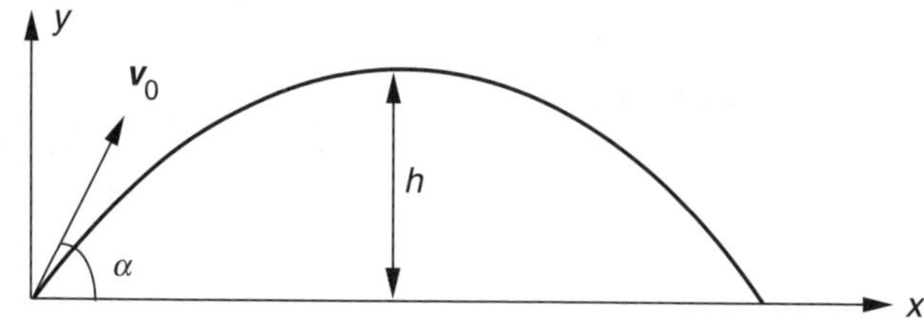

$$\boldsymbol{v} = v_0 \cos\alpha\,\hat{\boldsymbol{x}} + (v_0 \sin\alpha - gt)\,\hat{\boldsymbol{y}}$$

$$v^2 = v_0^2 - 2\,gy \qquad \begin{cases} x = v_0 t \cos\alpha \\ y = v_0 t \sin\alpha - \tfrac{1}{2}\,gt^2 \end{cases}$$

$$y = x \tan \alpha - \frac{g\,x^2}{2v_0^2\,\cos^2\alpha}$$

$$h = \frac{v_0^2}{2g}\,\sin^2\alpha$$

Linear momentum and impulse

$$\boldsymbol{p} = m\boldsymbol{v} \qquad \boldsymbol{S} = \boldsymbol{p}_2 - \boldsymbol{p}_1 = \int_{t_1}^{t_2} \boldsymbol{F}\,\mathrm{d}t$$

Work

$$W = \int \boldsymbol{F} \cdot \mathrm{d}\boldsymbol{r} \qquad W_2 - W_1 = \int_{t_1}^{t_2} \boldsymbol{F} \cdot \boldsymbol{v}\,\mathrm{d}t = \int_{\boldsymbol{r}_1}^{\boldsymbol{r}_2} \boldsymbol{F} \cdot \mathrm{d}\boldsymbol{r}$$

Kinetic energy

$$E_\mathrm{k} = \tfrac{1}{2}\,mv^2$$

Conservative force

$$\mathrm{curl}\,\boldsymbol{F} = \nabla \times \boldsymbol{F} = \boldsymbol{0}$$

$$\boldsymbol{F} = -\,\mathrm{grad}\,E_\mathrm{p} = -\,\nabla E_\mathrm{p}$$

Law of energy conservation (mechanical energy)

$$E = E_\mathrm{p} + E_\mathrm{k} = \text{constant}$$

Power

$$P = \boldsymbol{F} \cdot \boldsymbol{v} = \frac{\mathrm{d}W}{\mathrm{d}t}$$

Angular momentum

$$\boldsymbol{L} = \boldsymbol{r} \times \boldsymbol{p}$$

Torque (moment of force)

$$\boldsymbol{M} = \boldsymbol{r} \times \boldsymbol{F}$$

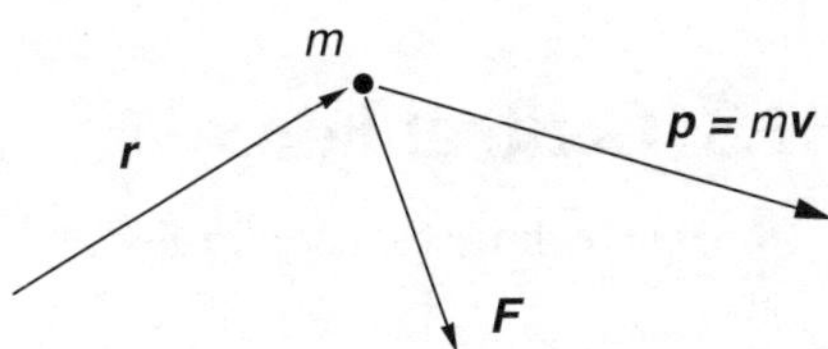

Law of momenta (Euler's second law)

$$\frac{\mathrm{d}\boldsymbol{L}^\mathrm{P}}{\mathrm{d}t} = \boldsymbol{M}^\mathrm{P}$$

1.2 Forces

Gravitational force (Newton's law of gravitation)

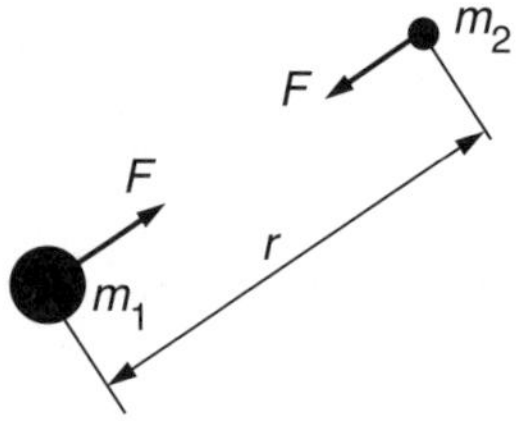

$$F = G\frac{m_1 m_2}{r^2}$$

In particular, force of gravity at sea level

$$\boldsymbol{F} = -m\,g\,\hat{z}$$

$$g = G\,\frac{M}{R^2}$$

M = mass of earth

R = radius of earth

See also Sec. F–11.1

Frictional force

$f \le \mu_s N$	at rest	μ_s = static frictional coefficient
$f = \mu N$	in motion	μ = sliding frictional coefficient
		N = normal force

Elastic (spring) force

$$F = -kx$$

k = spring constant

x = displacement from equilibrium

Centripetal force in circular motion

$$F_{cp} = -m\,\omega^2\,r = -\frac{m v^2}{r}$$

F_{cp} is directed towards the motion centre

See also Sec. F–1.6.

$$v = \frac{2\pi r}{T} = \omega r$$

v = speed = constant

$$a = -\frac{v^2}{r} = -\omega v = -\omega^2 r$$

a = acceleration

1.3 Central Force

(r, θ, z) = cylindrical coordinates

$$\boldsymbol{F} = F(r)\hat{\boldsymbol{r}}$$

Angular momentum

$$\boldsymbol{L}^0 = m\,r^2\,\dot{\theta}\,\hat{z} = L\,\hat{z} = \text{constant}$$

Twice areal velocity

$$h = |\boldsymbol{r} \times \boldsymbol{v}| = \frac{L}{m}$$

Effective potential energy

$$V_{\text{eff}} = E_{\text{p}} + \frac{L^2}{2m\,r^2}$$

Virial theorem for central force

$$\langle E_{\text{k}} \rangle = -\tfrac{1}{2}\,\langle \boldsymbol{F} \cdot \boldsymbol{r} \rangle = \tfrac{1}{2}\,\langle r\,\frac{\partial E_{\text{p}}}{\partial r} \rangle$$

Laws of motion

$$\dot{r}^2 = \frac{2}{m}\,(E - V_{\text{eff}})$$

$$\frac{\mathrm{d}r}{\mathrm{d}\theta} = \pm \frac{m\,r^2}{L}\sqrt{\frac{2}{m}(E - V_{\text{eff}})}$$

$$t_1 - t_0 = \frac{m}{L}\int_{\theta_0}^{\theta_1} r(\theta)^2\,\mathrm{d}\theta$$

$$m\ddot{r} = F(r) + \frac{L^2}{m\,r^3}$$

Binet's formula

$$\frac{\mathrm{d}^2}{\mathrm{d}\theta^2}\left(\frac{1}{r}\right) = -\,\frac{1}{r} - \frac{m\,r^2}{L^2}\,F(r)$$

In particular, if $\boldsymbol{F} = \dfrac{k}{r^2}\,\hat{\boldsymbol{r}}$

$$E_{\text{p}} = \frac{k}{r}$$

$$\frac{1}{r} = A \cos(\theta - \theta_0) - \frac{m\,k}{L^2}$$

An ellipse, if $E < 0$ and $k < 0$

A parabola, if $E = 0$ and $k < 0$

A hyperbola, if $k > 0$ or $E > 0$

Major axis

$$2\,a = \left|\frac{k}{E}\right|$$

1.4 Collisions

Impulse of force

$$\boldsymbol{S} = m\,\boldsymbol{v} - m\,\boldsymbol{v}_0 = \int \boldsymbol{F}\mathrm{d}t$$

Restitution coefficient and conservation of momentum in head-on collision

$$e = -\frac{v - v'}{v_0 - v_0'}$$

v_0 and v_0' are the initial velocities of the bodies

$$mv_0 + m'v_0' = mv + m'v'$$

v and v' are the final velocities of the bodies

Special case: $v_0' = 0,\ e = 1$

$$\frac{v}{v_0} = \frac{m - m'}{m + m'}$$

1.5 Kinematics

Plane polar coordinates

$$\boldsymbol{r} = r\,\hat{\boldsymbol{r}}$$

$$\dot{\boldsymbol{r}} = \dot{r}\,\hat{\boldsymbol{r}} + r\dot{\theta}\hat{\boldsymbol{\theta}}$$

$$\ddot{\boldsymbol{r}} = (\ddot{r} - r\,\dot{\theta}^2)\hat{\boldsymbol{r}} + (r\ddot{\theta} + 2\,\dot{r}\,\dot{\theta})\hat{\boldsymbol{\theta}}$$

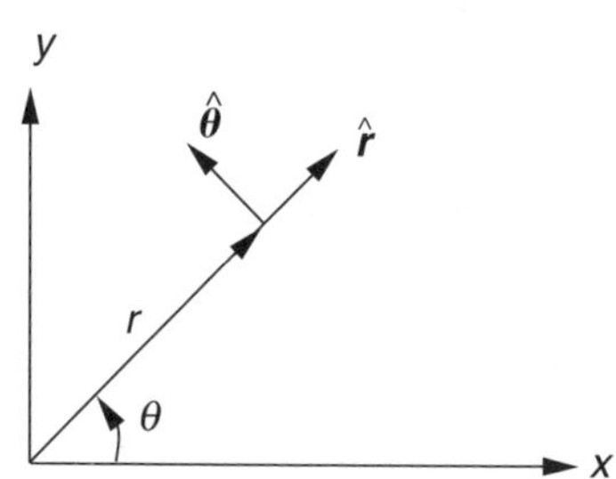

Special case: Circular motion

$r = \text{const}$

$\ddot{\boldsymbol{r}} = -r\dot{\theta}^2\hat{\boldsymbol{r}} + r\ddot{\theta}\hat{\boldsymbol{\theta}}$

$\text{radial acceleration} = -r\dot{\theta}^2 = -r\omega^2 = -\dfrac{v^2}{r}$

Cylindrical coordinates

$\boldsymbol{r} = \rho\hat{\boldsymbol{\rho}} + z\hat{z}$ See also Sec. M–10

$\dot{\boldsymbol{r}} = \dot{\rho}\hat{\boldsymbol{\rho}} + \rho\dot{\varphi}\hat{\boldsymbol{\varphi}} + \dot{z}\hat{z}$

$\ddot{\boldsymbol{r}} = (\ddot{\rho} - \rho\dot{\varphi}^2)\hat{\boldsymbol{\rho}} + (\rho\ddot{\varphi} + 2\dot{\rho}\dot{\varphi})\hat{\boldsymbol{\varphi}} + \ddot{z}\hat{z}$

Pure rotation round an axis

$\boldsymbol{\omega} = \dot{\varphi}\,\hat{z}$

$\dot{\boldsymbol{r}} = \boldsymbol{\omega} \times \boldsymbol{r}$

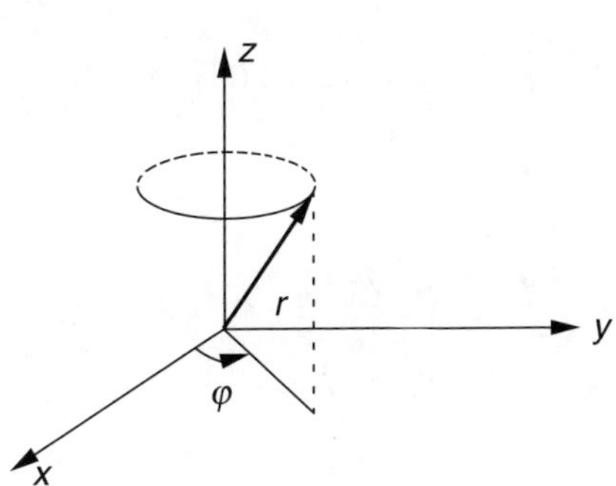

Spherical coordinates

See also Sec. M–10

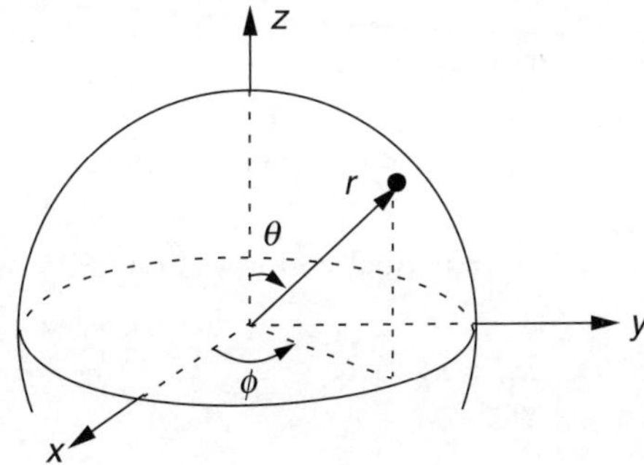

$\boldsymbol{r} = r\,\hat{\boldsymbol{r}}$

$\dot{\boldsymbol{r}} = \dot{r}\,\hat{\boldsymbol{r}} + r\,\dot{\theta}\,\hat{\boldsymbol{\theta}} + r\,\dot{\phi}\sin\theta\,\hat{\boldsymbol{\phi}}$

$$\ddot{\boldsymbol{r}} = (\ddot{r} - r\,\dot{\phi}^2\sin^2\theta - r\,\dot{\theta}^2)\hat{\boldsymbol{r}} + (r\,\ddot{\theta} + 2\,\dot{r}\,\dot{\theta} - r\,\dot{\phi}^2\sin\theta\cos\theta)\hat{\boldsymbol{\theta}} + (r\,\ddot{\phi}\sin\theta + 2\,\dot{r}\,\dot{\phi}\sin\theta + 2\,r\,\dot{\theta}\,\dot{\phi}\cos\theta)\hat{\boldsymbol{\phi}}$$

Planar motion, tangential and normal coordinates

Instantaneous radius of curvature

$$\rho = \frac{\mathrm{d}s}{\mathrm{d}\theta}$$

s = length of arc

θ = path angle

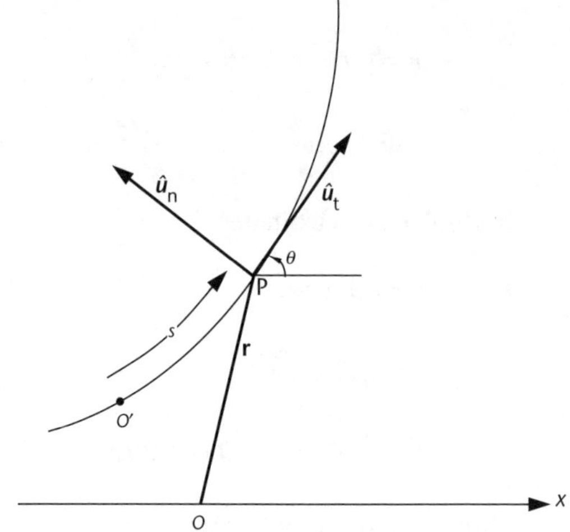

Velocity and acceleration relative to O

$$\boldsymbol{v} = v\hat{\boldsymbol{u}}_t = \dot{s}\,\hat{\boldsymbol{u}}_t$$

$$\boldsymbol{a} = a_t\hat{\boldsymbol{u}}_t + a_n\hat{\boldsymbol{u}}_n =$$

$$= \dot{v}\,\hat{\boldsymbol{u}}_t + v\dot{\theta}\,\hat{\boldsymbol{u}}_n$$

$$= \ddot{s}\,\hat{\boldsymbol{u}}_t + \frac{v^2}{\rho}\,\hat{\boldsymbol{u}}_n$$

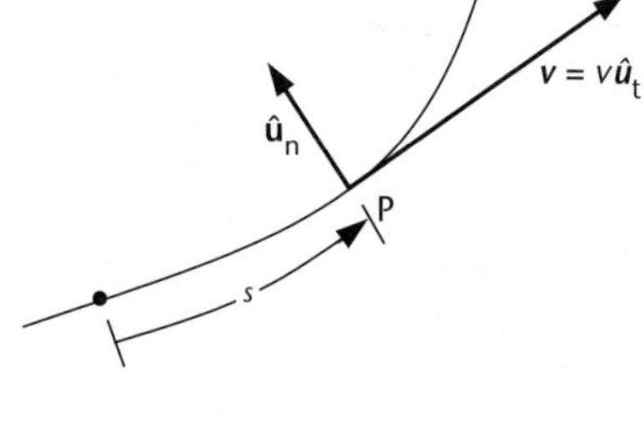

Unit vectors

$\hat{\boldsymbol{u}}_t$ = tangential vector, as in figure

$\hat{\boldsymbol{u}}_n$ = normal vector, as in figure

$\hat{\boldsymbol{u}}_p = \hat{\boldsymbol{u}}_t \times \hat{\boldsymbol{u}}_n$

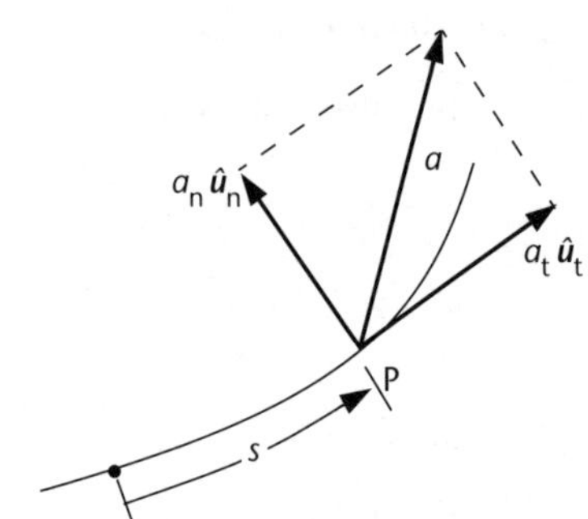

1.6 Relative Motion

Let $\boldsymbol{S}^*$ be a moving system where the origin has position vector $\boldsymbol{R}_0$ and where the axes are rotating at rotation velocity $\boldsymbol{\omega}$ with respect to an inertial system $\boldsymbol{S}$.

$$\boldsymbol{r} = \boldsymbol{r}^* + \boldsymbol{R}_0$$

$$\frac{\mathrm{d}}{\mathrm{d}t}\boldsymbol{r} = \frac{\mathrm{d}^*}{\mathrm{d}t}\boldsymbol{r}^* + \boldsymbol{\omega}\times\boldsymbol{r}^* + \dot{\boldsymbol{R}}_0$$

$$\frac{\mathrm{d}^2}{\mathrm{d}t^2}\boldsymbol{r} = \frac{\mathrm{d}^{2*}}{\mathrm{d}t^2}\boldsymbol{r}^* + 2\,\boldsymbol{\omega}\times\frac{\mathrm{d}^*}{\mathrm{d}t}\boldsymbol{r}^* + \dot{\boldsymbol{\omega}}\times\boldsymbol{r}^* + \boldsymbol{\omega}\times(\boldsymbol{\omega}\times\boldsymbol{r}^*) + \ddot{\boldsymbol{R}}_0$$

$\frac{\mathrm{d}}{\mathrm{d}t}$ is relative $\boldsymbol{S}$ and $\frac{\mathrm{d}^*}{\mathrm{d}t}$ is relative $\boldsymbol{S}^*$

Fictitious forces

Coriolis force $= -2\,m\,\boldsymbol{\omega}\times\frac{\mathrm{d}^*}{\mathrm{d}t}\boldsymbol{r}^*$ See also Sec. F – 11

Centrifugal force $= -\,m\,\boldsymbol{\omega}\times(\boldsymbol{\omega}\times\boldsymbol{r}^*)$

1.7 Mechanics of Particle Systems

Position of centre of mass, CM

$$\boldsymbol{R} = \frac{\Sigma_i m_i \boldsymbol{r}_i}{m} \qquad m = \Sigma_i\, m_i$$

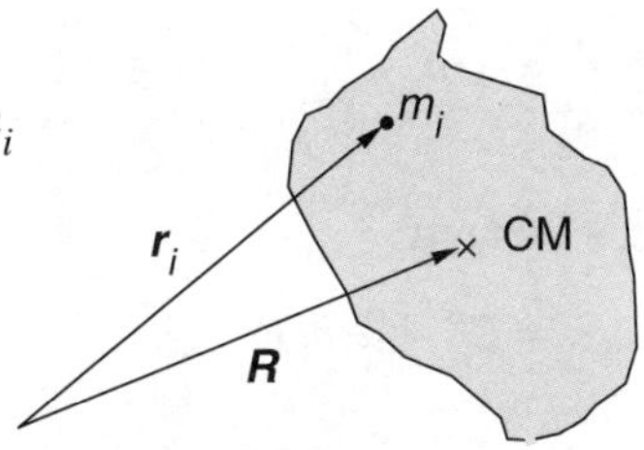

Law of motion of centre of mass

$$m\,\ddot{\boldsymbol{R}} = \boldsymbol{F}^{\mathrm{ext}} \qquad \boldsymbol{F}^{\mathrm{ext}} = \text{external force} = \Sigma_i \boldsymbol{f}_i$$

Two particles; reduced mass and law of motion

$$\mu = \frac{m_1\, m_2}{m_1 + m_2}$$

$$\mu(\ddot{\boldsymbol{r}}_1 - \ddot{\boldsymbol{r}}_2) = \boldsymbol{F}_1^{\mathrm{in}} + \mu\left(\frac{\boldsymbol{F}_1^{\mathrm{ext}}}{m_1} - \frac{\boldsymbol{F}_2^{\mathrm{ext}}}{m_2}\right)$$

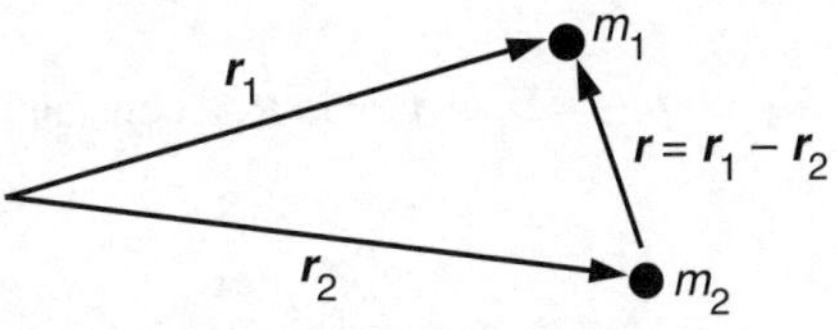

$\boldsymbol{F}_1^{\mathrm{in}}$ = the force with which particle 2 affects particle 1.

Laws of moments

$$\boldsymbol{L}^{\mathrm{p}} = \Sigma_i \boldsymbol{r}_i^{\mathrm{p}} \times m_i \dot{\boldsymbol{r}}_i^{\mathrm{p}}$$

$$\boldsymbol{L} = \boldsymbol{L}^* + \boldsymbol{R} \times m \dot{\boldsymbol{R}}$$

$\boldsymbol{L}^*$ = angular momentum relative to centre of mass

$$\frac{\mathrm{d}}{\mathrm{d}t} \boldsymbol{L}^{\mathrm{p}} = \boldsymbol{M}^{\mathrm{p}} - m(\boldsymbol{R} - \boldsymbol{r}_{\mathrm{p}}) \times \ddot{\boldsymbol{r}}_{\mathrm{p}}$$

In particular, if

1. the point P is moving at constant velocity, or
2. P = centre of mass, or
3. P is a point being accelerated towards the centre of mass

$$\frac{\mathrm{d}}{\mathrm{d}t} \boldsymbol{L}^{\mathrm{p}} = \boldsymbol{M}^{\mathrm{p}}$$

Kinetic energy

$$E_{\mathrm{k}} = \tfrac{1}{2} \Sigma_i m_i v_i^2$$

König's theorem

$$E_{\mathrm{k}} = E_{\mathrm{k}}^* + \tfrac{1}{2} m \dot{\boldsymbol{R}}^2$$

E_{k}^* = kinetic energy in centre-of-mass system

Conservative force

$$\boldsymbol{F} = -\operatorname{grad} E_{\mathrm{p}} = -\nabla E_{\mathrm{p}}$$

$$E = E_{\mathrm{p}} + E_{\mathrm{k}} = \text{constant}$$

Law of conservation of momentum in isolated system

$$\boldsymbol{p}_{\mathrm{tot}} = \Sigma_i m_i \boldsymbol{v}_i = \mathrm{m} \dot{\boldsymbol{R}} = \text{constant}$$

1.8 Mechanics of Particle Systems in Generalized Coordinates

Work in virtual displacement δr

$$\delta W = \sum_i^N \boldsymbol{F}_i \cdot \delta \boldsymbol{r}_i = \sum_\nu^n Q_\nu \delta q_\nu$$

q_ν = generalized coordinates

N = number of particles

n = number of generalized coordinates

Generalized force

$$Q_\nu = \Sigma_i \boldsymbol{F}_i^{\text{dir}} \cdot \frac{\partial \boldsymbol{r}_i}{\partial q_\nu}$$

$\boldsymbol{F}_i^{\text{dir}}$ = directly applied force, doing work in virtual displacements

For conservative forces

$$Q_\nu = -\frac{\partial E_\text{p}}{\partial q_\nu}$$

Lagrange's equations

$$\frac{\text{d}}{\text{d}t}\frac{\partial E_\text{k}}{\partial \dot{q}_\nu} - \frac{\partial E_\text{k}}{\partial q_\nu} = Q_\nu$$

The Lagrange function and equations for monogenic forces

$$\mathcal{L} = E_\text{k} - U$$

U = generalized potential

$$\frac{\text{d}}{\text{d}t}\frac{\partial \mathcal{L}}{\partial \dot{q}_\nu} - \frac{\partial \mathcal{L}}{\partial q_\nu} = 0$$

$\nu = 1, \dots n$

Generalized momentum (conjugated canonically to q_ν)

$$p_\nu = \frac{\partial \mathcal{L}}{\partial \dot{q}_\nu}$$

$$\dot{p}_\nu = \frac{\partial \mathcal{L}}{\partial q_\nu}$$

The Hamiltonian

$$H = \Sigma_\nu p_\nu \dot{q}_\nu - \mathcal{L} = E_\mathrm{k} + U - \Sigma_\nu \frac{\partial U}{\partial \dot{q}_\nu} \dot{q}_\nu$$

Hamilton's equations

$$\frac{\partial H}{\partial q_\nu} = -\dot{p}_\nu \qquad \frac{\partial H}{\partial p_\nu} = \dot{q}_\nu \qquad \frac{\partial H}{\partial t} = -\frac{\partial \mathcal{L}}{\partial t}$$

1.9 Rigid Body Mechanics

Velocity of arbitrary point P

$$\boldsymbol{v}_\mathrm{P} = \boldsymbol{v}_\mathrm{A} + \boldsymbol{\omega} \times \boldsymbol{r}_\mathrm{AP}$$

$\boldsymbol{\omega}$ = total angular velocity

Tensor of inertia

$$I_{ik} = \int (r^2 \delta_{ik} - r_i r_k)\, \mathrm{d}m \qquad \begin{cases} \delta_{ik} = 1 \text{ if } i = k \\ \delta_{ik} = 0 \text{ if } i \neq k \end{cases}$$

In particular

$$I_{xx} = \int (r^2 - x^2) \mathrm{d}m = \int (y^2 + z^2) \mathrm{d}m$$

$$I_{xy} = -\int xy\, \mathrm{d}m$$

Moments of inertia

$$I = \Sigma_i m_i r_i^2 = \int r^2\, \mathrm{d}m$$

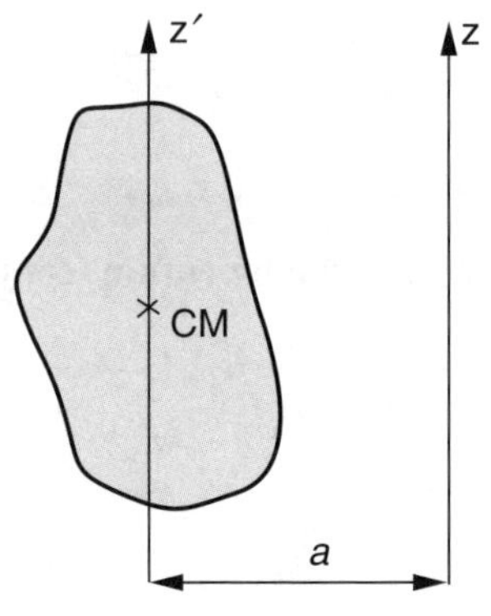

Radius of gyration

$$K = \sqrt{I/m}$$

Steiner's theorem

$$I_z = I^*_{z'} + m a^2$$

Generalization of Steiner's theorem

$$\boldsymbol{I} = \boldsymbol{I}^* + m\,(R^2\,\boldsymbol{1} - \boldsymbol{R}\,\boldsymbol{R})$$

$\boldsymbol{R}$ = (X, Y, Z) = position vector of centre of mass

$\boldsymbol{1}$ = unity tensor

$\boldsymbol{I}^*$ = inertial tensor relative to centre of mass

Particularly, moments and products of inertia for parallel axes

$$I_{xx} = I^*_{x'x'} + m(Y^2 + Z^2)$$

$$I_{xy} = I^*_{x'y'} - m\,X\,Y$$

Relation for moments of inertia of thin plane disc

$$I_z = I_x + I_y$$

x and y axes in the plane of the disc

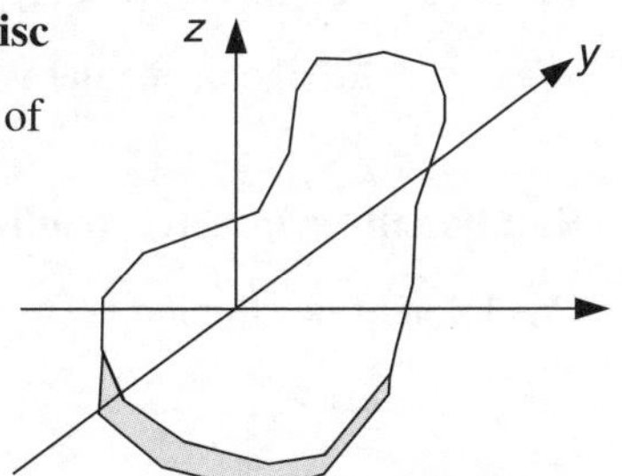

Rotational energy

$$E_\mathrm{k} = \tfrac{1}{2}\,\boldsymbol{\omega}\cdot\boldsymbol{I}\cdot\boldsymbol{\omega} = \tfrac{1}{2}\,\boldsymbol{\omega}\cdot\boldsymbol{L}$$

in particular, for principal systems

$$E_\mathrm{k} = \tfrac{1}{2}\,(I_{11}\,\omega_1^2 + I_{22}\,\omega_2^2 + I_{33}\,\omega_3^2)$$

Rotation round a fixed axis

$$E_\mathrm{k} = \tfrac{1}{2}\,I\,\omega^2$$

Rotational power in principal system

$$P = \boldsymbol{M}\cdot\boldsymbol{\omega}$$

Work for plane turning

$$W = \int |\boldsymbol{M}|\,\mathrm{d}\varphi$$

Angular momentum

$$\boldsymbol{L} = \boldsymbol{I}\cdot\boldsymbol{\omega} = \begin{pmatrix} I_{11} & I_{12} & I_{13} \\ I_{21} & I_{22} & I_{23} \\ I_{31} & I_{32} & I_{33} \end{pmatrix}\begin{pmatrix} \omega_1 \\ \omega_2 \\ \omega_3 \end{pmatrix}$$

In particular, for principal systems

$$\boldsymbol{L} = I_{11}\,\omega_1\,\hat{\boldsymbol{x}} + I_{22}\,\omega_2\,\hat{\boldsymbol{y}} + I_{33}\,\omega_3\,\hat{\boldsymbol{z}}$$

Torque in a system S* rotating with Ω

$$\boldsymbol{M}^{\mathrm{p}} = \boldsymbol{I}^{\mathrm{p}} \cdot \dot{\boldsymbol{\omega}} + \Omega \times (\boldsymbol{I}^{\mathrm{p}} \cdot \boldsymbol{\omega})$$

In particular, Euler's equations of motion in a corotating principal system

$$M_1 = I_{11}\,\dot{\omega}_1 - \omega_2\,\omega_3\,(I_{22} - I_{33})$$

$$M_2 = I_{22}\,\dot{\omega}_2 - \omega_3\,\omega_1\,(I_{33} - I_{11})$$

$$M_3 = I_{33}\,\dot{\omega}_3 - \omega_1\,\omega_2\,(I_{11} - I_{22})$$

$\boldsymbol{\omega}$ = total angular velocity of the body

Rotationally symmetric bodies rolling down a slope

Condition of no slipping

$$v = \omega\, r$$

Conservation of energy

$$\tfrac{1}{2}\, mv^2 + \tfrac{1}{2}\, I\,\omega^2 - mg\, x \sin\alpha = E$$

Acceleration

$$\frac{\mathrm{d}v}{\mathrm{d}t} = \frac{mg \sin\alpha}{(m + I/r^2)}$$

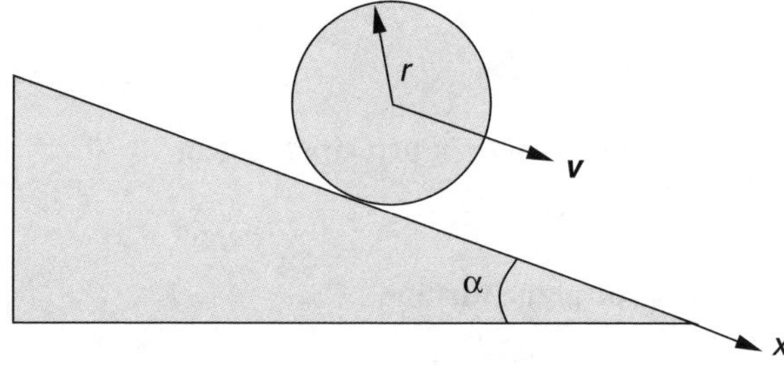

Spinning tops

$$\boldsymbol{M} = \boldsymbol{\omega}_{\mathrm{p}} \times \boldsymbol{L}$$

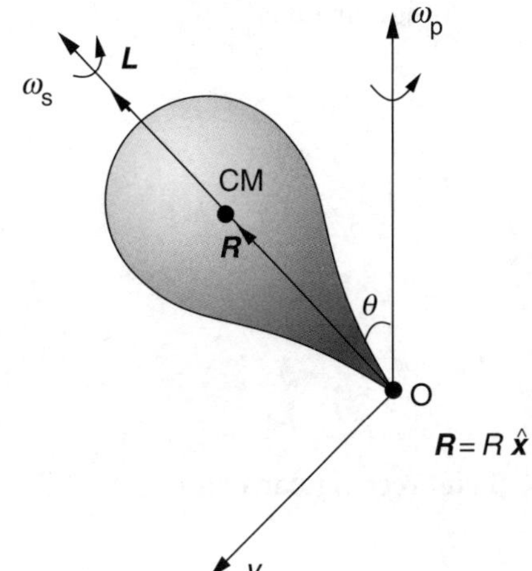

Velocity of precession

$$\omega_{\mathrm{p}} = \frac{m\,g\,R}{I\,\omega_{\mathrm{s}}}$$

Relation among the nutation θ, ω_{s}, and ω_{p}

$$mg\,R = I_x\,\omega_{\mathrm{p}}\,(\omega_{\mathrm{p}}\cos\theta + \omega_{\mathrm{s}}) - I_y\,\omega_{\mathrm{p}}^2\cos\theta$$

Compound pendulum

$$T = 2\pi\sqrt{\frac{I_{\mathrm{O}}}{mg\,R}}$$

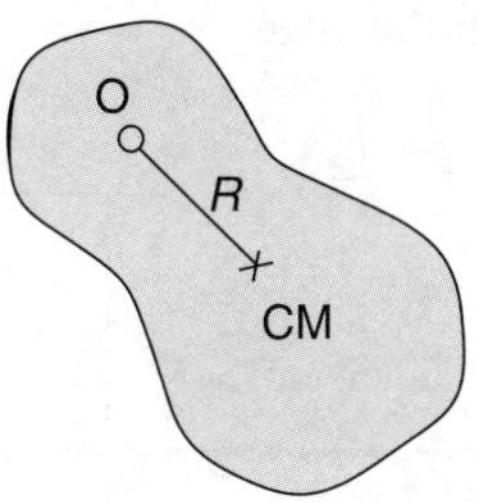

1.10 Moments of Inertia

1. Thin rod

$$I_x = \frac{1}{12}\,m\,a^2$$

$$I_{x'} = \frac{1}{3}\,m\,a^2$$

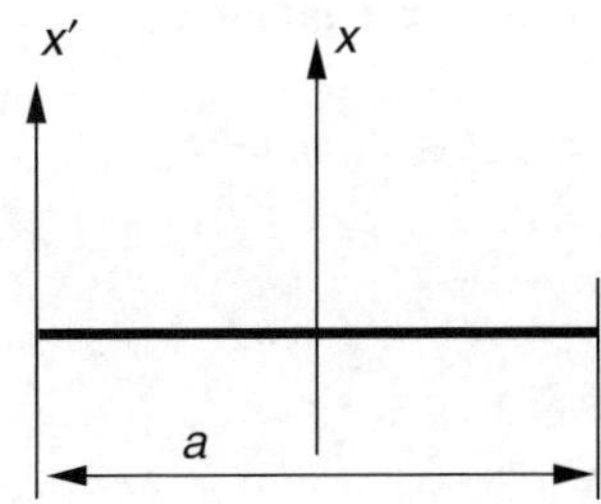

2. Rectangular block

$$I_x = \frac{1}{12} m (a^2 + b^2)$$

$$I_{x'} = \frac{1}{12} m (4 a^2 + b^2)$$

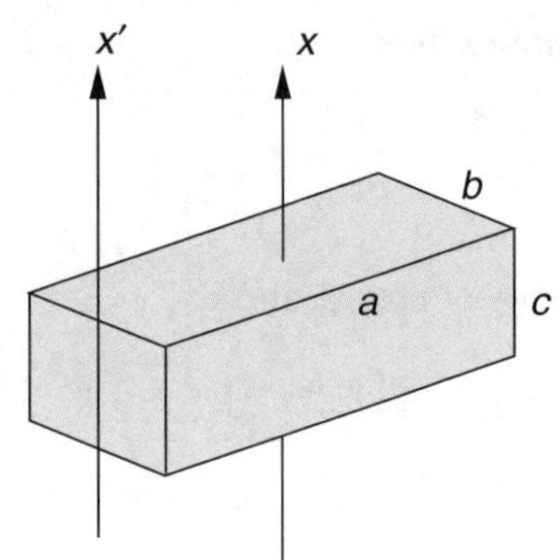

3. Thin, rectangular plate

$$I_x = \frac{1}{12} m a^2$$

$$I_z = \frac{1}{12} m (a^2 + b^2)$$

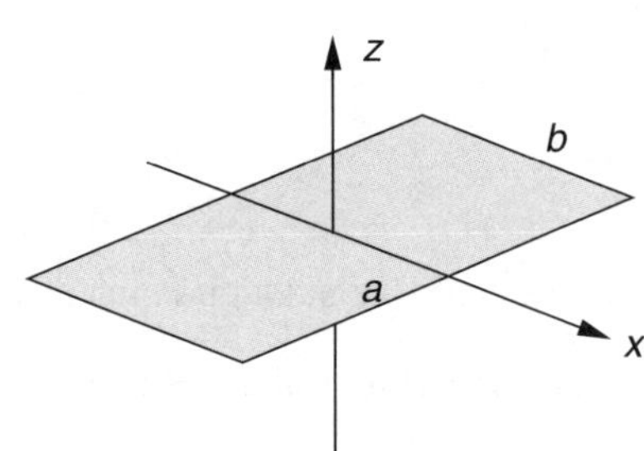

4. Triangular plate

$$I_x = \frac{m h^2}{6}$$

$$y_{CM} = \frac{h}{3}$$

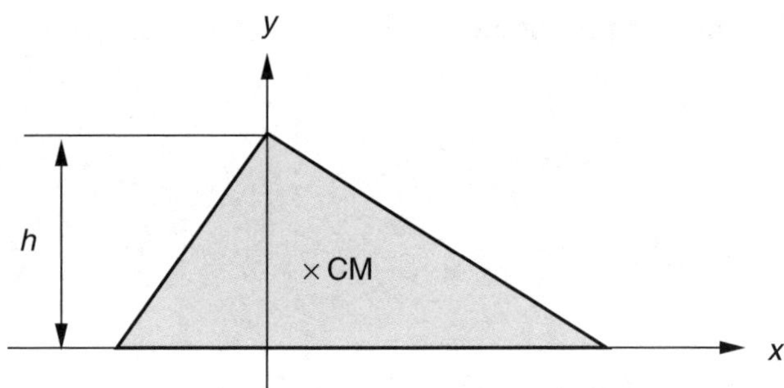

5. Isosceles triangle

$$I_x = \frac{1}{18} m h^2$$

$$I_y = \frac{1}{24} m b^2$$

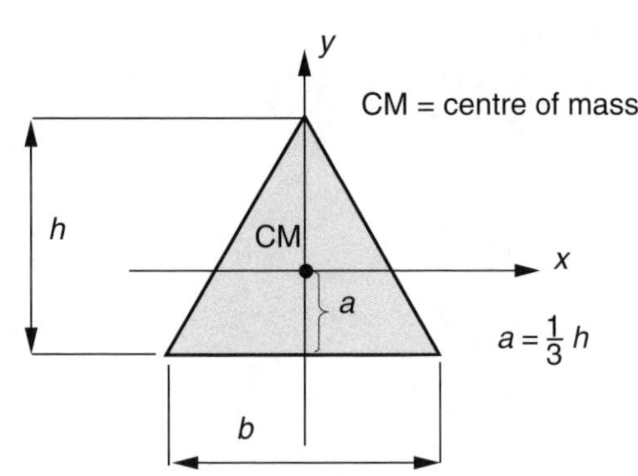

6. Cylinder

$$I_y = \frac{1}{2}\, m\, r^2$$

$$I_x = \frac{1}{12}\, m\, (h^2 + 3\, r^2)$$

$$I_{x'} = \frac{1}{12}\, m\, (4\, h^2 + 3\, r^2)$$

solid cylinder

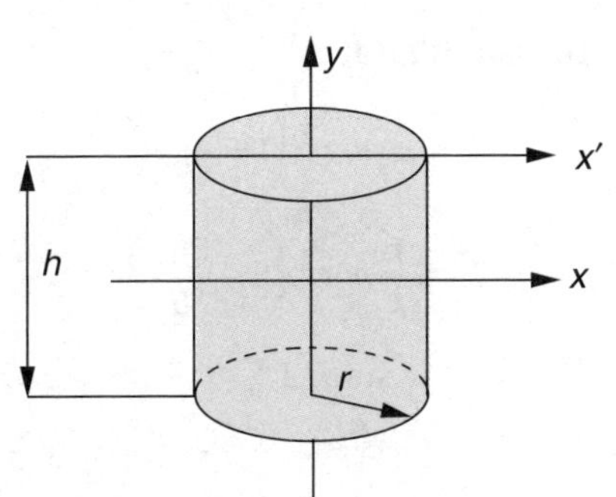

$$I_y = \frac{1}{2}\, m\, (r_1^2 + r_2^2)$$

$$I_x = \frac{1}{12}\, m\, (h^2 + 3\, r_1^2 + 3\, r_2^2)$$

$$I_{x'} = \frac{1}{12}\, m\, (4\, h^2 + 3\, r_1^2 + 3\, r_2^2)$$

cylinder shell of radii r_1 and r_2

7. Circular plate

$$I_x = I_y = \frac{1}{2}\, I_z = m\, \frac{r^2}{4}$$

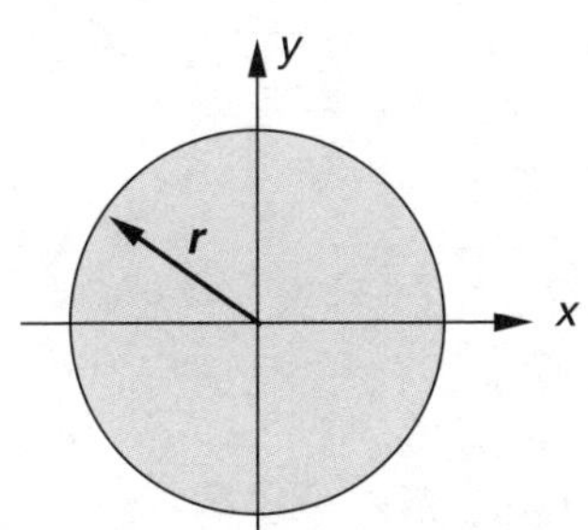

8. Circular ring

$$I_x = I_y = \frac{1}{2}\, I_z = \frac{1}{4}\, m\, (r_1^2 + r_2^2)$$

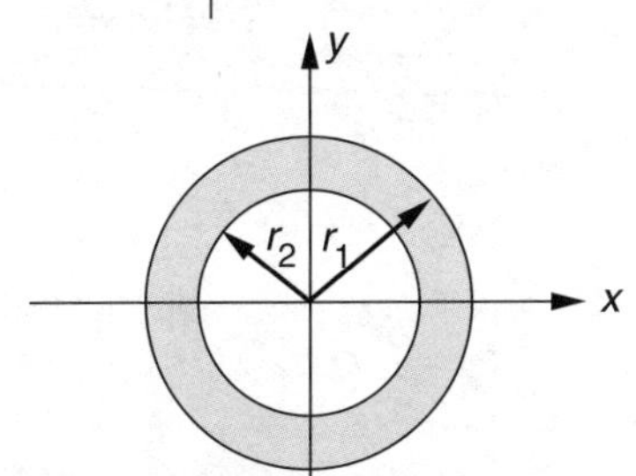

9. Ellipse

$$I_x = \frac{1}{4}\, m\, b^2$$

$$I_y = \frac{1}{4}\, m\, a^2$$

$$I_z = I_x + I_y$$

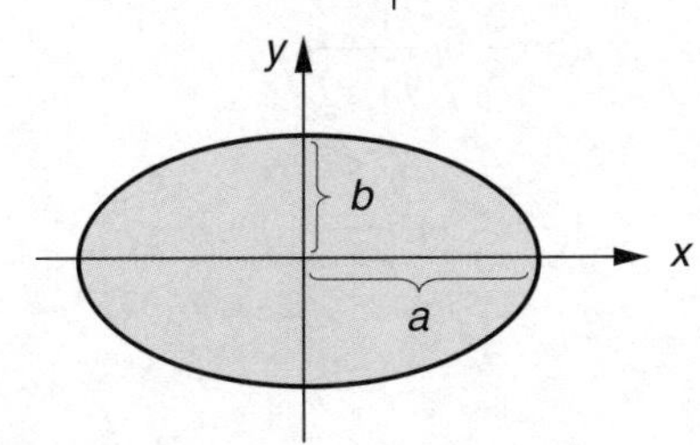

10. Circular arc

$$I_x = \frac{1}{2}\, m\, r^2 \left(1 + \frac{\sin\theta}{\theta}\right)$$

$$I_y = \frac{1}{2}\, m\, r^2 \left(1 - \frac{\sin\theta}{\theta}\right)$$

$$a = r\, \frac{\sin\theta/2}{\theta/2}$$

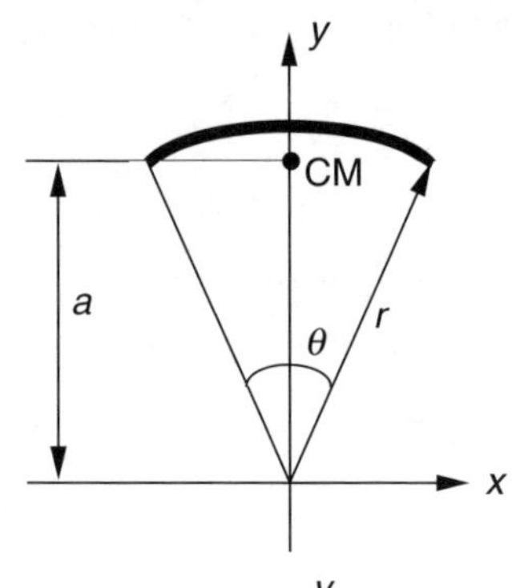

11. Circular sector

$$I_x = \frac{1}{4}\, m\, r^2 \left(1 + \frac{\sin\theta}{\theta}\right)$$

$$I_y = \frac{1}{4}\, m\, r^2 \left(1 - \frac{\sin\theta}{\theta}\right)$$

$$a = \frac{2}{3}\, r\, \frac{\sin\theta/2}{\theta/2}$$

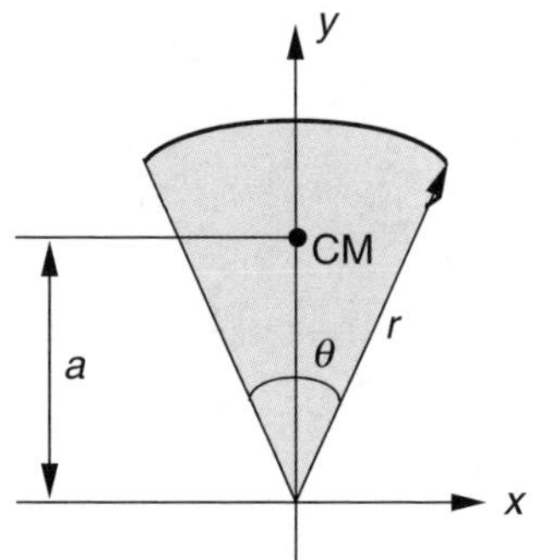

12. Sphere

$$I_x = \frac{2}{5}\, m\, r^2 \qquad I_{x'} = \frac{7}{5}\, m\, r^2$$

solid sphere

$$I_x = \frac{2}{3}\, m\, r^2 \qquad I_{x'} = \frac{5}{3}\, m\, r^2$$

spherical shell

$$I_x = \frac{2}{5}\, m\, \frac{r_1^5 - r_2^5}{r_1^3 - r_2^3} \qquad I_{x'} = \frac{2}{5}\, m\, \frac{r_1^5 - r_2^5}{r_1^3 - r_2^3} + m\, r_1^2$$

hollow sphere of outer radius r_1 and inner radius r_2

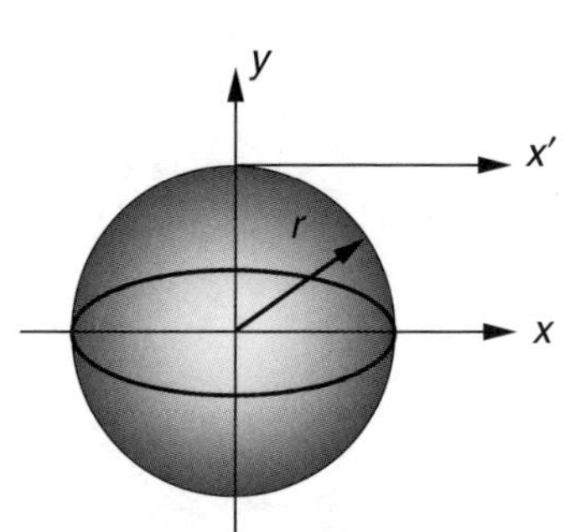

13. Ellipsoid

$$I_z = \frac{1}{5}\ m\,(a^2 + b^2)$$

$$I_{z'} = \frac{1}{5}\ m\,(6\,a^2 + b^2)$$

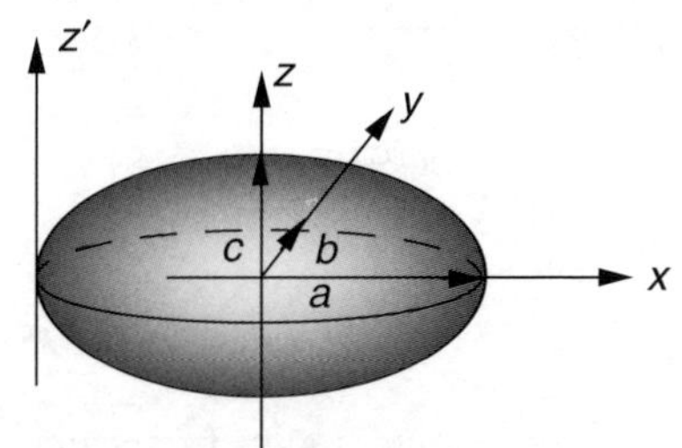

14. Cone

$$\left.\begin{aligned} I_x &= \frac{3}{80}\ m\,(4\,r^2 + h^2) \\ I_y &= \frac{3}{10}\ m\,r^2 \end{aligned}\right\}\ \text{solid cone}$$

$$I_y = \frac{1}{2}\ m\,r^2 \text{ open conical shell}$$

$$a = \begin{cases} \frac{1}{4}\,h & \text{solid cone} \\ \frac{1}{3}\,h & \text{open conical shell} \end{cases}$$

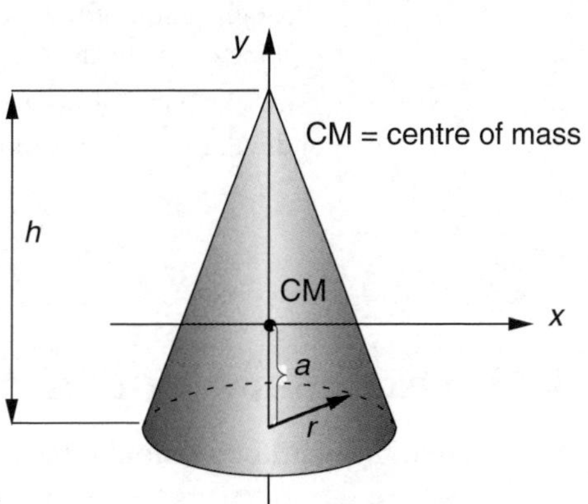

15. Spherical cap

$$I_y = \frac{1}{10}\ m\,h\,\frac{20\,r^2 - 15\,r\,h + 3\,h^2}{3\,r - h}\ \text{(solid cap)}$$

$$a = \begin{cases} \dfrac{3}{4}\,\dfrac{(2r-h)^2}{3r-h} & \text{(solid cap)} \\ r - \dfrac{1}{2}\,h & \text{(open cap)} \end{cases}$$

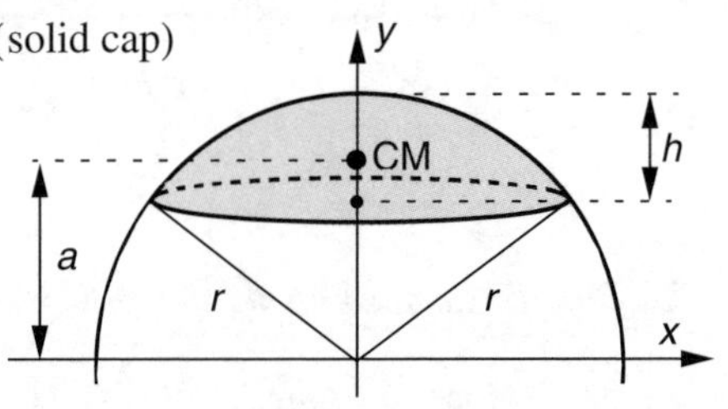

Centre of mass for a half sphere

$$a = \frac{3r}{8} \qquad \text{(solid half sphere)}$$

$$a = \frac{r}{2} \qquad \text{(half spherical shell)}$$

16. Torus

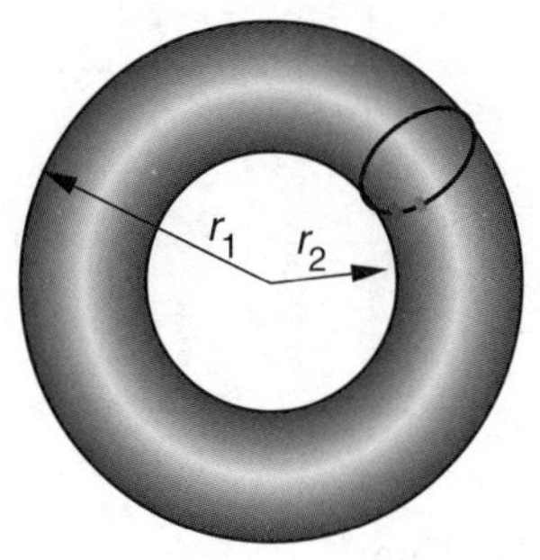

$$I_x = \frac{1}{4}\, m\,(9\, r_1^2 - 10\, r_1 r_2 + 5\, r_2^2)$$

$$I_y = \frac{1}{4}\, m\,(7\, r_1^2 - 6\, r_1 r_2 + 3\, r_2^2)$$

r_1 = outer radius from centre of mass

r_2 = inner radius from centre of mass

x = axis through center of mass and in the plane of the torus

y = axis through center of mass and perpendicular to plane of torus

1.11 Oscillatory Motion

Free oscillations without damping (harmonic oscillator)

Differential equation

$$\ddot{x} + \omega_0^2\, x = 0$$

Solution

$$x = A\cos(\omega_0\, t + \varphi)$$

$$x = A_1 \cos \omega_0\, t + A_2 \sin \omega_0\, t$$

$$x = A\, \mathrm{e}^{\mathrm{i}(\omega_0 t + \varphi)}$$

Example 1. Elastic (spring) force $F = -kx$ applied to mass m

$$T = \frac{2\pi}{\omega_0} = 2\pi\sqrt{\frac{m}{k}}$$

$$E_\mathrm{p} = \frac{1}{2}\, k\, x^2$$

$$E = \frac{1}{2}\, k\, A^2$$

Example 2. Simple pendulum (For compound pendulum, see F – 1.7)

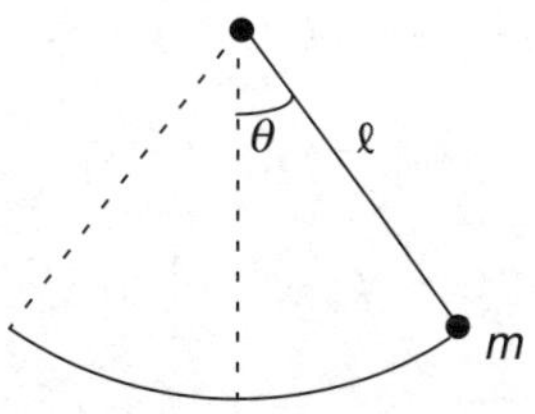

$$T = 2\pi\sqrt{\frac{\ell}{g}}$$

$$v^2 = 2g\,\ell\,(\cos\theta - \cos\theta_0)$$

$$S = mg(3\cos\theta - 2\cos\theta_0)$$

S = string tension

Example 3. Conical pendulum

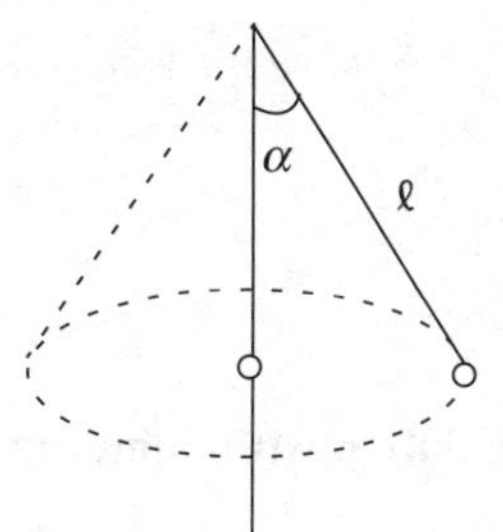

$$T = 2\pi\sqrt{\frac{\ell\cos\alpha}{g}}$$

$$v^2 = g\,\ell\,\tan\alpha\sin\alpha$$

$$S = \frac{mg}{\cos\alpha}$$

Example 4. Torsional pendulum

$$T = 2\pi\sqrt{\frac{I_0}{D}}$$

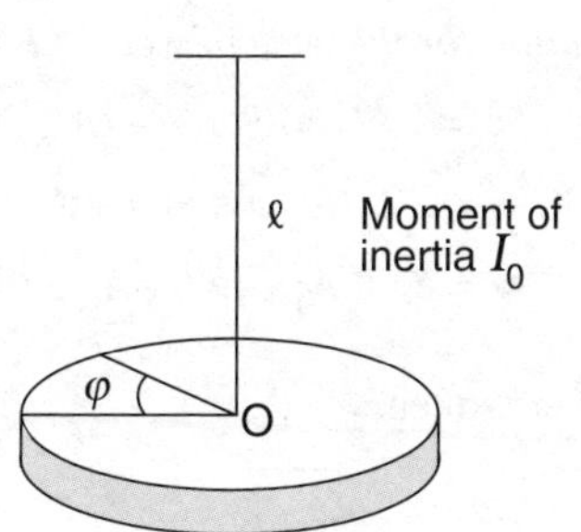

Restoring torque for a circular wire of radius r and length ℓ, being twisted an angle φ

$$M = \frac{\pi r^4}{2\,\ell}G\,\varphi = D\,\varphi$$

G = shear modulus of wire

D = torque constant

Superposed oscillations

$x_1 = A_1 \sin\omega_1 t$ $\qquad$ $x_2 = A_2 \sin\omega_2 t$

Amplitude for $x_1 + x_2$

$$A = (A_1{}^2 + A_2{}^2 + 2A_1A_2\cos(\omega_1 - \omega_2)t)^{1/2}$$

In particular, if $A_1 = A_2$

$$A = 2A_1\cos\tfrac{1}{2}(\omega_1 - \omega_2)t$$

Coupled oscillations

$$m_1 \ddot{x}_1 = -k_1 x_1 - k(x_1 - x_2)$$

$$m_2 \ddot{x}_2 = -k_2 x_2 + k(x_1 - x_2)$$

$$E_p = \frac{1}{2} k_1 x_1^2 + \frac{1}{2} k_2 x_2^2 + \frac{1}{2} k(x_1 - x_2)^2$$

Solution for $k_1 = k_2$ and $m_1 = m_2 = m$

$$x_1 = A_1 \sin(\omega_1 t + \varphi_1) + A_2 \sin(\omega_2 t + \varphi_2)$$

$$x_2 = A_1 \sin(\omega_1 t + \varphi_1) - A_2 \sin(\omega_2 t + \varphi_2)$$

$$\omega_1 = \sqrt{\frac{k_1}{m}} \qquad \omega_2 = \sqrt{\frac{2k + k_1}{m}}$$

Free oscillation with damping

Differential equation

$$\ddot{x} + 2\gamma \dot{x} + \omega_0^2 x = 0$$

Solution for the case of weak damping, $\gamma < \omega_0$

$$x = A\mathrm{e}^{-\gamma t} \cos(\omega_e t + \varphi)$$

$$x = \mathrm{e}^{-\gamma t} (A_1 \cos \omega_e t + A_2 \sin \omega_e t)$$

$$x = \mathrm{e}^{-\gamma t} (A_1 \mathrm{e}^{\mathrm{i}\omega_e t} + A_2 \mathrm{e}^{-\mathrm{i}\omega_e t})$$

Eigenfrequency

$$\omega_e = \sqrt{\omega_0^2 - \gamma^2}$$

Energy of oscillation

$$\langle E_k \rangle = \langle E_k \rangle_0 \, \mathrm{e}^{-2\gamma t}$$

Solution in the case of strong damping, $\gamma > \omega_0$

$$x = \mathrm{e}^{-\gamma t} (A_1 \mathrm{e}^{\omega_e' t} + A_2 \mathrm{e}^{-\omega_e' t})$$

$$\omega_e' = \sqrt{\gamma^2 - \omega_0^2}$$

Solution in the case of critical damping, $\gamma = \omega_0$

$$x = \mathrm{e}^{-\gamma t}(A_1 + A_2 t)$$

Forced Oscillations

Differential equation

$$\ddot{z} + 2\,\gamma\,\dot{z} + \omega_0^{\,2}\,z = \frac{F_0}{m}\,\mathrm{e}^{\mathrm{i}\omega t}$$

Particular solution

$$z = A\cos(\omega t - \varphi)$$

$$z = A\,\mathrm{e}^{\mathrm{i}(\omega t - \varphi)}$$

Amplitude of oscillation and phase displacement

$$A = \frac{F_0/m}{\sqrt{(\omega_0^2 - \omega^2)^2 + (2\,\gamma\,\omega)^2}} \approx \frac{A_0\,\gamma}{\sqrt{(\omega_0 - \omega)^2 + \gamma^2}}$$

$$\tan\varphi = \frac{2\,\gamma\,\omega}{\omega_0^2 - \omega^2}$$

Resonance amplitude, small damping

$$A_0 = \frac{F_0}{2\,m\,\omega_0\,\gamma}$$

Resonance amplification (sharpness of resonance, quality)

$$Q = \frac{\omega_0}{2\gamma}$$

Resonance frequency

$$\omega_r = \sqrt{\omega_0^2 - 2\,\gamma^2} \approx \omega_0$$

Resonance frequency when amplitude of applied function is proportional to ω (e.g. electric current in a series resonance circuit)

$$\omega_r = \omega$$

1.12 General Theory of Small Vibrations

Small vibrations about a stable equilibrium in u_0 for a particle which is free to move without friction along a curve $r(u)$, and acted on by a conservative force F

$$\frac{dE_p}{du} = -\boldsymbol{F} \cdot \frac{d\boldsymbol{r}}{du} \qquad \frac{dE_p}{du}(u_0) = 0$$

$$E_p \approx E_p(u_0) + \frac{1}{2}(u - u_0)^2 \alpha \qquad \alpha = \frac{d^2E_p}{du^2}(u_0) > 0$$

$$E_k \approx \frac{1}{2}\dot{u}^2 \beta \qquad \beta = m\left(\frac{d\boldsymbol{r}}{du}(u_0)\right)^2$$

Angular frequency of vibration

$$\omega \approx \sqrt{\frac{\alpha}{\beta}}$$

Equations for a holonomous system of particles acted on by conservative forces with time-independent constraints

$$E_k = \frac{1}{2}\dot{\boldsymbol{q}} \cdot \boldsymbol{T} \cdot \dot{\boldsymbol{q}} = \frac{1}{2}\Sigma_{\nu\mu} T_{\nu\mu} \dot{q}_\nu \dot{q}_\mu$$

q_ν = generalized coordinates

Components of tensor $\boldsymbol{T}$

$$T_{\nu\mu} = \Sigma_i m_i \frac{\partial \mathbf{r}_i}{\partial q_\nu} \cdot \frac{\partial \mathbf{r}_i}{\partial q_\mu}$$

$$E_p = \frac{1}{2}\boldsymbol{q} \cdot \mathbf{V} \cdot \boldsymbol{q} = \frac{1}{2}\Sigma_{\nu\mu} V_{\nu\mu} q_\nu q_\mu$$

Components of tensor $\boldsymbol{V}$

$$V_{\nu\mu} = \frac{\partial^2 E_p}{\partial q_\nu \partial q_\mu}$$

Equation of motion

$$\boldsymbol{T} \cdot \ddot{\boldsymbol{q}} + \boldsymbol{V} \cdot \boldsymbol{q} = \mathbf{0}$$

For the case of two degrees of freedom

$$T_{11}\ddot{q}_1 + T_{12}\ddot{q}_2 + V_{11}q_1 + V_{12}q_2 = 0$$

$$T_{21}\ddot{q}_1 + T_{22}\ddot{q}_2 + V_{21}q_1 + V_{22}q_2 = 0$$

1.13 Theory of Relativity and Relativistic Mechanics

The Lorentz transformation and its inverse (the x'-axis moves along the x-axis with speed v)

$$x' = \frac{x - vt}{\sqrt{1 - v^2/c^2}} \qquad x = \frac{x' + vt'}{\sqrt{1 - v^2/c^2}}$$

$$y' = y \qquad y = y'$$

$$z' = z \qquad z = z'$$

$$t' = \frac{t - \frac{v}{c^2}x}{\sqrt{1 - v^2/c^2}} \qquad t = \frac{t' + \frac{v}{c^2}x'}{\sqrt{1 - v^2/c^2}}$$

Addition of velocities ($u_x = \frac{\mathrm{d}x}{\mathrm{d}t}$, etc.)

$$u_{x'} = \frac{u_x - v}{1 - vu_x/c^2} \qquad u_x = \frac{u_{x'} + v}{1 + vu_{x'}/c^2}$$

$$u_{y'} = u_y \frac{\sqrt{1 - v^2/c^2}}{1 - vu_x/c^2} \qquad u_y = u_{y'} \frac{\sqrt{1 - v^2/c^2}}{1 + vu_{x'}/c^2}$$

$$u_{z'} = u_z \frac{\sqrt{1 - v^2/c^2}}{1 - vu_x/c^2} \qquad u_z = u_{z'} \frac{\sqrt{1 - v^2/c^2}}{1 + vu_{x'}/c^2}$$

Relative velocity

$$v_{\text{rel}}^2 = \frac{(\boldsymbol{v}_1 - \boldsymbol{v}_2)^2 - \frac{1}{c^2}\left|\boldsymbol{v}_1 \times \boldsymbol{v}_2\right|^2}{\left(1 - \frac{1}{c^2}\boldsymbol{v}_1 \cdot \boldsymbol{v}_2\right)^2}$$

Length contraction for fixed t

$$x_2 - x_1 = (x_2' - x_1')\sqrt{1 - v^2/c^2}$$

Time dilation for fixed x'

$$t_2 - t_1 = \gamma(t_2' - t_1') \qquad \gamma = \frac{1}{\sqrt{1 - v^2 / c^2}}$$

Total energy, relativistic momentum and velocity of a particle

$$E_{\text{tot}} = E_{\text{k}} + m\,c^2 = \gamma m\,c^2$$

$$E_{\text{tot}}^2 = (p\,c)^2 + (m\,c^2)^2$$

$$\boldsymbol{p} = \gamma m \boldsymbol{v}$$

$$p = \frac{1}{c}\sqrt{E_{\text{k}}^2 + 2\,E_{\text{k}}\,m\,c^2}$$

$$\boldsymbol{v} = \frac{c^2}{E_{\text{tot}}}\,\boldsymbol{p} \qquad v = c\,\sqrt{1 - (m\,c^2 / E_{\text{tot}})^2}$$

The quantity γm is often, but somewhat inadvertently, called “relativistic mass”. At zero velocity it equals m, which is referred to as “rest mass” and denoted m_0.

Total mass (Lorentz invariant mass) of a system of particles

$$M^2\,c^4 = \left(\sum_n E_{n\ \text{tot}}\right)^2 - \left(\sum_n \boldsymbol{p}_n\right)^2 c^2$$

Newton’s second law

$$\frac{\mathrm{d}}{\mathrm{d}t}\,\boldsymbol{p} = \boldsymbol{F}$$

1.14 General Relativity

In this section the metre is used as unit for time.
One metre of time = the time it takes light to travel one metre.
Thus $c = 1$ (dimensionless) and $G \approx 7.4242 \cdot 10^{-28}$ m/kg.

Notation

$$\gamma = \frac{1}{\sqrt{1-v^2}} \qquad \text{(Since } c = 1\text{)}$$

A four-vector

$$\Delta\vec{x} = (\Delta t, \Delta x, \Delta y, \Delta z) = (\Delta x^0, \Delta x^1, \Delta x^2, \Delta x^3) = \{\Delta x^\alpha\}$$

Bold characters denote three-vectors.
~ above a symbol denotes a one-form (a covector, a covariant vector).

Example The gradient $\tilde{\mathrm{d}}\phi = \left(\frac{\partial\phi}{\partial t}, \frac{\partial\phi}{\partial x}, \frac{\partial\phi}{\partial y}, \frac{\partial\phi}{\partial z}\right)$

Indices on vectors are up (V^α), on one-forms down (V_α).

An $\binom{M}{N}$ tensor is a linear function of M one-forms and N vectors into real numbers.

Derivatives: $\phi_{,\alpha} \equiv \partial_\alpha \phi \equiv \dfrac{\partial\phi}{\partial x^\alpha}$

The Einstein summation convention
Whenever an expression contains one index as a superscript and the same index as a subscript, a summation is implied over all values that the index can take.

Example The Lorentz transformation in Cartesian coordinates

$$\Delta x^{\bar{\alpha}} = \Lambda^{\bar{\alpha}}{}_{\beta} \Delta x^\beta \qquad \Lambda = \begin{pmatrix} \gamma & -v\gamma & 0 & 0 \\ -v\gamma & \gamma & 0 & 0 \\ 0 & 0 & 1 & 0 \\ 0 & 0 & 0 & 1 \end{pmatrix}$$

Invariant interval between any two events

$$\Delta s^2 = -(\Delta t)^2 + (\Delta x)^2 + (\Delta y)^2 + (\Delta z)^2$$

Metric η on Minkowski space in Cartesian coordinates

$$ds^2 = \eta_{\mu\nu} dx^\mu dx^\nu = -(dt)^2 + (dx)^2 + (dy)^2 + (dz)^2$$

$$V_\alpha = \eta_{\alpha\beta} V^\beta \qquad V^\beta = \eta^{\beta\alpha} V_\alpha \qquad \eta^{\beta\alpha} = \eta_{\alpha\beta} = \begin{pmatrix} -1 & 0 & 0 & 0 \\ 0 & 1 & 0 & 0 \\ 0 & 0 & 1 & 0 \\ 0 & 0 & 0 & 1 \end{pmatrix}$$

Four-velocity and four-momentum

$\vec{U} = \dfrac{\partial \vec{x}}{\partial \tau}$ — τ = proper time, $d\tau^2 = -ds^2$

m = rest mass

$\vec{p} = m\vec{U} = (E, p^1, p^2, p^3)$ — E = total energy

Exempel 1 A particle moves with velocity $\boldsymbol{v}$ in the x direction

$\vec{U} = (\gamma, v\gamma, 0, 0)$ $\qquad$ $\vec{p} = (m\gamma, mv\gamma, 0, 0)$

Example 2 Photons

$\vec{U}$ is not defined since $d\tau = 0$.

$\vec{p} \cdot \vec{p} = 0$ $\qquad$ $p^0 = E = h\nu$

Doppler shift for photons

A photon has frequency ν in frame $\mathcal{O}$ and moves in the x direction.

Frame $\bar{\mathcal{O}}$ moves with velocity v in the x direction relative $\mathcal{O}$.

$$\frac{\bar{\nu}}{\nu} = \frac{(1-v)}{\sqrt{1-v^2}}$$ (In units where $c = 1$)

Tensor algebra and tensor calculus in polar coordinates (r, θ)

$x = r\cos\theta$ $\qquad$ $r = (x^2 + y^2)^{1/2}$

$y = r\sin\theta$ $\qquad$ $\theta = \arctan\dfrac{y}{x}$

Basis vectors $\vec{e}_\alpha$

$$(\vec{e}_r, \vec{e}_\theta) = \begin{pmatrix} \dfrac{\partial x}{\partial r} & \dfrac{\partial y}{\partial r} \\ \dfrac{\partial x}{\partial \theta} & \dfrac{\partial y}{\partial \theta} \end{pmatrix} \begin{pmatrix} \vec{e}_x \\ \vec{e}_y \end{pmatrix}$$

$\vec{e}_r = \cos\theta\, \vec{e}_x + \sin\theta\, \vec{e}_y$ $\qquad$ $|\vec{e}_r|^2 = 1$

$\vec{e}_\theta = -r\sin\theta\, \vec{e}_x + r\cos\theta\, \vec{e}_y$ $\qquad$ $|\vec{e}_\theta|^2 = r^2$

One-forms

$$\tilde{\mathrm{d}}\theta = -\frac{1}{r}\sin\theta\,\tilde{\mathrm{d}}x + \frac{1}{r}\cos\theta\,\tilde{\mathrm{d}}y$$

$$\tilde{\mathrm{d}}r = \cos\theta\,\tilde{\mathrm{d}}x + \sin\theta\,\tilde{\mathrm{d}}y$$

Derivatives of basis vectors

$$\frac{\partial\vec{e}_r}{\partial r} = 0 \qquad \frac{\partial\vec{e}_r}{\partial\theta} = -\sin\theta\,\vec{e}_x + \cos\theta\,\vec{e}_y = \frac{1}{r}\,\vec{e}_\theta$$

$$\frac{\partial\vec{e}_\theta}{\partial r} = -\sin\theta\,\vec{e}_x + \cos\theta\,\vec{e}_y = \frac{1}{r}\,\vec{e}_\theta \qquad \frac{\partial\vec{e}_\theta}{\partial\theta} = -r\cos\theta\,\vec{e}_x - r\sin\theta\,\vec{e}_y = -\,r\vec{e}_r$$

Derivatives of any vector $\vec{V} = V^r\vec{e}_r + V^\theta\vec{e}_\theta$

$$\frac{\partial\vec{V}}{\partial r} = \frac{\partial V^r}{\partial r}\,\vec{e}_r + V^r\frac{\partial\vec{e}_r}{\partial r} + \frac{\partial V^\theta}{\partial r}\,\vec{e}_\theta + V^\theta\frac{\partial\vec{e}_\theta}{\partial r}$$

$$\frac{\partial\vec{V}}{\partial\theta} = \frac{\partial V^r}{\partial\theta}\,\vec{e}_r + V^r\frac{\partial\vec{e}_r}{\partial\theta} + \frac{\partial V^\theta}{\partial\theta}\,\vec{e}_\theta + V^\theta\frac{\partial\vec{e}_\theta}{\partial\theta}$$

Metric tensor

$$g_{\alpha\beta} = \mathbf{g}(\vec{e}_\alpha, \vec{e}_\beta) = \vec{e}_\alpha\cdot\vec{e}_\beta$$

Definition of the inverse $g^{\alpha\beta}$

$$g^{\alpha\nu}g_{\nu\beta} = \delta^\alpha_\beta \qquad \delta^\alpha_\beta = \begin{pmatrix} 1 & 0 & 0 & 0 \\ 0 & 1 & 0 & 0 \\ 0 & 0 & 1 & 0 \\ 0 & 0 & 0 & 1 \end{pmatrix}$$

Properties

$$V_\alpha = g_{\alpha\beta}V^\beta \qquad V^\alpha = g^{\alpha\beta}V_\beta$$

Connection coefficients (Christoffel symbol) Γ

$$\frac{\partial\vec{e}_\alpha}{\partial x^\beta} = \Gamma^\mu{}_{\alpha\beta}\vec{e}_\mu$$

General expressions in any coordinate system

$$\Gamma^\mu{}_{\alpha\beta} = \Gamma^\mu{}_{\beta\alpha}$$

$$\Gamma^{\mu}{}_{\alpha\beta} = \frac{1}{2} g^{\gamma\mu}(g_{\gamma\alpha,\beta} + g_{\gamma\beta,\alpha} - g_{\alpha\beta,\gamma})$$

Especially for polar coordinates

$$g^{rr} = 1,\ g^{r\theta} = g^{\theta r} = 0,\ g^{\theta\theta} = \frac{1}{r} \qquad g_{rr} = 1,\ g_{\theta\theta} = r^2,\ g_{r\theta} = g_{\theta r} = 0$$

$$\Gamma^{r}{}_{rr} = \Gamma^{\theta}{}_{rr} = 0 \qquad \Gamma^{r}{}_{r\theta} = 0,\ \Gamma^{\theta}{}_{r\theta} = \frac{1}{r}$$

$$\Gamma^{r}{}_{\theta r} = 0,\ \Gamma^{\theta}{}_{\theta r} = \frac{1}{r} \qquad \Gamma^{r}{}_{\theta\theta} = -r,\ \Gamma^{\theta}{}_{\theta\theta} = 0$$

Covariant derivative

$$\frac{\partial \vec{V}}{\partial x^{\beta}} = \frac{\partial V^{\alpha}}{\partial x^{\beta}} \vec{e}_{\alpha} + V^{\alpha} \frac{\partial \vec{e}_{\alpha}}{\partial x^{\beta}} = \frac{\partial V^{\alpha}}{\partial x^{\beta}} \vec{e}_{\alpha} + V^{\mu} \Gamma^{\alpha}{}_{\mu\beta} \vec{e}_{\alpha}$$

Symbols ; and ∇

$$V^{\alpha}{}_{;\beta} \equiv (V^{\alpha}{}_{,\beta} + V^{\mu}\Gamma^{\alpha}{}_{\mu\beta}) \qquad V^{\alpha}{}_{;\beta} = (\nabla_{\beta}\vec{V})^{\alpha} = (\nabla\vec{V})^{\alpha}{}_{\beta}$$

General expressions

$$\frac{\partial \vec{V}}{\partial x^{\beta}} = V^{\alpha}{}_{;\beta} \vec{e}_{\alpha} \qquad \nabla_{\beta}\tilde{V} = \mathbf{g}(\nabla_{\beta}\vec{V},\quad) \text{ if } \tilde{V} = \mathbf{g}(\vec{V},\quad)$$

$$\text{or } V_{\alpha;\beta} = g_{\alpha\mu} V^{\mu}{}_{;\beta}$$

Riemann curvature tensor R

$$R^{\alpha}{}_{\beta\mu\nu} \equiv \Gamma^{\alpha}{}_{\beta\nu,\mu} - \Gamma^{\alpha}{}_{\beta\mu,\nu} + \Gamma^{\alpha}{}_{\sigma\mu}\ \Gamma^{\sigma}{}_{\beta\nu} - \Gamma^{\alpha}{}_{\sigma\nu}\ \Gamma^{\sigma}{}_{\beta\mu}$$

$$R_{\alpha\beta\mu\nu} \equiv g_{\alpha\lambda} R^{\lambda}{}_{\beta\mu\nu} = \frac{1}{2}(g_{\alpha\nu,\beta\mu} - g_{\alpha\mu,\beta\nu} + g_{\beta\mu,\alpha\nu} - g_{\beta\nu,\alpha\mu})$$

Properties in a locally flat frame

$$R_{\alpha\beta\mu\nu} = -R_{\beta\alpha\mu\nu} = -R_{\alpha\beta\nu\mu} = R_{\mu\nu\alpha\beta}$$

$$R_{\alpha\beta\mu\nu} + R_{\alpha\nu\beta\mu} + R_{\alpha\mu\nu\beta} = 0$$

Bianchi identity

$$R_{\alpha\beta\mu\nu;\lambda} + R_{\alpha\beta\lambda\mu;\nu} + R_{\alpha\beta\nu\lambda;\mu} = 0$$

The Ricci tensor $R_{\mu\nu}$ and the Ricci scalar R

$$R_{\mu\nu} = R^{\sigma}{}_{\mu\sigma\nu} = g^{\sigma\lambda} R_{\mu\sigma\nu\lambda} \qquad R = g^{\mu\lambda} R_{\mu\lambda}$$

The Einstein tensor G and the contracted Bianchi identity

$$G_{\mu\nu} = R_{\mu\nu} - \frac{1}{2} g_{\mu\nu} R \qquad G^{\mu\nu}{}_{;\nu} = 0$$

Einstein's field equations

$$G_{\mu\nu} + \Lambda g_{\mu\nu} = 8\pi G T_{\mu\nu}$$

$T_{\mu\nu}$ = energy-momentum tensor, sometimes called stress-energy tensor

Λ = cosmological constant

Especially for a perfect fluid in a flat space

$T^{00} = \rho$ = energy density $\qquad T^{\mu\nu} = 0$ for $\mu \neq \nu$

$T^{ii} = p$ = pressure, for i = 1, 2, 3. $\qquad p = \alpha\rho$, where

$\alpha \approx 0$ for (non-relativistic) matter, $\alpha = 1/3$ for radiation (photons) and relativistic matter (neutrinos etc), and $\alpha = -1$ for vacuum energy

Geodesic equation of freely falling bodies

$$\ddot{x}^{\mu} + \dot{x}^{\nu}\dot{x}^{\lambda}\Gamma^{\mu}{}_{\nu\lambda} = 0 \qquad \text{where } \dot{x}^{\nu} = \frac{\mathrm{d}}{\mathrm{d}\tau}x^{\nu}$$

$$\frac{\mathrm{d}}{\mathrm{d}\tau}(g_{\mu\nu}\dot{x}^{\nu}) = \frac{1}{2} g_{\lambda\nu,\mu}\dot{x}^{\lambda}\dot{x}^{\nu}$$

Schwarzschild metric in vaccum outside a spherical object

$$\mathrm{d}s^2 = -\left(1 - \frac{2MG}{r}\right)\mathrm{d}t^2 + \frac{1}{\left(1 - \frac{2MG}{r}\right)}\mathrm{d}r^2 + r^2\,\mathrm{d}\theta^2 + r^2\sin^2\theta\,\mathrm{d}\phi^2$$

(r, θ, ϕ) = spherical coordinates, see Sec M–10

M = mass of object

G = gravitational constant

Effective potential outside the object

$$\tilde{V}^2(r) = \left(1 - \frac{2MG}{r}\right)\left(1 + \frac{\tilde{L}^2}{r^2}\right) \qquad \text{for particles, } \tilde{L} = \frac{p_{\phi}}{m}$$

m = rest mass of particle

$$V^2(r) = \left(1 - \frac{2MG}{r}\right)\frac{L^2}{r^2} \qquad \text{for photons, } L = p_{\phi}$$

Perihelion shift, in radians per orbit, for a nearly circular orbit

$$\Delta\phi_{\mathrm{p}} \approx \frac{6\pi MG}{r_0} \qquad r_0 = \text{radius of orbit} \gg 3MG$$

Deflection of light close to the object

$$\Delta\phi_{\mathrm{d}} \approx \frac{4MG}{b} \qquad b = \text{impact parameter} = r_{\min} \gg MG$$

Kruskal-Szekeres coordinates

$$\left.\begin{aligned} u &= \left(\frac{r}{2MG} - 1\right)^{1/2} \mathrm{e}^{r/4MG} \cosh\frac{t}{4MG} \\ v &= \left(\frac{r}{2MG} - 1\right)^{1/2} \mathrm{e}^{r/4MG} \sinh\frac{t}{4MG} \end{aligned}\right\} \quad \text{for } r > 2MG$$

$$\left.\begin{aligned} u &= \left(1 - \frac{r}{2MG}\right)^{1/2} \mathrm{e}^{r/4MG} \sinh\frac{t}{4MG} \\ v &= \left(1 - \frac{r}{2MG}\right)^{1/2} \mathrm{e}^{r/4MG} \cosh\frac{t}{4MG} \end{aligned}\right\} \quad \text{for } r < 2MG$$

Properties

$$\mathrm{d}s^2 = \frac{32(MG)^3}{r} \mathrm{e}^{r/2MG}(\mathrm{d}u^2 - \mathrm{d}v^2) + r^2\,\mathrm{d}\theta^2 + r^2 \sin^2\theta\,\mathrm{d}\phi^2$$

$$u^2 - v^2 = \left(\frac{r}{2MG} - 1\right)\mathrm{e}^{r/2MG} \qquad \frac{v}{u} = \tanh\left(\frac{t}{4MG}\right)$$

Kerr metric for a Kerr black hole with angular momentum L

$$\mathrm{d}s^2 = -\frac{(\Delta - a^2 \sin^2\theta)}{\rho^2}\,\mathrm{d}t^2 - 2a\frac{2MGr\sin^2\theta}{\rho^2}\,\mathrm{d}t\,\mathrm{d}\phi +$$

$$+ \frac{(r^2 + a^2)^2 - a^2\Delta\sin^2\theta}{\rho^2} \sin^2\theta\,\mathrm{d}\phi^2 + \frac{\rho^2}{\Delta}\,\mathrm{d}r^2 + \rho^2\,\mathrm{d}\theta^2$$

$$a = \frac{L}{M}$$

$$\Delta = r^2 - 2MGr + a^2$$

$$\rho^2 = r^2 + a^2\cos^2\theta$$

Robertson-Walker metrics for a homogeneous and isotropic universe

$$\mathrm{d}s^2 = -\mathrm{d}t^2 + R^2(t)\left(\frac{\mathrm{d}r^2}{1 - \kappa r^2} + r^2\,\mathrm{d}\theta^2 + r^2\sin^2\theta\,\mathrm{d}\phi^2\right)$$

$R(t)$ = scale factor of the universe

Example 1 $\kappa = 0$: a flat Euclidean space

Example 2 $\kappa = +1$: a closed, or spherical, Robertson-Walker metric

$$ds^2 = -dt^2 + R^2(t)\,(d\chi^2 + \sin^2\chi\, d\theta^2 + \sin^2\chi \sin^2\theta\, d\phi^2)$$

$$d\chi^2 = \frac{dr^2}{1 - r^2} \qquad r = \sin\chi$$

Example 3 $\kappa = -1$: an open, or hyperbolic, Robertson-Walker metric

$$ds^2 = -dt^2 + R^2(t)\,(d\chi^2 + \sinh^2\chi\, d\theta^2 + \sinh^2\chi \sin^2\theta\, d\phi^2)$$

$$d\chi^2 = \frac{dr^2}{1 + r^2} \qquad r = \sinh\chi$$

Friedmann's equation

$$\dot{R}^2 = \frac{8}{3}\pi G\rho R^2 + \frac{1}{3}\Lambda R^2 - \kappa \qquad R = R(t) = \text{scale factor of the universe}$$

More information:
http://scienceworld.wolfram.com/physics/FriedmannsEquation.html 

Friedmann's acceleration equation for $\kappa = 0$

$$\frac{\ddot{R}}{R} = -\frac{4\pi G}{3}(\rho + 3p) \qquad p = \alpha\rho = \text{pressure}$$

Derived equation

$$\dot{\rho} = -3\frac{\dot{R}}{R}(\rho + p)$$

Deceleration parameter q and Hubble's parameter H

$$q = -\frac{R\ddot{R}}{\dot{R}^2} \qquad H = \frac{\dot{R}}{R}$$

Cosmological redshift z

$$1 + z = \frac{\lambda(t)}{\lambda(t_0)} = \frac{R(t)}{R(t_0)}$$

Energy density when $\kappa = 0$

$$\rho = \rho_0 / R^{3(1+\alpha)}$$

Matter-dominated universe

$$\alpha \approx 0 \qquad \rho = \rho_0 / R^3 \qquad R \sim t^{2/3}$$

Radiation-dominated universe

$$\alpha = 1/3 \qquad \rho = \rho_0 / R^4 \qquad R \sim t^{1/2}$$

Vacuum-dominated universe

$$\alpha = -1 \qquad \rho = \rho_0 \qquad R \sim e^{\sqrt{\frac{1}{3}\Lambda t}} = e^{Ht}$$

More information on general relativity:
http://arxiv.org/abs/gr-qc/9712019
http://preposterousuniverse.com/grnotes/

1.15 Elasticity

Strain in a rod (Hooke's law)

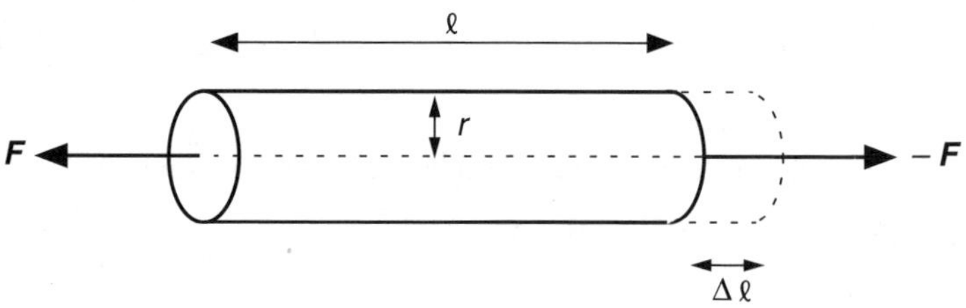

Young's modulus

$$Y = \frac{F/A}{\Delta \ell / \ell} = \frac{\text{normal stress}}{\text{strain}}$$

A = cross-sectional area

Work required to prolong a rod $\Delta \ell$

$$W = \frac{Y A (\Delta \ell)^2}{2 \ell}$$

Poisson's constant for transverse contraction

$$\nu = -\frac{\Delta r / r}{\Delta \ell / \ell} = (-1)\,\frac{\text{transverse strain}}{\text{ongitudinal strain}}$$

$$0 \le \nu \le \frac{1}{2}$$

Fractional change of volume at uni-axial loading

$$\frac{\Delta V}{V} = (1 - 2\nu)\,\frac{\Delta \ell}{\ell}$$

Shear

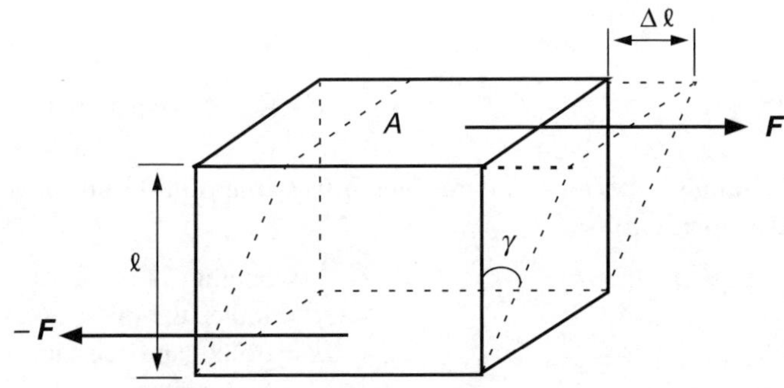

Shear modulus

$$G = \frac{F/A}{\Delta\ell/\ell} = \frac{F/A}{\tan\gamma} = \frac{\text{shear stress}}{\text{shear strain}}$$

A = area of application

Hydrostatic compression

Hydrostatic pressure

$$p = \frac{F}{A}$$ A = body area

Bulk modulus

$$B = -\frac{p}{\Delta V/V} = (-1)\,\frac{\text{hydrostatic pressure}}{\text{fractional volume change}}$$

Compressibility (isothermal)

$$\kappa_T = \frac{1}{B} = -\frac{1}{V}\left(\frac{\partial V}{\partial p}\right)_T$$ T = temperature

Relations among elastic constants for isotropic materials

$$G = \frac{Y}{2(1+\nu)}$$

$$B = \frac{Y}{3(1-2\nu)}$$

$$Y = \frac{9\,GB}{G+3B}$$

1.16 Fluid Mechanics

Pressure

$$p = \frac{F}{A}$$

F = total force perpendicular to area A

Vertical change in pressure for an incompressible liquid and for an ideal gas (barometer formula)

$$p = p_0 - \rho\, g\, h$$

$$p = p_0 \exp\left(-\frac{\mu\, g\, h}{k\, T}\right)$$

h = height
μ = molecular mass
k = Boltzmann constant
T = temperature

Bernoulli's theorem for an ideal fluid flowing at speed v

$$p + \frac{1}{2}\,\rho\, v^2 + \rho\, g\, h = \text{constant}$$

Rate of flow through an opening of area A

$$\frac{\mathrm{d}V}{\mathrm{d}t} = A\, v$$

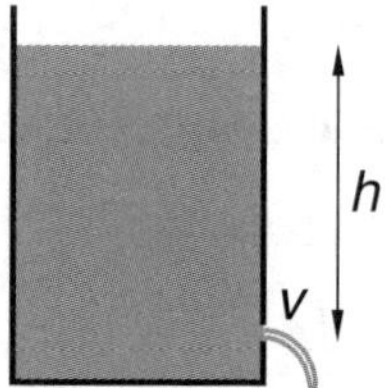

Velocity of liquid emerging from a container (Torricelli's principle)

$$v = \sqrt{2\, g\, h}$$

Differential equation for a Newtonian liquid flowing in the y-direction with speed $v_y = v_y(x,t)$

$$\frac{\partial v_y}{\partial t} = \frac{\eta}{\rho}\,\frac{\partial^2 v_y}{\partial x^2} + \frac{f}{\rho}$$

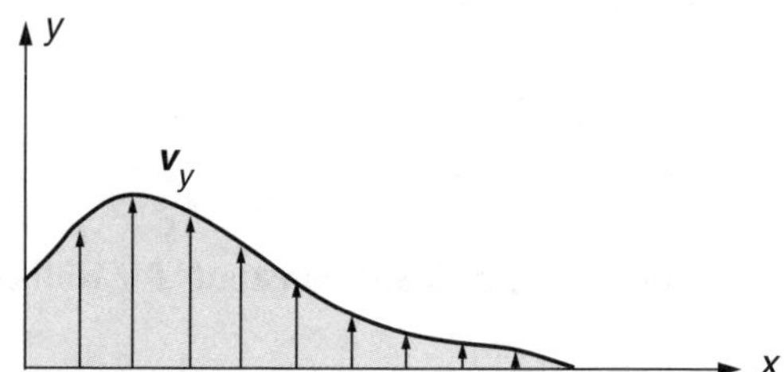

f = external force in y-direction per volume

η = coefficient of viscosity

Tangential tension between two layers

$$\tau(x) = -\,\eta\,\frac{\partial v_y}{\partial x}$$

Poiseuille's formula for stationary flow through a tube of radius R and length ℓ

$$\frac{dV}{dt} = \frac{\pi R^4}{8\eta}\frac{\Delta p}{\ell}$$

Δp = pressure difference between ends of tube

$\frac{dV}{dt}$ = discharged volume per time (rate of flow)

Rate of flow at radius r

$$v = \frac{\Delta p}{4\ell\eta}(R^2 - r^2)$$

Equations of continuity and energy for stationary flow

$$A_1 v_1 \rho_1 = A_2 v_2 \rho_2$$

$$q = w + (h_2 + \tfrac{1}{2}v_2^2 + gz_2) - (h_1 + \tfrac{1}{2}v_1^2 + gz_1)$$

q = supplied energy per mass

w = work done by system per mass between 1 and 2

h = enthalpy per mass

z = height

Inner friction and Archimedes' principle

$$\boldsymbol{f} = -K\eta\boldsymbol{v}$$

$K = 6\pi r$ for a sphere of radius r (Stokes' law)

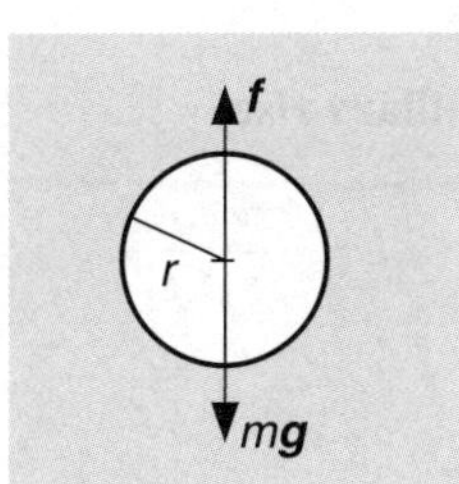

$$\boldsymbol{F}_{\text{buoyant}} = -m_f\boldsymbol{g} = -\rho_f V\boldsymbol{g}$$

$$m\frac{dv}{dt} = (m - m_f)g - K\eta v$$

m = mass of body

m_f = mass of the displaced fluid

$$v = \frac{(m-m_f)g}{K\eta}\left(1-\exp\left(-\frac{K\eta}{m}t\right)\right)$$

if initial speed $v_0 = 0$

Flow equations for compressible liquids

$$\frac{\mathrm{d}A}{A} = (M^2 - 1)\,\frac{\mathrm{d}v}{v}$$

$$\frac{1}{\rho}\,\mathrm{d}p + v\,\mathrm{d}v = 0$$

The Mach number

$$M = \frac{v}{c}$$ c = speed of sound

$$c^2 = \frac{\mathrm{d}p}{\mathrm{d}\rho}$$

Laplace's formula for the curvature pressure (capillary pressure)

$$p = \gamma\left(\frac{1}{R_1} + \frac{1}{R_2}\right)$$

γ = surface tension (capillary constant)

$R_{1,2}$ = curvature radii of the surface (largest and smallest radius, $R > 0$ for convex surface)

For a spherical surface

$$p = \frac{2\,\gamma}{R}$$

Capillary rise

$$h = \frac{2\,\gamma\cos\theta}{r\,\rho\,g}$$

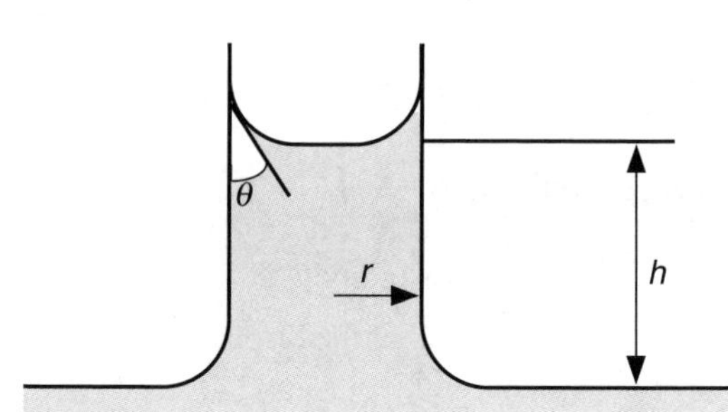

2 Thermal Physics

Quantity	Symbol	SI Unit
Number of particles	N	
Number of moles	n	
Temperature	T	K
Mass	m	kg
Area	A	m^2
Volume	V	m^3
Velocity	v	m/s
Work	W	J = Nm
Internal energy	U	J
Thermal energy	Q	J
Chemical potential (one particle)	μ	J
Entropy	S	J/K
Pressure	p	Pa = N/m^2
Heat capacity	C	J/K
Specific heat capacity	c	J/kg K
Density	ρ	kg/m^3
Thermal conductivity	λ	J/m s K
Diffusion coefficient	D	m^2/s

2.1 Kinetic Theory of Gases

Energy for each degree of freedom (law of equipartition of energy)

$$E = \frac{1}{2}\, kT$$

Collision number

$$n_s = \frac{1}{4}\frac{N}{V}\,\langle v\rangle = \frac{p}{\sqrt{2\pi m\, kT}}$$

Pressure

$$p = \frac{1}{3}\frac{N}{V}\, m\,\langle v^2\rangle$$

Mean free path of molecules in hard-sphere model

$$\ell = \frac{V}{\sqrt{2}\,\pi\, d^2 N} = \text{const}\cdot\frac{T}{p}$$

d = molecular diameter

Collision frequency of molecules

$$\Gamma = \frac{\langle v\rangle}{\ell}$$

Molecular distribution of speeds and energies for an ideal gas

$$v_{\text{mp}} = \sqrt{\frac{2kT}{m}} = \sqrt{\frac{2RT}{M}} \qquad E_{\text{mp}} = \frac{1}{2}\,kT \qquad U = N\langle E\rangle$$

$$\langle v\rangle = \sqrt{\frac{8kT}{\pi m}} = \sqrt{\frac{8RT}{\pi M}} \qquad \langle E\rangle = \frac{3}{2}\,kT$$

$$\sqrt{\langle v^2\rangle} = \sqrt{\frac{3kT}{m}} = \sqrt{\frac{3RT}{M}}$$

v_{mp} and E_{mp} refer to the most probable molecular value

M = molar mass = $m\,N_A$

Maxwell-Boltzmann distribution

$$n(v)\,\mathrm{d}v = C_1\, v^2\, \mathrm{e}^{-\frac{mv^2}{2kT}}\,\mathrm{d}v \qquad n(E)\,\mathrm{d}E = C_2\,\sqrt{E}\,\mathrm{e}^{-\frac{E}{kT}}\,\mathrm{d}E$$

$$C_1 = \frac{4N}{\sqrt{\pi}}\left(\frac{m}{2kT}\right)^{\frac{3}{2}} \qquad C_2 = \frac{2N}{\sqrt{\pi}}\left(\frac{1}{kT}\right)^{\frac{3}{2}}$$

$n(v)\,\mathrm{d}v$ = number of particles with velocities in interval $[v, v + \mathrm{d}v]$
$n(E)\,\mathrm{d}E$ = number of particles with energies in interval $[E, E + \mathrm{d}E]$
N = total number of particles
m = mass of each particle

See also Sec. 9.2 and 9.4.

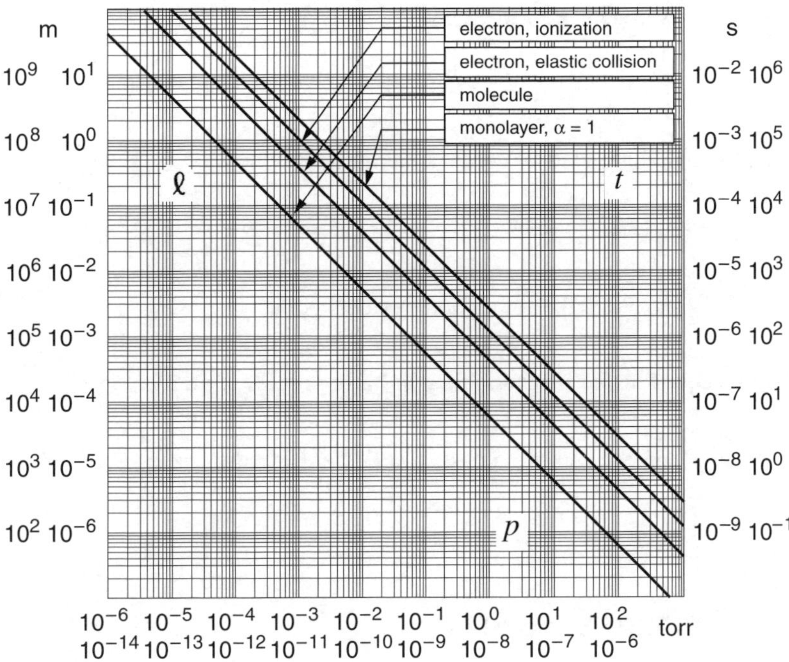

Mean Free Path, Time of Contamination

The diagram shows the mean free path ℓ *and contamination time* t *as functions of the pressure* p *at* 300 K *for molecules and electrons. The notation is explained as follows:*

Electron, ionization: *Mean free path of* 100 eV *electrons between ionizing collisions with air molecules.*

Electron, elastic collision: *Mean free path of* 100 eV *electrons between elastic collisions with air molecules.*

Molecule: *Weighted mean free path of oxygen and nitrogen molecules in air.*

Monolayer, $\alpha = 1$: *Time required for formation of contaminating nitrogen monolayer on a surface with sticking coefficient* $\alpha = 1$.

2.2 Equations of State

Ideal gas law

$$pV = n\,R\,T$$

$$pV = N\,k\,T$$

Van der Waals law for non-ideal gas

$$\left(p+\frac{a}{(V/n)^2}\right)(V-nb)=n\,R\,T$$

a and b are constants, see Section T – 1.8

Reduced equation (the law of corresponding states)

$$\left(\pi+\frac{3}{\phi^2}\right)(3\phi-1)=8\theta \qquad \pi=\frac{p}{p_c},\ \phi=\frac{V}{V_c},\ \theta=\frac{T}{T_c}$$

p_c, V_c and T_c are the values at the critical point.

Poisson's equation for a reversible adiabatic process of an ideal gas

$$pV^{\gamma}=\text{const},\quad TV^{\gamma-1}=\text{const},\quad Tp^{-(\gamma-1)/\gamma}=\text{const},\quad \gamma=\frac{C_p}{C_v}$$

2.3 Thermal Capacity

Some expressions for C_v and C_p in a closed system doing pressure-volume work

$$C_v=\left(\frac{\partial U}{\partial T}\right)_V$$

$$C_p=\left(\frac{\partial U}{\partial T}\right)_p+p\left(\frac{\partial V}{\partial T}\right)_p=\left(\frac{\partial H}{\partial T}\right)_p$$

H = enthalpy according to Section 2.5

$$C_v=T\left(\frac{\partial S}{\partial T}\right)_V \qquad C_p=T\left(\frac{\partial S}{\partial T}\right)_p$$

Relations between C_v and C_p

$$C_p-C_v=\frac{T\,V\,\beta^2}{\kappa_T}$$

β and κ_T defined as in Section 2.8.

$$C_p-C_v=\left(\left(\frac{\partial U}{\partial V}\right)_T+p\right)\left(\frac{\partial V}{\partial T}\right)_p$$

In particular, for an ideal gas

$$C_p=C_v+n\,R$$

Specific heat capacity

$$c_v = \frac{C_v}{m} \qquad c_p = \frac{C_p}{m}$$

The molar heat capacity is found by division by *n* instead.

2.4 Thermodynamic Relations

First Law of Thermodynamics

$\mathrm{d}U = đQ + đW$

$đQ$ = heat (thermal energy) flowing into the system

$đW$ = (work done on the system) – (work done by the system)

Example 1 Pressure-volume work

$đW = -p\,\mathrm{d}V$

Example 2 Stretching of elastic fibre

$đW = J\,\mathrm{d}L$

J = tension
L = length of fibre

Example 3 Expansion of liquid surface

$đW = \gamma\, đA$

γ = surface tension

Example 4 Magnetization *M* in a homogeneous magnetizing field *H*

$đW = \mu_0 V H\,\mathrm{d}M$

$M = (\mu_r - 1)H = \chi H$

μ_r = relative permeability

χ = magnetic susceptibility

V = volume of the specimen

Example 5 Polarization *P* in a homogeneous electric field *E*

$đW = V E\,\mathrm{d}P$

$P = (\varepsilon_r - 1)\varepsilon_0 E$

ε_r = relative dielectric constant

V = volume of the specimen

Internal energy of an ideal gas

$$U = \frac{n\,R\,T}{\gamma - 1} \qquad \gamma = \frac{C_p}{C_v}$$

Entropy

$S = k \ln P$ $\qquad$ P = thermodynamic probability

Principle of increasing entropy in an isolated system
(Second law of thermodynamics)

$$\mathrm{d}S \geq 0$$

Entropy change for a reversible process

$$\mathrm{d}S = \frac{đQ}{T}$$

$\mathrm{d}S = 0$ for a reversible and adiabatic process.

Central equation of thermodynamics (thermodynamic identity)

$\mathrm{d}U = T\,\mathrm{d}S - p\,\mathrm{d}V + \mu\,\mathrm{d}N$ $\qquad$ (pressure-volume work, $\mathrm{d}N = 0$ for a closed system)

Carnot cycle

Efficiency

$$\eta = 1 - \frac{T_2}{T_1}$$

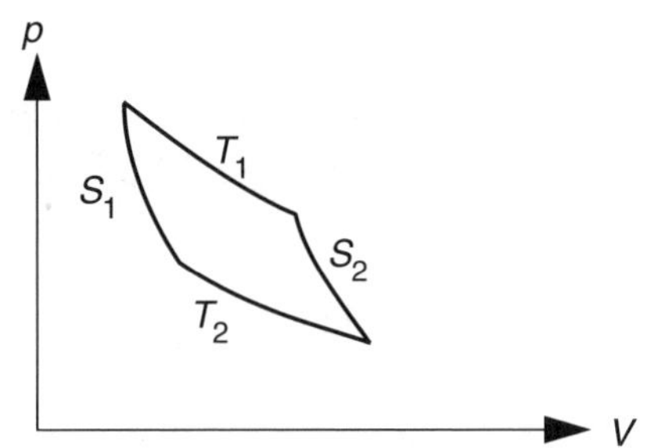

Heating factor for raising temperature to T_1

$$\varepsilon_\mathrm{h} = \frac{T_1}{T_1 - T_2}$$

Cooling factor for lowering temperature to T_2

$$\varepsilon_\mathrm{c} = \frac{T_2}{T_1 - T_2} \qquad \varepsilon_\mathrm{h} = 1 + \varepsilon_\mathrm{c}$$

Work done in Carnot cycle

$$W = Q_1 - Q_2 = (T_1 - T_2)(S_2 - S_1)$$

Work done on the system in adiabatic compression of ideal gas

$$W = \frac{n\,R}{\gamma - 1}\,(T_1 - T_2) \qquad \gamma = \frac{C_p}{C_v} \qquad T_1 > T_2$$

Otto cycle

Efficiency

$$\eta = 1 - \left(\frac{V_\mathrm{min}}{V_\mathrm{max}}\right)^{\gamma - 1}$$

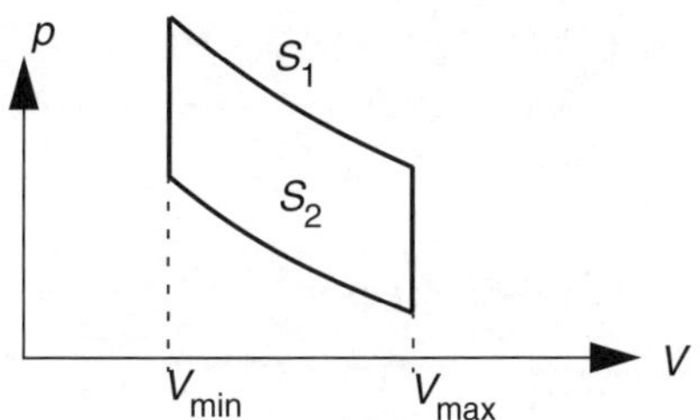

2.5 Thermodynamic Potentials

(one-component systems, pressure-volume work only)

Internal energy

$$U = TS - pV + \mu N$$

$$\mathrm{d}U = T\,\mathrm{d}S - p\,\mathrm{d}V + \mu\,\mathrm{d}N$$

(for a closed system, $\mathrm{d}N = 0$)

Enthalpy

$$H = U + pV$$

$$\mathrm{d}H = T\,\mathrm{d}S + V\,\mathrm{d}p + \mu\,\mathrm{d}N$$

Helmholz' free energy

$$F = U - TS$$

$$\mathrm{d}F = -S\,\mathrm{d}T - p\,\mathrm{d}V + \mu\,\mathrm{d}N$$

Gibbs' free energy

$$G = U - TS + pV$$

$$\mathrm{d}G = -S\,\mathrm{d}T + V\,\mathrm{d}p + \mu\,\mathrm{d}N$$

Grand canonical potential

$$\Omega = U - TS - \mu N$$

$$\mathrm{d}\Omega = -S\,\mathrm{d}T - p\,\mathrm{d}V - N\,\mathrm{d}\mu$$

Maxwell's relations

$$\left(\frac{\partial T}{\partial V}\right)_S = -\left(\frac{\partial p}{\partial S}\right)_V$$

$$\left(\frac{\partial T}{\partial p}\right)_S = \left(\frac{\partial V}{\partial S}\right)_p$$

$$\left(\frac{\partial S}{\partial V}\right)_T = \left(\frac{\partial p}{\partial T}\right)_V$$

$$\left(\frac{\partial S}{\partial p}\right)_T = -\left(\frac{\partial V}{\partial T}\right)_p$$

Euler's equation for the internal energy

$$U = TS - pV + \mu N + \ldots$$

Gibbs-Duhem relation

$$0 = S\,\mathrm{d}T - V\,\mathrm{d}p + N\,\mathrm{d}\mu + \ldots$$

Gibbs-Helmholtz equation

$$U = -T^2 \frac{\partial}{\partial T}\left(\frac{F}{T}\right)_V$$

2.6 Thermal Transport

Basic formula of calorimetrics

$$\mathrm{d}Q = c\, m\, \mathrm{d}T$$

$\mathrm{d}Q$ = absorbed heat

Melting and solidification

$$Q = mH_{\mathrm{fus}}$$

H_{fus} = Enthalpy of fusion (or latent heat of fusion)

Vaporization and condensation

$$Q = mH_{\mathrm{vap}}$$

H_{vap} = Enthalpy of vaporization (or latent heat of vaporization)

Fourier's law

$$\boldsymbol{G} = -\lambda \,\mathrm{grad}\, T$$

$\boldsymbol{G}$ = heat flow

Example Heat flow in a rod

$$\frac{\partial Q}{\partial t} = -A\lambda \frac{\mathrm{d}T}{\mathrm{d}x}$$

Normal component of heat flow

$$G_{\mathrm{n}} = -\frac{1}{A}\frac{\partial Q}{\partial t}$$

Newton's law for external heat transfer

$$G_{\mathrm{n}} = \alpha(T - T_0)$$

T_0 = temperature of the surroundings

α = heat transfer number

Continuity equation

$$\frac{\partial T}{\partial t} + \frac{\kappa}{\lambda}\,\mathrm{div}\,\boldsymbol{G} = \frac{\kappa h}{\lambda}$$

h = source density = heat generated per time and volume

Thermal conduction capacity

$$\kappa = \frac{\lambda}{c_v \rho}$$

Heat conduction equation

$$\frac{\partial T}{\partial t} = \kappa \nabla^2 T + \frac{\kappa h}{\lambda}$$

2.7 Diffusion

Continuity equation

$$\frac{\partial n}{\partial t} + \operatorname{div} \boldsymbol{j} = h$$

n = number of particles in unit volume

$\boldsymbol{j}$ = diffusion flow

h = source density = generated number of particles per unit time and volume

Fick's first law

$$\boldsymbol{j} = -D \operatorname{grad} n$$

Fick's second law (diffusion equation)

$$\frac{\partial n}{\partial t} = D\,\nabla^2 n + h$$

2.8 Special Relations

Coefficient of thermal volume expansion

$$\beta = \frac{1}{V}\left(\frac{\partial V}{\partial T}\right)_p \qquad V - V_0 \approx \beta(T - T_0)$$

Relation between coefficients of linear (α) and volume (β) expansion

$$\beta \approx 3\alpha$$

Coefficient of isothermal compressibility

$$\kappa_T = -\frac{1}{V}\left(\frac{\partial V}{\partial p}\right)_T$$

Enthalpy (heat) of vaporization (Clausius-Clapeyron formula)

$$L = \int đQ = T(V_2 - V_1)\frac{\mathrm{d}p}{\mathrm{d}T}$$

Joule-Thomson coefficient

$$\left(\frac{\partial T}{\partial p}\right)_H = \frac{1}{C_p}\left(T\left(\frac{\partial V}{\partial T}\right)_p - V\right)$$

Curie's law of magnetic susceptibility in a paramagnetic material

$$\chi = \frac{C}{T}$$

C = Curie constant according to Section 10.8.

Temperature dependence of the relative permittivity

$$\varepsilon_{\mathrm{r}} = A + \frac{B}{T}$$

A and B are constants

Relations between partial derivatives for deriving thermodynamic formulae

$$\left(\frac{\partial x}{\partial y}\right)_z = \frac{1}{\left(\frac{\partial y}{\partial x}\right)_z}$$

$$\left(\frac{\partial x}{\partial y}\right)_v = \frac{\left(\frac{\partial x}{\partial u}\right)_v}{\left(\frac{\partial y}{\partial u}\right)_v}$$

when $x = x(u, v)$ $y = y(u, v)$

$$\left(\frac{\partial x}{\partial u}\right)_v = \left(\frac{\partial x}{\partial y}\right)\left(\frac{\partial y}{\partial u}\right)_v$$

when $x = x[y(u, v), v]$

$$\left(\frac{\partial x}{\partial y}\right)_z = -\frac{\left(\frac{\partial z}{\partial y}\right)_x}{\left(\frac{\partial z}{\partial x}\right)_y} = -\frac{\left(\frac{\partial x}{\partial z}\right)_y}{\left(\frac{\partial y}{\partial z}\right)_x}$$

when each of the three variables x, y, and z may be expressed in terms of the two others, so that $x = x(y, z)$, $y = y(x, z)$ and $z = z(x, y)$

Coefficients of viscosity, self-diffusion and thermal conductivity according to the diffusion approximation

(the formulae hold only for gases and give only the order of η, D and λ over a limited temperature range)

$$\eta \approx \frac{1}{\pi d^2}\sqrt{\frac{m k T}{\pi}} \approx \frac{D N m}{V}$$

m and d are the molecular mass and linear "diameter", respectively

$$D \approx \frac{1}{3} \langle v \rangle \ell$$

$\langle v \rangle$ for an ideal gas is given in Section 2.1

ℓ = mean free path

$$\lambda \approx \frac{3D k N}{2V} \approx \frac{3k \eta}{2m}$$

3 Electromagnetic Theory

Quantity	Symbol	SI Unit
Position vector	$\boldsymbol{r}$	m
Frequency	f	$\mathrm{Hz} = \mathrm{s}^{-1}$
Angular frequency	ω	rad/s
Phase difference	φ	rad
Length	ℓ, a, r	m
Area	A, S	m^2
Velocity	$\boldsymbol{v}\ (v)$	m/s
Force	$\boldsymbol{F}\ (F)$	$\mathrm{N} = \mathrm{kg\ m/s}^2$
Electric current	I	A
Electric current density	$\boldsymbol{j}\ (j)$	$\mathrm{A/m}^2$
Electric charge	Q	C = As
Charge density	ρ	$\mathrm{C/m}^3$
Surface charge	σ	$\mathrm{C/m}^2$
Electric potential and voltage*	V	V = Nm/As
Energy	W	J = Nm = VC
Energy density	w	$\mathrm{J/m}^3$
Power	P	W = VA = J/s
Electric field	$\boldsymbol{E}\ (E)$	V/m = N/C
Polarization	$\boldsymbol{P}$	$\mathrm{C/m}^2$
Electric displacement field	$\boldsymbol{D}$	$\mathrm{C/m}^2$
Relative permittivity	ε_r	
Resistance	R	Ω = V/A
Complex impendance	Z	Ω
Reactance	X	Ω
Capacitance	C	F = C/V = s/Ω
Inductance	L	H = Vs/A = Ωs
Electric conductivity	σ	A/Vm
Magnetic flux density	$\boldsymbol{B}\ (B)$	$\mathrm{T} = \mathrm{Vs/m}^2$
Magnetic flux	$\boldsymbol{\Phi}$	$\mathrm{Wb} = \mathrm{T\ m}^2 = \mathrm{Vs}$
Magnetizing field	$\boldsymbol{H}\ (H)$	A/m
Magnetization	$\boldsymbol{M}$	A/m
Number of turns	N	
Relative permeability	μ_r	
Magnetic vector potential	$\boldsymbol{A}$	T m

* Recommended symbols are V or φ for electric potential and U or V for potential difference. We have adopted the symbol V for both quantities, in accordance with British and US practice.

3.1 The Coulomb Field

Coulomb's law

$$F = \frac{Q_1\, Q_2}{4\pi\, \varepsilon_0\, r^2}$$

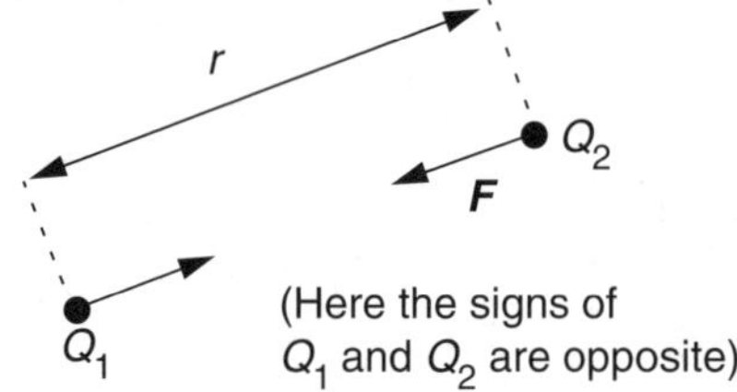

Field intensity and potential

$$\boldsymbol{E} = \boldsymbol{F}/Q \qquad \boldsymbol{E} = \frac{1}{4\pi\, \varepsilon_0}\int \frac{\rho}{r^2}\frac{\boldsymbol{r}}{r}\,\mathrm{d}v \qquad v = \text{volume}$$

$\boldsymbol{F}$ = force on a charge Q in the electric field $\boldsymbol{E}$

$$\boldsymbol{E} = -\,\mathrm{grad}\, V \qquad V_{\mathrm{A}} - V_{\mathrm{B}} = -\int_{\mathrm{B}}^{\mathrm{A}} \boldsymbol{E}\cdot \mathrm{d}\boldsymbol{r}$$

$$V = \frac{1}{4\pi\, \varepsilon_0}\int \frac{\rho}{r}\,\mathrm{d}v$$

Uniform field

$$\boldsymbol{E} = \frac{V_2 - V_1}{a}\,\hat{\boldsymbol{x}}$$

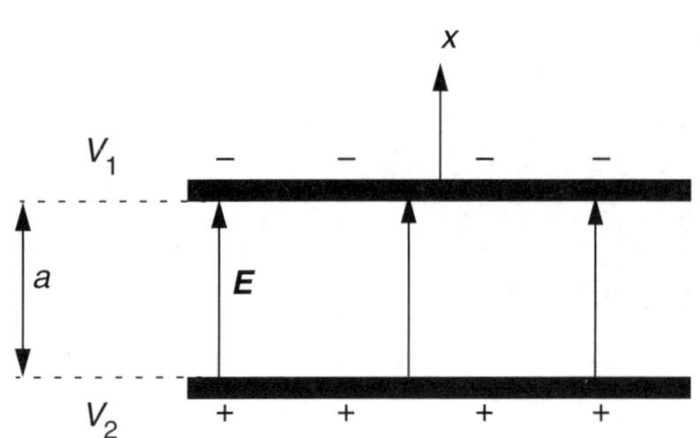

Field around an electric point charge and outside a spherical charge distribution

$$E = \frac{1}{4\pi\, \varepsilon_0}\frac{Q}{r^2} \qquad \boldsymbol{E} = \frac{Q}{4\pi\, \varepsilon_0\, r^3}\boldsymbol{r}$$

$$V = \frac{1}{4\pi\, \varepsilon_0}\frac{Q}{r}$$

r = distance from centre of charge distribution

Field inside a spherical, uniform charge distribution

$$E_r = \frac{\rho\, r}{3\, \varepsilon_0} = \frac{Q\, r}{4\pi\, \varepsilon_0\, a^3}$$

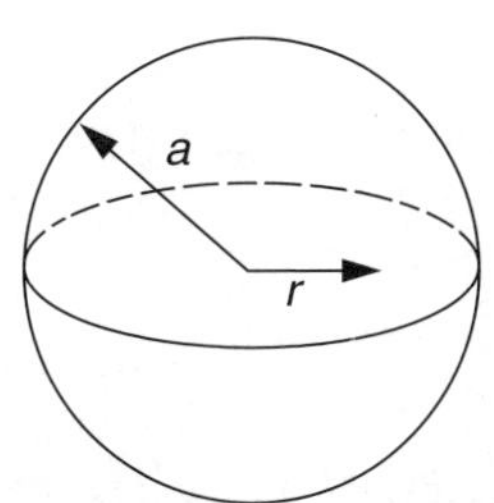

$$V = \frac{Q}{8\pi\, \varepsilon_0\, a^3}\, (3a^2 - r^2)$$

Q = total charge of sphere

Field on axis of circular, uniform charge distribution

$$E_z = \frac{\sigma z}{2\, \varepsilon_0}\left(\frac{1}{|z|} - \frac{1}{\sqrt{z^2+a^2}}\right)$$

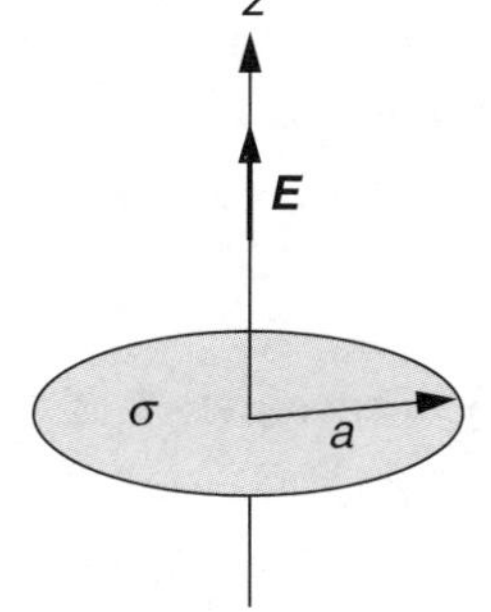

$$V = \frac{\sigma}{2\, \varepsilon_0}\, (\sqrt{z^2+a^2} - |z|)$$

Surface charge density

$$\sigma = \frac{Q}{\pi\, a^2} \qquad Q = \text{total disc charge}$$

Field from an infinitely long conductor

$$E = \frac{\lambda}{2\pi\, \varepsilon_0\, a}$$

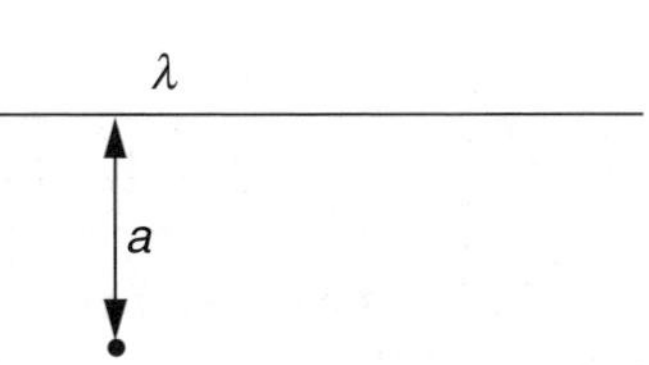

$$V = -\,\frac{\lambda}{2\pi\, \varepsilon_0}\, \ln \frac{a}{a_0}$$

λ = linear charge density
a_0 = distance from conductor where $V = 0$

Poisson's equation

$$\nabla^2 V = -\,\rho / \varepsilon_0$$

Gauss' law of flux of a field vector

$$\Phi_E = \oint_S \boldsymbol{E}\, \mathrm{d}\boldsymbol{S} = Q / \varepsilon_0$$

Q = total charge within the closed surface S.

Alternative formulation of Gauss' law in polarization-free space

$$\operatorname{div} \boldsymbol{E} = \rho / \varepsilon_0$$

Irrotation of Coulomb field

$$\operatorname{curl} \boldsymbol{E} = \boldsymbol{0}$$

3.2 Motion of Charged Particles

Lorentz force

$$\boldsymbol{F} = Q(\boldsymbol{E} + \boldsymbol{v} \times \boldsymbol{B})$$

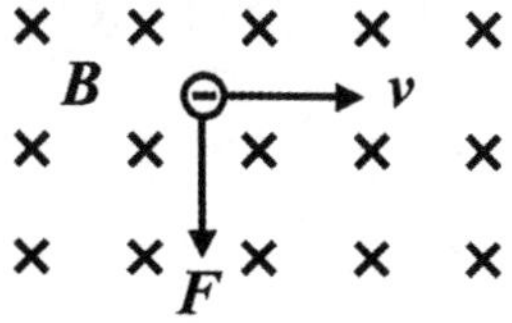

Electric current

$$I = \frac{\mathrm{d}Q}{\mathrm{d}t}$$

Direction of force $\boldsymbol{F}$ on a moving negatively charged particle in a magnetic field.

Density of current

$$j = n\, Q\, \langle v \rangle \qquad j = I/A$$

n = number of charges Q per unit volume

$\langle v \rangle$ = drift velocity of charge carriers

Total energy of one particle moving in an electric field (non-relativistic case)

$$W_{\text{tot}} = \tfrac{1}{2} m\, v^2 + Q\, V$$

m = particle mass

Drift velocity of charged particle acted on by a magnetic field and a non-magnetic force

$\boldsymbol{B}$ = magnetic flux density
$\boldsymbol{F}$ = non-magnetic force
q = particle charge
m = particle mass
$\boldsymbol{u}$ = drift velocity
$\boldsymbol{v}$ = particle velocity

Indices $\perp$ and // mean "component perpendicular to magnetic field" and "component parallel to magnetic field", respectively.

$$\boldsymbol{u}_{\perp} = -\frac{1}{qB^2} \boldsymbol{B} \times \left(\boldsymbol{F} - \frac{1}{B} \frac{m v_{\perp}^2}{2} \operatorname{grad} B - m \frac{\mathrm{d}\boldsymbol{u}}{\mathrm{d}t} \right)$$

$$\text{For } \boldsymbol{u}_{//}: \boldsymbol{B} \cdot \left(\boldsymbol{F} - \frac{1}{B} \frac{m v_{\perp}^2}{2} \operatorname{grad} B - m \frac{\mathrm{d}\boldsymbol{u}}{\mathrm{d}t} \right) = 0$$

Special cases

1. $\boldsymbol{F} = q\boldsymbol{E}$; $\operatorname{grad} B \equiv \boldsymbol{0}$, $\quad \dfrac{\mathrm{d}\boldsymbol{u}}{\mathrm{d}t} = \boldsymbol{0}$

$$\boldsymbol{u}_{\perp} = -\frac{1}{B^2} \boldsymbol{B} \times \boldsymbol{E} \quad \text{(indep. of } q, \boldsymbol{v}\text{)}$$

2. $\boldsymbol{F} = \dfrac{\mathrm{d}\boldsymbol{u}}{\mathrm{d}t} = \boldsymbol{0}$

$$\boldsymbol{u}_{\perp} = \frac{m v_{\perp}^2}{2qB^3} \boldsymbol{B} \times \operatorname{grad} B$$

3. $\dfrac{\mathrm{d}\boldsymbol{u}}{\mathrm{d}t}$ dominated by centrifugal acceleration

$$\boldsymbol{u}_{\perp} = -\frac{1}{qB^2} \boldsymbol{B} \times \left[\boldsymbol{F} - \frac{m}{B} \left(\frac{v_{\perp}^2}{2} + v_{//}^2 \right) \operatorname{grad} B \right]$$

Generalized momentum

$$\boldsymbol{p} = m\,\boldsymbol{v} + Q\,\boldsymbol{A}$$

Generalized potential of Lorentz force

$$V_{\text{gen}} = Q(V - \boldsymbol{v} \cdot \boldsymbol{A})$$

Radiation at a point P from an accelerated charge ($v \ll c_0$)

$$\boldsymbol{E}(\boldsymbol{r}, t_{\text{P}}) = -\frac{Q}{4\pi\varepsilon_0 c_0^2 r} \dot{\boldsymbol{v}}_{\perp}(t_0)$$

$$\boldsymbol{B} = \frac{1}{c_0} (\hat{\boldsymbol{r}} \times \boldsymbol{E})$$

Poynting's vector

$$\boldsymbol{S} = \frac{1}{\mu_0} \boldsymbol{E} \times \boldsymbol{B}$$

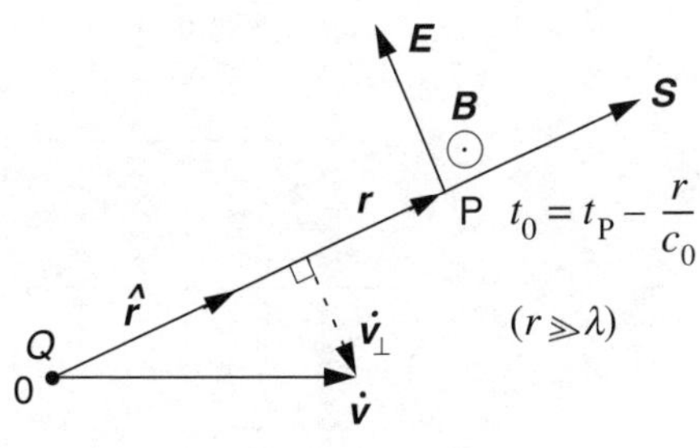

See also Sec. F–5.3.

3.3 The Quasi-Stationary *B*-field

Ampere-Laplace law of the magnetic field from a closed loop

$$\boldsymbol{B} = \frac{\mu_0 I}{4\pi} \oint_{\mathrm{L}} \frac{\boldsymbol{u}_{\mathrm{T}} \times \boldsymbol{u}_r}{r^2} \, \mathrm{d}\ell$$

$$\mathrm{d}\boldsymbol{B} = \frac{\mu_0 I}{4\pi} \frac{\boldsymbol{u}_{\mathrm{T}} \times \boldsymbol{u}_r}{r^2} \, \mathrm{d}\ell$$

$\boldsymbol{u}_{\mathrm{T}}$ and $\boldsymbol{u}_r$ = unit vectors

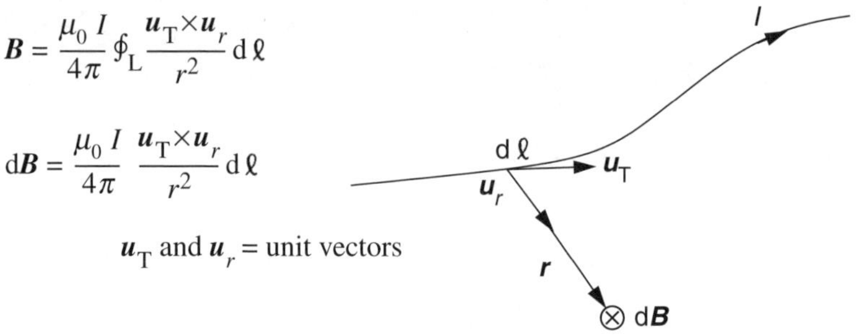

Example 1. The field at distance *d* from a long, straight conductor (Biot-Savart formula)

$$B = \frac{\mu_0}{4\pi} \frac{2I}{d}$$

Example 2. Field on axis of circular loop (short solenoid)

$$B = \frac{\mu_0}{4\pi} \frac{2NIA}{r^3}$$

Example 3. Field on axis of thin solenoid

$$B = \mu_0 \frac{NI}{2\ell} (\cos \alpha_1 + \cos \alpha_2)$$

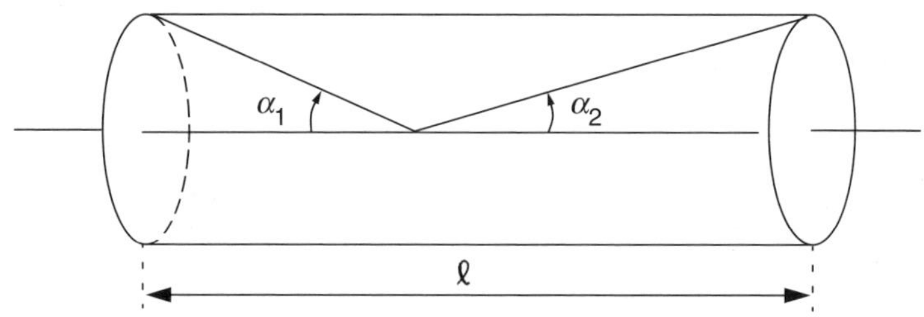

In particular, inside a long solenoid or a toroidal solenoid

$$B = \mu_0 \frac{N\,I}{\ell}$$

Exempel 4. Field from a charged particle in uniform motion

$$\boldsymbol{B} = \frac{\mu_0}{4\pi} \frac{Q}{r^2}\, \boldsymbol{v} \times \boldsymbol{u}_r$$

$$\boldsymbol{B} = \mu_0\, \varepsilon_0\, \boldsymbol{v} \times \boldsymbol{E}$$

Ampere's law (law of circulation)

$\oint_{\mathrm{L}} \boldsymbol{B}\, \mathrm{d}\boldsymbol{r} = \mu_0 I$ $\qquad$ $\oint_{\mathrm{L}} \boldsymbol{H}\, \mathrm{d}\boldsymbol{r} = I$

$\mathrm{curl}\, \boldsymbol{B} = \mu_0 \boldsymbol{j}$ $\qquad$ I = current inside the closed loop L

In particular, inside a conductor of uniform current density

$$B = \frac{\mu_0}{4\pi} \frac{2\,r}{a^2}\, I$$

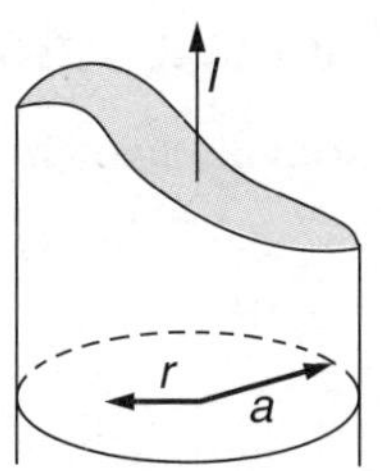

Gauss' law of the magnetic flux

$\Phi_B = \oint_{\mathrm{S}} \boldsymbol{B}\, \mathrm{d}\boldsymbol{S} = 0$ $\qquad$ $\mathrm{div}\, \boldsymbol{B} = 0$

Magnetic vector potential *A*

$\boldsymbol{B} = \mathrm{curl}\, \boldsymbol{A}$ $\qquad$ $\boldsymbol{A} = \frac{\mu_0}{4\pi} \int \frac{\boldsymbol{j}}{r}\, \mathrm{d}V$

In particular, for a uniform ***B***-field

$$\boldsymbol{A} = \tfrac{1}{2}\, \boldsymbol{B} \times \boldsymbol{r}$$

Magnetic force on an electric conductor

$$dF = B\,I \sin\theta\, dr$$

$$d\boldsymbol{F} = I\, d\boldsymbol{r} \times \boldsymbol{B}$$

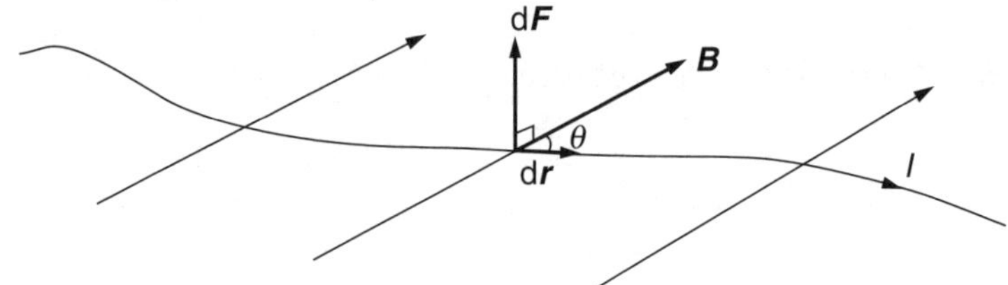

In particular the mutual force on parallel conductors, each of length ℓ

$$F = \frac{\mu_0}{4\pi}\,\frac{2\,\ell}{a}\,I_1\,I_2$$

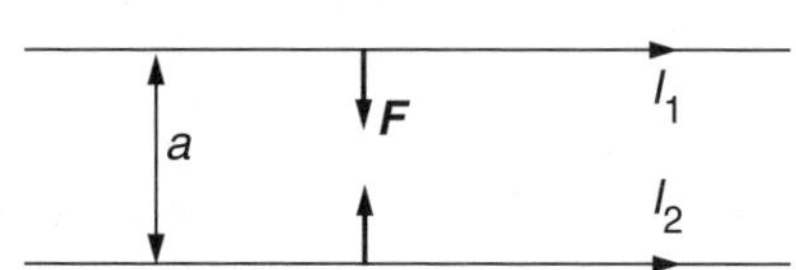

3.4 Electric and Magnetic Dipoles

Electric dipole moment for charges $+Q$ and $-Q$ at a distance d from each other

$$\boldsymbol{p} = Q\,\boldsymbol{d}$$

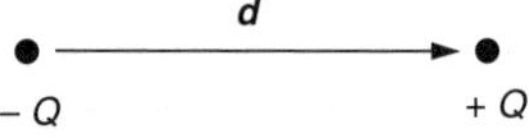

Magnetic dipole moment of a small current loop

$$\boldsymbol{m} = I\,A\,\boldsymbol{u}_N \qquad \boldsymbol{u}_N = \text{unit vector}$$

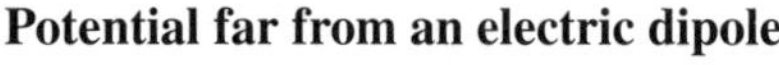

Potential far from an electric dipole

$$V = \frac{1}{4\pi\,\varepsilon_0}\,\frac{p\cos\theta}{r^2}$$

$$V = \frac{1}{4\pi\,\varepsilon_0}\,\frac{\boldsymbol{p}\cdot\boldsymbol{r}}{r^3}$$

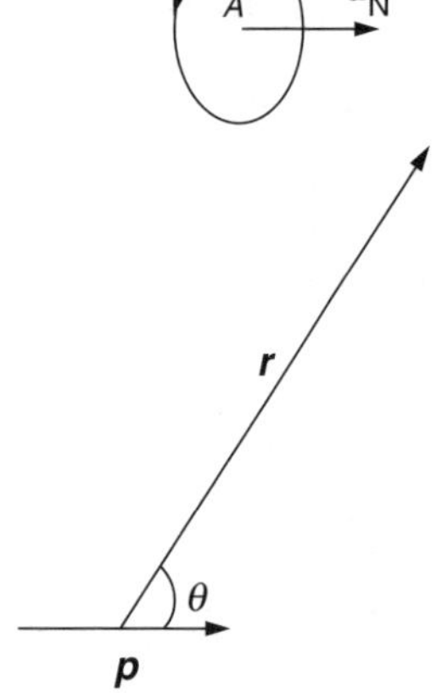

Electric field

$$\boldsymbol{E} = -\,\text{grad}\,V$$

In particular, in spherical polar coordinates

$$E_r = -\frac{\partial V}{\partial r} \qquad E_\theta = -\frac{1}{r}\frac{\partial V}{\partial \theta}$$

Field far from a magnetic dipole

$$B_r = \frac{\mu_0}{4\pi}\frac{2m\cos\theta}{r^3}$$

$$B_\theta = \frac{\mu_0}{4\pi}\frac{m\sin\theta}{r^3}$$

Vector potential of a magnetic dipole

$$\boldsymbol{A} = \frac{\mu_0}{4\pi}\frac{\boldsymbol{m}\times\boldsymbol{r}}{r^3}$$

Torque on dipole in external field

$$\tau_e = p\,E\sin\alpha \qquad \tau_m = m\,B\sin\alpha$$

$$\boldsymbol{\tau}_e = \boldsymbol{p}\times\boldsymbol{E} \qquad \boldsymbol{\tau}_m = \boldsymbol{m}\times\boldsymbol{B}$$

Potential energy of a dipole in external field

$$W_e = -\boldsymbol{p}\cdot\boldsymbol{E} = -\,p\,E\cos\alpha$$

$$W_m = -\,\boldsymbol{m}\cdot\boldsymbol{B} = -\,m\,B\cos\alpha$$

Force on dipole in external field

$$\boldsymbol{F} = (\boldsymbol{p}\cdot\nabla)\boldsymbol{E} \qquad \boldsymbol{F} = (\boldsymbol{m}\cdot\nabla)\,\boldsymbol{B} + \boldsymbol{m}\times(\nabla\times\boldsymbol{B})$$

In particular, if dipole is directed along x-axis, and if curl $\boldsymbol{B} = \boldsymbol{0}$.
(Force directed towards stronger field.)

$$\boldsymbol{F} = p_x\frac{\partial}{\partial x}\boldsymbol{E} \qquad \boldsymbol{F} = m_x\frac{\partial}{\partial x}\boldsymbol{B}$$

Electric dipole moment of an uncharged body with a charge distribution

$$\boldsymbol{p} = \int_{\upsilon} \boldsymbol{r}\,\rho\,\mathrm{d}\upsilon$$

υ = volume of body

General definition of magnetic dipole moment

$$\boldsymbol{m} = \tfrac{1}{2}\int \boldsymbol{r}\times\boldsymbol{j}\,\mathrm{d}\upsilon$$

3.5 Dielectric and Magnetic Media

Relations between *E*, *D*, and *P*

$$\boldsymbol{D} = \varepsilon_0\,\boldsymbol{E} + \boldsymbol{P} = \varepsilon_r\,\varepsilon_0\,\boldsymbol{E}$$

$$\boldsymbol{P} = (\varepsilon_r - 1)\varepsilon_0\,\boldsymbol{E} = \chi_e\,\varepsilon_0\,\boldsymbol{E}$$

χ_e = electric susceptibility (sometimes χ_e is given such that $\boldsymbol{P} = \chi_e\,\boldsymbol{E}$)

$\boldsymbol{P}$ = dipole moment per unit volume = polarization

$\varepsilon_r\,\varepsilon_0$ = permittivity

$\boldsymbol{D}$ = electric displacement = electric flux density

Surface charge density from polarization

$$\sigma_p = \boldsymbol{P}\cdot\boldsymbol{u}_N$$

$\boldsymbol{u}_N$ = normal vector

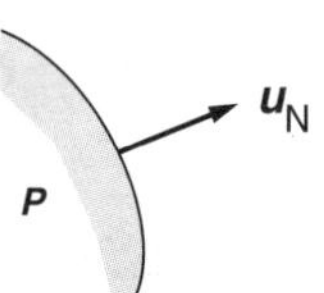

Relations between *B*, *H* and *M*

$$\boldsymbol{B} = \mu_0(\boldsymbol{H} + \boldsymbol{M}) = \mu_r\,\mu_0\,\boldsymbol{H}$$

$$\boldsymbol{M} = (\mu_r - 1)\boldsymbol{H} = \chi_m\,\boldsymbol{H}$$

χ_m = magnetic susceptibility

$\boldsymbol{M}$ = dipole moment per volume = magnetization

Gauss' law of polarizable space

$$\text{div}\,\boldsymbol{E} = \frac{1}{\varepsilon_0}\,(\rho_f - \text{div}\,\boldsymbol{P})$$

ρ_f = charge density other than that of polarization

$-\text{div}\,\boldsymbol{P} = \rho_p$ = charge density of polarization

Field vector components at interfaces

$$E_{1T} - E_{2T} = 0$$

$$D_{1N} - D_{2N} = \sigma_f$$

$$H_{1T} - H_{2T} = 0$$

$$B_{1N} - B_{2N} = 0$$

σ_f = surface charge density other than that of polarization

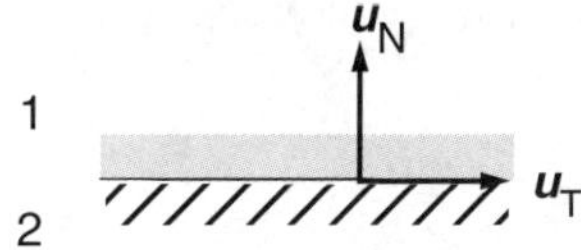

In particular, the refraction of the $\boldsymbol{B}$-field

$$\frac{\tan\alpha_1}{\tan\alpha_2} = \frac{\mu_{r_1}}{\mu_{r_2}}$$

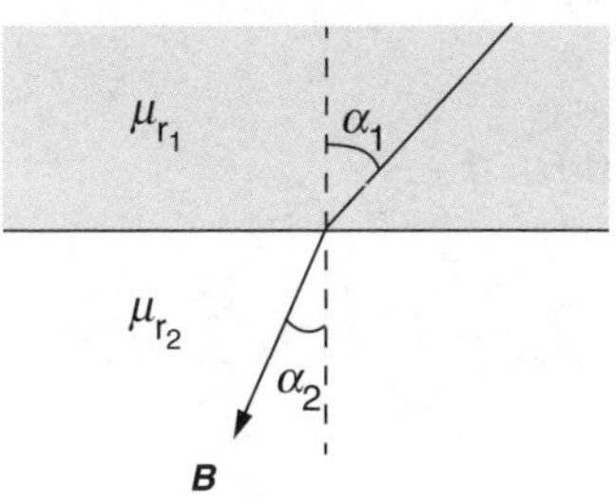

Maxwell stresses

$$T_e = \tfrac{1}{2}\,\varepsilon_0\,E^2$$

$$T_m = \frac{1}{2\mu_0}\,B^2$$

$$\boldsymbol{T} = \boldsymbol{T}_e + \boldsymbol{T}_m$$

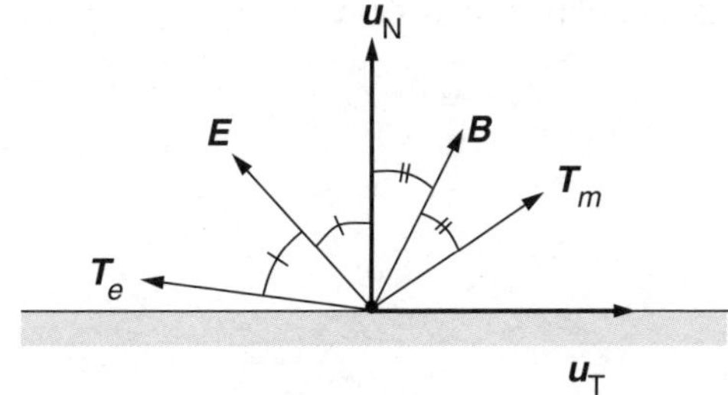

Real force

$$\text{d}\boldsymbol{F} = \boldsymbol{T}\,\text{d}A$$

3.6 The Electromagnetic Field

Maxwell's equations

I $\oint_S \boldsymbol{D}\,\mathrm{d}\boldsymbol{S} = Q_f$ — Q_f = total free charge inside surface S

$\operatorname{div} \boldsymbol{D} = \rho_f$ — ρ_f = charge density other than that of polarization

$\operatorname{div} \boldsymbol{E} = \rho/\varepsilon_0$ (in polarization-free space)

II $\oint_S \boldsymbol{B}\,\mathrm{d}\boldsymbol{S} = 0$

$\operatorname{div} \boldsymbol{B} = 0$

III $\oint_L \boldsymbol{E}\,\mathrm{d}\boldsymbol{r} = -\dfrac{\mathrm{d}}{\mathrm{d}t}\displaystyle\int_S \boldsymbol{B}\,\mathrm{d}\boldsymbol{S}$

$$\operatorname{curl} \boldsymbol{E} = -\frac{\partial \boldsymbol{B}}{\partial t}$$

IV $\oint_L \boldsymbol{H}\,\mathrm{d}\boldsymbol{r} = I + \dfrac{\mathrm{d}}{\mathrm{d}t}\displaystyle\int_S \boldsymbol{D}\,\mathrm{d}\boldsymbol{S}$ — $\boldsymbol{D}$ = electric displacement = electric flux density

$$\operatorname{curl} \boldsymbol{H} = \boldsymbol{j} + \frac{\partial \boldsymbol{D}}{\partial t}$$

$$\operatorname{curl} \boldsymbol{B} = \mu_0 \boldsymbol{j} + \varepsilon_0 \mu_0 \frac{\partial \boldsymbol{E}}{\partial t} \quad \text{(when } \boldsymbol{P} = \boldsymbol{M} = \boldsymbol{0}\text{)}$$

Law of charge conservation

$$\oint_S \boldsymbol{j}\,\mathrm{d}\boldsymbol{S} + \frac{\mathrm{d}}{\mathrm{d}t}\oint_S \boldsymbol{D}\,\mathrm{d}\boldsymbol{S} = 0$$

$$\operatorname{div} \boldsymbol{j} + \frac{\partial \rho}{\partial t} = 0$$

Energy density in an electric and magnetic field respectively when ε_r and μ_r are constant

$$w_e = \frac{1}{2}\,\boldsymbol{D}\cdot\boldsymbol{E} = \frac{1}{2}\,\varepsilon_r \varepsilon_0 E^2$$

$$w_m = \frac{1}{2}\,\boldsymbol{H}\cdot\boldsymbol{B} = \frac{1}{2}\,\frac{1}{\mu_r \mu_0}\,B^2$$

In particular, total electrostatic energy of a chage distribution

$$W = \frac{1}{2}\int \boldsymbol{D}\cdot\boldsymbol{E}\,\mathrm{d}v = \frac{1}{2}\int \rho_{\mathrm{f}} V\,\mathrm{d}v \qquad v = \text{volume}$$

Relation between ε_0 and μ_0

$$\varepsilon_0\,\mu_0 = 1/c_0^2$$

Poynting vector (power flux density)

$$\boldsymbol{S} = \boldsymbol{E}\times\boldsymbol{H} = c_0^2\,\varepsilon_0\,\boldsymbol{E}\times\boldsymbol{B}$$

Effective penetration depth (skin effect)

$$\delta = 1/\sqrt{\pi\,\mu_{\mathrm{r}}\,\mu_0\,\sigma\,f} \qquad \sigma = \text{electric conductivity}$$

Vector field (at point P) from a point charge Q_1 with velocity $\boldsymbol{v}_0$

$$\boldsymbol{E} = \frac{Q_1}{4\pi\,\varepsilon_0}\left(-\,\mathrm{grad}\,\frac{1}{r_1} - \frac{1}{c_0^2}\,\frac{\partial}{\partial t}\left(\frac{\boldsymbol{v}_0}{r_1}\right)\right) = -\,\mathrm{grad}\,V - \frac{\partial \boldsymbol{A}}{\partial t}$$

$$\boldsymbol{B} = \frac{1}{c_0}\,\frac{\boldsymbol{r}_0}{r_0}\times\boldsymbol{E} = \mathrm{curl}\,\boldsymbol{A}$$

Scalar field r_1

$$r_1 = r\sqrt{1 - \frac{v_0^2}{c_0^2}\,\sin^2\theta}$$

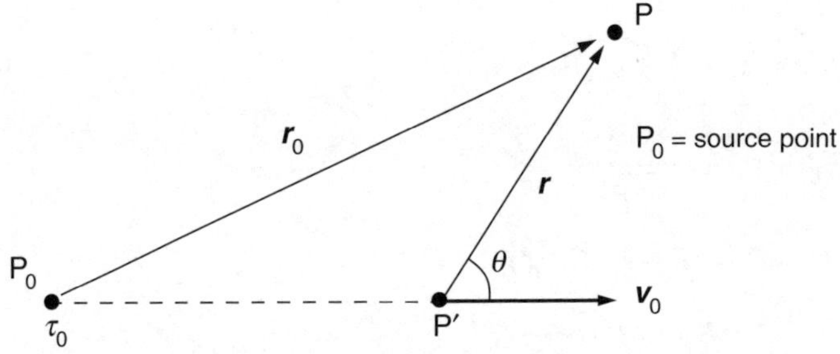

At time t the particle is found at P′

Example 1. Uniform motion

$$\boldsymbol{E} = \frac{Q_1}{4\pi\,\varepsilon_0}\,\frac{\boldsymbol{r}}{r_1^3}\left(1-\frac{v_0^2}{c_0^2}\right)$$

$$\boldsymbol{B} = \frac{\mu_0\,Q_1}{4\pi}\,\frac{\boldsymbol{v}_0\times\boldsymbol{r}}{r_1^3}\left(1-\frac{v_0^2}{c_0^2}\right)$$

Example 2. Stationary or quasi-stationary particle

$$\boldsymbol{E} = \frac{Q_1}{4\pi\,\varepsilon_0}\,\frac{\boldsymbol{r}}{r^3}$$

$$\boldsymbol{B} = \frac{\mu_0}{4\pi}\,Q_1\,\frac{\boldsymbol{v}_0\times\boldsymbol{r}}{r^3} = \frac{\mu_r\,\varepsilon_r}{c_0^2}\,\boldsymbol{v}_0\times\boldsymbol{E}$$

Relation between field time t and source time τ

$$\tau = t - r/c_0$$

3.7 Relativity and Electromagnetism

Transformation to a system S′, moving with a velocity v along the x-axis of the system S (both origins coincide at time $t = t' = 0$)

$$\rho' = \frac{\rho - v\,j_x/c_0^2}{\sqrt{1-\beta^2}} = \frac{\rho - \boldsymbol{v}\cdot\boldsymbol{j}/c_0^2}{\sqrt{1-\beta^2}} \qquad \beta = \frac{v}{c_0}$$

$$j'_{x'} = \frac{j_x - v\,\rho}{\sqrt{1-\beta^2}} = \frac{(\boldsymbol{j}-\rho\,\boldsymbol{v})_x}{\sqrt{1-\beta^2}}$$

$$j'_{y',z'} = j_{y,z} = (\boldsymbol{j} - \rho\,\boldsymbol{v})_{y,z}$$

$$E'_{x'} = E_x = (\boldsymbol{E} + \boldsymbol{v}\times\boldsymbol{B})_x$$

$$E'_{y',z'} = \frac{(\boldsymbol{E}+\boldsymbol{v}\times\boldsymbol{B})_{y,z}}{\sqrt{1-\beta^2}}$$

$$B'_{x'} = B_x = (\boldsymbol{B} - \boldsymbol{v}\times\boldsymbol{E}/c_0^2)_x$$

$$B'_{y',z'} = \frac{(\boldsymbol{B}-\boldsymbol{v}\times\boldsymbol{E}/c_0^2)_{y,z}}{\sqrt{1-\beta^2}}$$

3.8 The Magnetic Circuit

Circulation law of a toroidal solenoid with small air gap δ

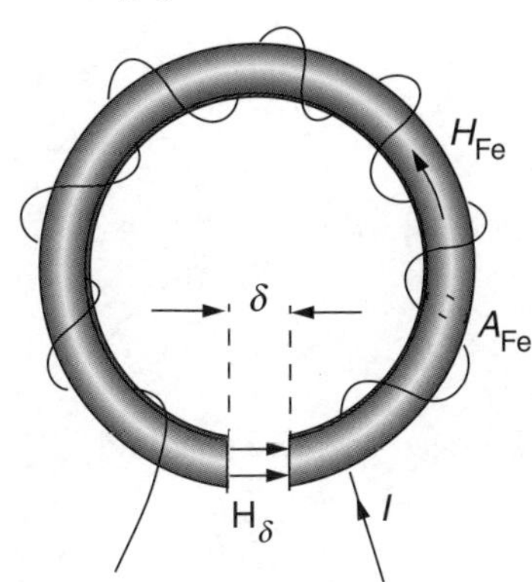

$$H_{Fe}\,\ell + H_\delta\,\delta = N\,I$$

ℓ = length of solenoid

Magnetizing field in the solenoid if the leak can be neglected

$$H_{Fe} = \frac{N\,I}{\ell+\delta} - \frac{\delta}{\ell+\delta} M_{Fe}$$

$$\frac{\delta}{\ell+\delta} = \text{demagnetization factor}$$

Magnetic flux

$$\Phi_B = B_{Fe} A_{Fe} = B_\delta A_\delta$$

A = cross-section area

Example 1.

$$B_\delta = \mu_r\,\mu_0 \frac{NI}{\ell - \delta + \mu_r \delta}$$

Example 2. Permanent magnet ($I = 0$)

$$H_{Fe}\,\ell + H_\delta\,\delta = 0$$

$$B_{Fe} = -\,H_{Fe}\,\frac{\mu_0\,\ell}{\delta}\,\frac{A_\delta}{A_{Fe}}$$

Reluctance and permeance of a magnetic "conductor" of length ℓ and area A

$$R_m = \ell / \mu_r\,\mu_0\,A$$
$$\Lambda = 1/R_m$$

"Ohm's law" of magnetic circuit

$$N\,I = R_m\,\Phi_B$$

Magnetic "pole mass" inside a closed surface S

$$Q_m = \oint_S \boldsymbol{H}\,\mathrm{d}\boldsymbol{S}$$

Magnetic "pole density"

$$\rho_m = -\,\mathrm{div}\,\boldsymbol{M}$$

3.9 Induction and Inductance

Law of induction (Faraday-Henry law)

$$\varepsilon = -\frac{\mathrm{d}\Phi_B}{\mathrm{d}t}$$

ε = induced electromotive force, emf

$$\oint_{\mathrm{L}} \boldsymbol{E}\,\mathrm{d}\boldsymbol{r} = -\frac{\mathrm{d}}{\mathrm{d}t}\int_{\mathrm{S}} \boldsymbol{B}\,\mathrm{d}\boldsymbol{S}$$

Induced emf in a solenoid

$$\varepsilon = -N\,\frac{\mathrm{d}\Phi_B}{\mathrm{d}t}$$

Lenz's law

The direction of the induced current is such as to create a magnetic field which opposes the change of magnetic flux.

Definition of inductance L

$$N\,\Phi_B = L\,I$$

Potential drop over a pure inductor

$$V_L = L\,\frac{\mathrm{d}I}{\mathrm{d}t}$$

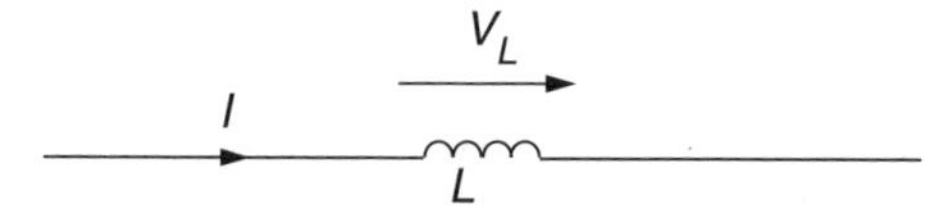

Inductance in a long straight or toroidal solenoid of length ℓ

$$L = \mu_0\,\frac{N^2 A}{\ell}$$

Inductance of a coaxial cable

$$L = \mu_{r_2} \mu_0 \frac{\ell}{2\pi} \ln \frac{r_2}{r_1} + \frac{\mu_{r_1} \mu_0 \ell}{8\pi}$$

μ_{r_1} = relative permeability of inner conductor

(The thickness of the outer conductor has been neglected.)

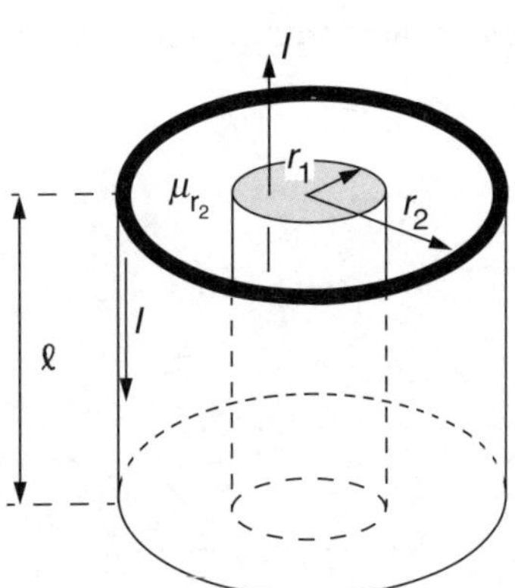

Equivalent inductance of inductors in series and in parallel

$$L_s = \Sigma_n L_n$$

$$\frac{1}{L_p} = \Sigma_n \frac{1}{L_n}$$

Energization and deenergization of an inductor

$$V = R\,I + L\,\frac{dI}{dt}$$

$$I = I_0 \left(1 - e^{-\frac{R}{L}t}\right)$$

$$I = I_0\, e^{-\frac{R}{L}t}$$

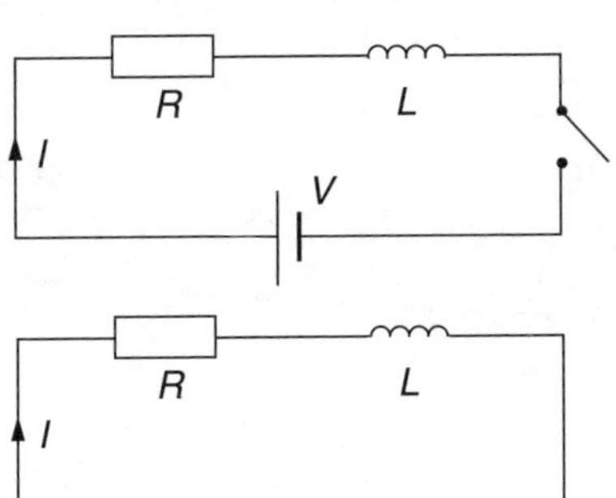

Energy stored in solenoidal magnetic field

$$W_L = \frac{1}{2} L I^2$$

Energy density

$$w_m = \frac{1}{2\mu_r \mu_0} B^2$$

Definition of mutual inductance M

$$N_2\,\Phi_{12} = M\,I_1$$

$$N_1\,\Phi_{21} = M\,I_2$$

Φ_{12} = flux generated by coil 1 affecting coil 2 (in most cases $\approx \Phi_1$)

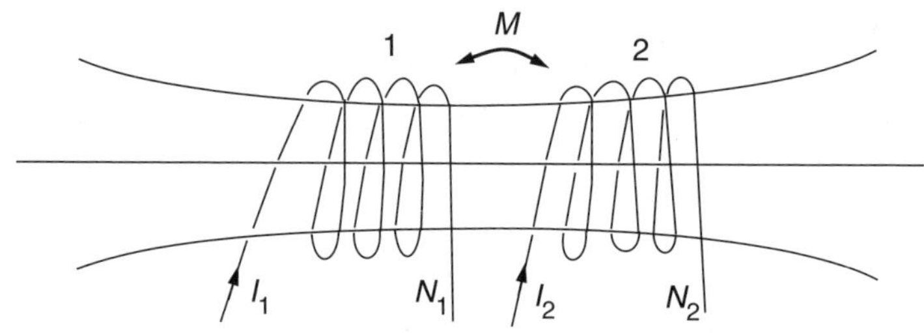

Coupling factor

$$k = \frac{M}{\sqrt{L_1\,L_2}} \qquad |k| < 1$$

Example. Air transformer

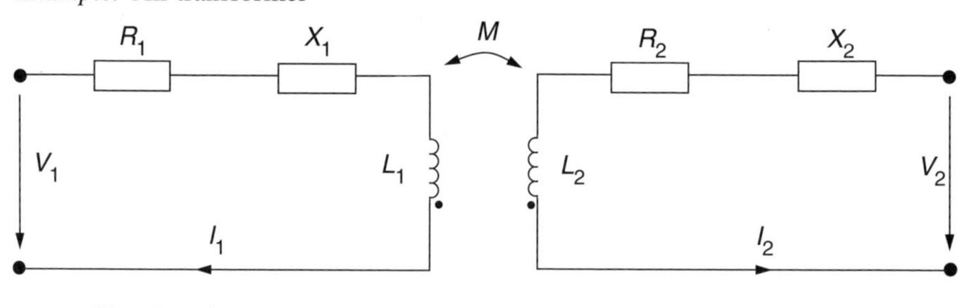

$$Z_1 = R_1 + \mathrm{j}\,X_1 \qquad Z_2 = R_2 + \mathrm{j}\,X_2$$

$$V_1 = (Z_1 + \mathrm{j}\,\omega\,L_1)I_1 + \mathrm{j}\,\omega\,M\,I_2 \qquad V_2 = (Z_2 + \mathrm{j}\,\omega\,L_2)I_2 + \mathrm{j}\,\omega\,M\,I_1$$

Ideal transformer

Equivalent circuit

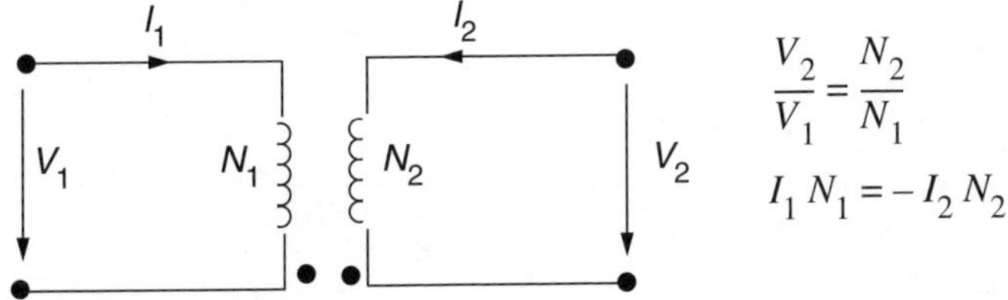

$$\frac{V_2}{V_1} = \frac{N_2}{N_1}$$

$$I_1 N_1 = - I_2 N_2$$

Real transformer

Equivalent circuit

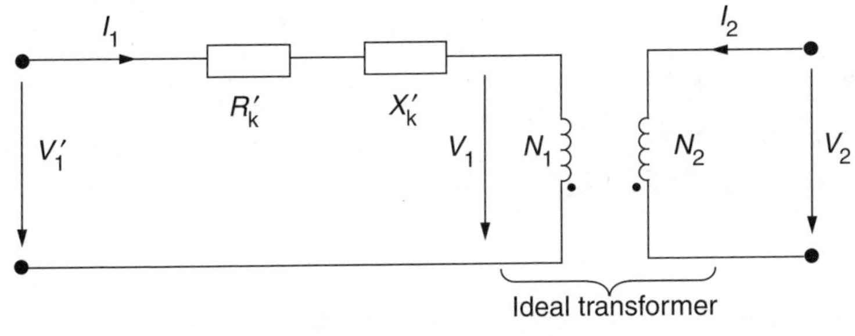

$$R'_k = R_1 + \left(\frac{N_1}{N_2}\right)^2 R_2 \qquad X'_k = X_1 + \left(\frac{N_1}{N_2}\right)^2 X_2$$

Efficiency of transformer

$$\eta = \frac{P_2}{P_2 + P_0 + P_b}$$

$$P_2 = V_2 I_2 \cos \varphi_2 = x \cdot S_{2M} \cos \varphi_2$$

x = load factor, i.e. $I_2 = x\, I_{2M}$

I_{2M} = rated current of secondary

$S_{2M} = V_{2M} I_{2M}$ = rated power of secondary

P_0 = no-load loss

$P_b = x^2 P_{bM}$ = load loss

$P_{bM} = R'_k I_{1M}^2$ = maximum copper loss

I_{1M} = rated current of primary

3.10 Capacitance

Definition of capacitance

$$C = Q/V$$

Parallel-plate capacitor

$$C = \varepsilon_r \varepsilon_0 \frac{A}{a}$$

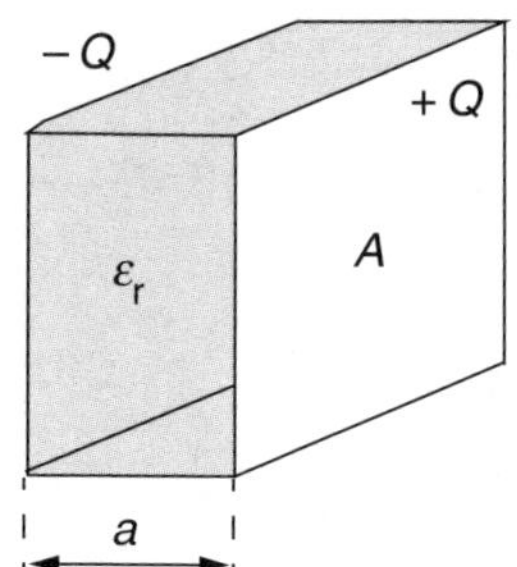

Mutual force on plates

$$F = \frac{dW_C}{da} = \frac{Q^2}{2\varepsilon_r \varepsilon_0 A}$$

Cylinder capacitor

$$C = \varepsilon_r \varepsilon_0 \frac{2\pi \ell}{\ln(r_2/r_1)}$$

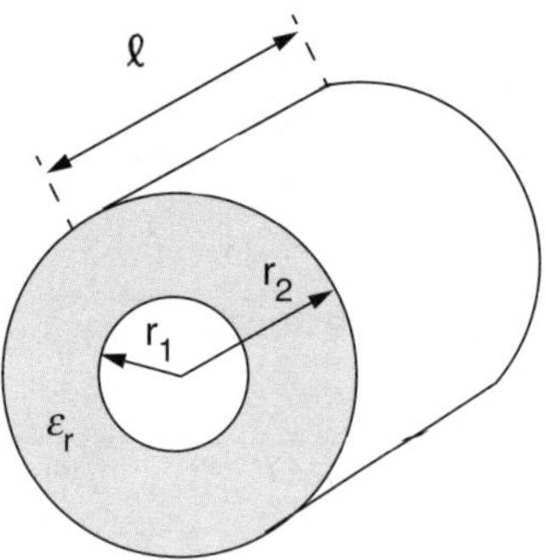

Spherical capacitor

$$C = \varepsilon_r \varepsilon_0 \frac{4\pi r_1 r_2}{r_2 - r_1}$$

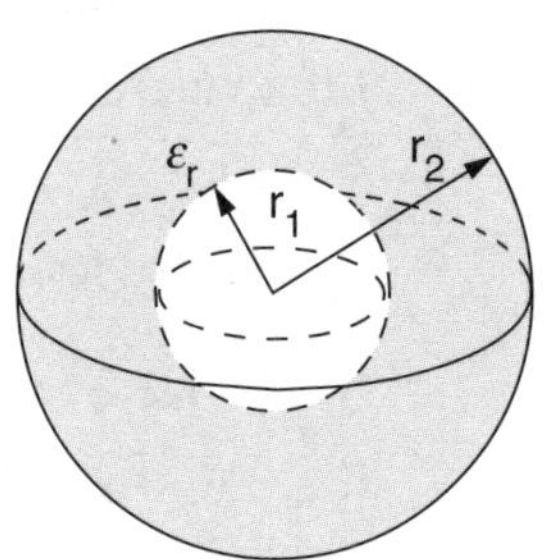

Leakage conductance

$$\frac{1}{R} = \frac{\sigma\, C}{\varepsilon_{\mathrm{r}}\, \varepsilon_0}$$

σ = conductivity of dielectric

Resultant capacitance in series and parallel arrangements

$$\frac{1}{C_s} = \Sigma_n \frac{1}{C_n}$$

$$C_p = \Sigma_n\, C_n$$

Charging and discharging of a capacitor through a resistor

$$V = V_C + R\, C\, \frac{\mathrm{d}\, V_C}{\mathrm{d}t}$$

$$V_C = V(1 - \mathrm{e}^{-t/RC})$$

$$V_C = V_0\, \mathrm{e}^{-t/RC}$$

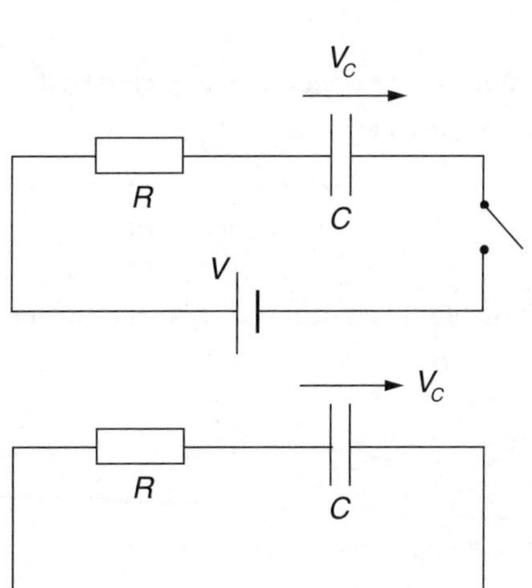

Energy of electric field in capacitor

$$W_C = \tfrac{1}{2}\, C\, V^2 = \tfrac{1}{2}\, Q^2/C = \tfrac{1}{2}\, Q\, V$$

Energy density

$$w_e = \tfrac{1}{2}\, \varepsilon_{\mathrm{r}}\, \varepsilon_0\, E^2 = \tfrac{1}{2}\, D\, E$$

3.11 The Electric Circuit

Ohm's law

$$V = R\,I$$

$$\boldsymbol{j} = \sigma \boldsymbol{E}$$

Electric conductivity of a conductor

$$\sigma = \frac{\ell}{R\,A}$$

ℓ = length of conductor

A = cross-section area

Resistivity and conductance

$$\rho = \frac{1}{\sigma} = \frac{R\,A}{\ell}$$

$$G = 1/R$$

Temperature dependence of resistivity

$$\rho = \rho_0(1 + \alpha(T - T_0))$$

α = temperature coefficient

T = temperature

Equivalent resistance of resistors in series and in parallel

$$R_s = \Sigma_n R_n$$

$$\frac{1}{R_p} = \Sigma_n \frac{1}{R_n}$$

In particular, two parallel resistors

$$R_p = \frac{R_1 R_2}{R_1 + R_2}$$

The same rule applies for impedances Z in series and parallel.

Voltage division

$$V_2 = \frac{R_2}{R_1 + R_2} V_0$$

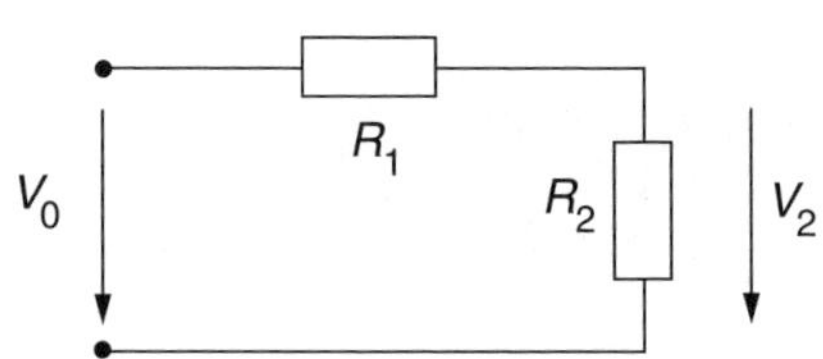

Current division

$$I_2 = \frac{R_1}{R_1 + R_2} I$$

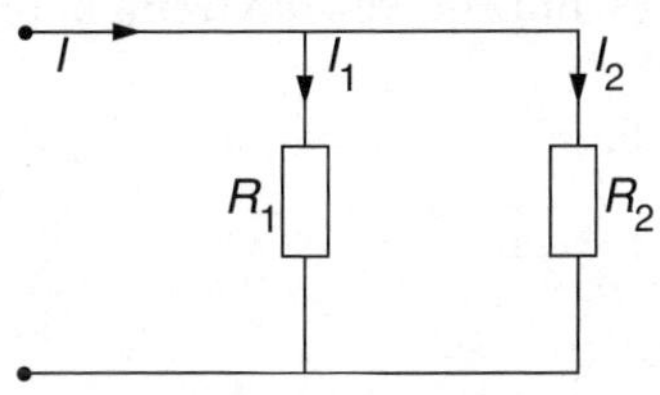

Kirchhoff's laws

$\Sigma_i I_i = 0$ at a junction

$\Sigma_i V_i = 0$ round a closed path

Joule's law for thermal power

$$P = V I = R I^2 = V^2/R$$

Transformation between Y and delta connections

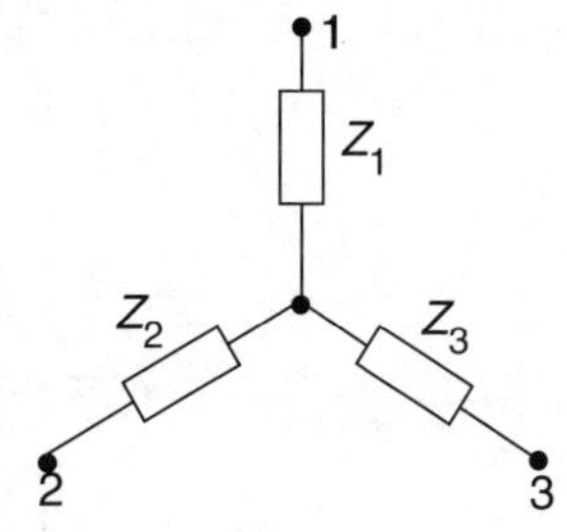

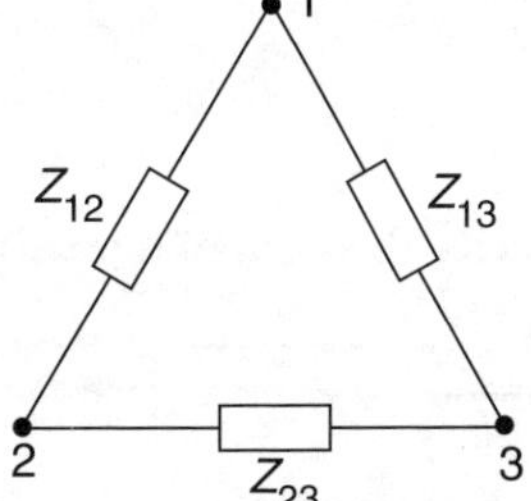

$$Z_1 = \frac{Z_{12}\, Z_{13}}{Z_{12} + Z_{13} + Z_{23}}$$

$$Z_{12} = Z_1\, Z_2 \left(\frac{1}{Z_1} + \frac{1}{Z_2} + \frac{1}{Z_3}\right)$$

Thevenin equivalent circuit of an active two-terminal network

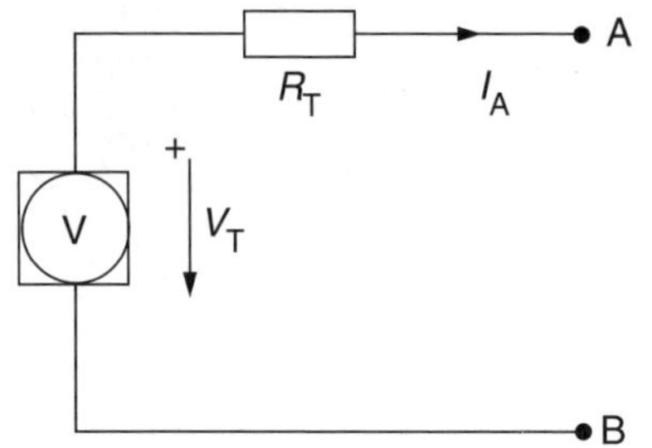

V_T = open-circuit potential difference between A and B ($I_A = 0$)

R_T = resistance between A and B when all voltage supplies are short-circuited and all current supplies are inactive.

Norton equivalent of an active two-terminal network

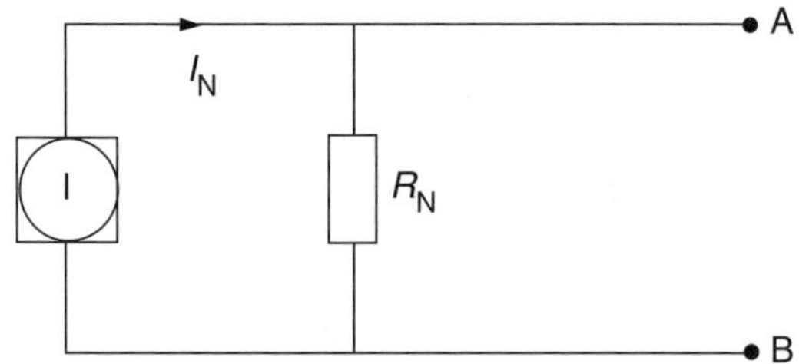

I_N = current through A – B when output is short-circuited

$R_N = R_T$

Reciprocity theorem of a passive circuit

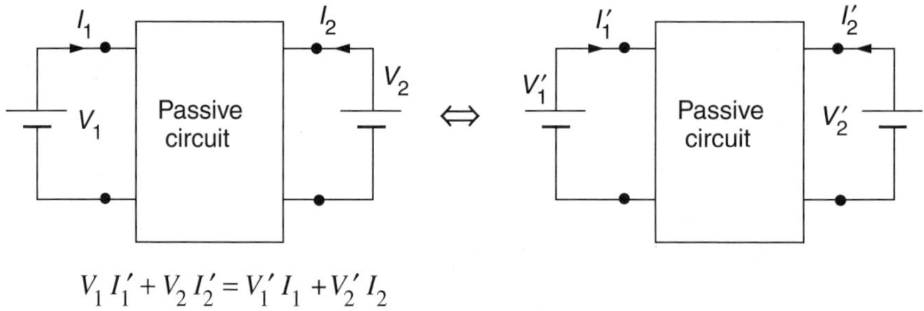

$$V_1 I_1' + V_2 I_2' = V_1' I_1 + V_2' I_2$$

Miller's theorem

I

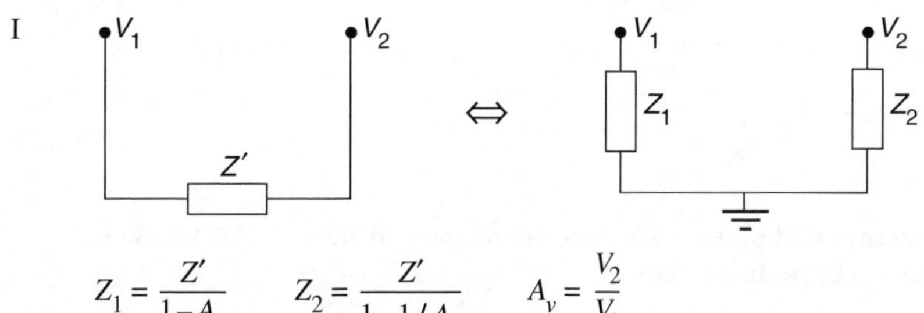

$$Z_1 = \frac{Z'}{1-A_v} \qquad Z_2 = \frac{Z'}{1-1/A_v} \qquad A_v = \frac{V_2}{V_1}$$

II

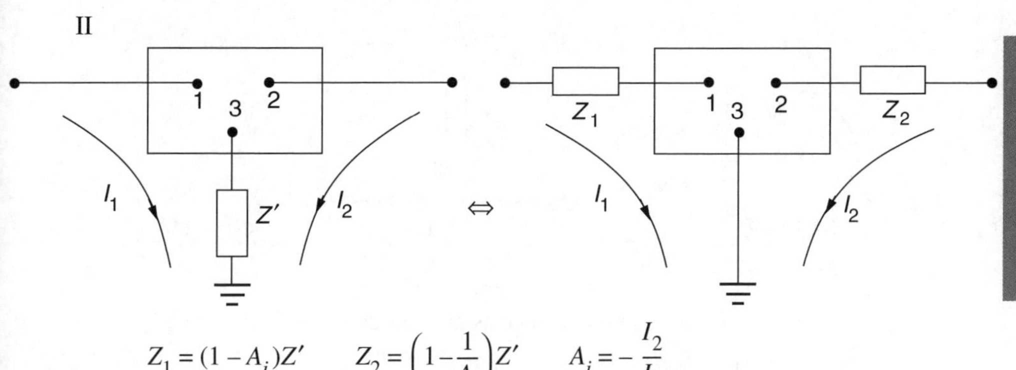

$$Z_1 = (1 - A_i)Z' \qquad Z_2 = \left(1 - \frac{1}{A_i}\right)Z' \qquad A_i = -\frac{I_2}{I_1}$$

3.12 Alternating (AC) Currents

Relations between frequency, angular frequency, and period T

$$f = \frac{1}{T} = \frac{\omega}{2\pi}$$

Impendance and phase difference of voltage in different (ideal) circuit elements relative to current

$Z = R$ (R) $\varphi = 0°$

$Z = \mathrm{j}\,\omega L$ (L) $\varphi = +90°$

$Z = -\mathrm{j}/\omega C$ (C) $\varphi = -90°$

$Z = \mathrm{j}\,\omega M$ (M) $\varphi = +90°$

M = mutual inductance as in Section 3.9

$\mathrm{j} = \sqrt{-1}$

Overall impendance

$Z = R + \mathrm{j}\,X$ $\qquad \varphi = \arctan \frac{X}{R}$

Admittance

$Y = 1/Z = G + \mathrm{j}\,B$ $\qquad G$ = conductance
B = susceptance

Effective value

$$I = \left(\frac{1}{T} \int_0^T i^2 \, \mathrm{d}t \right)^{1/2} \qquad V = \left(\frac{1}{T} \int_0^T v^2 \, \mathrm{d}t \right)^{1/2}$$

In particular,

$$i = \hat{\imath} \sin \omega t \quad \Rightarrow \quad I = \hat{\imath} / \sqrt{2}$$

$$v = \hat{v} \sin \omega t \quad \Rightarrow \quad V = \hat{v} / \sqrt{2}$$

Effective power

$P = V\,I\,\cos\varphi = R\,I^2$ $\qquad$ $\cos\varphi$ = power factor

Reactive power

$Q = V\,I\sin\varphi = X\,I^2$

Apparent power

$S = V\,I$

General definition of quality factor

$$Q_0 = \omega_0\,\frac{\langle W_L\rangle + \langle W_C\rangle}{P} \quad \text{at resonance}$$

ω_0 = angular frequency at resonance

$\langle W_L\rangle$ and $\langle W_C\rangle$ are time mean values of stored energy in inductors and capacitors.

$\langle W_L\rangle = \frac{1}{2}\,L\,I^2$ $\qquad$ $\langle W_C\rangle = \frac{1}{2}\,C\,V_C{}^2$

"Bridge"

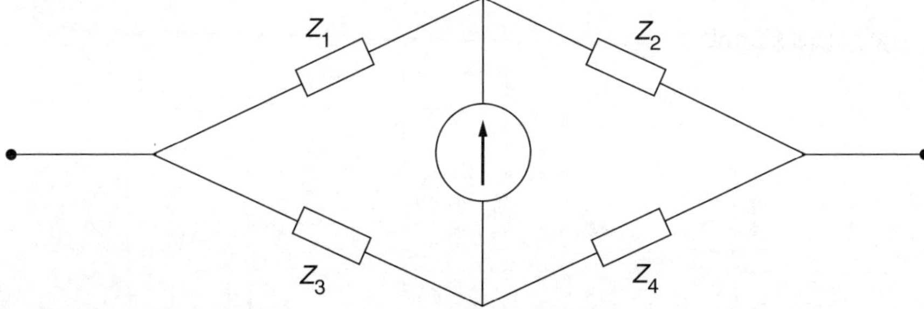

Condition of balance

$Z_1\,Z_4 = Z_2\,Z_3$

Power adjustment

$$P_y = \frac{R_y V^2}{(R_s+R_y)^2+(X_s+X_y)^2}$$

$Z_s = R_s + j X_s$, $Z_y = R_y + j X_y$, V

Condition for maximum power in Z_y

$$R_y = R_s \text{ and } X_y = -X_s \Rightarrow P_{max} = \frac{V^2}{4 R_s}$$

if R_y and X_y may be chosen independently

$$|Z_y| = |Z_s| \Rightarrow P_{max} = \frac{V^2 R_y}{|Z_s+Z_y|^2} = \frac{V^2 \cos \varphi_y}{2|Z_s|(1+\cos(\varphi_y-\varphi_s))}$$

if only the magnitude of Z_y may be varied (e.g. a resistor).

$$\varphi_y = \arctan \frac{X_y}{R_y} \qquad \varphi_s = \arctan \frac{X_s}{R_s}$$

$$R_y = |Z_y| \cos \varphi_y \qquad R_s = |Z_s| \cos \varphi_s$$

3.13 Series and Parallel Circuits

The Series Circuit

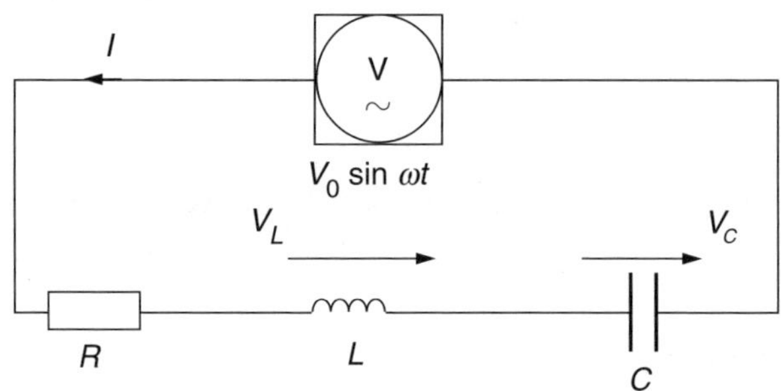

$$L \frac{d^2 I}{dt^2} + R\frac{dI}{dt} + \frac{1}{C} I = \omega V_0 \cos \omega t$$

Stationary solution

$$I = I_0 \sin(\omega t - \varphi)$$

$$I_0 = \frac{V_0}{\sqrt{R^2 + (\omega L - 1/\omega C)^2}}$$

$$\varphi = \arctan \frac{\omega L - 1/\omega C}{R}$$

Quality

$$Q_0 = \frac{1}{R}\sqrt{\frac{L}{C}} = \frac{\omega_0 L}{R} = \frac{1}{\omega_0 RC}$$

Resonant frequency

$$f_0 = \frac{\omega_0}{2\pi} = \frac{1}{2\pi}\,\frac{1}{\sqrt{LC}}$$

Impedance

$$Z = R + \mathrm{j}(\omega L - 1/\omega C) = R(1 + \mathrm{j}\,a) \qquad \mathrm{j} = \sqrt{-1}$$

Normalized impendance

$$a = Q_0\left(\frac{\omega}{\omega_0} - \frac{\omega_0}{\omega}\right)$$

Half-power points ω_1 and ω_2 (when $a = \pm 1$)

$$\frac{\omega_{1,2}}{\omega_0} = \frac{\pm 1}{2\,Q_0} + \sqrt{\frac{1}{4\,Q_0^2} + 1}$$

$$Q_0 = \frac{\omega_0}{\omega_1 - \omega_2}$$

$$\omega_1\,\omega_2 = \omega_0^2$$

Band with (in Hz)

$$B = \frac{\omega_1 - \omega_2}{2\pi} = \frac{\omega_0}{2\pi\,Q_0} = \frac{1}{2\pi}\,\frac{R}{L}$$

Formulae for $\omega \approx \omega_0$, $Q_0 \gg 1$

$$V_L \approx \mathrm{j}\,Q_0\,V$$

$$V_C \approx -\mathrm{j}\,Q_0\,V$$

$$Z \approx R(1 + \mathrm{j}\,2\delta\,Q_0)$$

$$\delta = \frac{\omega - \omega_0}{\omega_0} = \text{relative frequency difference}$$

Switching off a power source (Oscillating circuit)

I. $\omega_0 > \gamma$

$$\omega_e = \sqrt{\omega^2 - \gamma^2} \qquad \omega_0 = 1/\sqrt{LC}$$

$$I = I_0\, e^{-\gamma t} \sin(\omega_e t + \alpha) \qquad \gamma = R/2L$$

Damping ratio

$$K = e^{\gamma T} \qquad T = \frac{2\pi}{\omega_e}$$

Logarithmic decrement

$$\Lambda = \ln K = \frac{\pi R}{\omega_e L}$$

II. $\omega_0 < \gamma$

$$I = A\, e^{\beta_1 t} + B\, e^{\beta_2 t}$$

$$\beta_{1,2} = -\gamma \pm \sqrt{\gamma^2 - \omega_0^2}$$

III. $\omega_0 = \gamma = -\beta$

$$I = e^{\beta t}(A + Bt)$$

A and B are constants

The Parallel Circuit

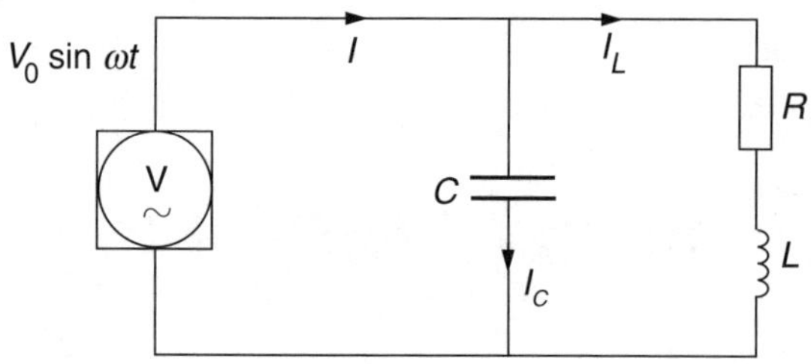

Resonant frequency

$$\omega_0' = \sqrt{\frac{1}{LC} - \frac{R^2}{L^2}} = \omega_0 \sqrt{1 - \frac{1}{Q_0^2}}$$

Impendance

$$Z = \frac{R + \mathrm{j}\,\omega L}{1 - \omega^2 LC + \mathrm{j}\omega RC} \qquad \mathrm{j} = \sqrt{-1}$$

In particular, if $\omega = \omega_0'$

$$Z = R\,Q_0^2 = \frac{L}{RC}$$

Quality of the parallel circuit

$$Q_0' = \frac{L}{R}\,\omega_0' = Q_0\sqrt{1 - \frac{1}{Q_0^2}}$$

Band width (in Hz)

$$B = \frac{\omega_1 - \omega_2}{2\pi} \approx \frac{\omega_0}{2\pi\,Q_0}$$

Formulae for $\omega = \omega_0'$, Q_0 ¡ 1

$$I_L \approx -\mathrm{j}\,Q_0\,I \qquad I_C \approx \mathrm{j}\,Q_0\,I$$

Approximate equivalent circuit when Q_0 ¡ 1

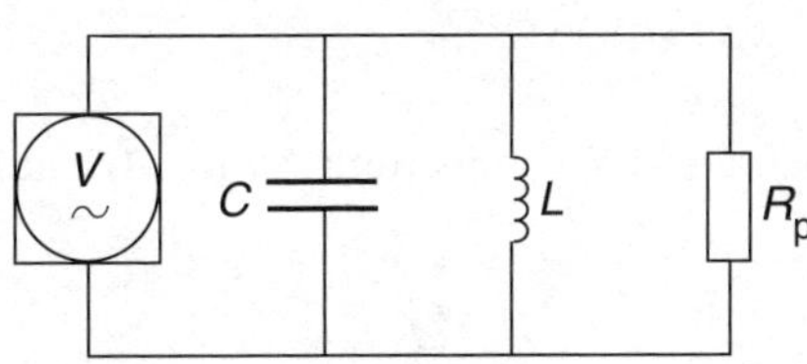

$$R_\mathrm{p} = \frac{\omega_0^2 L^2}{R}\,Q_0^2 R$$

$$Z \approx \frac{R_\mathrm{p}}{1 + \mathrm{j}\,2\delta\,Q_0}$$

$$\delta = \frac{\omega - \omega_0}{\omega_0}$$ = relative frequency difference

3.14 Three-Phase Current

Voltage phasor diagram

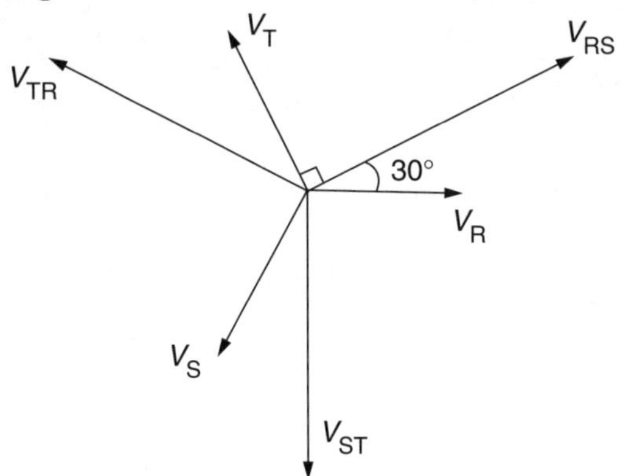

Line voltages V_{RS}, V_{ST}, and V_{TR} are denoted by V_h

Phase voltages V_R, V_S, and V_T are denoted by V_p

Relationship between line and phase voltages

$$V_h = \sqrt{3}\ V_p$$

Delta and Y connections, symmetrical load

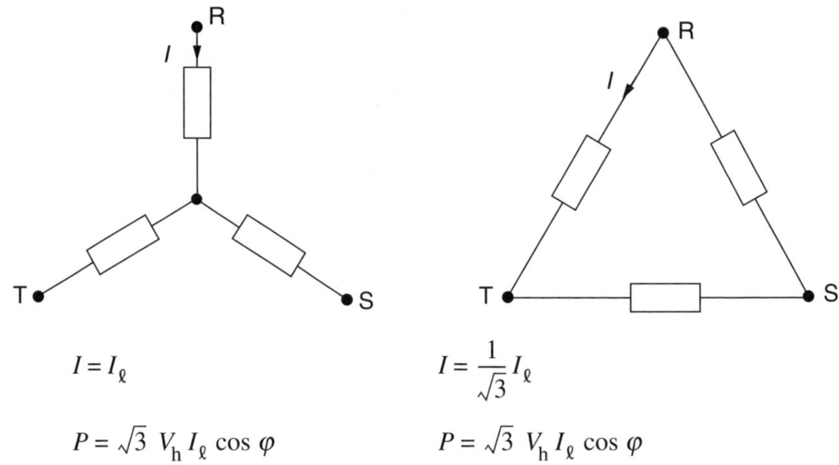

$I = I_\ell$ $\qquad\qquad$ $I = \dfrac{1}{\sqrt{3}} I_\ell$

$P = \sqrt{3}\ V_h I_\ell \cos\varphi$ $\qquad\qquad$ $P = \sqrt{3}\ V_h I_\ell \cos\varphi$

I_ℓ = phase current (line current effective value). See also Sec. 3.11.

Definition of short-circuit power

$S_{SC} = \sqrt{3}\, V_h I_{SC}$ $\qquad$ I_{SC} = short-circuit current

Voltage drop in transmission lines

$$V_{1h} - V_{2h} = \sqrt{3}\, I_\ell \,(R \cos \varphi_2 + X \sin \varphi_2)$$

R and X = resistance and reactance of transmission lines (per phase)

$\cos \varphi_2$ = load power factor

Power loss

$$P = 3\, R\, I_\ell^{\,2}$$

Measuring the power without a neutral by the two-wattmeter method

$$P = P_1 + P_2$$

$$\tan \varphi = \sqrt{3}\, \frac{P_2 - P_1}{P_2 + P_1}$$

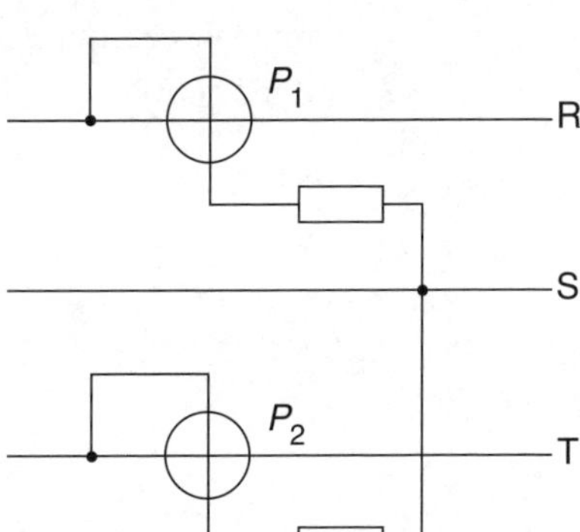

3.15 Rotating Electric Machines

Mechanical power

$P = M\omega$ $\qquad$ M = torque

DC machines

Induced emf in rotor (armature)

$\varepsilon = k_1\, n\, \Phi_\delta$ $\qquad$ n = number of turns

Φ_δ = air-gap flux

Rotor voltage

$$V_a = \varepsilon + R_a I_a + V_c$$

R_a =resistance in armature winding
I_a =armature current
V_c =brush voltage drop

Torque

$$M = k_2 I_a \Phi_\delta$$

Air-gap flux

$$\Phi_\delta = \Phi_{\delta 0}\left(1 + \frac{k_s I_a - k_a |I_a|}{I_{an}}\right)$$

k_s = series winding constant (if no series winding, set $k_s = 0$)
k_a = armature reaction constant
I_{an} = rated current

Leakage flux

$$\Phi_{\delta 0} \approx k_3 I_m + \Phi_r$$

I_m = magnetization current
Φ_r = residual flux

The series motor

$$M \approx k_4 I_a^2$$

Generator operation

$$I_a < 0 \qquad V_c < 0$$

The Asynchronous Motor

Flux r.p.m. (synchronous r.p.m.)

$$n_1 = \frac{60 f_0}{p}$$

f_0 = applied voltage frequency
p = number of pole pairs

Lag

$$s = \frac{n_1 - n}{n_1}$$

n = r.p.m. of motor

Induced emf in rotor

$$\varepsilon_2 = s\,\varepsilon_{20}$$

ε_{20} = emf with stationary rotor

Rotor frequency

$$f_2 = s\,f_0$$

Motor torque

$$M \approx M_{\text{max}} \frac{2\,s\,s_{\text{m}}}{s_{\text{m}}^2 + s^2}$$

$$s_{\text{m}} = \frac{R_2 + R_{\text{o}}}{X_{20}}$$

$$M_{\text{max}} = k_5 \frac{V_1^2}{X_{20}}$$

s_{m} = lag at maximum torque
R_2 = resistance of rotor winding (per phase)
R_{o} = outer resistance of rotor winding (per phase)
X_{20} = reactance of rotor winding (per phase) with stationary rotor
V_1 = applied voltage

In particular, when $s \ll s_{\text{m}}$

$$M = k_6\,s\,V_1^2$$

Efficiency of motor

$$\eta = \frac{P_1 - P_{\text{p}}}{P_1} = \frac{P_2}{P_1}$$

P_1 = input electric power
P_2 = output mechanical power

$$P_{\text{p}} = P_0 + P_{\text{b1}} + P_{\text{b2}}$$

P_0 = leakage loss
P_{b1} = power loss in stator
$P_{\text{b2}} \approx s\,(P_1 - P_0 - P_{\text{b1}})$ ≈ power loss in rotor

4 Electronics

Quantity	Symbol	SI Unit
Temperature	T	K
Frequency	f	$\mathrm{Hz} = \mathrm{s}^{-1}$
Angular frequency	ω	rad/s
Current	I, i	A = C/s
Voltage	V, v	V = Nm/C
Resistance	R, r	Ω = V/A
Capacitance	C	F = C/V
Inductance	L	H = Vs/A
Complex impedance	Z	Ω
Power	P	W = VA
Gain (amplification factor)	A	

4.1 The PN Diode

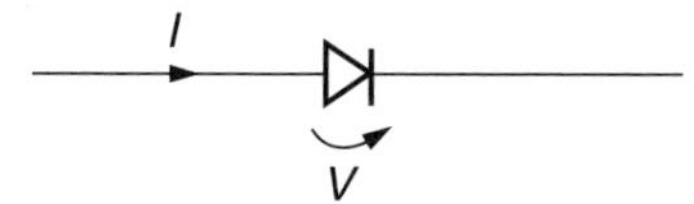

Diode equation

$$I = I_0 \left(\exp\left(\frac{V}{\eta V_T} \right) - 1 \right)$$

I_0 = saturation reverse current

η = ideality factor

$\eta_{\mathrm{Ge}} = 1$

$\eta_{\mathrm{Si}} \approx 2$

$$V_T = \frac{kT}{e} \approx 26 \text{ mV at room temperature}$$

k = Boltzmann constant

e = absolute value of electron charge

Temperature dependence of reverse current

$$I_0 \approx I_{01}\, 2^{\frac{T-T_1}{10}}$$

Differential diode resistance r_e

$$\frac{1}{r_\text{e}} = \frac{\mathrm{d}I}{\mathrm{d}V}$$

Capacitance of a reverse-biased diode

$$C = \frac{\text{constant}}{\sqrt{-V}}$$

4.2 The Bipolar Transistor

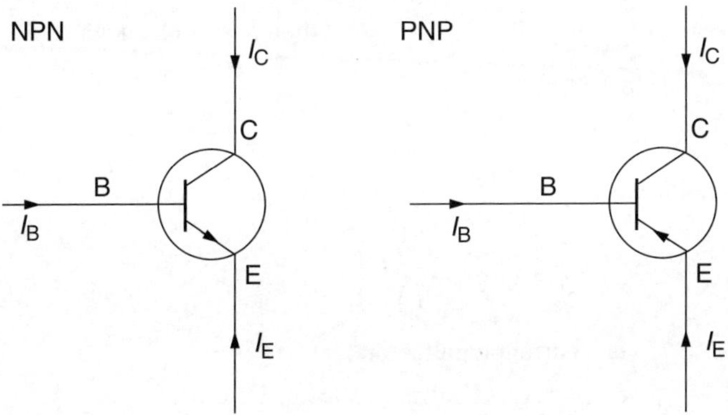

Low frequency equivalent circuit (CE configuration)

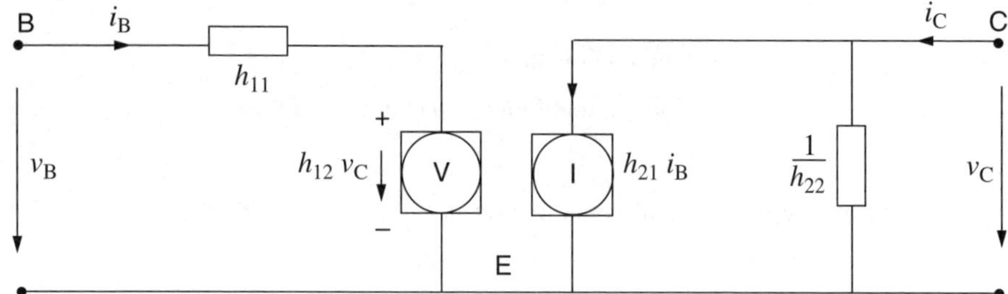

Hybrid parameters

$$h_{11} = h_{ie} = \left.\frac{\partial V_B}{\partial I_B}\right|_{V_{CE}} = \left.\frac{v_B}{i_B}\right|_{v_C = 0}$$

$$h_{12} = h_{re} = \left.\frac{\partial V_B}{\partial V_{CE}}\right|_{I_B} = \left.\frac{v_B}{v_C}\right|_{i_B = 0}$$

$$h_{21} = h_{fe} = \left.\frac{\partial I_C}{\partial I_B}\right|_{V_{CE}} = \left.\frac{i_C}{i_B}\right|_{v_C = 0}$$

$$h_{22} = h_{oe} = \left.\frac{\partial I_C}{\partial V_{CE}}\right|_{I_B} = \left.\frac{i_C}{v_C}\right|_{i_B = 0}$$

Signal equations

$$v_B = h_{11}\,i_B + h_{12}\,v_C$$

$$i_C = h_{22}\,i_B + h_{22}\,v_C$$

Often h_{12} can be neglected

Current gain

$$I_C = -\,\alpha I_E + I_{CO}\left(1 - \exp\frac{\pm V_{CB}}{\eta\,V_T}\right)$$

α = current gain factor

I_{CO} = reverse current collector-base

η and V_T are defined in Section 4.1

The sign of V_{CB} is chosen such that in the active region

$$I_C = -\,\alpha I_E + I_{CO}$$

Current gain in CE (common-emitter) configuration

$$I_C = (1 + \beta)\, I_{CO} + \beta I_B$$

$$\beta = \frac{\alpha}{1-\alpha}$$

For weak currents

$$\beta = \frac{I_C - I_{CB0}}{I_B + I_{CB0}}$$

I_{CB0} = collector current when $I_E = 0$

DC case when $|I_B| \gg |I_{CB0}|$

$$\beta_{DC} = h_{FE} = \frac{I_C}{I_B}$$

For signals

$$\beta = h_{fe} = \left.\frac{\partial I_C}{\partial I_B}\right|_{V_{CE}}$$

Condition of saturation

$$\beta_{DC}\, I_B > I_C$$

Thermal resistance

$$T_j - T_A = \Theta\, P_D$$

T_j = junction temperature

T_A = ambient temperature

Θ = thermal resistance in °C/W

P_D = power dissipated in crystal

4.3 The Field Effect Transistor (FET)

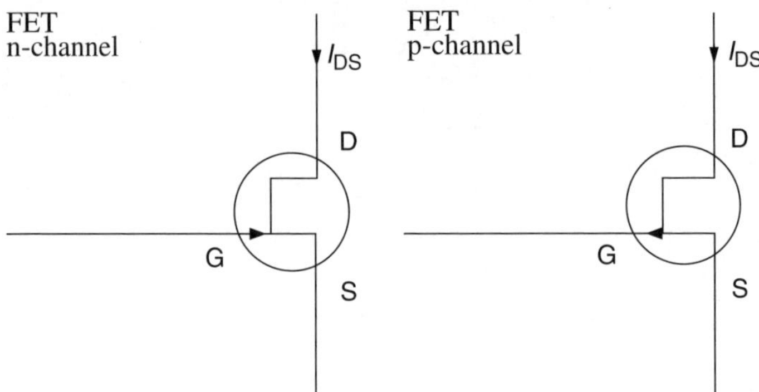

Low frequency equivalent circuit

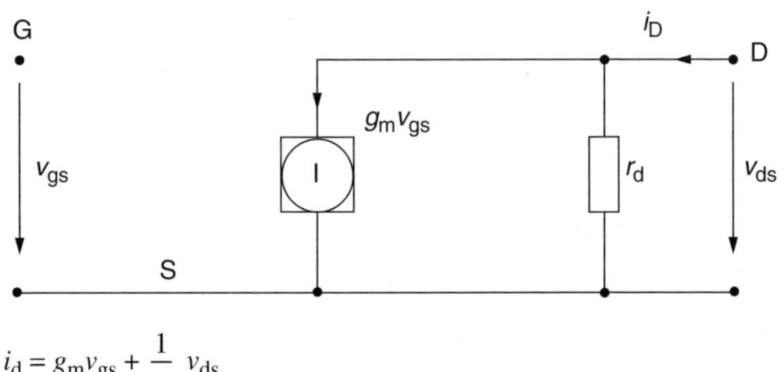

$$i_d = g_m v_{gs} + \frac{1}{r_d} v_{ds}$$

g_m = transistor transconductance

r_d= transistor drain resistance

Drain current

$$I_{DS} = I_{DSS}\left(1 - \frac{V_{GS}}{V_P}\right)^2$$

I_{DSS} = drain current of saturation when $V_{GS} = 0$

V_P = pinch-off voltage

Open circuit amplification factor

$$\mu = -\left.\frac{\partial V_{DS}}{\partial V_{GS}}\right|_{I_D} = -\left.\frac{v_{ds}}{v_{gs}}\right|_{i_d = 0} = r_d\, g_m$$

Transconductance

$$g_m = \left.\frac{\partial I_D}{\partial V_{GS}}\right|_{V_{DS}} = \left.\frac{i_d}{v_{gs}}\right|_{v_{ds}=0} = g_{m0}\left(1 - \frac{V_{GS}}{V_P}\right) = -\frac{2}{V_P}\sqrt{I_{DSS} I_{DS}}$$

$$g_{m0} = -\frac{2\,I_{DSS}}{V_P} = g_m \text{ when } V_{GS} = 0$$

Map of input/output relations for transistors

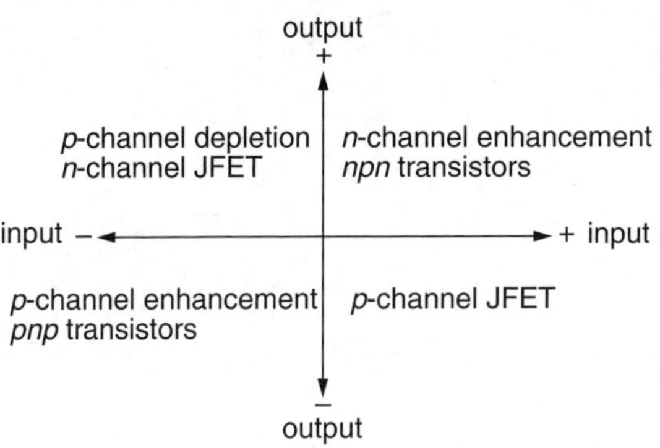

The MOS field-effect transistor

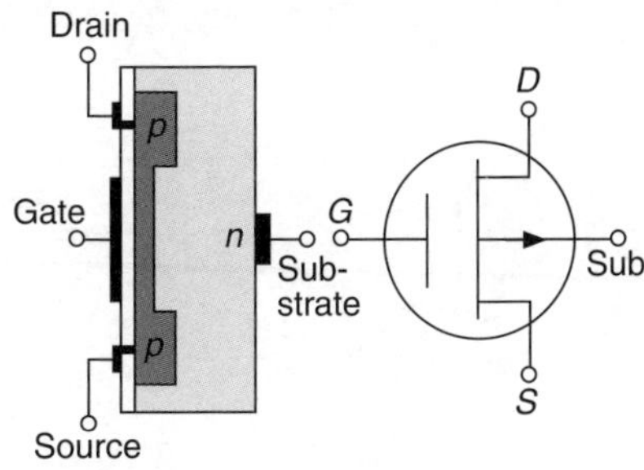

p-channel depletion mode MOSFET

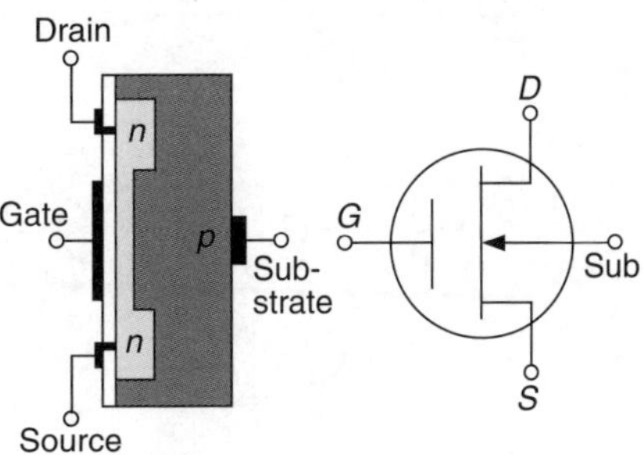

n-channel depletion mode MOSFET

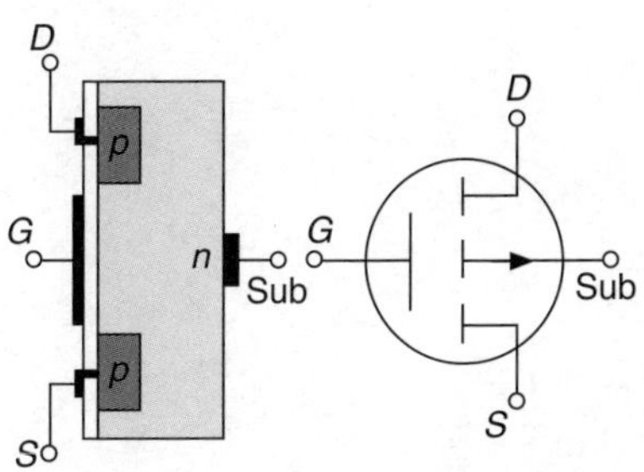

p-channel enhancement mode MOSFET

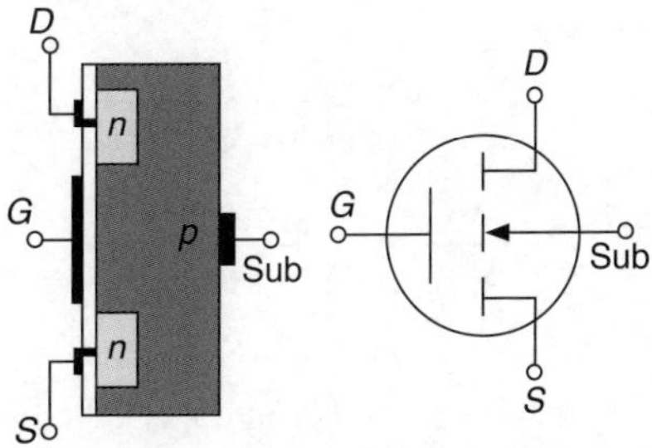

n-channel enhancement mode MOSFET

4.4 The Transistor as Amplifier

The Common-Emitter Amplifier with an Emitter Resistance

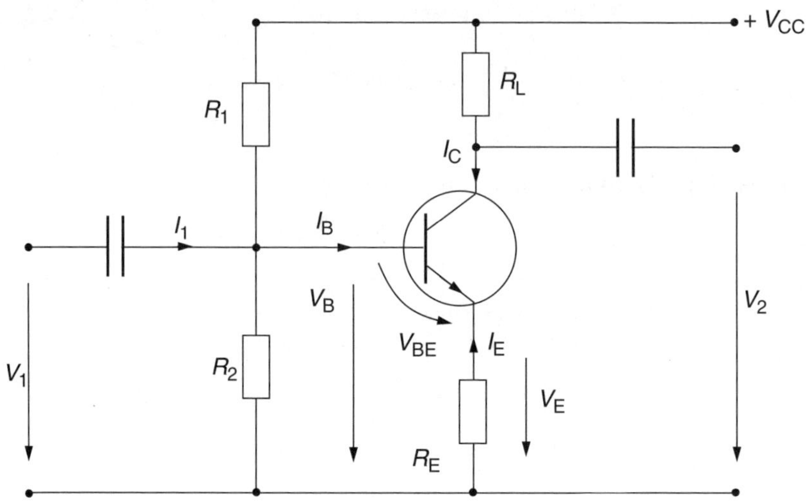

Operating point (quiescent point, Q point)

$$V_B = \frac{R_2(V_{CC} - R_1\, I_B)}{R_1 + R_2}$$

Stability factors

$$S_0 = \frac{\partial I_C}{\partial I_{CO}} = (1 + h_{FE})\,\frac{1 + R_P/R_E}{1 + h_{FE} + R_P/R_E} \approx 1 + \frac{R_P}{R_E}$$

$$R_P = \frac{R_1 \cdot R_2}{R_1 + R_2}$$

$$S_1 = \frac{I_C}{V_{BE}} = \frac{-h_{FE}}{R_P + (1 + h_{FE})\,R_E} \approx -\frac{1}{R_E}$$

$$S_2 = \frac{I_C}{h_{FE}} = \frac{I_C\, S_0}{h_{FE}\,(1 + h_{FE})}$$

$$\frac{\Delta I_C}{\Delta h_{FE}} = \frac{I_{C2} - I_{C1}}{h_{FE2} - h_{FE1}} = \frac{I_{C1}\, S_{02}}{h_{FE1}\,(1 + h_{FE2})}$$

$S_{02} = S_0$ when $I_C = I_{C2}$ and $h_{FE} = h_{FE2}$

h_{FE} and I_{CO} have been defined in Sec. 4.2

$$\Delta I_C = S_0 \Delta I_{CO} + S_1 \Delta V_{BE} + S_2 \Delta h_{FE}$$

Approximate low-frequency current and voltage gain of CE-stage (see also Sec. 4.2)

$$A_i = -\frac{i_C}{i_1} \approx -h_{fe}\frac{R_P}{R_B + R_P} \qquad R_B = h_{ie} + (1 + h_{fe})R_E$$

$$A_v = \frac{v_2}{v_1} = A_i\frac{R_L}{R_{in}}$$

Input resistance

$$R_{in} = \frac{v_1}{i_1} \approx \frac{R_P \cdot R_B}{R_P + R_B}$$

Approximate emitter follower current and voltage gain

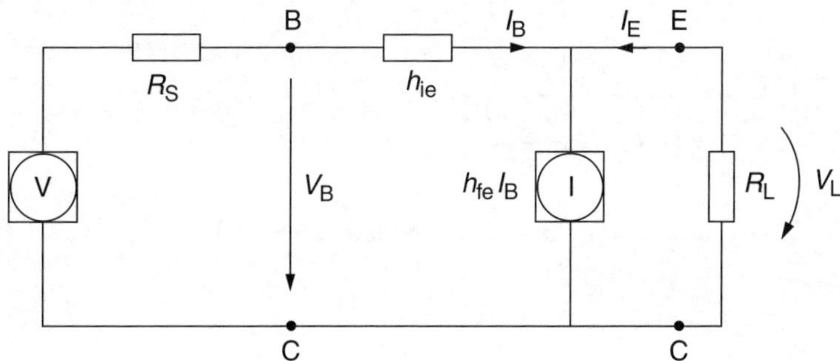

$$A_i = -\frac{i_E}{i_B} \approx 1 + h_{fe}$$

$$A_v = \frac{v_L}{v_B} \approx \frac{(1 + h_{fe})R_L}{Z_i}$$

$$\text{where } Z_i = \frac{v_B}{i_B} \approx h_{ie} + R_L(1 + h_{fe})$$

Field-effect transistor as amplifier

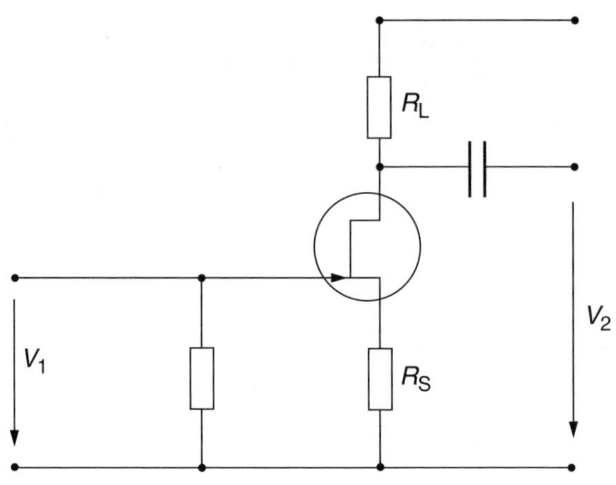

$$A_v = \frac{v_2}{v_1} = \frac{g_m R_L}{1 + g_m R_S + \dfrac{R_S + R_L}{r_d}}$$

g_m and r_d have been defined in Sec. 4.3.

CE-stage equivalent circuit and hybrid-π parameters for high frequencies

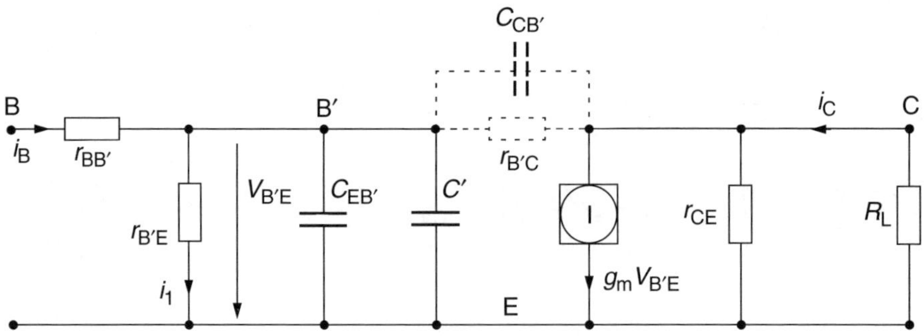

$$C' = C_{CB'}\left(1 + g_m \frac{R_L\, r_{CE}}{R_L + r_{CE}}\right)$$

$C_{CB'}$ = collector-base capacitance

$$g_m \approx \frac{|I_C|}{\eta\, V_T}$$

η and V_T are defined in Sec. 4.1

Higher cut-off frequency of current gain (3 dB limit)

$$f_\beta = \frac{1}{2\pi\, r_{B'E}(C_{EB'} + C')}$$

Current gain with short-circuited output

$$\beta = h_{fe} = \frac{i_C}{i_1} = \frac{\beta_0}{1 + \mathrm{j}\dfrac{f}{f_\beta}}$$

$\beta_0 = g_m\, r_{B'E}$ $\qquad\qquad$ $\mathrm{j} = \sqrt{-1}$

The gain-bandwidth product (transition frequency)

$f_T \approx f_\beta \beta_0$ $\qquad\qquad$ $f_T \approx f\beta$ when $f > f_\beta$

f_T = the frequency when $\beta = 1$, output short-circuited

Relations among hybrid parameters

$$h_{ie} = r_{BB'} + r_{B'E}$$

$$\frac{1}{r_{CE}} = h_{oe} - \frac{1 + h_{fe}}{r_{B'C}}$$

$$h_{re} = \frac{R_{B'E}}{r_{B'E} + r_{B'C}} \approx \frac{r_{B'E}}{r_{B'C}}$$

h_{ie}, h_{re}, h_{fe}, and h_{oe} are defined in Sec. 4.2

4.5 Filters and Cut-Off Frequencies

Gain in dB

$$A(\text{dB}) = 10 \lg \frac{P_2}{P_1} = 20 \lg \left|\frac{V_2}{V_1}\right|$$

(V_1 and V_2 to be measured over identical resistances)

High-pass filter

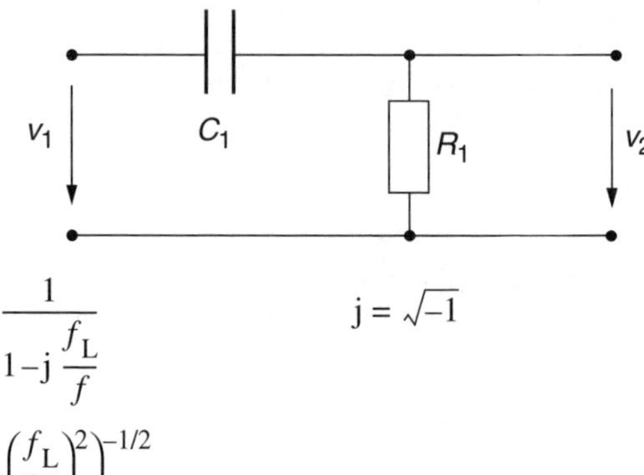

$$A = \frac{v_2}{v_1} = \frac{1}{1 - \text{j}\,\dfrac{f_\text{L}}{f}} \qquad \text{j} = \sqrt{-1}$$

$$|A| = \left(1 + \left(\frac{f_\text{L}}{f}\right)^2\right)^{-1/2}$$

Cut-off frequency (3 dB)

$$f_\text{L} = \frac{1}{2\pi R_1 C_1}$$

Phase shift of v_2

$$\theta = \arctan \frac{f_\text{L}}{f}$$

Low-pass filter

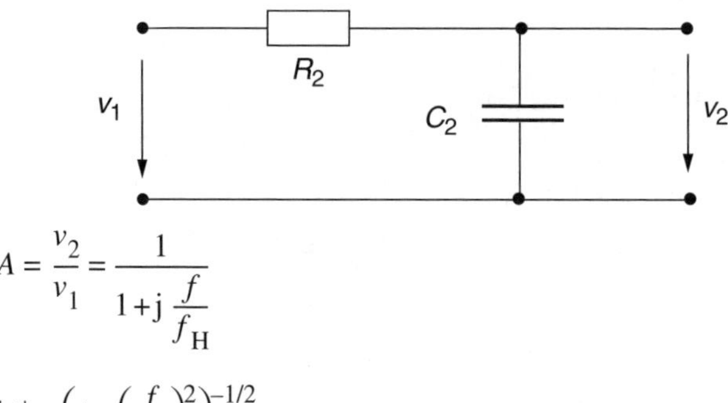

$$A = \frac{v_2}{v_1} = \frac{1}{1 + \text{j}\,\dfrac{f}{f_\text{H}}}$$

$$|A| = \left(1 + \left(\frac{f}{f_\text{H}}\right)^2\right)^{-1/2}$$

Cut-off frequency (3 dB)

$$f_H = \frac{1}{2\pi R_2 C_2}$$

Phase shift of v_2

$$\theta = -\arctan \frac{f}{f_H}$$

Step response of low-pass filter

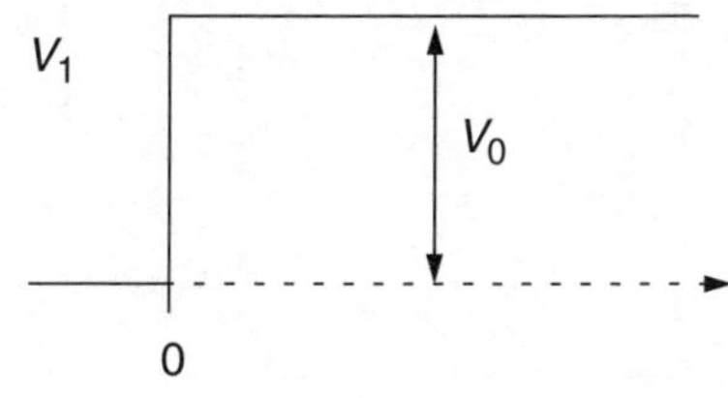

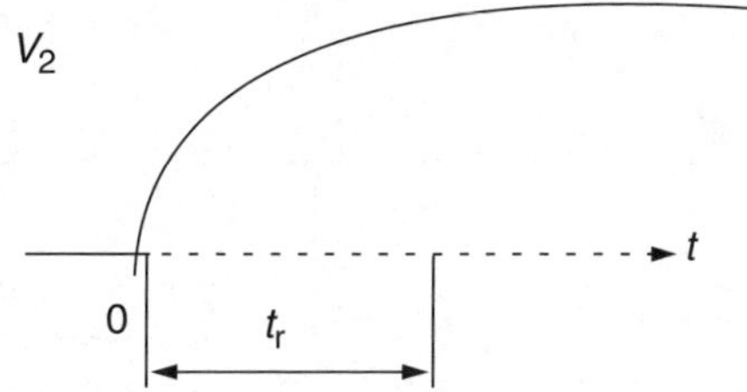

$$V_2(t) = V_0 \left(1 - e^{-\frac{t}{R_2 C_2}}\right)$$

Rise time

$$t_r = t_{90\%} - t_{10\%} = 2.20\, R_2 C_2 = \frac{0.35}{f_H}$$

Cascading transistor amplifier of *n* independent stages

$$2 = \left(1 + \left(\frac{f_H^*}{f_{H1}}\right)^2\right) \cdot \left(1 + \left(\frac{f_H^*}{f_{H2}}\right)^2\right) \cdot \ldots \cdot \left(1 + \left(\frac{f_H^*}{f_{Hn}}\right)^2\right)$$

$$2 = \left(1 + \left(\frac{f_{L1}}{f_L^*}\right)^2\right) \cdot \left(1 + \left(\frac{f_{L2}}{f_L^*}\right)^2\right) \cdot \ldots \cdot \left(1 + \left(\frac{f_{Ln}}{f_L^*}\right)^2\right)$$

f_L^* and f_H^* are the lower and higher cut-off frequency (3 dB) of the cascading amplifier.

With an error < 10 %, the rise time is

$$t_r^* \approx 1.1 \sqrt{t_{r1}^2 + t_{r2}^2 + \ldots + t_{rn}^2}$$

Butterworth filter

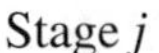

Stage j

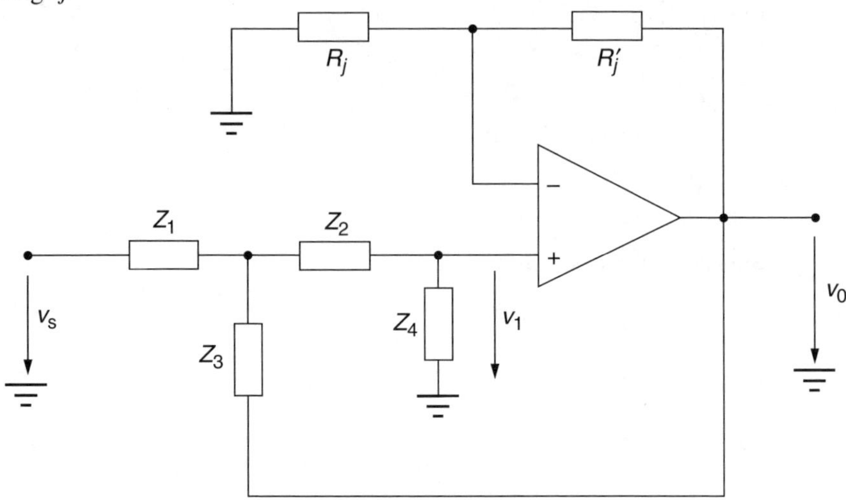

Low-pass version

$$Z_1 = Z_2 = R$$

$$Z_3 = Z_4 = \frac{1}{\mathrm{j}\omega C}$$

High-pass version

$$Z_1 = Z_2 = \frac{1}{\mathrm{j}\omega C}$$

$$Z_3 = Z_4 = R$$

Total cut-off (3 dB)

$$f_0 = \frac{\omega_0}{2\pi} = \frac{1}{2\pi RC}$$

Normalized Butterworth polynomials

n	$B_n(s)$
1	$s + 1$
2	$s^2 + 1.414\, s + 1$
3	$(s + 1)\,(s^2 + s + 1)$
4	$(s^2 + 0.765\, s + 1)\,(s^2 + 1.848\, s + 1)$
$2m$	$\prod_{j=1}^{m} (s^2 + 2k_j\, s + 1)$
$2m + 1$	$(s + 1) \prod_{j=1}^{m} (s^2 + 2k_j\, s + 1)$

Damping factors

$$k_j = \cos\,\theta_j$$

$$\theta_j = \begin{cases} \dfrac{j\pi}{n} & j = 1, 2, \ldots \dfrac{n-1}{2} \quad \text{for odd } n \\[2ex] \dfrac{j\pi}{n} + \dfrac{\pi}{2n} & j = 1, 2, \ldots \dfrac{n-2}{2} \quad \text{for even } n \end{cases}$$

Adjustment of the gain of each stage

$$A_j = 3 - 2k_j$$

$$A_j = \frac{v_0}{v_1} = 1 + \frac{R_j'}{R_j}$$

Total gain of n-th order low-pass filter

$$A = \frac{v_0}{v_s} = \frac{\prod_{j=1}^{m} A_j}{B_n\left(\dfrac{s}{\omega_0}\right)}$$

s = complex angular frequency

Total gain of n-th order high-pass filter

$$A = \frac{v_0}{v_s} = \frac{\prod_{j=1}^{m} A_j}{B_n\left(\dfrac{\omega_0}{s}\right)}$$

4.6 Feedback

Effective gain in feedback systems

$$A_{\text{eff}} = \frac{A}{1 - \beta A}$$

A = gain without feedback
β = reverse transmission factor

Condition of oscillation (the Barkhausen criterion)

$$|\beta A| = 1$$

$$\arg\,(\beta A) = 0$$

Phase and amplitude margins

$$\Phi_m = \pi - \arg(\beta A) \qquad \text{when } |\beta A| = 1$$

$$M = \frac{1}{|\beta A|} \qquad \text{when } I_m(\beta A) = 0$$

Ideal feedback operational amplifier

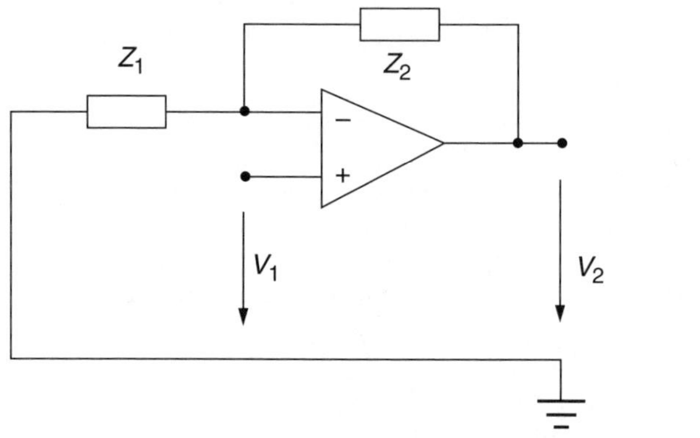

$$\frac{V_2}{V_1} = \left(1 + \frac{Z_2}{Z_1}\right)$$

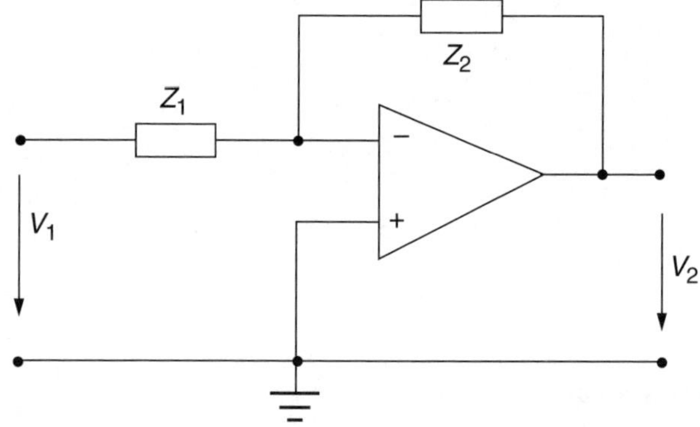

$$\frac{V_2}{V_1} = -\frac{Z_2}{Z_1}$$

In particular,

$$V_2 = -\frac{1}{RC}\int V_1 \, dt \qquad \text{when } Z_1 = R \text{ and } Z_2 = \frac{1}{j\omega C}$$

$$V_2 = -RC\frac{dV_1}{dt} \qquad \text{when } Z_1 = \frac{1}{j\omega C} \text{ and } Z_2 = R$$

llator

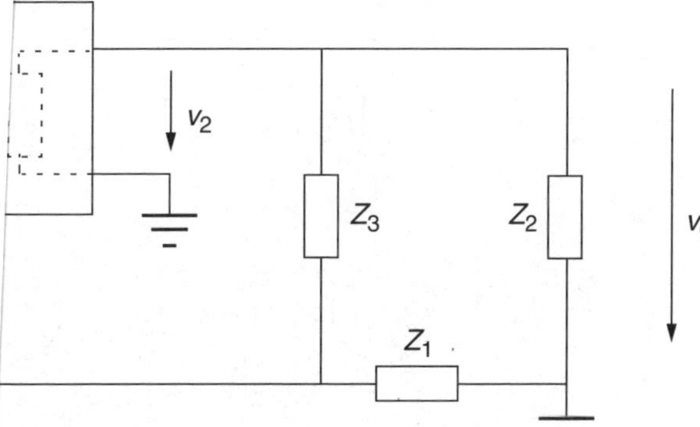

$$A_0 = \frac{v_2}{v_1}, \text{ open circuit } (A_0 < 0)$$

$$\beta = \frac{Z_1}{Z_1 + Z_3}$$

$$A = \frac{v_0}{v_1} = \frac{A_0 Z_2 (Z_1 + Z_3)}{Z_2(Z_1 + Z_3) + R_0(Z_1 + Z_2 + Z_3)}$$

Oscillation condition

$$Z_1 + Z_2 + Z_3 = 0, \qquad -\frac{A_0 Z_1}{Z_2} = 1$$

The Hartley oscillator

$$Z_1 = \mathrm{j}\omega L_1 \qquad Z_2 = \mathrm{j}\omega L_2 \qquad Z_3 = \frac{1}{\mathrm{j}\omega C_3}$$

The Colpitt oscillator

$$Z_1 = \frac{1}{\mathrm{j}\omega C_1} \qquad Z_2 = \frac{1}{\mathrm{j}\omega C_2} \qquad Z_3 = \mathrm{j}\omega L_3$$

4.7 The Differential Amplifier

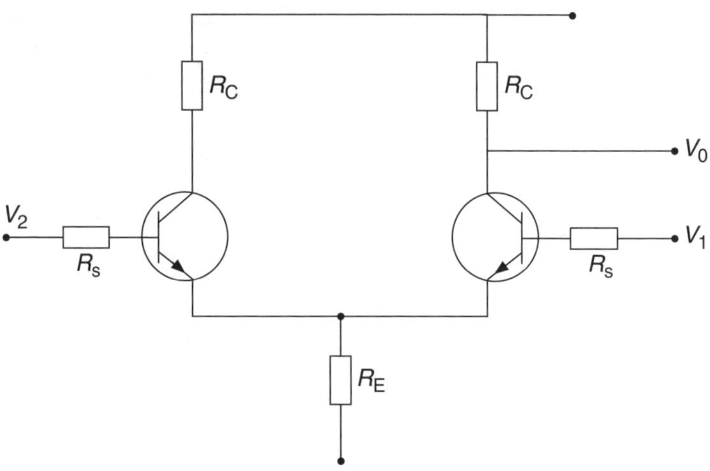

$$V_0 = A_D(V_2 - V_1) + A_C \frac{1}{2}(V_1 + V_2)$$

Differential gain

$$A_D = \frac{h_{fe}\, R_C}{R_s + h_{ie}} \cdot \frac{1}{2}$$

Common gain

$$A_C = \frac{(2h_{oe}R_E - h_{fe})\, R_C}{2R_E(1 + h_{fe}) + (R_s + h_{ie})\,(2h_{oe}R_E + 1)}$$

$$A_C \approx -\frac{1}{2}\,\frac{R_C}{R_E}$$

h_{ie}, h_{fe} and h_{oe} have been defined in Sec. 4.2

Common-mode rejection ratio, CMRR

$$\rho = \left|\frac{A_D}{A_C}\right|$$

Bias current

$$I_B = \frac{I_C}{h_{FE}}$$

h_{FE} is defined in Sec. 4.2

4.8 Transmission Lines

L = Line inductance per metre
C = Line capacitance per metre

Characteristic impedance of the line

$$Z_0 = \sqrt{\frac{L}{C}}$$

Delay per metre

$$T_0 = \sqrt{LC}$$

Wave velocity

$$v = \frac{1}{T_0}$$

Reflection coeffient

$$\Gamma = \frac{Z_L - Z_0}{Z_L + Z_0}$$

Z_L = impedance of load

Standing-wave relation

$$SWR = \frac{1 + |\Gamma|}{1 - |\Gamma|}$$

4.9 Digital Circuits and Boolean Algebra

Boolean algebra

		AND	OR	NOT	NAND	NOR	EXCLU-SIVE OR
x	y	$x \cdot y$	$x + y$	$\overline{x}$	$\overline{x \cdot y}$	$\overline{x + y}$	$x \oplus y$
1	1	1	1	0	0	0	0
1	0	0	1	0	1	0	1
0	1	0	1	1	1	0	1
0	0	0	0	1	1	1	0

See also table T–3.4.

de Morgan's laws

$$\overline{x+y} = \overline{x} \cdot \overline{y}$$

$$\overline{x \cdot y} = \overline{x} + \overline{y}$$

Consensus

$$x \cdot y + \overline{x} \cdot z = x \cdot y + \overline{x} \cdot z + y \cdot z$$

$$(x + y) \cdot (\overline{x} + z) = (x + y) \cdot (\overline{x} + z) \cdot (y + z)$$

Absorption

$$x + x \cdot y = x$$

$$x \cdot (x + y) = x$$

Modulo 2 addition and its inverse

$$x \oplus y = x \cdot \overline{y} + \overline{x} \cdot y$$

$$\overline{x \oplus y} = x \cdot y + \overline{x} \cdot \overline{y}$$

Function table for SR, JK, T, and D flip-flops

q	**q⁺**	**S**	**R**	**J**	**K**	**T**	**D**
0	0	0	0	0	–	0	0
0	1	1	0	1	–	1	1
1	0	0	1	–	1	1	0
1	1	0	0	–	0	0	1

q^+ = result

5 Waves

Quantity	Symbol	SI Unit
Positon vector	$\boldsymbol{r}$ (r)	m
Period	T	s
Wave length	λ	m
Frequency	ν	$\mathrm{Hz} = \mathrm{s}^{-1}$
Angular frequency	ω	rad/s
Phase velocity of waves	c	m/s
Velocity	$\boldsymbol{v}$ (v)	m/s
Wave number	σ	m^{-1}
Circular wave number	k	rad/m
Area	A	m^2
Volume	V	m^3
Force	$\boldsymbol{F}$ (F)	$\mathrm{N} = \mathrm{kg\ m/s}^2$
Density	ρ	$\mathrm{kg/m}^3$
Energy	W	J = Nm
Energy density	w	$\mathrm{J/m}^3$
Intensity, irradiance	I	$\mathrm{W/m}^2 = \mathrm{J/m}^2\ \mathrm{s}$
Electric field	$\boldsymbol{E}$ (E)	V/m = N/As
Relative dielectric constant	ε_r	
Refractive index	n	
Magnetic flux density	$\boldsymbol{B}$ (B)	$\mathrm{T} = \mathrm{Vs/m}^2$
Relative permeability	μ_r	

5.1 Wave Motion

The differential equation (wave equation) of wave propagation in one dimension and its general solution

$$\frac{\partial^2 \xi}{\partial x^2} = \frac{1}{c^2}\,\frac{\partial^2 \xi}{\partial t^2}$$

$$\xi = f(x + ct) + g(x - ct)$$

Solution in the case of harmonic waves

$$\xi = \xi_0 \sin(k(x - ct) - \delta)$$

$$\xi = \xi_0 \, e^{i(k(x - ct) - \delta)}$$

Fundamental relations

$$\lambda = \frac{2\pi}{k} = Tc$$

$$c = \lambda \nu = \frac{\omega}{k}$$

$$T = 1/\nu = \frac{2\pi}{kc}$$

$$\omega = 2\pi\nu = kc$$

$$\sigma = 1/\lambda$$

Phase of the wave when $\delta = 0$

$$\phi = k(x - ct) = kx - \omega t = 2\pi\left(\frac{x}{\lambda} - \nu t\right) = 2\pi\left(\frac{x}{\lambda} - \frac{t}{T}\right) = 2\pi\nu\left(\frac{x}{c} - t\right) = 2\pi(\sigma x - \nu t)$$

Longitudinal elastic wave (sound wave) in a rod

$$c_\ell = \sqrt{\frac{Y}{\rho}}$$

Y = Young's modulus of elasticity

$$\frac{F}{A} = Y\frac{\partial \xi}{\partial x}$$

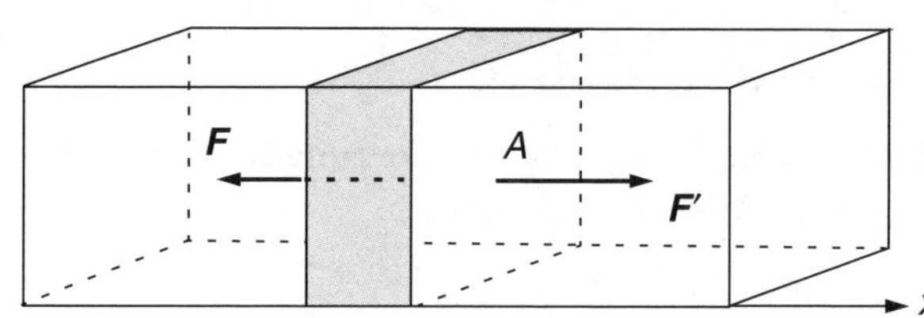

Transverse elastic wave in a rod

$$c_t = \sqrt{\frac{G}{\rho}}$$

G = shear modulus

$$\frac{F}{A} = G\frac{\partial \xi}{\partial x}$$

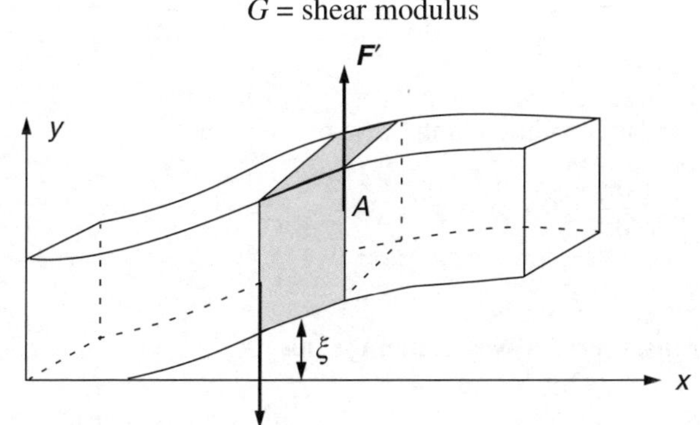

Longitudinal wave in a spring

$$c = \sqrt{\frac{k\ell}{\rho_\ell}}$$

ρ_ℓ = linear density

k = spring constant

$$F = k\,\Delta\ell$$

ℓ = length of spring

Transverse wave in a string

$$\rho_\ell \frac{\partial^2 \xi}{\partial t^2} = \frac{\partial}{\partial x}\left(F\frac{\partial \xi}{\partial x}\right) + f_y$$

F = tension

f_y = external force per unit length in the direction of disturbance

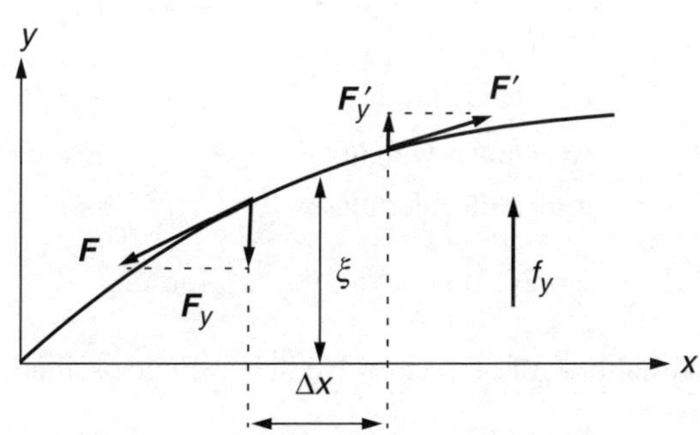

In particular, in the case $f_y = 0$, F = constant

$$c = \sqrt{\frac{F}{\rho_\ell}} \qquad \frac{F'_y - F_y}{\Delta x} = F\,\frac{\partial^2 \xi}{\partial x^2}$$

Boundary condition in the case of loose end

$$\frac{\partial \xi}{\partial x} = 0$$

Transerverse waves in membrane

$$c = \sqrt{\frac{\tau}{\sigma_0}}$$

$\tau =$ tension per unit length
$\sigma_0 =$ surface density

Pressure wave in gas (sound wave)

$$\rho - \rho_0 = -\,\rho_0\,\frac{\partial \xi}{\partial x}$$

$$p - p_0 = -\,B\,\frac{\partial \xi}{\partial x}$$

p = pressure

Bulk modulus

$$B = \gamma p = \rho_0 \left(\frac{\partial p}{\partial \rho}\right)_0 \qquad \gamma = \frac{C_p}{C_v}$$

Phase velocity

$$c = \sqrt{\frac{B}{\rho_0}} = \sqrt{\frac{\gamma R\, T_0}{M}} = \sqrt{\frac{\gamma k\, T_0}{\mu}} = \alpha\sqrt{T_0}$$

$T_0 =$ temperature
$M =$ molar weight
$\mu =$ molecular mass

$k =$ Boltzmann constant
$R =$ universal gas constant

$\alpha = 20.055$ m s^{-1} $K^{-1/2}$ for air

Amplitude of pressure wave in harmonic oscillation case

$$P_0 = c\omega\,\rho_0\,\xi_0$$

Velocity potential Φ for sound waves when curl $\boldsymbol{v} = 0$

$\boldsymbol{v} = -\operatorname{grad} \Phi$ — $\boldsymbol{v} = \boldsymbol{v}(x, y, z, t)$ = velocity field of gas

$$\nabla^2 \Phi = \frac{1}{c^2} \frac{\partial^2 \Phi}{\partial t^2}$$

$$\rho_0 \frac{\partial \Phi}{\partial t} = c^2 (\rho - \rho_0)$$

Pressure wave in liquid

$$c = \sqrt{\frac{B}{\rho}}$$ B = bulk modulus

Electromagnetic wave

$$c = 1/\sqrt{\varepsilon_r \varepsilon_0 \mu_r \mu_0}$$

Water waves

$c = \sqrt{\dfrac{g\lambda}{2\pi} + \dfrac{2\pi S}{\rho\lambda}}$ for $d \gtrsim \lambda/4$ — S = surface tension (≈ 0.074 N/m), ρ = density of the water

$c = \sqrt{gd}$ for $d \lesssim \lambda/12$ — d = depth of water

For tsunami waves $T \approx 10^3$ s and d $\ll \lambda$
For gravitational waves $\omega^2 = gk \tanh(kd)$

Differential equation for wave propagation in three dimensions

$$\nabla^2 \xi = \frac{1}{c^2} \frac{\partial^2 \xi}{\partial t^2}$$

Solution for plane harmonic waves

$\xi = \xi_0 \sin(\boldsymbol{k} \cdot \boldsymbol{r} - \omega t - \delta)$ — $\boldsymbol{k}$ = wave vector

$$\xi = \xi_0 \, e^{i(\boldsymbol{k} \cdot \boldsymbol{r} - \omega t - \delta)}$$

Solution in the case of spherical harmonic waves

$\xi = \dfrac{a}{r} \sin(\boldsymbol{k} \cdot \boldsymbol{r} - \omega t - \delta)$ — a = constant

5.2 Superposition of Waves

Beats

$$\nu = |\nu_1 - \nu_2|$$

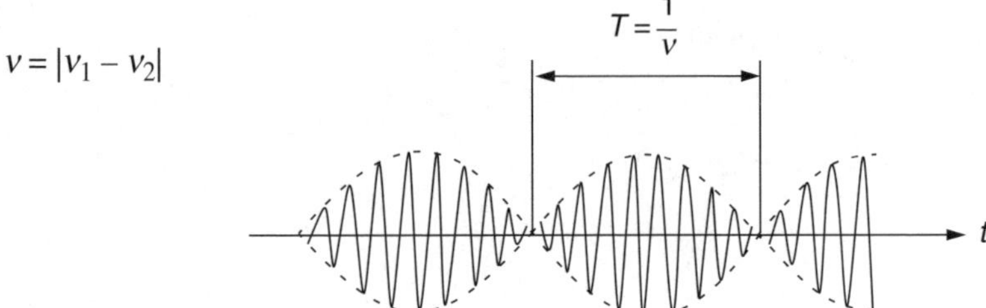

Phase velocity and group velocity

$$c = \frac{\omega}{k}$$

$$v_g = \frac{\partial \omega}{\partial k} = \frac{\partial \nu}{\partial \sigma} = c + k\,\frac{\partial c}{\partial k}$$

Standing wave in one dimension

$$\xi = f(x) \cdot g(t) = (a \sin kx + b \cos kx) \sin(\omega t - \delta)$$

a and b are constants

Allowed frequencies in three-dimensional case (rectangular cavity)

$$\nu = \frac{c}{2}\sqrt{\left(\frac{m_1}{\ell_1}\right)^2 + \left(\frac{m_2}{\ell_2}\right)^2 + \left(\frac{m_3}{\ell_3}\right)^2}$$

m_1, m_2 and m_3 are integers

ℓ_1, ℓ_2 and ℓ_3 are lengths of edges

Number of modes of oscillations between ν and $\nu + \mathrm{d}\nu$ in a body of volume V

$$N(\nu)\mathrm{d}\nu = 4\pi V \nu^2 \left(\frac{1}{c_\ell^3} + \frac{2}{c_t^3}\right)\mathrm{d}\nu$$

c_ℓ and c_t = phase velocities of longitudinal and transverse waves. In case there are no longitudinal (transverse) waves, one must set $1/c_\ell^3 = 0$ ($1/c_t^3 = 0$).

5.3 Energy Transport

Power of a longitudinal elastic wave

$$\frac{\mathrm{d}W}{\mathrm{d}t} = -F\,\frac{\partial\xi}{\partial t} = -YA\,\frac{\partial\xi}{\partial x}\,\frac{\partial\xi}{\partial t}$$

The symbols are taken from Sec. 5.1

In particular, energy density and intensity of a harmonic wave

$$\xi = \xi_0 \sin(kx - \omega t)$$

$$\langle w\rangle = \frac{1}{2}\,\rho\,\omega^2\,\xi_0^2$$

$$I = \frac{\langle\frac{\mathrm{d}W}{\mathrm{d}t}\rangle}{A} = c\langle w\rangle$$

Example Harmonic pressure wave

$$\langle w\rangle = \frac{P_0^2}{2c^2\rho_0} \qquad I = \frac{P_0^2}{2c\,\rho_0}$$

P_0 = amplitude as in Sec. 5.1

Acoustic intensity in decibel

$$B = 10\lg(I/I_0) \qquad I_0 = 10^{-12}\ \mathrm{Wm}^{-2}$$

Intensity of a spherical harmonic pressure wave $p - p_0 = \frac{P_0}{r}\sin(kr - \omega t)$

$$I = c\langle w\rangle = \frac{P_0^2}{2c\,\rho_0\,r^2} = \frac{I_0}{r^2}$$

(r is taken to be dimensionless)

Power of a transverse wave

$$\frac{\mathrm{d}W}{\mathrm{d}t} = F_y\,\frac{\partial\xi}{\partial t}$$

The symbols are taken from Sec. 5.1

Plane electromagnetic waves

$E = cB$

E

k

B

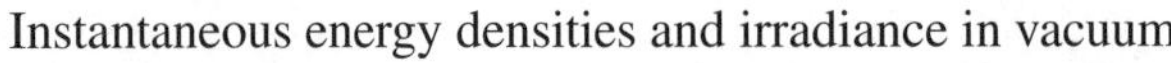

Instantaneous energy densities and irradiance in vacuum

$w_E = \frac{1}{2}\,\varepsilon_0\,E^2$ $\qquad$ $w_B = \frac{1}{2\mu_0}\,B^2$

$w = w_E + w_B = \varepsilon_0\,E^2$ $\qquad$ $I = c_0\langle w\rangle$

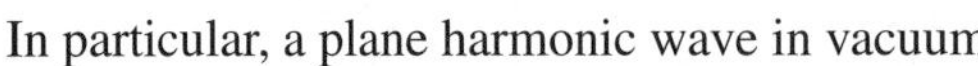

In particular, a plane harmonic wave in vacuum

$E = E_0 \sin\,(kx - \omega t)$

$I = c_0\langle w\rangle = \frac{1}{2}\,c_0\,\varepsilon_0\,E_0^2$

Irradiance of a plane harmonic wave in a dielectric media with refractive index n

$I = c\,\langle w\rangle = \frac{1}{2}\,c\,\varepsilon_r\,\varepsilon_0\,E_0^2 = \frac{1}{2}\,n\,c_0\,\varepsilon_0\,E_0^2$

Poynting's vector

$\boldsymbol{S} = \boldsymbol{E} \times \boldsymbol{H}$ $\qquad$ $\langle|\boldsymbol{S}|\rangle = I$

$\boldsymbol{H} = \frac{1}{\mu_r\mu_0}\,\boldsymbol{B}$ = magnetizing field

Momentum and angular momentum per volume in vacuum

$\boldsymbol{p} = \varepsilon_0\,\boldsymbol{E} \times \boldsymbol{B}$ $\qquad$ $\langle p\rangle = \frac{\langle w\rangle}{c_0} = \frac{I}{c_0^2}$

$\boldsymbol{L} = \boldsymbol{r} \times \boldsymbol{p}$

Radiation pressure

$P_s = cp\,\cos^2\theta$ $\quad$ at total absorption

$P_s = 2\,cp\,\cos^2\theta$ $\quad$ at total reflection

$P_s = \frac{1}{3}\,pc$ $\quad$ at total isotropic absorption

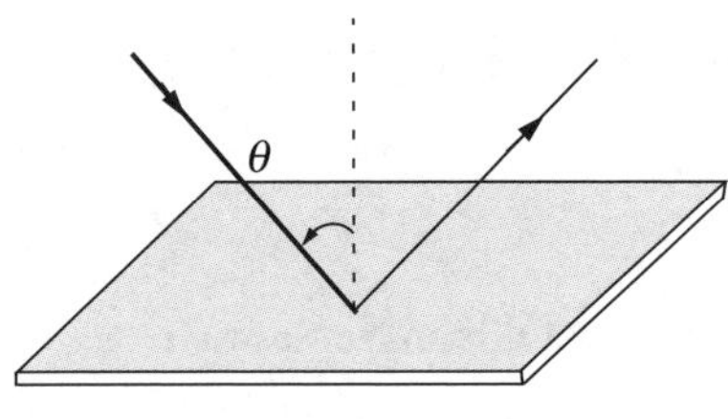

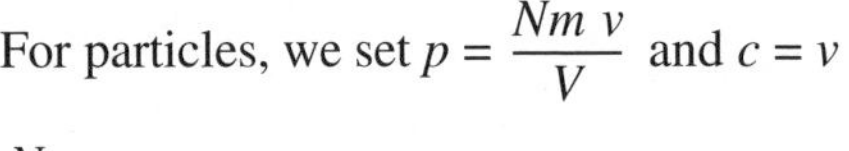

For particles, we set $p = \frac{Nm\,v}{V}$ and $c = v$

$\frac{N}{V}$ = number of molecules, of mass m and velocity v, per volume

Radiation from an accelerated charge Q ($v \ll c_0$)

$$|S| = \frac{\mu_0 Q^2 \dot{v}^2 \sin^2 \theta}{16\pi^2 c_0 r^2}$$

$$\frac{dW}{dt} = \frac{\mu_0 Q^2 \dot{v}^2}{6\pi c_0} = \frac{Q^2 \dot{v}^2}{6\pi \varepsilon_0 c_0^3}$$

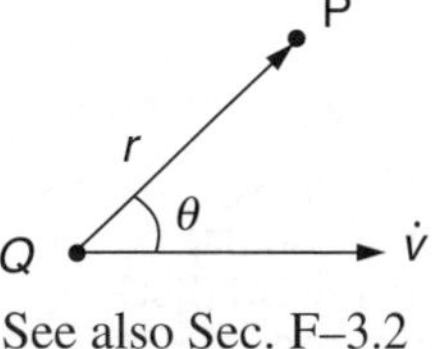

See also Sec. F–3.2

Irradiance and total radiated power from an oscillating electric dipole $Q z_0 \sin \omega t$

$$I = \frac{\mu_0 Q^2 z_0^2 \omega^4 \sin^2\theta}{32\pi^2 c_0 r^2}$$

$$\left\langle \frac{dW}{dt} \right\rangle = \frac{\mu_0 Q^2 z_0^2 \omega^4}{12\pi c_0}$$

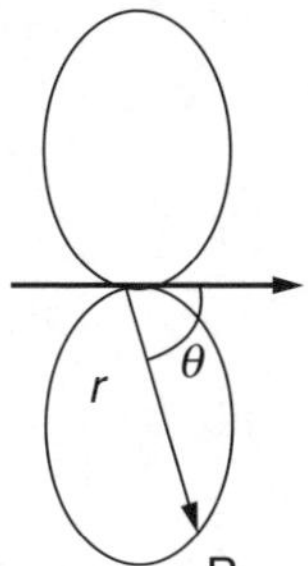

Total power radiated from electron synchrotron or storage ring

$$P = 88.5 \frac{I E^4}{R} = 26.5\, IBE^3$$

P = power	expressed in units of W
I = electron current	" " " " mA
E = electron energy	" " " " GeV
R = radius of electron orbit	" " " " m
B = magnetic flux density in bending magnet	" " " " T

Characteristic wavelength, characteristic frequency, and characteristic photon energy for synchrotron radiation

$$\lambda_c = 0.56\, R/E^3 = 1.86/BE^2$$

$$\nu_c = 540\, E^3/R = 162\, BE^2$$

$$\varepsilon_c = 2.2\, E^3/R = 0.67\, BE^2$$

λ_c	expressed in units of nm
ν_c	" " " " PHz
ε_c	" " " " keV
E	" " " " GeV
R	" " " " m
B	" " " " T

Spectral curve for synchrotron radiation

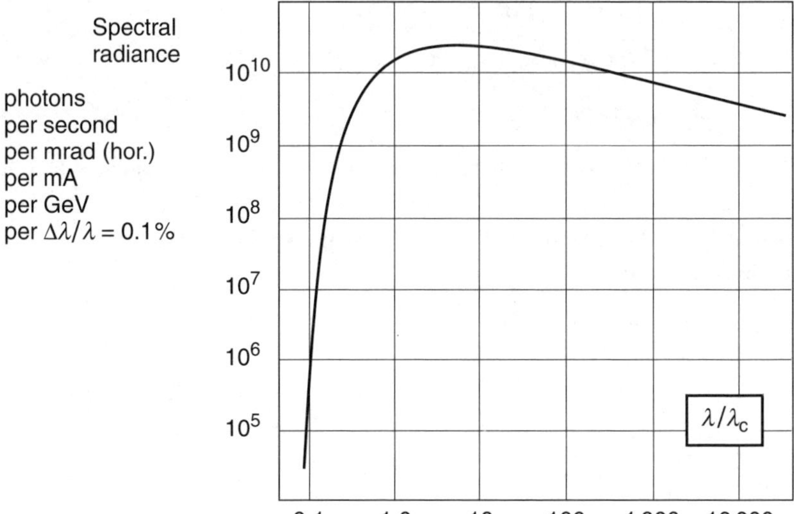

5.4 Doppler Effect

Mechanical waves

Sender S and observer O moving in a straight line ($c > v_S$, $c > v_O$)

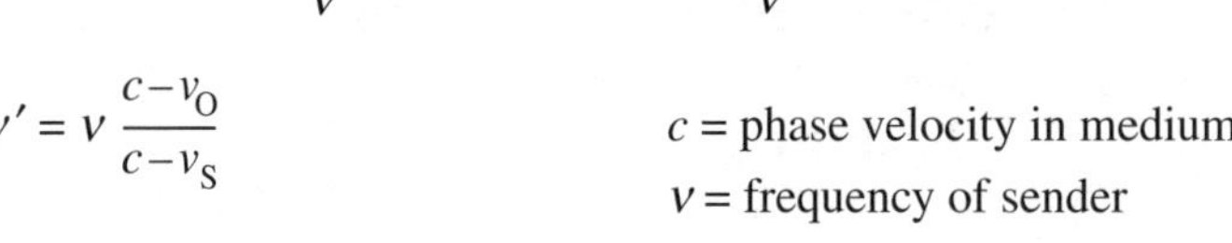

$$\nu' = \nu \frac{c - v_O}{c - v_S}$$

c = phase velocity in medium
ν = frequency of sender
v_S = velocity of sender
v_O = velocity of observer

S at rest and O moving at speed v_O *relative to medium* ($v_O \ll c$)

$$\nu' = \nu \left(1 - \frac{v_O}{c} \cos\theta\right)$$

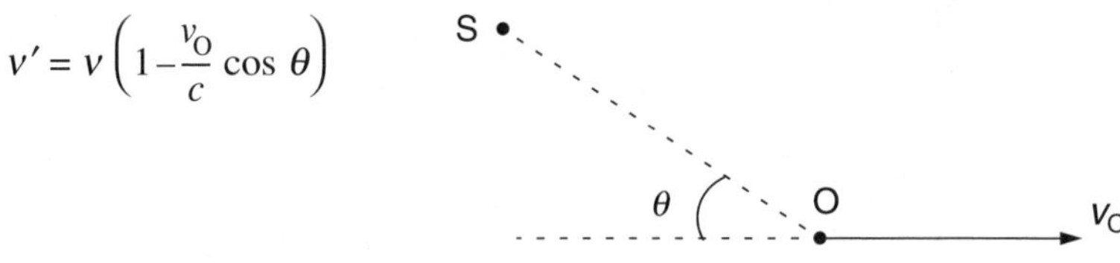

O at rest and S moving at speed v_S relative to medium ($v_S \ll c$)

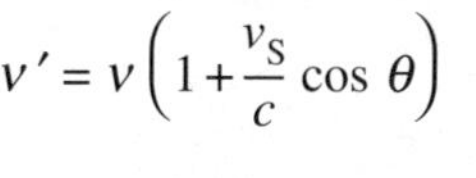

$$\nu' = \nu\left(1 + \frac{v_S}{c}\cos\theta\right)$$

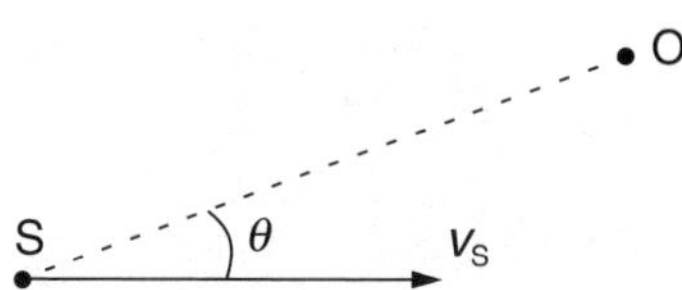

Wave length as observed by O

$$\lambda' = c/\nu'$$

Electromagnetic waves

An observer O′ moving at speed $\mathbf{v} = v\hat{\mathbf{x}}$ relative to the frame of the sender. O′ measures the quantities

$$\omega' = \omega\,\frac{1 - \dfrac{v}{c_0}\cos\theta}{\sqrt{1 - v^2/c_0^2}}$$

$$p'_x = \frac{p_x - \dfrac{vE}{c_0^2}}{\sqrt{1 - v^2/c_0^2}}$$

$$E' = \frac{E - vp_x}{\sqrt{1 - v^2/c_0^2}}$$

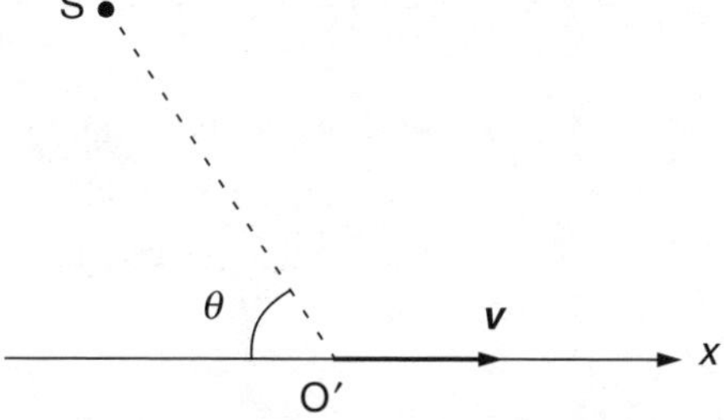

p = wave momentum (sender's frame)

$p_x = p\cos\theta$

$p'_x = p'\cos\theta'$

In particular, when sender S and observer O′ are moving rectilinearly from each other

$$\cos\theta = 1 \qquad k' = \frac{k - \dfrac{\omega v}{c_0^2}}{\sqrt{1 - v^2/c_0^2}}$$

p, k, ω, E and θ = quantities measured by an observer O at rest relative to S and stationed close to O′

Relation between θ and θ' (aberration)

$$\cos\theta' = \frac{\cos\theta - \dfrac{v}{c_0}}{1 - \dfrac{v}{c_0}\cos\theta}$$

When a sender S is moving at velocity **v** *in the observer's frame, O′ will find the frequency*

$$\nu' = \nu \frac{\sqrt{1 - v^2 / c_0^2}}{1 + \frac{v}{c_0} \cos \theta'}$$

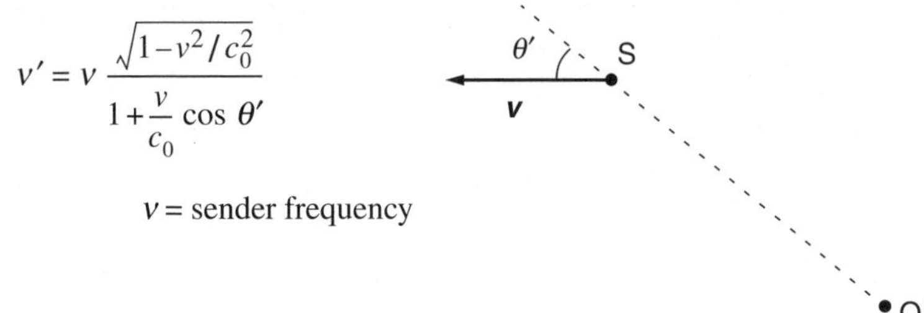

ν = sender frequency

Doppler broadening of spectral lines

$$\frac{\Delta\nu}{\nu} = \frac{4\pi}{c_0} \sqrt{\frac{2RT \ln 2}{M}} \approx 7.16 \cdot 10^{-7} \sqrt{\frac{T}{M}}$$

M = molecular weight (kg/mol)
T = temperature (kelvin)
R = molar gas konstant, see Sec. CU–1.1

5.5 Refraction, Absorption, Dispersion, and Reflection

Index of refraction

$$n = \frac{c_0}{c} = \sqrt{\varepsilon_r \mu_r} \approx \sqrt{\varepsilon_r} \qquad \varepsilon_r(\omega) \approx 1 + \frac{Ne^2}{m_e \varepsilon_0} \left(\sum_i \frac{f_i}{\omega_i^2 - \omega^2} \right)$$

f_i = oscillator strength
ω_i = resonance angular frequency
N = number of electrons per volume

Optical path length

$$L = \int n \, dx$$

x = geometrical path

Wavelength in medium of refractive index n

$$\lambda_m = \lambda / n$$

λ = vacuum wavelength

Absorption of electromagnetic radiation

$$\frac{\partial I}{\partial x} = -\mu I(x)$$

$\mu = \mu(\omega)$ = linear coefficient of absorption

$$I = I_0 \, e^{-\mu x}$$

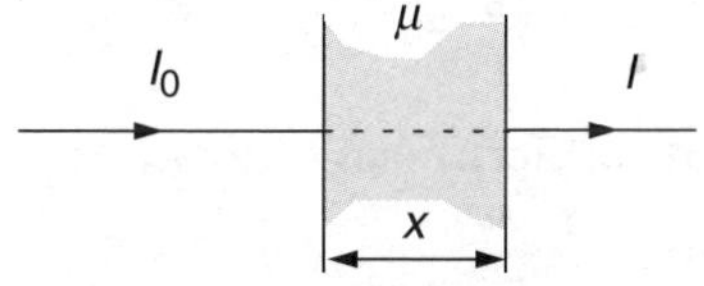

Coefficient of absorption in metals when $\sigma_{e1} \gg \varepsilon_r \varepsilon_0 \omega$ (skin effect)

$$\mu = \sqrt{\frac{1}{2}\mu_r \mu_0 \sigma_{e1} \omega}$$

σ_{e1} = electric conductivity

Complex index of refraction

$$n^* = n - i\,\frac{\mu c_0}{2\omega}$$

Dispersion

$$D = \frac{\partial n}{\partial \lambda}$$

$D < 0$ at normal dispersion

Relative dispersion (relative colour dispersion, dispersive power, Abbe number)

$$D_{rel} = \frac{n_F - n_C}{n_D - 1}$$

n_F, n_C and n_D = refractive indices for Fraunhofer's F-, D-, and C-lines (486, 589, and 656 nm)

Cauchy's dispersion formula

$$n = a + \frac{b}{\lambda^2}$$

a and *b* are constants

Cauchy formula for calculating the refractive index n of dry air, given the wavelength λ in vacuum in μm. (The pressure is assumed to be 1 atm.)

15 °C: $(n-1)\cdot 10^7 = 2726.43 + 12.288/\lambda^2 + 0.3555/\lambda^4$

0 °C: $(n-1)\cdot 10^7 = 2875.66 + 13.412/\lambda^2 + 0.3777/\lambda^4$

Group velocity of electromagnetic waves

$$v_g = \frac{\partial\omega}{\partial k} = \frac{\partial\nu}{\partial\sigma} = \frac{c_0}{n+\left(\omega\dfrac{\partial n}{\partial\omega}\right)}$$

Reflection of acoustic waves at normal incidence towards a dividing surface

$$\frac{I_R}{I_0} = \left(\frac{Z_1-Z_2}{Z_1+Z_2}\right)^2$$

I_R = reflected intensity

Acoustic impedance

$$Z = \rho c$$

Coefficients of reflection and transmission at the junction of two strings

$$R = \frac{\sqrt{\rho_{\ell 1}}-\sqrt{\rho_{\ell 2}}}{\sqrt{\rho_{\ell 1}}+\sqrt{\rho_{\ell 2}}}$$

$\rho_{\ell 1}$ and $\rho_{\ell 2}$ = linear densities
Wave is appearing from string 1

$$T = \frac{2\sqrt{\rho_{\ell 1}}}{\sqrt{\rho_{\ell 1}}+\sqrt{\rho_{\ell 2}}}$$

5.6 Polarization

Fresnel's formulae

Coefficients of reflection and transmission when $\boldsymbol{E}$ is parallel to the plane of incidence

$$R_p = \frac{E_{pr}}{E_{p0}} = \frac{n_1\cos\theta_2 - n_2\cos\theta_1}{n_1\cos\theta_2 + n_2\cos\theta_1} = -\frac{\tan(\theta_1-\theta_2)}{\tan(\theta_1+\theta_2)}$$

$$T_p = \frac{E_{pt}}{E_{p0}} = \frac{2n_1\cos\theta_1}{n_1\cos\theta_2 + n_2\cos\theta_1} = \frac{2\cos\theta_1\sin\theta_2}{\sin(\theta_1+\theta_2)\cos(\theta_1-\theta_2)}$$

Notations and sense of $\boldsymbol{E}$-vectors taken from the figure illustrating Snell's law in Sec. 5.7. If $\mu_r \neq 1$, see below.

Coefficients of reflection and transmission when $\boldsymbol{E}$ is perpendicular to plane of incidence

$$R_{\perp} = \frac{n_1 \cos\theta_1 - n_2 \cos\theta_2}{n_1 \cos\theta_1 + n_2 \cos\theta_2} = -\frac{\sin(\theta_1 - \theta_2)}{\sin(\theta_1 + \theta_2)}$$

$$T_{\perp} = \frac{2n_1 \cos\theta_1}{n_1 \cos\theta_1 + n_2 \cos\theta_2} = \frac{2\cos\theta_1 \sin\theta_2}{\sin(\theta_1 + \theta_2)}$$

In particular,

$$R_{\mathrm{p}} = R_{\perp} = \frac{n_1 - n_2}{n_1 + n_2} \quad \text{when } \theta_1 = \theta_2 = 0°$$

For media with $\mu_{\mathrm{r}} \neq 1$, n should be replaced by $\frac{n}{\mu_{\mathrm{r}}}$ in the formulae above.

Reflected and transmitted intensity

$$I_{\mathrm{R}} = R_{\mathrm{p}}^2\, I_{0\mathrm{p}} + R_{\perp}^2\, I_{0\perp} \qquad I_0 = I_{0\mathrm{p}} + I_{0\perp} = \text{incident intensity}$$

$$I_{\mathrm{T}} = \frac{n_2}{n_1}(T_{\mathrm{p}}^2 I_{0\mathrm{p}} + T_{\perp}^2 I_{0\perp})$$

$$R^2 + \frac{n_2}{n_1} T^2 = 1 \qquad R^2 = R_{\mathrm{p}}^2 + R_{\perp}^2 \qquad T^2 = T_{\mathrm{p}}^2 + T_{\perp}^2$$

Brewster angle $\theta_{1\mathrm{p}}$ (angle of polarization)

$$\tan\theta_{1\mathrm{p}} = n_2/n_1$$

$$\theta_1 + \theta_2 = 90° \qquad R_{\mathrm{p}} = 0$$

Malus' law

$$I = I_0 \cos^2 \theta$$

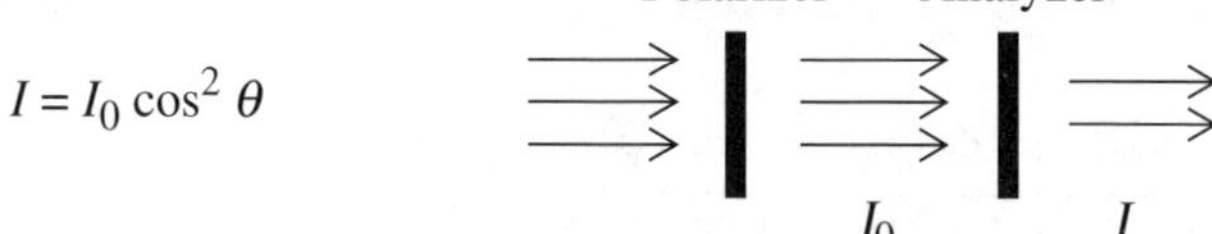

I = transmitted intensity of a linearly polarized wave incident on a polaroid

θ = angle between direction of polarization and orientation of the analyzer

Degree of polarization

$$P = \frac{I_{max} - I_{min}}{I_{max} + I_{min}}$$

Retardation plates

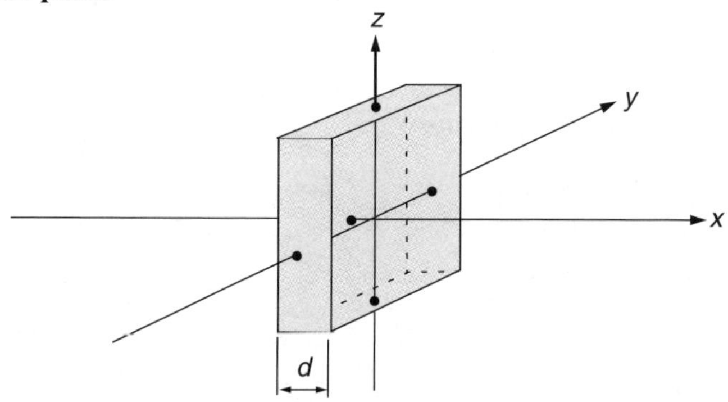

Phase difference between y- and z-components after plate with its optical axis parallel to y-axis

$$\varphi = \frac{2\pi d}{\lambda}(n_o - n_{eo})$$

n_o and n_{eo} = refractive indices of ordinary and extraordinary waves

Example 1. Plane-polarized wave becomes elliptically polarized

$$d = \frac{\lambda}{4}\frac{(1+2m)}{|n_o - n_{eo}|} \qquad m = 0, 1, 2, \ldots$$

Example 2. "Reflection" in the optical axis

$$d = \frac{\lambda}{2}\frac{(1+2m)}{|n_o - n_{eo}|} \qquad m = 0, 1, 2, \ldots$$

5.7 Ray Optics or Geometrical Optics

Snell's law of refraction

$$n_1 \sin \theta_1 = n_2 \sin \theta_2$$

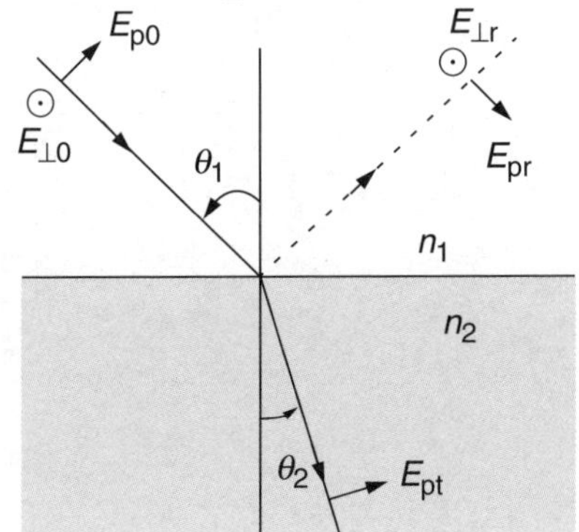

Critical angle of total reflection θ_{1c}

$$\sin \theta_{1c} = n_2 / n_1$$

Minimum deviation in prism

$$\theta_{i1} = \theta_{i2} = \theta_i$$

$$\delta_m = 2\theta_i - \alpha$$

$$\sin \frac{\alpha + \delta_m}{2} = n \sin \frac{\alpha}{2}$$

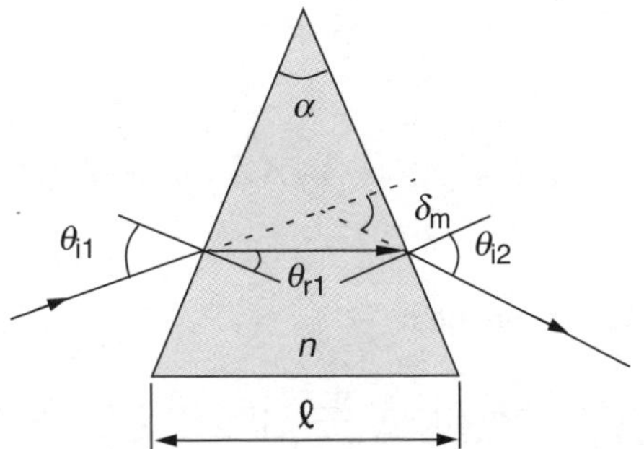

Dispersion of prism

$$D = \frac{\partial \delta}{\partial \lambda} = \frac{\partial n}{\partial \lambda} \frac{\sin \alpha}{\cos \theta_{i2} \cos \theta_{r1}}$$

In particular, at minimum deviation

$$D = \frac{\partial n}{\partial \lambda} \frac{2 \sin (\alpha/2)}{\cos \frac{1}{2} (\delta_m + \alpha)}$$

Resolving power of prism

$$R = \frac{\lambda}{\Delta\lambda} = -\ \ell \frac{\partial n}{\partial \lambda}$$

$\Delta\lambda$ = least difference in wavelength for two lines to be resolved

Two examples

$$d \cos \theta_r = a \sin(\theta_i - \theta_r)$$

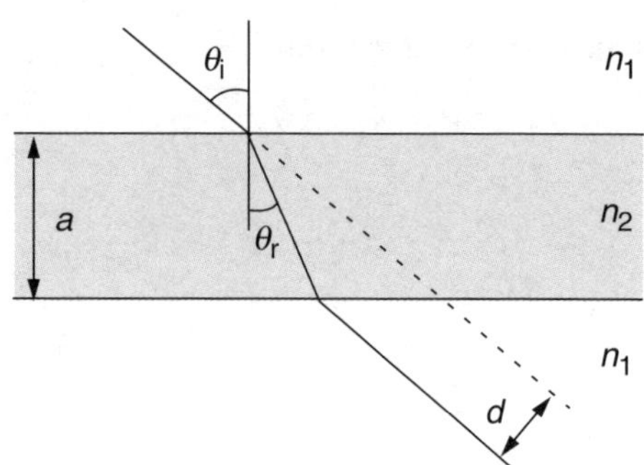

$$n_1^2 = n_2^2 - \sin^2 \alpha$$

(This is a way to measure n_1 or n_2. θ is slightly less than 90° and $n_1 < n_2$)

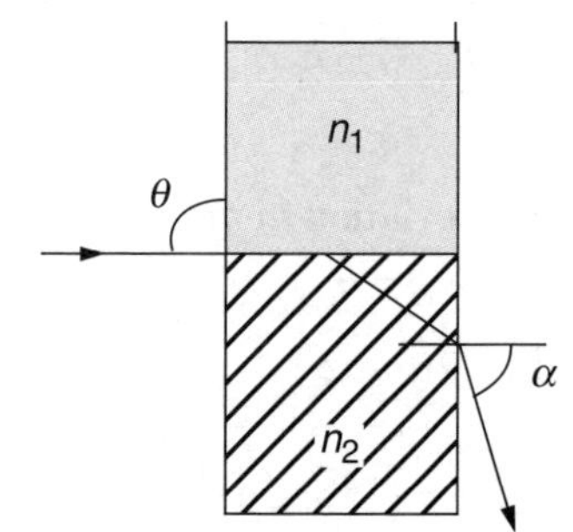

Mirror formulae

$$f = r\left(1 - \frac{1}{2 \cos \theta}\right)$$

For small θ

$$\frac{1}{a_1} + \frac{1}{a_2} = \frac{2}{r} = \frac{1}{f}$$

Linear magnification

$$M = -\frac{a_2}{a_1}$$

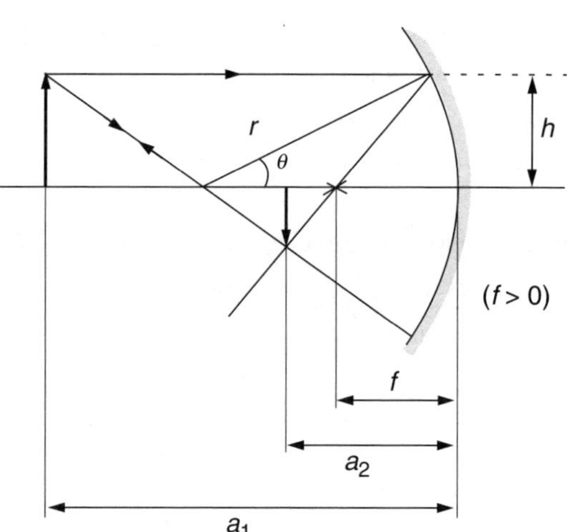

Non-central rays

$$\frac{1}{a_1}+\frac{1}{a_2}=\frac{2}{r}+\frac{h^2}{r}\left(\frac{1}{r}-\frac{1}{a_1}\right)^2 \qquad h = r\sin\theta$$

a_1 and a_2 are positive for real object and real image.
f and r are positive for a concave mirror and negative for a convex one.

Refraction at a spherical surface

$$\frac{n_1}{a_1}+\frac{n_2}{a_2}=\frac{n_2-n_1}{r}$$

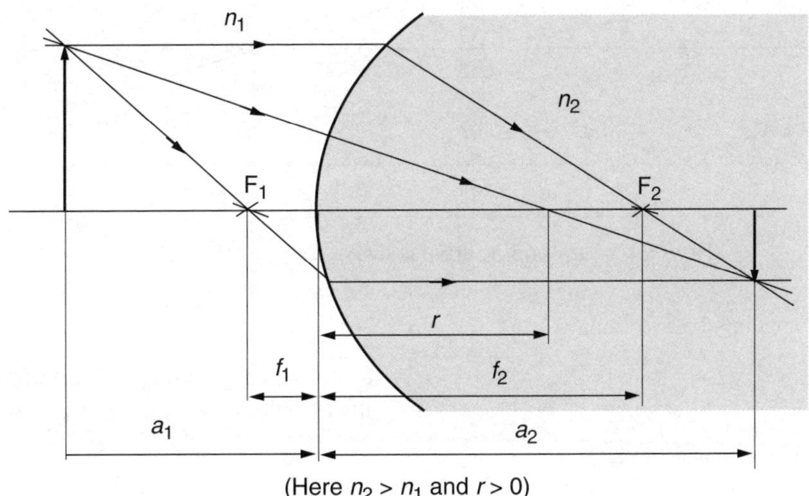

(Here $n_2 > n_1$ and $r > 0$)

Focal lengths

$$f_1 = \frac{n_1 r}{n_2-n_1} \qquad f_2 = \frac{n_2 r}{n_2-n_1}$$

Linear magnification

$$M = -\frac{n_1 a_2}{n_2 a_1}$$

a_1 and a_2 are positive for real object and real image.

Focal length of a thin lens (lensmaker's formula)

$$\frac{1}{f} = (n-1)\left(\frac{1}{r_1}+\frac{1}{r_2}\right)$$

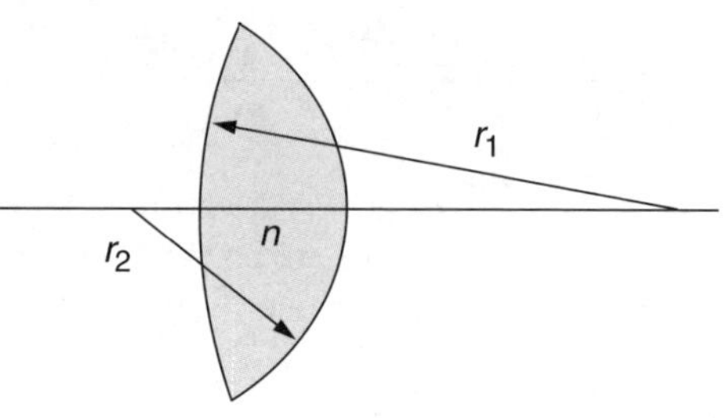

(Here $r_1 > 0$ and $r_2 > 0$)

Thin lenses

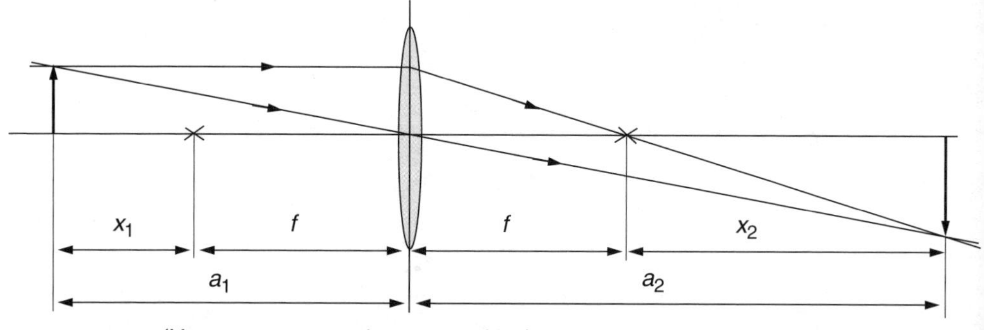

(Here x_1, x_2, a_1, and a_2 are positive)

Newton's lens formula

$$x_1\, x_2 = f^2$$

a_1 and a_2 are positive, if object and image both real.

Gauss' lens formula

$$\frac{1}{a_1}+\frac{1}{a_2}=\frac{1}{f}$$

Linear magnification

$$M = -\frac{a_2}{a_1}$$

Diopter number

$$D = \frac{1}{f}$$ when f is given in metre

System of two thin lenses when distances are measured from the lenses

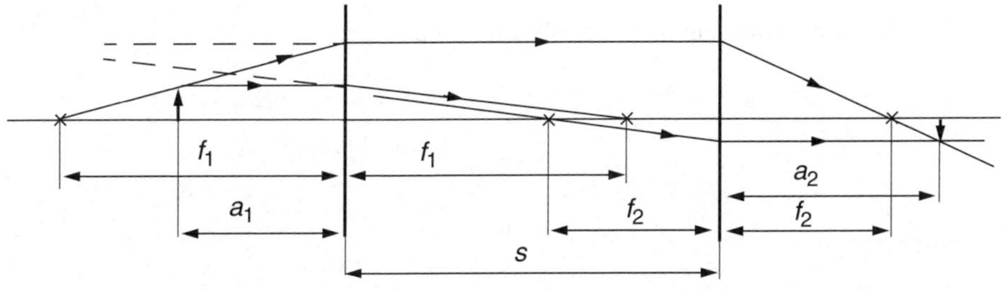

$$s = \frac{a_1 f_1}{a_1 - f_1} + \frac{a_2 f_2}{a_2 - f_2}$$

Linear magnification

$$M = \frac{f_1}{f_2} \left(\frac{a_2 - f_2}{a_1 - f_1} \right)$$

Focal length and position of main planes in thick lens system

$$\frac{1}{f} = (n - 1) \left(\frac{1}{r_1} + \frac{1}{r_2} - \frac{n-1}{n} \; \frac{d}{r_1 r_2} \right)$$

$$k_1 = \frac{n-1}{n} \; \frac{d}{r_2} \; f$$

$$k_2 = \frac{n-1}{n} \; \frac{d}{r_1} \; f$$

(Refractive index equal on both sides of the lens)

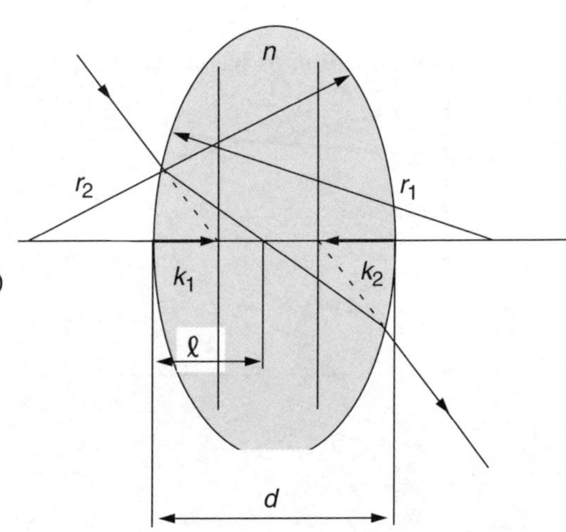

r_1 and r_2 = surface radii of curvature (here $r_1 > 0$ and $r_2 > 0$)

Position of optical centre

$$\ell = \frac{r_1 d}{r_1 + r_2}$$

5.8 Optical Instruments

Angular magnification of a magnifying glass

$$F = \frac{\sigma}{f}$$

σ = distance of clear vision

f = focal length

Microscope

Angular magnification

$$F = \frac{\Delta \sigma}{f_1 f_2}$$

Limit of resolution

$$s_r = \frac{1.2\,\lambda}{2n \sin \alpha}$$

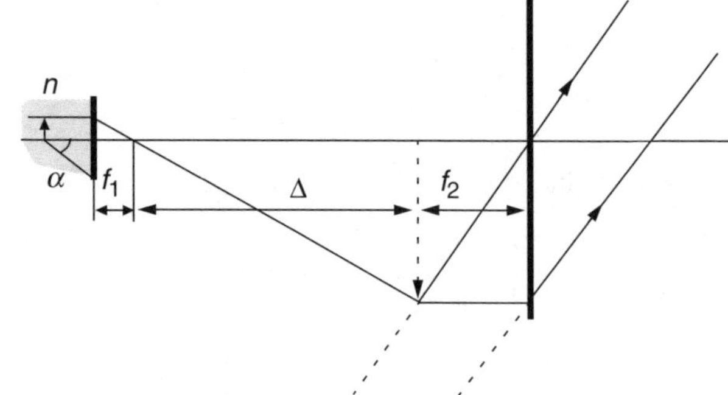

n = refractive index of medium in front of objective

2α = aperture angle according to figure

Angular magnification of Kepler telescope

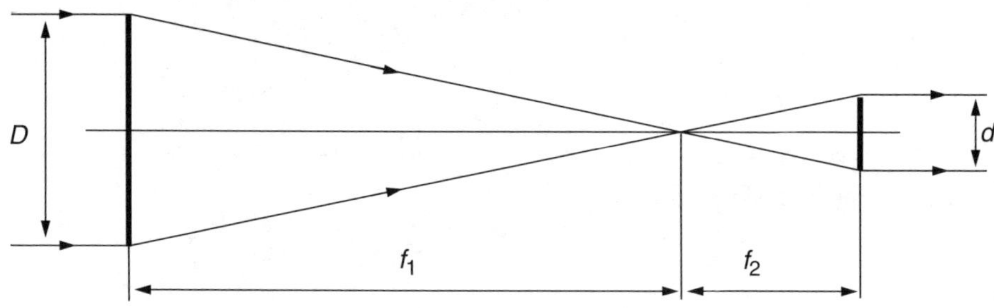

$$F = \frac{f_1}{f_2} = \frac{D}{d}$$

5.9 Interference

Interference of two coherent rays

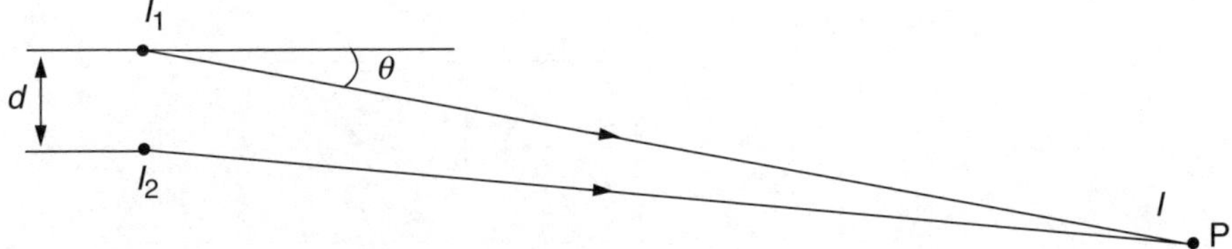

$$I = I_1 + I_2 + 2\sqrt{I_1 I_2}\ \cos\delta$$

Phase difference at interference point P

$$\delta = \frac{2\pi}{\lambda}\ d \sin\theta$$

Interference of N coherent rays

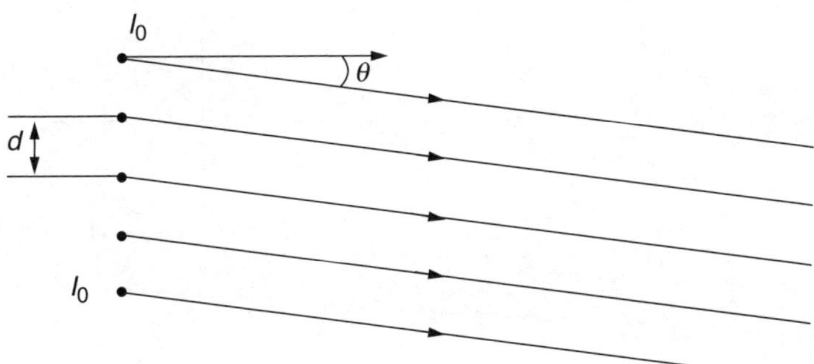

$$I = I_0 \left(\frac{\sin(N\delta/2)}{\sin(\delta/2)}\right)^2 \qquad \delta = \frac{2\pi}{\lambda}\ d \sin\theta$$

Principal maxima

$$d \sin\theta = m\,\lambda \qquad m = 0, \pm 1, \pm 2, \ldots$$

Minima

$$N\,d \sin\theta = m'\lambda \qquad m' \text{ integers}$$

$$m' \neq 0, \pm N, \pm 2N, \ldots$$

Interference in thin films

Phase difference of the two rays

$$\delta = \frac{2\pi}{\lambda} 2dn_2 \cos\theta_2 + \pi$$

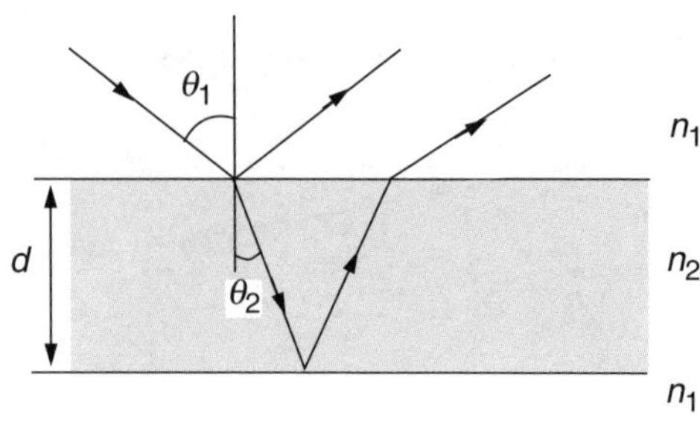

Condition of minimum of Newton's rings

$$2nd = m\,\lambda$$

$$r^2 n = mR\lambda$$

$$m = 0, 1, 2, \ldots$$

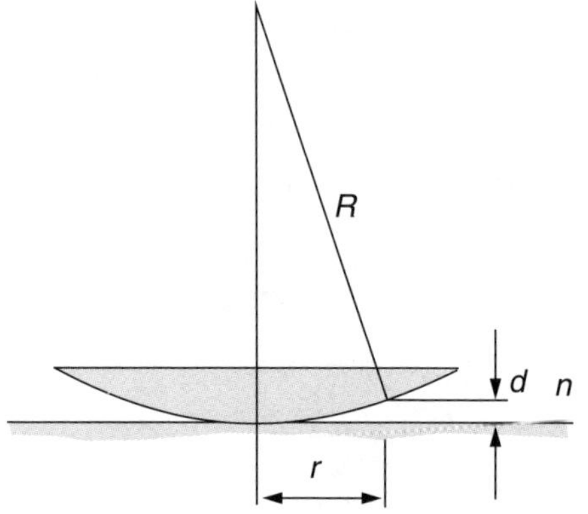

Fabry-Perot interferometer

Free Spectral Range (FSR)

$\Delta\nu_{FSR} = c_0/2nd$ plane mirrors

$\Delta\nu_{FSR} = c_0/4nd$ confocal mirrors

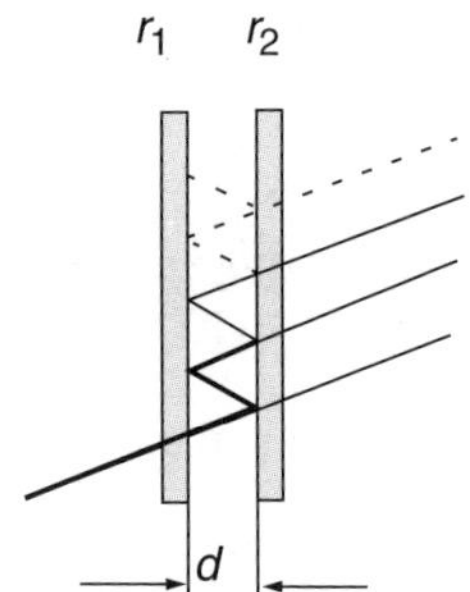

Resolving power

$$R = \frac{\lambda}{|\Delta\lambda|} = \frac{\nu}{|\Delta\nu|} \approx \frac{m\pi(r_1 r_2)^{1/4}}{1 - \sqrt{r_1 r_2}} \qquad m = \frac{2d}{\lambda}$$

r_i = reflectance

Finesse = $\Delta\nu_{FSR}/\Delta\nu$

Visibility

$$V = \frac{I_{max} - I_{min}}{I_{max} + I_{min}}$$

Time and length of coherence

$$\tau = 1/\Delta\nu$$

$\Delta\nu$ and $\Delta\lambda$ = widths of wave package frequency and wavelength

$$\ell = c_0\tau = \lambda^2/\Delta\lambda$$

5.10 Diffraction

Kirchhoff's formula for the contribution from the surface element dA of a wave front to the field at a point P

$$d\psi = a \exp i\omega t \; \frac{\exp(-ikR)}{R} \; \frac{(1+\cos\theta)}{2} \; dA$$

a is a constant

$$k = \frac{2\pi}{\lambda}$$

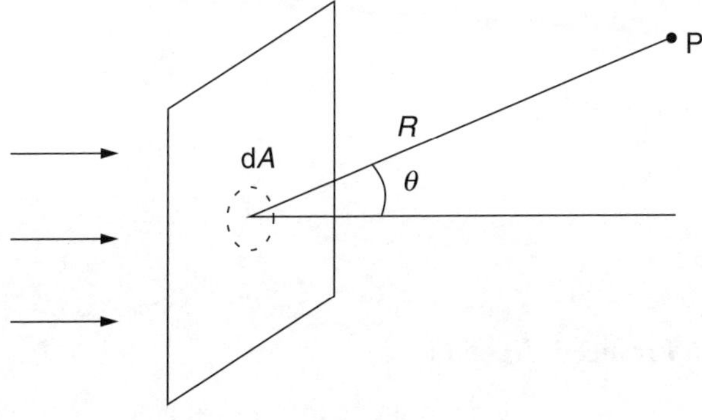

Intensity in Fraunhofer diffraction by a single slit (normal incidence)

$$I = I_0 \left(\frac{\sin(\beta/2)}{\beta/2} \right)^2$$

$$\beta = \frac{2\pi}{\lambda}\, b \sin\theta$$

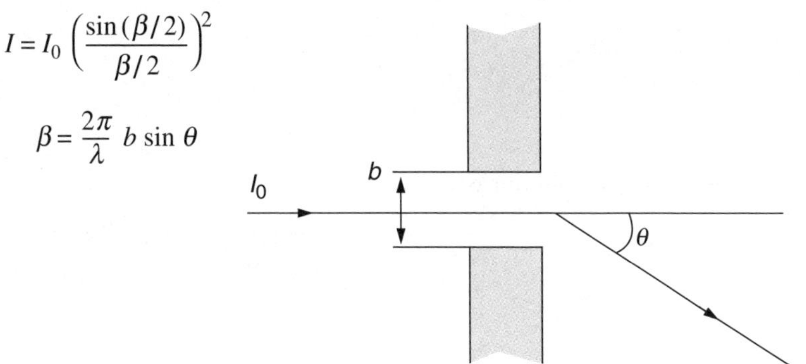

Condition of intensity minima

$$b \sin\theta = m\lambda \qquad m = \pm 1, \pm 2, \ldots$$

Resolving power, according to Rayleigh

$$\sin\varphi_r = \frac{\lambda}{b}$$

The angle θ_m to the first minimum in diffraction through a circular opening of diameter D

$$\sin\theta_m = 1.22\,\frac{\lambda}{D}$$

Limit of resolution, according to Rayleigh

$$\varphi_r = \theta_m$$

Area of a Fresnel zone in Fresnel diffraction

$$A = \pi d \lambda \qquad d = \text{distance to centre of hole}$$

Condition of extremum intensity

$$r^2 = m\lambda d \qquad m = 1,3,5, \ldots \Rightarrow I \approx 4\,I_0$$

$$m = 2,4,6, \ldots \Rightarrow I \approx 0$$

Fraunhofer diffraction in N slits at normal incidence

$$I = I_0 \left(\frac{\sin(\beta/2)}{\beta/2} \right)^2 \left(\frac{\sin(N\delta/2)}{\sin(\delta/2)} \right)^2$$

$$\beta = \frac{2\pi}{\lambda} b \sin\theta$$

$$\delta = \frac{2\pi}{\lambda} d \sin\theta$$

d = grating constant

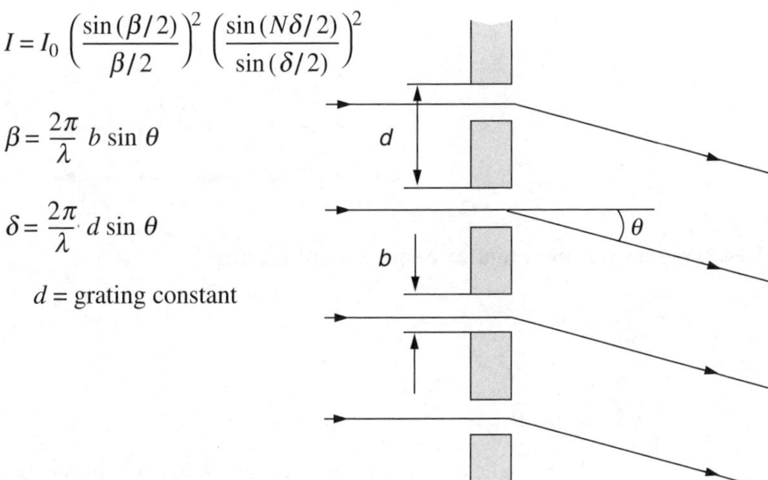

Condition of interference maximum (grating formula)

$$d \sin\theta = m\lambda \qquad m = 0, \pm 1, \pm 2, \ldots$$

Condition of diffraction minimum

$$b \sin\theta = m\lambda \qquad m = \pm 1, \pm 2, \ldots$$

Grating formula at oblique incidence

$$d(\sin\theta + \sin i) = m\lambda$$

$$m = 0, \pm 1, \pm 2, \ldots$$

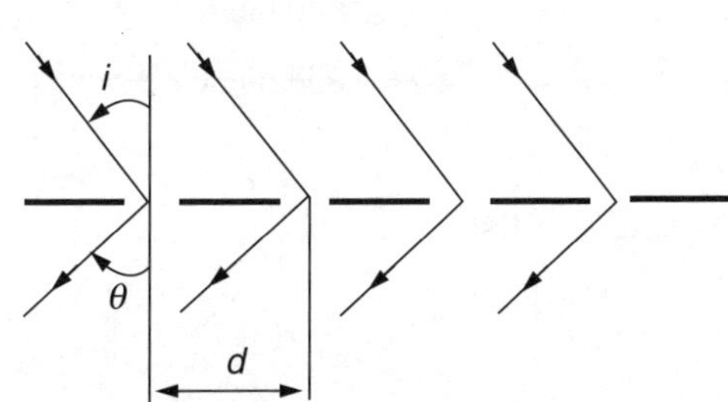

For reflection gratings

$$d(\sin\theta - \sin i) = m\lambda$$

$$m = 0, \pm 1, \pm 2, \ldots$$

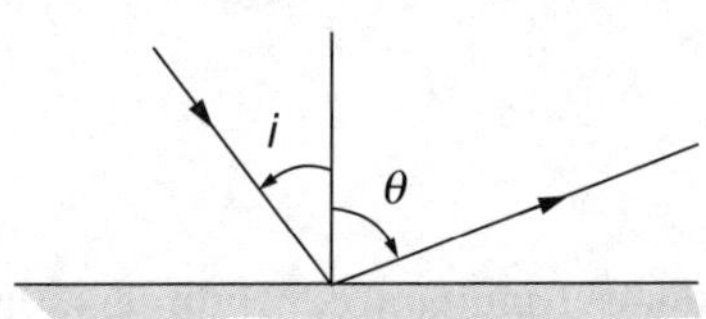

Bragg's law (condition of interference maximum)

$$2d \sin \theta = m \lambda$$

$m = 0,1,2, \ldots$

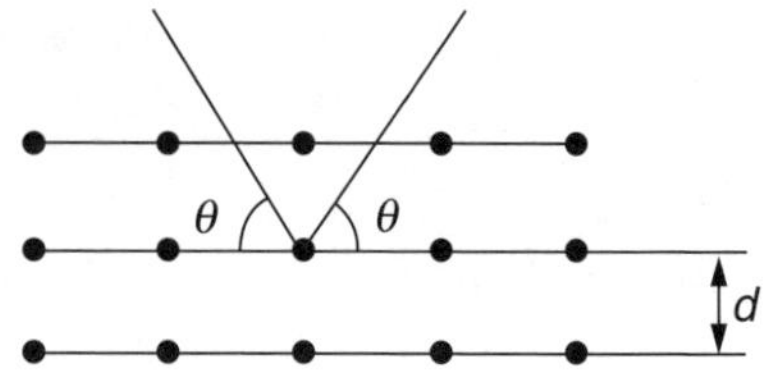

Resolving power and angular dispersion of gratings

$$R = \frac{\lambda}{\Delta\lambda} = Nm$$

$$D = \frac{\partial\theta}{\partial\lambda} = \frac{m}{d \cos \theta}$$

$\Delta\lambda$ =smallest difference in wavelength for two lines to be resolved

5.11 Radiation Laws and Photometry

Luminous flux (unit: lumen)

$$\phi = \int I \, d\Omega = \frac{dW}{dt}$$

$I =$ intensity of light (unit: candela) = radiated power per solid angle

$d\Omega =$ solid angle element

Quantity of light

$$Q = \int \phi \, dt$$

Illumination, illuminance (unit: lux = lumen/m²)

$$E = \frac{d\phi}{dA}$$

Radiance (luminance, brightness, light density)

$$L = I/A$$

Spectral radiance

$$R = \frac{\mathrm{d}L}{\mathrm{d}\lambda}$$

Lambert's cosine law

$$I_\varphi = I_0 \cos \varphi = eA \cos \varphi$$

e = emissivity of surface A

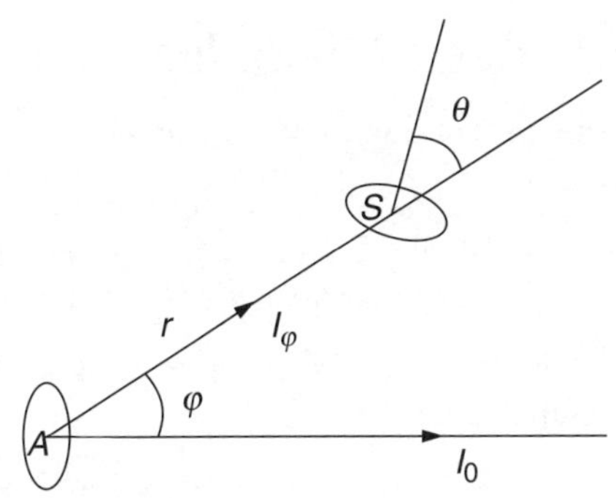

Illumination of surface S (law of illumination)

$$E_S = \frac{I_\varphi \cos \theta}{r^2} = \frac{I_0 \cos \varphi \cos \theta}{r^2}$$

Kirchhoff's law

$$\frac{e}{a} = \text{constant}$$

a = absorption power

Visible light, sensitivity of the human eye (absolute spectral luminous efficacy)

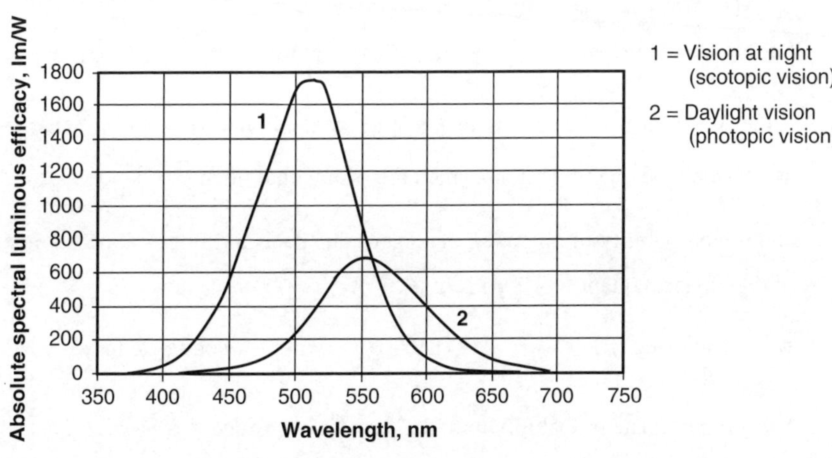

Response of the human eye to radiation of a given wavelength

6 Quantum Mechanics

6.1 Basic Formulae

Principles of quantum mechanics

1. To every physical quantity $A(\boldsymbol{r}, \boldsymbol{p})$, which is a function of the position and momentum of a particle, there corresponds a hermitian quantum operator, obtained by replacing $\boldsymbol{p}$ by $-i\hbar\nabla$; that is, $\mathbf{A}(\boldsymbol{r}, -i\hbar\nabla)$.

   For other quantities than momentum there are other operators according to the following table.

Quantity	Classical definition	Quantum operator
Position	$\boldsymbol{r}$	$\boldsymbol{r}$
Momentum	$\boldsymbol{p}$	$-i\hbar\nabla$
Angular momentum	$\boldsymbol{r} \times \boldsymbol{p}$	$-i\hbar\, \boldsymbol{r} \times \nabla$
Kinetic energy	$p^2/2m$	$-(\hbar^2/2m)\nabla^2$
Total energy	$p^2/2m + E_p(\boldsymbol{r})$	$-(\hbar^2/2m)\nabla^2 + E_p(\boldsymbol{r})$

2. The only possible values that can be obtained when a physical quantity $A(\boldsymbol{r}, \boldsymbol{p})$ is measured are the eigenvalues (proper values) of the quantum operator $\mathbf{A}(\boldsymbol{r}, -i\hbar\nabla)$. In mathematical terms:

   $$\mathbf{A}(\boldsymbol{r}, -i\hbar\nabla)\psi = a\psi$$

   Solutions are eigenvalues $a_1, a_2, a_3, \ldots$ and eigenfunctions $\psi_1, \psi_2, \psi_3, \ldots$

3. If the state of a system corresponds to a wave function $\Phi(\boldsymbol{r}) = \sum_n c_n \psi_n$ where ψ_n are normalized eigenfunctions which are orthogonal, then $c_n = \int_{\substack{All \\ space}} \psi_n{}^* \Phi \, dV$

   and the probability of obtaining the eigenvalue a_n as a result of a measurement of the physical quantity $A(\boldsymbol{r}, \boldsymbol{p})$ is $|c_n|^2$.

   ψ is normalized if $1 = \int_{\substack{All \\ space}} \psi^* \psi \, dV$ $\quad \psi^*$ = complex conjugate to ψ

   Special case: The probability to find a particle in volume V is $\int_V \psi^* \psi \, dV$.

4. The time evolution of a physical system is described by the equation

$$i\hbar \frac{\partial \Psi}{\partial t} = \mathbf{H}\,\Psi$$

where **H** is the hamiltonian operator of the system.

Expectation value

$$\langle A \rangle = \int_{\substack{All\\space}} \Phi^* \, \mathbf{A} \, \Phi \, dV \qquad \Phi(\boldsymbol{r}) = \text{normalized wave function} = \sum_n c_n \psi_n$$

Definition of a hermitian operator Q

$$\int (\mathbf{Q}\,\psi)^* \, \phi \, dV = \int \psi^* \, \mathbf{Q} \, \phi \, dV$$

ψ and ϕ satisfy physical boundary conditions

The Schrödinger equation

$$\mathbf{H}\Psi(\boldsymbol{r}, t) = i\hbar \frac{\partial}{\partial t} \Psi(\boldsymbol{r}, t)$$

Stationary states

$$\mathbf{H}\,\psi(\boldsymbol{r}) = E\,\psi(\boldsymbol{r})$$

$$\Psi(\boldsymbol{r}, t) = \psi(\boldsymbol{r}) \cdot \phi(t) \qquad \phi(t) = e^{-\frac{i}{\hbar}Et}$$

Relativistic Schrödinger equation for one particle (Klein-Gordon equation)

$$-\hbar^2 \frac{\partial^2 \Psi(\boldsymbol{r},t)}{\partial t^2} = -\hbar^2 c_0^2 \nabla^2 \Psi(\boldsymbol{r},t) + m^2 c_0^4 \Psi(\boldsymbol{r},t)$$

The Hamiltonian of one particle

$$\mathbf{H} = -\frac{\hbar^2}{2m} \nabla^2 + E_{\mathrm{p}}(\boldsymbol{r}) \qquad E_{\mathrm{p}}(\boldsymbol{r}) = \text{potential energy}$$

Probability flux

$$S = -\frac{i\hbar}{2m} (\Psi^* \nabla \Psi - \Psi \nabla \Psi^*)$$

$$S = \frac{1}{m} \operatorname{Re}(\Psi^* \, \mathbf{p} \, \Psi) \qquad \mathbf{p} = -i\hbar\nabla$$

In particular, for a plane wave

$$\psi(\boldsymbol{r}) = A\ \mathrm{e}^{\frac{\mathrm{i}}{\hbar}\boldsymbol{p}\cdot\boldsymbol{r}}$$

$$\boldsymbol{S} = |A|^2\,\frac{\boldsymbol{p}}{m}$$

Equation of continuity

$$\frac{\partial\rho}{\partial t} + \operatorname{div}\boldsymbol{S} = 0 \qquad\qquad \rho = \Psi^*\,\Psi$$

Commutator of operators P and Q

$$[\mathbf{P}, \mathbf{Q}] = \mathbf{PQ} - \mathbf{QP}$$

In particular,

$$[\mathbf{p}_x, x]\Psi(x) = -\,\mathrm{i}\hbar\;\Psi(x) \qquad\qquad \mathbf{p}_x = -\,\mathrm{i}\hbar\,\frac{\partial}{\partial x}$$

Square of the average deviation

$$(\Delta Q)^2 = \langle(\mathbf{Q} - \langle\mathbf{Q}\rangle)^2\rangle = \langle\mathbf{Q}^2\rangle - \langle\mathbf{Q}\rangle^2$$

Uncertainty relation

$$\Delta P\,\Delta Q \geq \tfrac{1}{2}\,|\langle[\mathbf{P}, \mathbf{Q}]\rangle|$$

P and **Q** are hermitian operators

In particular,

$$\Delta x\,\Delta p_x \geq \tfrac{1}{2}\hbar$$

$$\Delta E\,\Delta t \geq \tfrac{1}{2}\hbar$$

Heisenberg's equation of motion

$$\frac{\mathrm{d}}{\mathrm{d}t}\,\langle Q\rangle = \langle\frac{\mathrm{i}}{\hbar}\,[\mathbf{H}, \mathbf{Q}] + \frac{\partial\mathbf{Q}}{\partial t}\rangle$$

Ehrenfest's theorem

$$m\,\frac{\mathrm{d}}{\mathrm{d}t}\,\langle\boldsymbol{r}\rangle = \langle\boldsymbol{p}\rangle$$

$$\frac{\mathrm{d}}{\mathrm{d}t}\,\langle\boldsymbol{p}\rangle = -\,\langle\operatorname{grad} E_{\mathrm{p}}\rangle$$

6.2 Angular Momentum

Orbital angular momentum operator

$$\mathbf{L} = -\mathrm{i}\,\hbar\,\boldsymbol{r} \times \nabla$$

In particular, $\mathbf{L}_z$ in Cartesian and spherical polar coordinates

$$\mathbf{L}_z = x\,\mathbf{p}_y - y\,\mathbf{p}_x = -\mathrm{i}\hbar\,\frac{\partial}{\partial\varphi}$$

Uncertainty relation

$$\Delta L_x\,\Delta L_y \geq \tfrac{1}{2}\hbar\,L_z$$

Commutator relations of angular momenta

$$[\mathbf{j}_x, \mathbf{j}_y] = \mathrm{i}\hbar\mathbf{j}_z \qquad [\mathbf{j}_y, \mathbf{j}_z] = \mathrm{i}\hbar\mathbf{j}_x \qquad [\mathbf{j}_z, \mathbf{j}_x] = \mathrm{i}\hbar\mathbf{j}_y$$

$$[\mathbf{j}^2, \mathbf{j}_z] = [\mathbf{H}, \mathbf{j}_z] = [\mathbf{H}, \mathbf{j}^2] = 0$$

$\mathbf{H}$ = Hamiltonian as in Sec. 6.1, central force only

Eigenvalue relations of angular momentum j with eigenfunctions Ψ_{jm}

$$\mathbf{j}^2\,\Psi_{jm} = j(j+1)\hbar^2\,\Psi_{jm}$$

$$\mathbf{j}_z\,\Psi_{jm} = m\,\hbar\,\Psi_{jm}$$

$$m = -j, -j+1, \ldots, j-1, j$$

Example 1. Orbital angular momentum

$$\mathbf{L}^2 Y_{\ell m} = \ell(\ell+1)\hbar^2\,Y_{\ell m}$$

Length of orbital angular momentum vector

$$|\boldsymbol{L}| = \sqrt{\ell(\ell+1)}\,\hbar$$

$$|L_z| = |m|\hbar$$

Spherical harmonics (see also Sec. 6.4)

$$Y_{\ell m}(\theta, \varphi) = N_{\ell m}\,P_\ell^{|m|}(\cos\theta)\mathrm{e}^{\mathrm{i}m\varphi}$$

$P_\ell^{|m|}(\cos\theta)$ = associated Legendre polynomials

Example 2. Electron spin

$$\mathbf{S}^2 \chi_{s\, m_s} = s(s+1)\hbar^2 \chi_{s\, m_s} \qquad s = \tfrac{1}{2}$$

$$\chi_{\frac{1}{2}\,\frac{1}{2}} = \alpha \qquad \chi_{\frac{1}{2}\,-\frac{1}{2}} = \beta$$

Length of electron spin vector

$$|S| = \frac{\sqrt{3}}{2}\hbar$$

Shift operators

$$\mathbf{j}_+ = \mathbf{j}_x + \mathrm{i}\mathbf{j}_y \qquad \mathbf{j}_+ \Psi_{jm} = \sqrt{j(j+1) - m(m+1)}\,\hbar\,\Psi_{jm+1}$$

$$\mathbf{j}_- = \mathbf{j}_x - \mathrm{i}\mathbf{j}_y \qquad \mathbf{j}_- \Psi_{jm} = \sqrt{j(j+1) - m(m-1)}\,\hbar\,\Psi_{jm-1}$$

Addition theorem for vector operators

$$\mathbf{J} = \mathbf{J}_1 + \mathbf{J}_2$$

has eigenfunctions Ψ_{jm} where

$$j = |j_1 - j_2|,\ |j_1 - j_2| + 1,\ \ldots,\ |j_1 + j_2|$$

$$m = -j, -j+1, \ldots, j$$

Total number of eigenfunctions

$$j = \sum_{|j_1 - j_2|}^{j_1 + j_2} (2j+1) = (2j_1 + 1)(2j_2 + 1)$$

Example 1. Total angular momentum

$$\mathbf{J} = \mathbf{L} + \mathbf{S} \qquad \mathbf{S} = \text{spin operator}$$

$$\mathbf{J}^2 = \mathbf{L}^2 + \mathbf{S}^2 + 2\mathbf{L}\cdot\mathbf{S} = \mathbf{L}^2 + \mathbf{S}^2 + 2\mathbf{L}_z\mathbf{S}_z + \mathbf{L}_+\mathbf{S}_- + \mathbf{L}_-\mathbf{S}_+$$

Example 2. Addition of two spins $\frac{1}{2}$

$\mathbf{S} = \mathbf{S}_1 + \mathbf{S}_2$

Eigenfunction $\chi_{S\,M_s}$	Parity π	M_S	S	$\mathbf{S}_1 \cdot \mathbf{S}_2$	
$\alpha_1\,\alpha_2$	1	1	1	$\frac{1}{4}\hbar^2$	Triplet
$\frac{1}{\sqrt{2}}(\alpha_1\beta_2 + \beta_1\alpha_2)$	1	0	1	$\frac{1}{4}\hbar^2$	Triplet
$\beta_1\,\beta_2$	1	– 1	1	$\frac{1}{4}\hbar^2$	Triplet
$\frac{1}{\sqrt{2}}(\alpha_1\beta_2 - \beta_1\alpha_2)$	– 1	0	0	$-\frac{3}{4}\hbar^2$	Singlet

α_k and β_k eigenfunctions of $\mathbf{S}_k^2$ and $\mathbf{S}_{kz}$ $\quad k = 1,2$

One-electron eigenfunctions of $\mathbf{J}^2$, $\mathbf{J}_z$, $\mathbf{L}^2$ and $\mathbf{S}^2$

$$\Psi_{j,m_j} = \left(\frac{j-m_j}{2j}\right)^{1/2} Y_{\ell\,m_\ell}\,\beta + \left(\frac{j+m_j}{2j}\right)^{1/2} Y_{\ell\,m_\ell-1}\,\alpha$$

when $j = \ell + \frac{1}{2}$

$$\Psi_{j,m_j} = \left(\frac{j+1+m_j}{2(j+1)}\right)^{1/2} Y_{\ell\,m_\ell}\,\beta - \left(\frac{j+1-m_j}{2(j+1)}\right)^{1/2} Y_{\ell\,m_\ell-1}\,\alpha$$

when $j = \ell - \frac{1}{2}$

where $m_j = m_\ell - \frac{1}{2}$, $\quad m_j \neq \ell + \frac{1}{2}$

$$\Psi_{\ell+\frac{1}{2},\,\ell+\frac{1}{2}} = Y_{\ell\,\ell}\,\alpha$$

Pauli spin matrices σ_x, σ_y and σ_z for spin $\frac{1}{2}$

$$S_x = \frac{1}{2}\hbar\,\sigma_x \qquad \sigma_x = \begin{pmatrix} 0 & 1 \\ 1 & 0 \end{pmatrix}$$

$$S_y = \frac{1}{2}\hbar\,\sigma_y \qquad \sigma_y = \begin{pmatrix} 0 & -\mathrm{i} \\ \mathrm{i} & 0 \end{pmatrix}$$

$$S_z = \frac{1}{2}\hbar\,\sigma_z \qquad \sigma_z = \begin{pmatrix} 1 & 0 \\ 0 & -1 \end{pmatrix}$$

In particular, for $\alpha = \begin{pmatrix} 1 \\ 0 \end{pmatrix}$ and $\beta = \begin{pmatrix} 0 \\ 1 \end{pmatrix}$

$$S_x\alpha = \frac{1}{2}\hbar\,\beta \qquad S_x\beta = \frac{1}{2}\hbar\,\alpha$$

$$S_y\alpha = \frac{1}{2}\,\mathrm{i}\hbar\,\beta \qquad S_y\beta = -\frac{1}{2}\,\mathrm{i}\hbar\,\alpha$$

$$S_z\alpha = \frac{1}{2}\hbar\,\alpha \qquad S_z\beta = -\frac{1}{2}\hbar\,\beta$$

6.3 Special Solutions of the Schrödinger Equation

Free particle in one-dimensional box

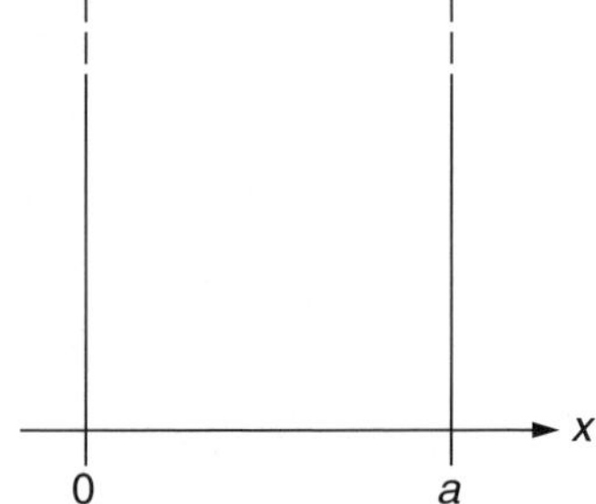

$$E_n = \frac{n^2\,\pi^2\,\hbar^2}{2\,m\,a^2}$$

$$n = 1, 2, \ldots$$

$$\psi_n(x) = \sqrt{\frac{2}{a}}\,\sin\frac{n\,\pi\,x}{a}$$

Solutions when the potential energy E_p is constant

$$\psi(x) = A\,\mathrm{e}^{\mathrm{i}kx} + B\,\mathrm{e}^{-\mathrm{i}kx},\; k = \frac{1}{\hbar}\sqrt{2m(E-E_\mathrm{p})} \qquad \text{when } E > E_\mathrm{p}$$

$$\psi(x) = C\,\mathrm{e}^{\alpha x} + D\,\mathrm{e}^{-\alpha x},\; \alpha = \frac{1}{\hbar}\sqrt{2m(E_\mathrm{p}-E)} \qquad \text{when } E < E_\mathrm{p}$$

E = particle energy

Linear harmonic oscillator

$$E_\mathrm{p}(x) = \frac{1}{2}k\,x^2 = \frac{1}{2}m\,\omega^2\,x^2$$

$$E_n = \left(n+\frac{1}{2}\right)\hbar\omega \qquad n = 0, 1, 2, \ldots$$

$$\psi_n(x) = N_n\,H_n(\alpha x)\,\mathrm{e}^{-\frac{1}{2}\alpha^2x^2}$$

$$\alpha = \sqrt{\frac{m\,\omega}{\hbar}} = \left(\frac{\sqrt{km}}{\hbar}\right)^{1/2} \qquad H_n(x) = \text{Hermite polynomial}$$

$$\psi_0(x) = \left(\frac{\alpha}{\sqrt{\pi}}\right)^{1/2} \exp(-\tfrac{1}{2}\,\alpha^2 x^2)$$

$$\psi_1(x) = \left(\frac{\alpha}{2\sqrt{\pi}}\right)^{1/2} 2\alpha x \exp(-\tfrac{1}{2}\,\alpha^2 x^2)$$

$$\psi_2(x) = \left(\frac{\alpha}{8\sqrt{\pi}}\right)^{1/2} (4\alpha^2 x^2 - 2) \exp(-\tfrac{1}{2}\,\alpha^2 x^2)$$

$$\psi_3(x) = \left(\frac{\alpha}{48\sqrt{\pi}}\right)^{1/2} (8\alpha^3 x^3 - 12\alpha x) \exp(-\tfrac{1}{2}\,\alpha^2 x^2)$$

Transmission and reflection coefficients of a potential step

$E > E_p$

$$k_1 = \frac{1}{\hbar}\sqrt{2m\,E}$$

$$k_2 = \frac{1}{\hbar}\sqrt{2m(E-E_p)}$$

$$T = \frac{|\boldsymbol{S}_{\text{transmitted}}|}{|\boldsymbol{S}_{\text{incident}}|} = \frac{4k_1\,k_2}{(k_1+k_2)^2}$$

$$R = \frac{|\boldsymbol{S}_{\text{reflected}}|}{|\boldsymbol{S}_{\text{incident}}|} = \frac{(k_1-k_2)^2}{(k_1+k_2)^2}$$

S = probability flux

$$T + R = 1$$

Transmission coefficient of a barrier of height E_p and width a

$$T = \left[1+\frac{E_p^2 \sinh^2\alpha a}{4E(E_p-E)}\right]^{-1} \qquad (E < E_p)$$

$$T = \left[1+\frac{E_p^2 \sin^2\alpha a}{4E(E-E_p)}\right]^{-1} \qquad (E > E_p)$$

$$\alpha = \frac{\sqrt{2m|E_p-E|}}{\hbar}$$

Density of one-particle states in a box of volume V

$$g(E) = \frac{dN}{dE} = \frac{4\pi V(2m^3)^{1/2}}{h^3} E^{1/2}$$ N = number of energy levels

$$g(p) = \frac{dN}{dp} = \frac{4\pi V}{h^3} p^2$$ p = momentum

$$g(\nu) = \frac{dN}{d\nu} = \frac{4\pi V}{c^3} \nu^2$$ ν = frequency
c = phase velocity

6.4 Hydrogenic Atoms

Schrödinger equation in spherically polar coordinates

$$\left(-\frac{\hbar^2}{2m}\frac{1}{r}\frac{\partial^2}{\partial r^2}r\cdot + \frac{L^2}{2mr^2} + E_p(r)\right)\psi(\boldsymbol{r}) = E\,\psi(\boldsymbol{r})$$

$$E_p(r) = -\frac{Z e^2}{4\pi\varepsilon_0 r}$$ m = reduced mass $\approx m_e$
Z = number of protons in nucleus

Solution

$$E_n = -\frac{m e^4}{8\varepsilon_0^2 h^2}\frac{Z^2}{n^2} = -E_H\frac{Z^2}{n^2}$$ $n = 1, 2, \ldots$

$$\Psi_{n\ell m}(\rho,\theta,\varphi) = N_{n\ell}\, e^{-\frac{1}{2}\rho}\, \rho^{\ell}\, L_{n+\ell}^{2\ell+1}(\rho)\, Y_{\ell m}(\theta,\varphi)$$

$$\rho = \frac{2Zr}{n a_0}$$ a_0 = Bohr radius

$L_{n+\ell}^{2\ell+1}(\rho)$ = associated Laguerre polynomial of degree $n - \ell - 1$

Spherical harmonics

ℓ	m	$Y_{\ell m}(\theta,\varphi) = P_\ell^{\|m\|}(\cos\theta)\, e^{im\varphi}$
0	0	$Y_{00} = 1/\sqrt{4\pi}$
1	0	$Y_{10} = \sqrt{3/4\pi}\cos\theta$
1	±1	$Y_{1\pm1} = \mp\sqrt{3/8\pi}\sin\theta\, e^{\pm i\varphi}$
2	0	$Y_{20} = \sqrt{5/16\pi}\,(3\cos^2\theta - 1)$
2	±1	$Y_{2\pm1} = \mp\sqrt{15/8\pi}\sin\theta\cos\theta\, e^{\pm i\varphi}$
2	±2	$Y_{2\pm2} = \sqrt{15/32\pi}\sin^2\theta\, e^{\pm 2i\varphi}$

$P_\ell^m(x)$ = associated Legendre functions, see M–8 and M–12

Angular dependence of orbitals

Orbital	Angular dependence	Orbital	Angular dependence
s	Y_{00}	$d_{x^2-y^2}$	$\frac{1}{\sqrt{2}}(Y_{22} + Y_{2-2})$
p_x	$\frac{1}{\sqrt{2}}(Y_{11} + Y_{1-1})$	d_{xz}	$-\frac{1}{\sqrt{2}}(Y_{21} - Y_{2-1})$
p_y	$-\frac{i}{\sqrt{2}}(Y_{11} - Y_{1-1})$	d_{z^2}	Y_{20}
p_z	Y_{10}	d_{yz}	$\frac{i}{\sqrt{2}}(Y_{21} + Y_{2-1})$
		d_{xy}	$-\frac{i}{\sqrt{2}}(Y_{22} - Y_{2-2})$

Radial wave functions of hydrogenic atoms

$$\rho = \frac{2\,Z\,r}{n\,a_0} \qquad a_0 = \text{Bohr radius}$$

n	ℓ	$R_{n\ell}(r)$
1	0	$R_{10}(r) = 2\left(\frac{Z}{a_0}\right)^{3/2} \mathrm{e}^{-\rho/2}$
2	0	$R_{20}(r) = \frac{1}{2\sqrt{2}}\left(\frac{Z}{a_0}\right)^{3/2} (2-\rho)\mathrm{e}^{-\rho/2}$
2	1	$R_{21}(r) = \frac{1}{2\sqrt{6}}\left(\frac{Z}{a_0}\right)^{3/2} \rho\mathrm{e}^{-\rho/2}$
3	0	$R_{30}(r) = \frac{1}{9\sqrt{3}}\left(\frac{Z}{a_0}\right)^{3/2} (6-6\rho+\rho^2)\mathrm{e}^{-\rho/2}$
3	1	$R_{31}(r) = \frac{1}{9\sqrt{6}}\left(\frac{Z}{a_0}\right)^{3/2} \rho(4-\rho)\mathrm{e}^{-\rho/2}$
3	2	$R_{32}(r) = \frac{1}{9\sqrt{30}}\left(\frac{Z}{a_0}\right)^{3/2} \rho^2\,\mathrm{e}^{-\rho/2}$

Expectation values of r^{k}

Definition of expectation value can be found in Sec. F–6.1.

$$\langle r\rangle = \frac{1}{2}\,[3n^2 - \ell(\ell+1)]\left(\frac{a_0}{Z}\right)$$

$$\langle r^2\rangle = \frac{1}{2}\,[5n^2 + 1 - 3\ell(\ell+1)]n^2\left(\frac{a_0}{Z}\right)^2$$

$$\langle r^{-1}\rangle = \frac{1}{n^2}\left(\frac{Z}{a_0}\right)$$

$$\langle r^{-2}\rangle = \frac{2}{(2\ell+1)\,n^3}\left(\frac{Z}{a_0}\right)^2$$

$$\langle r^{-3}\rangle = \frac{1}{\ell\,(\ell+\frac{1}{2})\,(\ell+1)\,n^3}\left(\frac{Z}{a_0}\right)^3$$

Most probable radial distance (“radius” of the orbital)

$$r \approx \frac{n^2}{Z}\,a_0$$

Slater orbital

$$\Psi_{n\ell m}(r,\theta,\varphi) = N_{n\ell}\, r^{n^*-1} \exp\left(-\frac{Z_{\text{eff}}\, r}{n^*\, a_0}\right) Y_{\ell m}(\theta,\varphi)$$

Effective principal quantum number n^*

n	1	2	3	4	5	6
n^*	1	2	3	3.7	4.0	4.2

Effective nuclear charge

$$Z_{\text{eff}} = Z - \sum_{\substack{\text{other}\\ \text{electrons}}} s_i$$

The electrons are grouped according to

(1 s), (2 s 2 p), (3 s 3 p), (3 d), (4 s 4 p), (4 d), (4 f), (5 s 5 p)

Rules for determining screening factors

1. $s_i = 0$ for higher-group electrons
2. $s_i = 0.35$ for same-group electrons (0.30 for 1 s)
3. $s_i = 0.85$ for s and p electrons with principal quantum number one unit less than the electron under consideration
4. $s_i = 1$ for other electrons

6.5 Perturbation Theory

Time-independent perturbation theory of non-degenerate states

Equation and perturbation term $\mathbf{H}'$

$$\mathbf{H}\,\Psi_n = E_n\,\Psi_n \qquad \mathbf{H} = \mathbf{H}^0 + \mathbf{H}' \qquad \langle \mathbf{H}' \rangle \ll \langle \mathbf{H}^0 \rangle$$

Solution

$$\Psi_n = \Psi_n^0 + \Psi_n^{(1)} + \Psi_n^{(2)} + \Psi_n^{(3)} + \dots$$

$$E_n = E_n^0 + E_n^{(1)} + E_n^{(2)} + E_n^{(3)} + \dots$$

Ψ_k^0, $k = 1, 2, \dots, n, \dots$ are orthonormal eigenfunctions of $\mathbf{H}^0$ with eigenvalue E_k^0: $\mathbf{H}^0\,\Psi_k^0 = E_k^0\,\Psi_k^0$
$k = 1, 2, \dots$

First order perturbation terms

$$E_n^{(1)} = \int (\Psi_n^0)^* \, \mathbf{H}' \, \Psi_n^0 \, \mathrm{d}V$$

Ψ^* = complex conjugate to Ψ

$$\Psi_n^{(1)} = -\sum_{k \neq n} \frac{\int (\Psi_k^0)^* \, \mathbf{H}' \, \Psi_n^0 \, \mathrm{d}V}{E_k^0 - E_n^0} \Psi_k^0$$

Second order terms of perturbation of energy

$$E_n^{(2)} = \int (\Psi_n^0)^* \, \mathbf{H}' \, \Psi_n^{(1)} \, \mathrm{d}V$$

6.6 Scattering Theory

Differential cross section

$$\frac{\mathrm{d}\sigma}{\mathrm{d}\Omega} = \frac{N(\theta, \varphi)}{N_0}$$

N_0 = number of incident particles per time and area

$N(\theta,\varphi)$ = number of scattered particles per time in the solid angle element $\mathrm{d}\Omega = \sin\theta \, \mathrm{d}\theta \, \mathrm{d}\varphi$

Total scattering cross section

$$\sigma = \int \frac{\mathrm{d}\sigma}{\mathrm{d}\Omega} \, \mathrm{d}\Omega$$

Elastic scattering in centre-of-mass system

Incident wave and particle energy

$$\psi_0 = \mathrm{e}^{\mathrm{i}\boldsymbol{k} \cdot \boldsymbol{r}} = \mathrm{e}^{\frac{\mathrm{i}}{\hbar}\boldsymbol{p} \cdot \boldsymbol{r}}$$

$$E = \frac{\hbar^2 k^2}{2\mu} = \frac{p^2}{2\mu}$$

μ = reduced mass

$\boldsymbol{p}$ = momentum in c.m. system

Total wave function

$$\psi = \psi_0 + \psi_S$$

ψ_S = scattered wave

Differential equation

$$(\nabla^2 + k^2)\psi_S = \frac{2\mu}{\hbar^2} E_p(\boldsymbol{r})\psi \qquad E_p(\boldsymbol{r}) = \text{scattering potential}$$

Solution when $E_p(\boldsymbol{r}) = 0$ *outside the volume* V'.

Integral equation of elastic scattering

$$\psi_S(\boldsymbol{r}) = -\frac{1}{4\pi}\frac{2\mu}{\hbar^2}\int \frac{e^{ikR}}{R} E_p(\boldsymbol{r}')\,\psi(\boldsymbol{r}')dV'$$

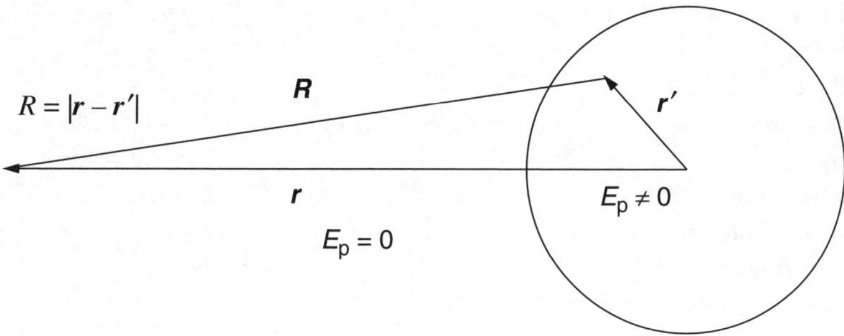

In particular, when $R \gg r'$

$$\psi_S = \frac{e^{ikr}}{r} f(\theta,\varphi)$$

Scattering amplitude

$$f(\theta,\varphi) = -\frac{1}{4\pi}\frac{2\mu}{\hbar^2}\int \exp\left(-\frac{ik\boldsymbol{r}\cdot\boldsymbol{r}'}{r}\right) E_p(\boldsymbol{r}')\,\psi(\boldsymbol{r}')dV'$$

Differential scattering cross section

$$\frac{d\sigma}{d\Omega} = |f(\theta,\varphi)|^2$$

First Born approximation

$$\psi(\boldsymbol{r}') \approx \psi_0(\boldsymbol{r}') = e^{i\boldsymbol{k}\cdot\boldsymbol{r}'}$$

can be inserted as approximate solution into the integral equation of elastic scattering.

7 Atomic and Molecular Physics

Quantity	Symbol	SI Unit
Mass	m, M	kg
Velocity	$\boldsymbol{v}$ (v)	m/s
Wavelength	λ	m
Frequency	ν	$\mathrm{Hz} = \mathrm{s}^{-1}$
Momentum	$\boldsymbol{p}$ (p)	Ns = kg m/s
Energy	E	$\mathrm{J} = \mathrm{Nm} = \mathrm{kg\ m^2/s^2}$
Kinetic energy	E_k	J
Orbital angular momentum	$\boldsymbol{L}$	Js
Spin angular momentum	$\boldsymbol{S}$	Js
Number of protons in nucleus	Z	

7.1 Wave Properties of Particles

de Broglie's relation

$$\lambda = \frac{h}{p}$$

λ = de Broglie wavelength

Momentum

$$p = \hbar\, k = \frac{h}{\lambda}$$

$\boldsymbol{k}$ = wave vector

Group velocity and phase velocity

$$v_g \equiv \frac{d\omega}{dk} = \frac{dE}{dp} = v$$

$$v_p = \frac{\omega}{k} = \frac{E}{p}$$

Dispersion law and de Broglie wavelength of a relativistic particle

$$E^2 = p^2 c_0^2 + E_0^2 \qquad E_0 = m_0 c_0^2$$

$$\lambda = \frac{h c_0}{(E_k^2 + 2 E_k E_0)^{1/2}}$$

Dispersion law and de Broglie wavelength of a non-relativistic particle

$$E = \frac{p^2}{2m} + E_0$$

$$\lambda = \frac{h}{(2m E_k)^{1/2}}$$

Heisenberg's uncertainty principle

$$\Delta x \, \Delta p_x \geq \frac{1}{2} \hbar$$

$$\Delta E \, \Delta t \geq \frac{1}{2} \hbar$$

See also Sec. F – 6.1

Δt is the time required to observe an energy to within the uncertainty ΔE.

7.2 The Photon

Energy and momentum

$$E = \hbar \omega = h \nu$$

$$p = \frac{h \nu}{c_0} n$$

n = refractive index

$$\boldsymbol{p} = \hbar \boldsymbol{k} \qquad |\boldsymbol{k}| = \frac{2\pi}{\lambda} = \frac{\omega n}{c_0}$$

$\boldsymbol{k}$ = wave vector

Photoelectric effect

$E_{k,max} = h\,\nu - \phi_0$ — ϕ_0 = work function

$\sigma_p \propto \dfrac{Z^5}{(h\nu)^{7/2}}$ — σ_p = cross section of photoelectric effect

Compton effect

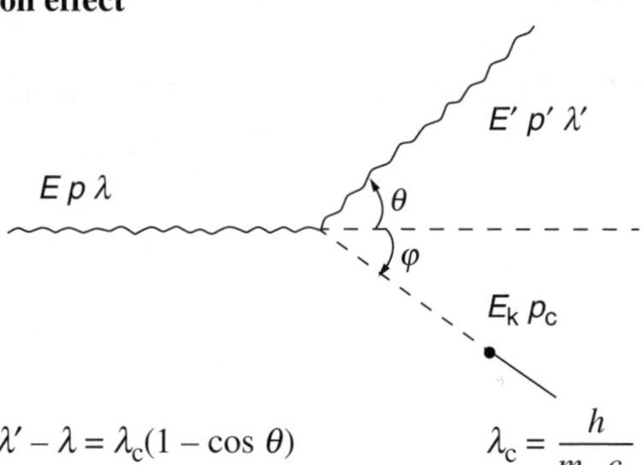

$\lambda' - \lambda = \lambda_c(1 - \cos\theta)$ — $\lambda_c = \dfrac{h}{m_0\,c_0}$ = Compton wavelength of the particle

$$E' = h\,\nu' = \frac{E}{1+\alpha(1-\cos\theta)}$$

$$\alpha = \frac{E}{m_0\,c_0^2}$$

$$E_k = E - E' = c_0\sqrt{p_c^2 + m_0^2\,c_0^2} - m_0\,c_0^2$$

$$\cos\theta = 1 - \frac{2}{(1+\alpha)^2\tan^2\varphi + 1}$$

$$\sin\varphi = \frac{p'\sin\theta}{p_c} = \frac{p'\sin\theta\cos\varphi}{p - p'\cos\theta}$$

$\sigma_c \propto \dfrac{Z}{E}$ — σ_c = cross section of Compton effect

Pair production

$\gamma \rightarrow e^- + e^+$ — (Must occur near a nucleus)

$E_{k,e^+} \approx E_{k,e^-} \approx \dfrac{1}{2}\,h\,\nu - m_e\,c_0^2$ — $E_{nucleus} \approx 0$

$\sigma_p \propto h\,\nu\,Z^2$ when $h\,\nu > 2m_e\,c_0^2$

σ_p = cross section of pair production

See also Sec. F–8.9

7.3 The Bohr Model

Quantization condition

$$2\,\pi\,r = n\,\frac{h}{p}$$

Total energy

$$E_n = -\,\frac{m\,e^4\,Z^2}{8\,\varepsilon_0^2\,h^2\,n^2} = -\,E_{\mathrm{H}}\,\frac{Z^2}{n^2} = -\,R\,h\,c_0\,\frac{Z^2}{n^2}$$

m = reduced mass $\approx m_{\mathrm{e}}$

$$R = R_\infty\left(1 - \frac{m_{\mathrm{e}}}{M}\right)$$

R_∞ = Rydberg constant

M = atomic mass

Balmer formula of line spectra (Rydberg's formula)

$$\frac{1}{\lambda} = R\,Z^2\left(\frac{1}{n_1^2} - \frac{1}{n_2^2}\right)$$

$n_2 = n_1 + 1,\ n_1 + 2,\ n_1 + 3,\ \ldots$

Radius of electron orbit

$$r_n = \frac{\varepsilon_0\,h^2\,n^2}{\pi\,m\,e^2\,Z} = \frac{n^2}{Z}\,a_0$$

a_0 = Bohr radius

Velocity of electron in orbit

$$v_n = \frac{Z\,e^2}{2\,\varepsilon_0\,h\,n}$$

7.4 Electron Structure of Atoms

Shells and subshells

n	1	2	3	4	5
Symbol	K	L	M	N	O

In shell n there is room for $2\,n^2$ electrons

ℓ	0	1	2	3	4	5
Symbol	s	p	d	f	g	h

In subshell ℓ there is room for $2(2\ell + 1)$ electrons

$$\ell = 0, 1, 2, \ldots, (n-1)$$

A subshell configuration is described as $n\ell^x$ where x = number of electrons

Term (level)

$${}^{2S+1}\mathrm{L}_J^{M_J}$$

S = atomic spin

$\mathrm{L} = \mathrm{S}, \mathrm{P}, \mathrm{D} \ldots$ = symbol of the atomic L quantum number

J = fine structure levels, $J = |L - S|, |L - S| + 1, \ldots, L + S$

M_J = magnetic quantum number = $M_L + M_S$

$$M_J = -J, -J + 1, \ldots, J - 1, J$$

$2S + 1$ = multiplicity

Spin-orbit interaction in single-electron systems

$$E = E_{n\ell} + E_{\mathrm{SO}}$$

$$E_{\mathrm{SO}} = a\,\boldsymbol{L} \cdot \boldsymbol{S}$$

$$a \approx \frac{Z}{8\pi\,\varepsilon_0}\left(\frac{e}{m\,c_0}\right)^2 \langle r^{-3}\rangle$$

m = reduced mass $\approx m_{\mathrm{e}}$

$$\boldsymbol{L} \cdot \boldsymbol{S} = \tfrac{1}{2}\,\hbar^2\,(j(j+1) - \ell(\ell+1) - s(s+1))$$

$$s = \tfrac{1}{2} \text{ for a single electron}$$

$$\langle r^{-3} \rangle = \frac{Z^3}{\ell\,(\ell+\frac{1}{2})\,(\ell+1)\,n^3\,a_0^3} \qquad a_0 = \text{Bohr radius}$$

Pauli's exclusion principle

The total wave function of a system of electrons must be antisymmetric. ⇒

No two electrons in an atom may have the same set of quantum numbers.

Terms of configurations p^x and d^x

Electron configuration	Terms				
p^0 , p^6	^{1}**S**				
p^1 , p^5		^{2}**P**			
p^2 p^4	^{1}**S,D**		^{3}**P**		
p^3		^{2}**P,D**		^{4}**S**	
d^0 , d^{10}	^{1}**S**				
d^1 , d^9		^{2}**D**			
d^2 , d^8	^{1}**S,D,G**		^{3}**P,F**		
d^3 , d^7		^{2}**D,P,D,F,G,H**		^{4}**P,F**	
d^4 , d^6	^{1}**S,D,G,S,D,F,G,I**		^{3}**P,F,P,D,F,G,H**		^{5}**D**
d^5		^{2}**D,P,D,F,G,H,S,D,F,G,I**		^{4}**P,F,D,G**	^{6}**S**

Spin-orbit interaction in many-electron systems in Russell-Saunders case (pure LS-coupling)

$$E_{SO} = a'\,\boldsymbol{L} \cdot \boldsymbol{S} = \tfrac{1}{2}\,a'\,\hbar^2\,(J(J+1) - L(L+1) - S(S+1))$$

Interval rules, the second is called Landé's rule of intervals

$$E_J - E_{J-1} = a'\,\hbar^2\,J \qquad a' = \text{a substance constant}$$

$$\frac{E_{J+1} - E_J}{E_J - E_{J-1}} = \frac{J+1}{J}$$

Hund's rules, giving the quantum numbers with the lowest LS-coupling energy

1. Maximum spin quantum number S
2. Maximum orbital angular momentum quantum number L (compatible with maximum S)
3. $J = |L - S|$ if less than half of the subshell has been occupied (normal fine structure, $a' > 0$)
   $J = L + S$ if more than half of the subshell has been occupied (inverted fine structure, $a' < 0$)

Coupling of angular momenta, Clebsch-Gordan (Wigner) coefficients

Coefficients for the addition of any angular momentum j_1 to angular momenta $j_2 = \frac{1}{2}$ and $j_2 = 1$

$(j_1 \, \tfrac{1}{2} \, m_1 m_2 | j_1 \, \tfrac{1}{2} \, j \, m)$

$j =$	$m_2 = \frac{1}{2}$	$m_2 = -\frac{1}{2}$
$j_1 + \frac{1}{2}$	$\sqrt{\dfrac{j_1 + m + \frac{1}{2}}{2j_1 + 1}}$	$\sqrt{\dfrac{j_1 - m + \frac{1}{2}}{2j_1 + 1}}$
$j_1 - \frac{1}{2}$	$-\sqrt{\dfrac{j_1 - m + \frac{1}{2}}{2j_1 + 1}}$	$\sqrt{\dfrac{j_1 + m + \frac{1}{2}}{2j_1 + 1}}$

$(j_1 1 \, m_1 m_2 | j_1 1 \, j \, m)$

$j =$	$m_2 = 1$	$m_2 = 0$	$m_2 = -1$
$j_1 + 1$	$\sqrt{\dfrac{(j_1+m)(j_1+m+1)}{(2j_1+1)(2j_1+2)}}$	$\sqrt{\dfrac{(j_1-m+1)(j_1+m+1)}{(2j_1+1)(j_1+1)}}$	$\sqrt{\dfrac{(j_1-m)(j_1-m+1)}{(2j_1+1)(2j_1+2)}}$
j_1	$-\sqrt{\dfrac{(j_1+m)(j_1-m+1)}{2j_1(j_1+1)}}$	$\dfrac{m}{\sqrt{j_1(j_1+1)}}$	$\sqrt{\dfrac{(j_1-m)(j_1+m+1)}{2j_1(j_1+1)}}$
$j_1 - 1$	$\sqrt{\dfrac{(j_1-m)(j_1-m+1)}{2j_1(2j_1+1)}}$	$-\sqrt{\dfrac{(j_1-m)(j_1+m)}{j_1(2j_1+1)}}$	$\sqrt{\dfrac{(j_1+m+1)(j_1+m)}{2j_1(2j_1+1)}}$

Further information:
http://electron6.phys.utk.edu/qm2/modules/m4/clebsch.htm

Atom in weak magnetic field, Zeeman effect

$$E = E_{n\ell} + E_{SO} + E_B$$

$$E_B = \mu_B \, g_J \, M_J \, B \qquad \mu_B = \frac{e\hbar}{2m_e} = \text{Bohr magneton } (e > 0)$$

B = magnetic flux density

The Landé factor

$$g_J = 1 + \frac{J(J+1) - L(L+1) + S(S+1)}{2J(J+1)}$$

Selection rules for electron transitions

$\Delta M_J = 0, \pm 1$ (not $0 \curvearrowright 0$ if $\Delta J = 0$)

Atom in a strong magnetic field, Paschen-Back effect

$$E_B \approx \mu_B \, B(M_L + 2\,M_S)$$

$$E_{SO} = a' \hbar^2 M_S M_L$$

Selection rules for electron transitions

$\Delta M_L = 0, \pm 1 \qquad \Delta M_S = 0$

Relativistic correction to kinetic energy and spin-orbit interaction energy in single-electron atoms

$$E_r = -|E_n| \frac{(\alpha Z)^2}{n^2} \left(\frac{n}{j + \frac{1}{2}} - \frac{3}{4} \right)$$

E_n = total energy as in Sec. 7.3
α = fine structure constant, see Sec. CU – 1.1

Electron orbital and spin magnetic dipole moments

$$\mu_\ell = -\frac{e}{2m_e} \boldsymbol{L} = -\frac{\mu_B}{\hbar} \boldsymbol{L}$$

$$\mu_s = \frac{g_e \, \mu_B}{\hbar} \boldsymbol{S} = -\gamma_e \, \boldsymbol{S} \approx -2 \frac{e}{2m_e} \boldsymbol{S}$$

g_e = electron g factor = – 2 · 1.001 159 652 186

$\gamma_e = \dfrac{e}{2m_e} g_e$ = electron gyromagnetic ratio

Splitting of electron energy levels due to spin-orbit interaction

$$E = \begin{cases} E_n + \frac{1}{2}\, a\,\ell\,\hbar^2 & \text{when } j = \ell + \frac{1}{2} \\ E_n - \frac{1}{2}\, a(\ell + 1)\hbar^2 & \text{when } j = \ell - \frac{1}{2} \end{cases}$$

$$a = \frac{|E_n|\,(\alpha Z)^2}{\hbar^2 n\,\ell\,(\ell + 1)\,(\ell + \frac{1}{2})}$$

7.5 Electronic Transitions

Selection rules

Single-electron systems

$\Delta\ell = \pm 1$ $\qquad\qquad$ $\Delta m_\ell = 0, \pm 1$

Many-electron systems (pure LS coupling)

$\Delta L = \pm 1$

$\Delta S = 0$

$\Delta J = 0, \pm 1$ (not $0 \curvearrowright 0$)

X-ray emission

K_{α_1} : L_{III} (= 2 $p_{3/2}$) $\rightarrow$ K (= 1 $s_{1/2}$)

K_{α_2} : L_{II} (= 2 $p_{1/2}$) $\rightarrow$ K (= 1 $s_{1/2}$)

K_β : $M_{III,II}$ (= 3 p) $\rightarrow$ K (= 1 s)

Auger effect

Approximate kinetic energy of the ejected electron in the transition $K\,L_i\,L_j$

$$E_k \approx E_b(K) - E_b(L_i) - E_b(L_j)$$

$E_b(L_i)$ = binding energy of electron in shell L_i

(No selection rule operative)

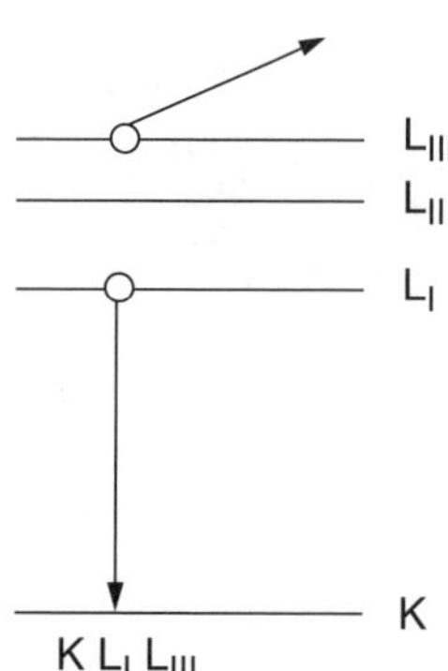

Energy of emitted (absorbed) photon

$$h\nu = \Delta E \,(\mp)\, \frac{(\Delta E)^2}{2\,M\,c_0^2} \qquad (\Delta E \ll M\,c_0^2)$$

M = mass of emitting (absorbing) atom or system

Number of atoms in different energy levels according to Maxwell-Boltzmann statistics

$$\frac{N_i}{N_j} = \frac{g_i}{g_j}\, e^{-\Delta E/kT}$$

$\Delta E = E_i - E_j$

N_i = number of atoms in level i

k = Boltzmann's constant

T = temperature

g_i = statistical weight of level i

Einstein's coefficients *A* and *B*

$$g_i\,B_{ij} = g_j\,B_{ji}$$

$$A_{ji} = \frac{8\pi\,h\,\nu^3}{c_0^3}\,B_{ji}$$

B_{ij} = coefficient of absorption

B_{ji} = coefficient of stimulated emission

A_{ji} = coefficient of spontaneous emission

$$A_{ji} = \frac{16\pi^2\nu^3}{3h\,\varepsilon_0 c_0^3}\,|\mu_{ji}|^2$$

μ_{ji} = electric dipole moment

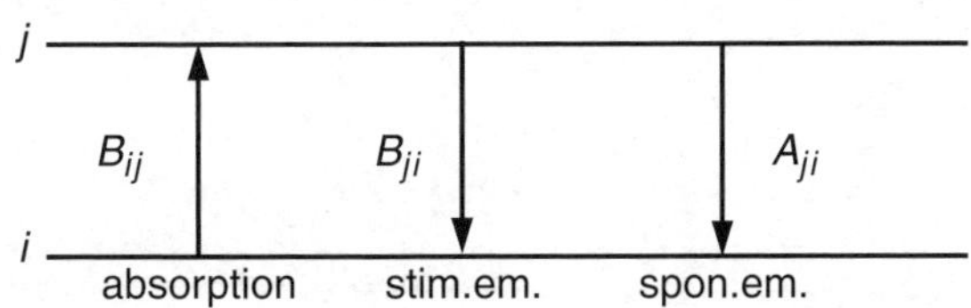

Lifetimes

$$\tau_{\mathrm{sp}} = \frac{1}{\Sigma_i\,A_{ji}} = \frac{1}{A}$$

sp = spontaneous (emission)

$$\frac{1}{\tau} = \frac{1}{\tau_{\mathrm{sp}}} + \frac{1}{\tau_{\mathrm{nr}}}$$

nr = non-radiative (transitions)

$$\phi = \frac{\tau}{\tau_{\mathrm{sp}}}$$

ϕ = fluorescence yield

Line profiles

Normalized Lorentzian profile (area = 1, natural line shape)

$$g(\omega) = \frac{1}{\pi}\,\frac{\gamma/2}{(\omega-\omega_0)^2+(\gamma/2)^2}$$

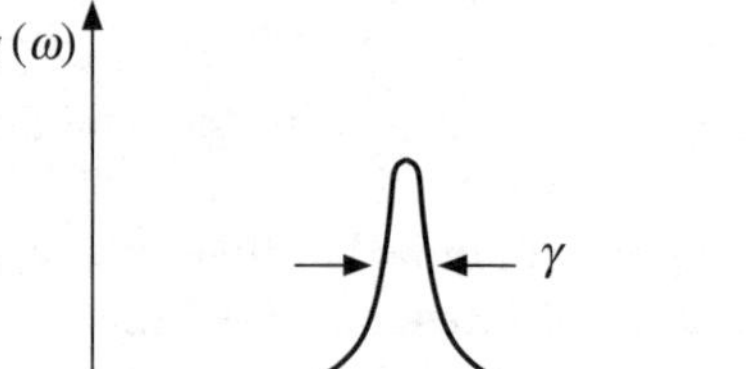

Normalized Doppler profile (area = 1, Gaussian line shape)

$$g(\omega) = \frac{2}{(\Delta\omega)_D}\sqrt{\frac{\ln 2}{\pi}}\exp\left\{-4\ln 2\frac{(\omega-\omega_0)^2}{(\Delta\omega)_D^2}\right\}$$

Doppler width

$$(\Delta\omega)_D = \frac{\omega_0}{c_0}\sqrt{\ln 2\cdot 8\,kT/m}$$

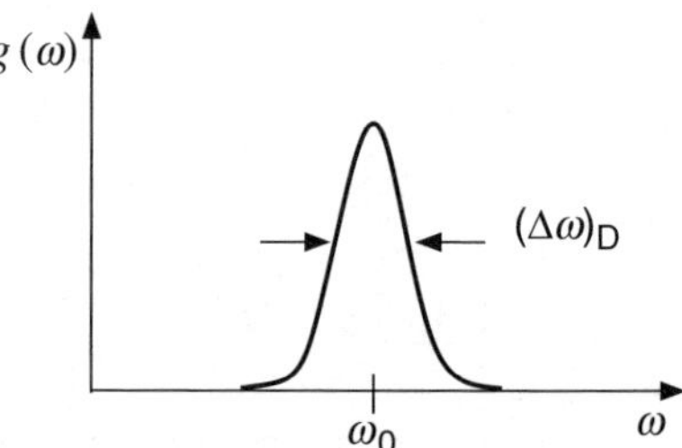

Transition probability (per time)

$\Lambda = \frac{1}{\tau}$ — τ = mean life

$\Gamma\,\tau = \hbar$ — Γ = width of level

$\frac{dN_\gamma}{dt} = \Lambda\,N_{ex}$ — $\frac{dN_\gamma}{dt}$ = number of emitted quanta per time

N_{ex} = number of excited electrons in the shell

7.6 Diatomic Molecules

Rotational energy of a rigid molecule

$$E_r = \frac{r(r+1)}{2\,I}\,\hbar^2 = Br(r+1) = \frac{1}{2}\,I\,\omega_{rot}^2$$

r = rotational quantum number
B = rotational constant
ω_{rot} = angular velocity

Moment of inertia

$$I = \mu R_0^2$$

$$\mu = \text{reduced mass} = \frac{M_1 M_2}{M_1 + M_2}$$

R_0 = bond length

In particular,

$$E_{r+1} - E_r = \frac{\hbar^2}{I}(r+1) = 2B(r+1)$$

Selection rule for rotational transitions

$$\Delta r = \pm 1$$

Vibrational energy of harmonically vibrating molecule

$$E_v = \left(v + \frac{1}{2}\right)\hbar\,\omega_0$$

v = vibrational quantum number

Selection rule for transitions between vibrational levels in harmonically oscillating molecule

$$\Delta v = \pm 1$$

Morse potential

$$E_p(R) = D_e(1 - e^{-\alpha(R-R_0)})^2$$

D_e, α, and R_0 are characteristic constants of each molecule.

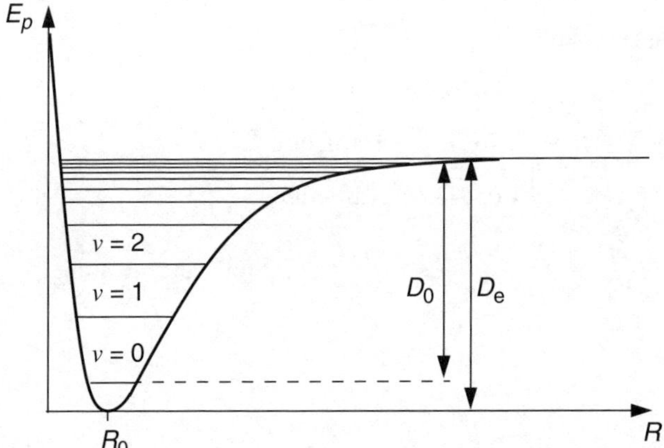

Dissociation energy

$$D_0 = D_e - \tfrac{1}{2}\,\hbar\,\omega_0$$

$\tfrac{1}{2}\,\hbar\,\omega_0$ = zero-point vibrational energy

Vibrational energy of a molecule in a Morse potential

$$E_v = (v + \tfrac{1}{2})\hbar\,\omega_0 - (v + \tfrac{1}{2})^2\,\frac{\hbar^2\,\omega_0^2}{4\,D_e}$$

v = vibrational quantum number

$$\omega_0 = \sqrt{\frac{k}{\mu}} \qquad k = 2\,D_e\,\alpha^2$$

Potential energy of an ion pair, ionically bonded

$$E_p(R) = -\frac{Z_1\,Z_2\,e^2}{4\pi\varepsilon_0\,R} + E_{rep}(R) \qquad e = \text{electron charge}$$

$$E_{rep}(R) = \lambda\,e^{-R/\rho} \quad \text{or} \quad E_{rep}(R) = \frac{b}{R^n},\ n \approx 9$$

λ, ρ, and b are constants, $\rho \approx 0.1\,R_0$

$$\frac{d}{dR}\,E_p(R) = 0 \quad \text{for} \quad R = R_0$$

Energy needed to create two neutral atoms

$$E_e = -\,E_p(R_0) - (\text{IP} - \text{EA})$$

IP = ionization energy of the positive ion
EA = electron affinity of the negative ion

Madelung constant

$$\alpha = \sum_j \frac{\pm 1}{p_j} = \begin{cases} 1.7476, \text{ NaCl structure} \\ 1.7627, \text{ CsCl structure} \\ 1.6381, \text{ Zn S structure} \end{cases}$$

$$p_j = \frac{R_j}{R_0} \qquad R_j = \text{distance to atom } j$$

8 Nuclear and Subnuclear Physics

Quantity	Symbol	SI Unit
Number of protons	Z	
Mass number (integer)	A	
Mass	m, M	kg
Decay constant	λ	s^{-1}
Nuclear radius	R	m
Velocity	$\boldsymbol{v}$ (v)	m/s
Momentum	$\boldsymbol{p}$ (p)	Ns = kg m/s
Nuclear spin	$\boldsymbol{I}$	Ns
Nuclear spin quantum number	I	
Energy	E	J = Nm = kg m^2/s^2
Kinetic energy of particle x	E_x	J
Q value	Q	J
Density	ρ	kg/m^3
Dose equivalent	H	Sv = J/kg
Absorbed dose	D	Gy = J/kg

8.1 Radioactivity and Nuclear Mass

Radioactive decay law

$$dN = -N\lambda\, dt$$
$$N = N_0\, e^{-\lambda t}$$

λ = disintegration constant

N = number of nuclei at time t

N_0 = number of nuclei at time $t = 0$

Activity

$$n = -\frac{dN}{dt} = \lambda N$$

Half-life and mean life

$$t_{1/2} = \frac{\ln 2}{\lambda}$$

$$\tau = \frac{1}{\lambda}$$

Power

$$P = n\,Q$$

Q = Q value as in Sec. 8.2

Decay law with continuous contribution U (e.g. from irradiation as in Sec. 8.3)

$$\frac{\mathrm{d}N}{\mathrm{d}t} = U - \lambda N$$

$$N = \frac{U}{\lambda}\,(1 - \mathrm{e}^{-\lambda t})$$

Decay A → B → C

$$N_\mathrm{B} = \frac{\lambda_\mathrm{A}}{\lambda_\mathrm{B} - \lambda_\mathrm{A}}\,N_{\mathrm{A},0}\,(\mathrm{e}^{-\lambda_\mathrm{A} t} - \mathrm{e}^{-\lambda_\mathrm{B} t})$$

$N_{\mathrm{A},0}$ = number of A particles at time $t = 0$

In particular, if $\lambda_\mathrm{A} \ll \lambda_\mathrm{B}$

$$\lambda_\mathrm{A} N_\mathrm{A} \approx \lambda_\mathrm{B}\,N_\mathrm{B}$$

Weizsäcker's semi-empirical nuclear mass formula

$$M = M_0 + M_1 + M_2 + M_3 + M_4 + \delta$$

Total rest mass of particles

$$M_0 = Z\,m_\mathrm{p} + (A - Z)m_\mathrm{n}$$

Nucleon binding energy

$$M_1 = -\,a_1\,A \qquad a_1 \approx 0.016919\ \mathrm{u}$$

Surface tension

$$M_2 = a_2\,A^{2/3} \qquad a_2 \approx 0.019114\ \mathrm{u}$$

Coulomb repulsion

$$M_3 = a_3 \frac{Z^2}{A^{1/3}} \qquad a_3 \approx 0.0007626 \text{ u}$$

Pairing of nucleons

$$M_4 = a_4 \frac{(A-2Z)^2}{A} \qquad a_4 \approx 0.02544 \text{ u}$$

Odd-even effect

$$\delta = \begin{cases} -f(A) & A \text{ and } Z \text{ even} \\ 0 & A \text{ odd} \\ +f(A) & A \text{ even, } Z \text{ odd} \end{cases}$$

$$f(A) \approx a_5 A^{-3/4} \qquad a_5 \approx 0.036 \text{ u}$$

Stable isotopes

$$\frac{dM}{dZ} = 0 \qquad Z \approx \frac{A}{2 + \frac{a_3}{2a_4} A^{2/3}}$$

Packing fraction

$$f = \frac{M-A}{A}$$

Total electrostatic energy in nucleus (Coulomb energy)

$$E_C \approx \frac{3}{5} \cdot \frac{e^2 Z(Z-1)}{4\pi \varepsilon_0 R} \qquad e = \text{fundamental charge}$$

Nuclear radius

$$R = r_0 A^{1/3}$$

$$r_0 = \begin{cases} (1.20 \pm 0.03) \cdot 10^{-15} \text{ m} & (\text{“charge radius”}) \\ (1.4 \pm 0.1) \cdot 10^{-15} \text{ m} & (\text{“nuclear force radius”}) \end{cases}$$

8.2 Nuclear Reactions

Nuclear reaction X(x,y)Y

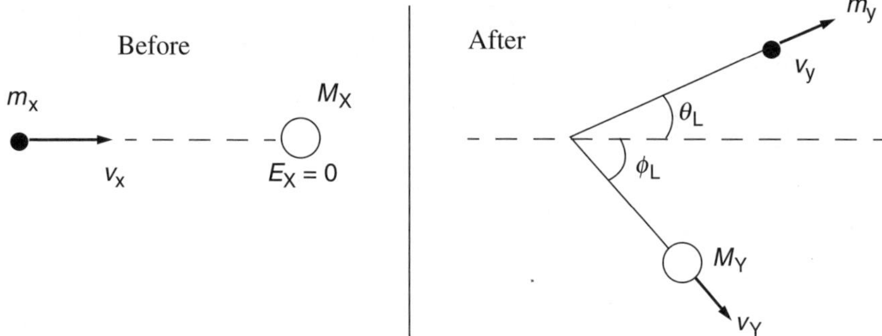

v_x, v_y, and v_Y	velocities in lab system
v_x', v_X', v_y', and v_Y'	velocities in centre-of-mass (c.m.) system
E_x, E_y, and E_Y	kinetic energies in lab system
E_x', E_X', E_y', and E_Y'	kinetic energies in c.m. system
θ_L, ϕ_L, θ_c, and ϕ_c	angles in lab and c.m. systems

Nuclear binding energy

$$E_b = (Z\,m_p + (A - Z)m_n - M_{nucleus})\,c_0^2$$

$$E_b = (Z\,m_H + (A - Z)m_n - M_{atom})\,c_0^2$$

$m_H\,c_0^2$ = mass of hydrogen atom = 1.0078250 u

$(m_H\,c_0^2 = 938.783\text{ MeV})$

Q value

$$Q_0 = (M_{before} - M_{after})\,c_0^2 = E_Y + E_y - E_x + E^*$$

$$Q_0 = E_b(Y) + E_b(y) - E_b(X) - E_b(x)$$

$$Q = Q_0 - E^* = E_y\left(1 + \frac{m_y}{M_Y}\right) - E_x\left(1 - \frac{m_x}{M_Y}\right) - \frac{2\sqrt{m_x\,m_y\,E_x\,E_y}}{M_Y}\cos\theta_L$$

Q_0 = Q value when the nuclei are in their ground states

E^* = Excitation energy of the product nucleus

Energy available in c.m. system

$$W \approx \sqrt{2E_x M_X c_0^2} \quad (E_x \gg m_x c_0^2, E_x \gg M_X c_0^2)$$

Solutions to the Q value formula

$$\sqrt{E_y} = a \pm \sqrt{a^2+b}$$

$$a = \frac{\sqrt{m_x m_y E_x}}{M_Y+m_y} \cos \theta_L$$

$$b = \frac{E_x(M_Y-m_x)+M_Y Q}{M_Y+m_y}$$

$$\sqrt{E_x} = -c + \sqrt{c^2+d}$$

$$c = \frac{\sqrt{m_x m_y E_y}}{M_Y-m_x} \cos \theta_L$$

$$d = \frac{E_y(M_Y+m_y)-M_Y Q}{M_Y-m_x}$$

Threshold energy

$$(E_x)_{\min} = -Q \frac{m_x+M_X+m_y+M_Y}{2M_X} \approx -Q\left(1+\frac{m_x}{M_X}\right)$$

Transforming between the lab and centre-of-mass systems

$$v_x' = v_x - v_c = \frac{M_X}{m_x+M_X} v_x$$

v_c = speed of centre of mass before reaction

$$v_X' = -v_c = -\frac{m_x}{m_x+M_X} v_x$$

$$E_x' = \left(\frac{M_X}{m_x+M_X}\right)^2 E_x$$

$$E'_X = \frac{m_x M_X}{(m_x + M_X)^2} E_x$$

$$E'_y + E'_Y = Q + E'_X + E'_x = Q + \frac{M_X}{m_x + M_X} E_x$$

$$E'_y = \frac{M_Y}{m_y + M_Y} (E'_y + E'_Y)$$

$$E'_Y = \frac{m_y}{m_y + M_Y} (E'_y + E'_Y)$$

$$\tan \theta_L = \frac{\sin \theta_c}{\gamma + \cos \theta_c}$$

$$\gamma = \frac{v_c}{v'_y}$$

In particular, if $\frac{Q}{c_0^2} \ll 1$ u

$$\gamma \approx \sqrt{\frac{m_x m_y}{M_X M_Y} \cdot \frac{E_x}{E_x + Q\left(1 + \frac{m_x}{M_X}\right)}}$$

Kinetic energy of c.m. before and after reaction

$$(E_c)_{\text{before}} = \frac{m_x}{M_X + m_x} E_x$$

$$(E_c)_{\text{after}} = \frac{m_x}{M_Y + m_y} E_x$$

Compound nucleus excitation energy

$$E_{exc} = E_b(\text{compound}) - E_b(\text{X}) - E_b(\text{x}) + E_a$$

$$E_a = \frac{M_X}{m_x + M_X} E_x$$

8.3 Irradiation of Thin Foils

Number of target nuclei per volume

$$n = \frac{\rho N_A}{M}$$

N_A = Avogadro's number

M = atomic mass

Sensitive part of irradiated area A

$$f = n \sigma x$$

σ = cross section

x = thickness of foil

Mean free path and coefficient of absorption

$$\ell = \frac{1}{n \sigma} = \frac{1}{\Sigma}$$

Σ = macroscopic cross section

$$\mu = n \sigma$$

Number of particles passing through the foil

$$N = N_0 \, e^{-n\sigma x}$$

Yield (number of nuclear reactions per time)

$$U = \phi f A = \phi n \sigma V = \phi \sigma N_k$$

N_k = number of target nuclei in volume V

The flux

$$\phi = \frac{1}{A} \frac{dN_0}{dt}$$

N_0 = number of particles impinging on area A

8.4 Nuclear Spin

Length of nuclear spin vector *I*

$$|\boldsymbol{I}| = \sqrt{I(I+1)}\,\hbar$$

I = nuclear spin quantum number

Total atomic angular momentum (spin)

$$\boldsymbol{F} = \boldsymbol{J} + \boldsymbol{I}$$

$\boldsymbol{J}$ = total electronic angular momentum

Selection rules for transitions

$$\Delta F = 0, \pm 1 \qquad \text{(not } 0 \curvearrowright 0\text{)}$$

Nuclear magnetic dipole moment

$$\boldsymbol{\mu}_I = \frac{1}{\hbar}\, g_I\, \mu_N\, \boldsymbol{I}$$

$$\mu_z = \frac{1}{\hbar}\, g_I\, \mu_N\, I_z = g_I\, \mu_N\, m_I$$

$m_I = -I,\ -I+1, \ldots, I-1, I$

$\mu_N = \dfrac{e\,\hbar}{2m_p}$ = nuclear magneton ($e > 0$)

g_I = g-factor of nucleus

Energy splitting in external magnetic field *B*

$$E_B = -\,\boldsymbol{\mu}_I \cdot \boldsymbol{B} = -\,\mu_z\, B$$

8.5 Gamma Rays and Internal Conversion

Radiative energy flux

$$I = h\,\nu\,\phi \qquad \phi = \text{photon flux}$$

$$I = I_0\, e^{-\mu x}$$

μ = linear absorption coefficient

x = thickness of absorber

Absorption coefficients

$$\mu_{\mathrm{m}} = \frac{\mu}{\rho} = \text{mass absorption coefficient}$$

$$\mu_{\mathrm{e}} = \frac{A}{N_{\mathrm{A}}\,Z}\mu_{\mathrm{m}} = \text{electronic absorption coefficient}$$

N_{A} = Avogadro's number

$$\mu_{\mathrm{a}} = Z\,\mu_{\mathrm{e}} = \frac{A}{N_{\mathrm{A}}}\mu_{\mathrm{m}} = \text{atomic absorption coefficient}$$

Half-thickness

$$x_{1/2} = \frac{\ln 2}{\mu} = \frac{\ln 2}{\rho\,\mu_{\mathrm{m}}}$$

Weisskopf's estimate of 2^L-multipole transition probabilities (L = photon angular momentum)

$$\lambda_{EL} = \frac{4.4\cdot 10^{21}(L+1)}{L((2L+1)!!)^2}\left(\frac{3}{L+3}\right)^2\left(\frac{E_\gamma}{197}\right)^{2L+1}R^{2L}$$

$$\lambda_{ML} = \frac{1.9\cdot 10^{21}(L+1)}{L((2L+1)!!)^2}\left(\frac{3}{L+3}\right)^2\left(\frac{E_\gamma}{197}\right)^{2L+1}R^{2L-2}$$

when λ is measured in s^{-1}, E_γ in MeV, and R in 10^{-15} m

(λ_{E1} = probability of electric dipole transition)

(λ_{M2} = probability of magnetic quadrupole transition)

Selection rules of transitions

$$|I_{\mathrm{before}} - I_{\mathrm{after}}| \le L \le I_{\mathrm{before}} + I_{\mathrm{after}}$$

$\pi_{\mathrm{before}}\cdot\pi_{\mathrm{after}} = (-1)^L$ for EL transitions

$\pi_{\mathrm{before}}\cdot\pi_{\mathrm{after}} = (-1)^{L+1}$ for ML transitions

I = nuclear spin quantum number

π = parity

Conversion coefficient

$$\alpha = \frac{N_e}{N_\gamma} = \frac{\lambda_e}{\lambda_\gamma}$$

N_e = number of conversion electrons
N_γ = number of gamma quanta

Mean life

$$\tau = \frac{1}{\lambda} = \frac{1}{\lambda_e + \lambda_\lambda}$$

λ_e = internal conversion transition probability

Partial life

$$\tau_\gamma = \tau(1 + \alpha)$$

Level width

$$\Gamma = \frac{\hbar}{\tau}$$

Recoil energy of nucleus

$$E_R = \frac{1}{2}\, E_\gamma \frac{v_{nucleus}}{c_0} = \frac{E_\gamma^2}{2M\, c_0^2}$$

Mössbauer effect

$$\Gamma = \frac{1}{2}\, E_\gamma \frac{\Delta v}{c_0}$$

Dose

$$D = \frac{k_R\, C\, k_A\, t}{d^2}$$

k_R = dose rate from diagram
C = strength of source
d = distance
k_A = fraction penetrating a radiation shield

Dose rate

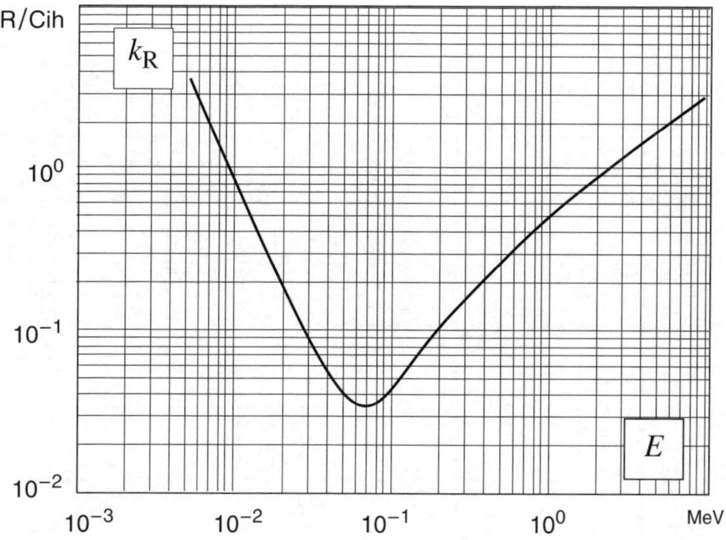

Dose rate at a distance of 1 metre from a localized gamma emitter as a function of gamma energy. The intervening medium is assumed to be air or another non-absorbing material.

Half-thickness

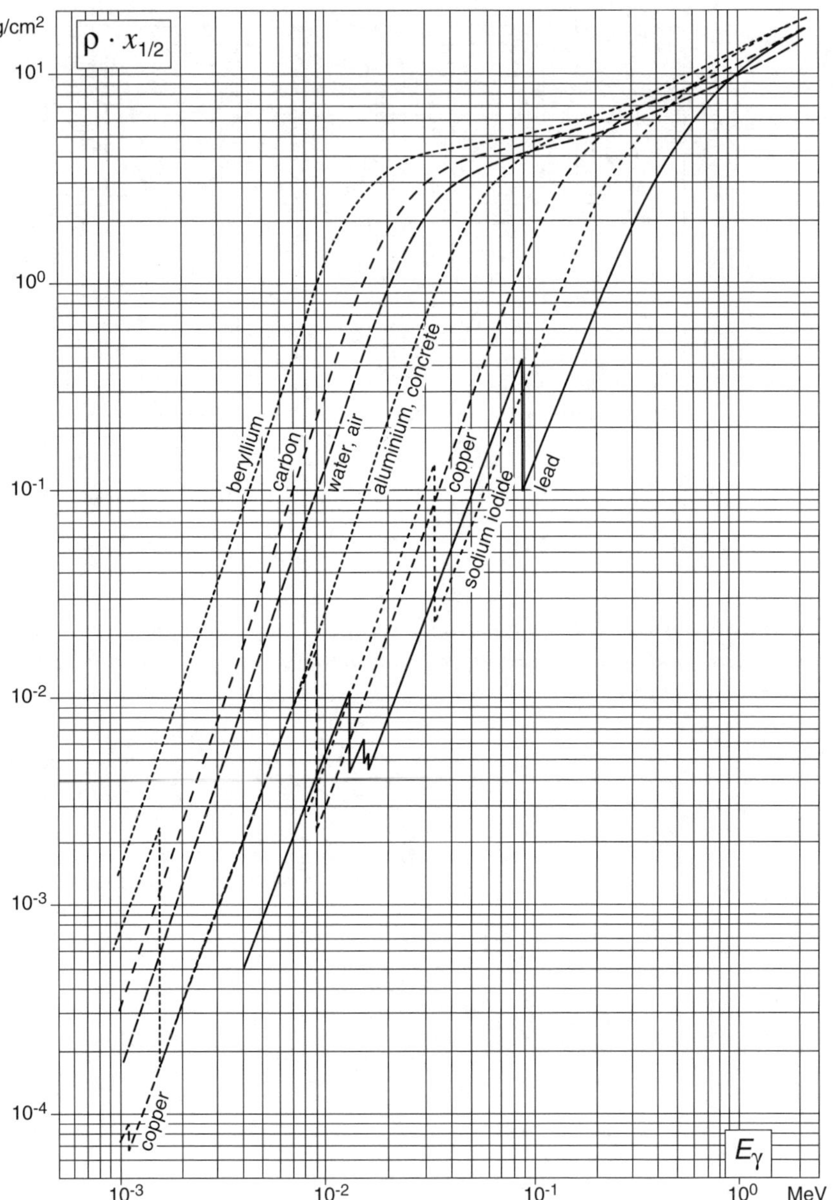

Half-thickness $\rho \cdot x_{1/2}$ *of various materials being irradiated with photons of energy* E_γ.

Density of concrete = 2.3 g/cm^3.

8.6 Alpha and Beta Decay

Kinetic energy of α-particle in decay X → Y + α

$$E_\alpha = \frac{M_\text{Y}}{M_\text{Y}+m_\alpha}\, Q$$

Q = Q value as in Sec. 8.2

Recoil energy of nucleus

$$E_\text{R} = \frac{m_\alpha}{M_\text{Y}+m_\alpha}\, Q$$

Geiger-Nuttal law

$$\log \lambda = A \log R_\alpha + B$$

A and B are constants

Range

$$R_\alpha \approx \text{constant} \cdot v^3$$

Particularly, in air

$R_\alpha \approx 0.318\ E_\alpha^{3/2}$ at 15 °C if R_α is measured in cm and E_α in MeV

Rutherford scattering formula

$$\sigma(\theta) = \left(\frac{q_\text{x}\, q_\text{n}}{2E_\text{x} \cdot 4\pi\, \varepsilon_0}\right)^2 \frac{1}{(1-\cos\,\theta)^2}$$

$\sigma(\theta)$ = cross section of scattering an angle θ, when a particle x of charge q_x is scattered by a nucleus of charge q_n

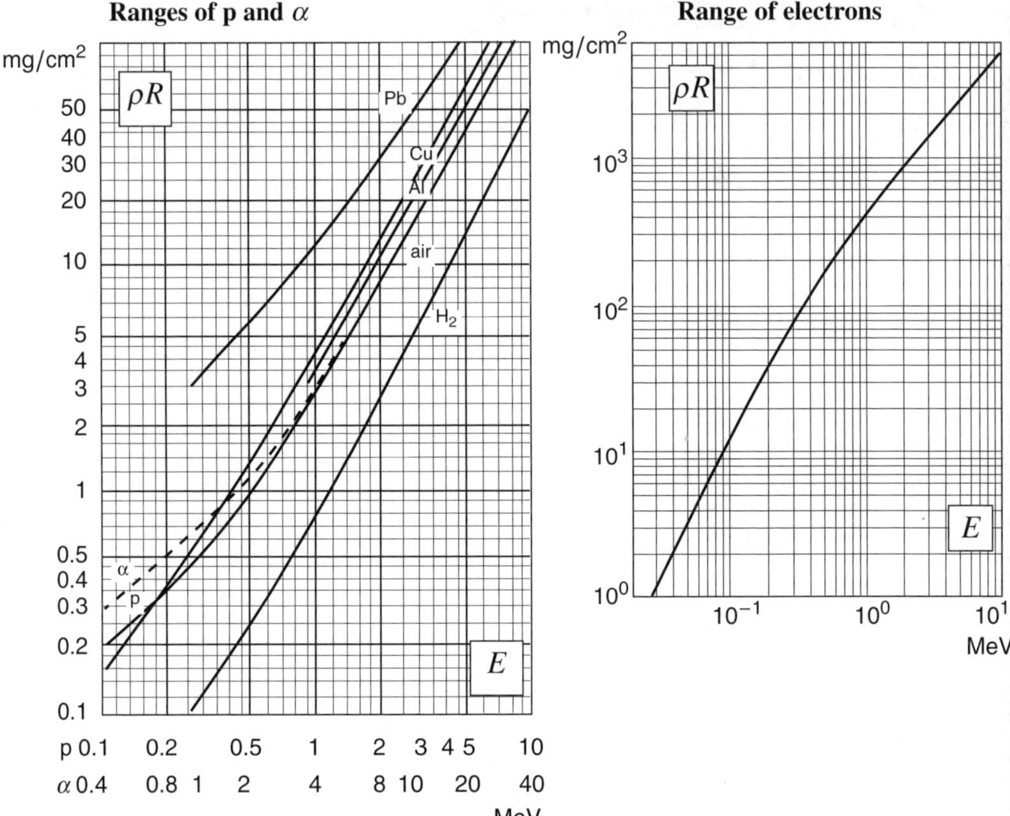

Range R of protons, α-particles, and electrons of kinetic energy E in a material of density ρ.

Range of electrons

$$R \approx \frac{1}{\rho}\,(0.530\,E - 0.106)$$

when $3 \lesssim E \lesssim 20$ and R is measured in cm, ρ in $\mathrm{g/cm^3}$ and E in MeV

β-decay X $\rightarrow$ Y + Σ y$_i$

Example 1. Electron emission

$$n \rightarrow p + e^- + \overline{\nu}$$

$$Q_{\beta^-} = E_b(Y) - E_b(X) + \Delta m\, c_0^2$$

$$\Delta m\, c_0^2 = (m_n - m_p - m_e)\, c_0^2 = 0.78232 \text{ MeV}$$

E_b = binding energy

Example 2. Positron emission

$$p \rightarrow n + e^+ + \nu$$

$$Q_{\beta^+} = E_b(Y) - E_b(X) - \Delta m\, c_0^2 - 2m_e\, c_0^2$$

Example 3. Electron capture

$$p + e^- = n + \nu$$

$$Q_{EC} = E_b(Y) - E_b(X) - \Delta m\, c_0^2$$

Maximum nuclear recoil energy

$$(E_R)_{max} = \frac{p_e^2}{2M} = \frac{E_k^2 + 2E_k\, m_e\, c_0^2}{2M\, c_0^2}$$

$E_k \approx Q$

E_k = total kinetic energy of both particles

8.7 Shell Model and Collective Model

Energy levels from shell model

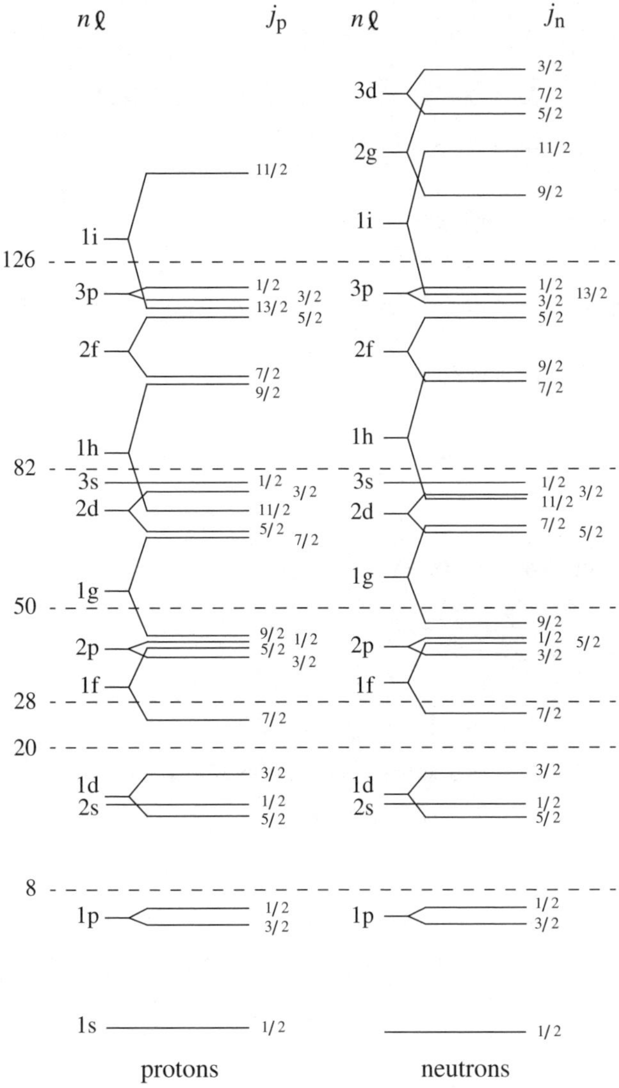

$|j_p - j_n| \le I \le j_p + j_n$ for unpaired nucleons

Every level holds $2(2j + 1)$ nucleons

Parity $\pi = (-1)^{\ell}$

For a numerical value of ℓ, see Sec. 7.4

Magnetic moment (in nuclear magnetons) from shell model (Schmidt estimate)

$$\mu_I = \left(j - \frac{1}{2}\right) g_{\ell} + \frac{1}{2} g_s \qquad \text{if } j = \ell + \frac{1}{2}$$

$$\mu_I = \frac{j}{j+1}\left(\left(j + \frac{3}{2}\right) g_{\ell} - \frac{1}{2} g_s\right) \qquad \text{if } j = \ell - \frac{1}{2}$$

For a proton $g_{\ell} = 1$ and $g_s = 5.5855$

For a neutron $g_{\ell} = 0$ and $g_s = -3.8263$

Rotational energy from collective model

$$E_r = \frac{\hbar^2}{2J}\left(I(I+1) - K^2\right) = \frac{1}{2} J\omega^2$$

For even A and even Z: $K = 0$, $I = 0^+, 2^+, 4^+, \ldots$

For odd A, K is half-integer, and $I = K, K + 1, \ldots$

J = moment of inertia
ω = angular velocity

8.8 Neutron Physics, Fission and Fusion

Elastic neutron scattering

E_0 = incident neutron kinetic energy (in lab system)
E_n = scattered neutron kinetic energy
θ_L = scattering angle in lab system
θ_c = scattering angle in centre-of-mass system
M = nuclear mass

$$\frac{E_n}{E_0} = \frac{M^2 + m_n^2 + 2Mm_n \cos\theta_c}{(M+m_n)^2} \approx \frac{A^2 + 1 + 2A \cos\theta_c}{(A+1)^2}$$

$$\cos\theta_L = \frac{A\cos\theta_c + 1}{(A^2 + 1 + 2A\cos\theta_c)^{1/2}}$$

Average energy

$$\langle E_n \rangle = f E_0 = \frac{A^2+1}{(A+1)^2} E_0$$

Average energy after x collisions

$$\langle E_n \rangle_x = f^x E_0$$

Recoil energy of nucleus

$$E_R = \frac{4m_n M}{(m_n + M)^2} E_0 \cos^2\varphi$$

φ = angle of nucleus in lab system

Fission yield

$$Y(A) = \frac{\text{number of nuclei of mass number } A \text{ formed in fission}}{\text{total number of fissions}}$$

$$\Sigma Y(A) \approx 2$$

Energy distribution of fission fragments $\mathbf{X + n \rightarrow Y_1 + Y_2 + k \cdot n}$

$$E_1 M_1 \approx E_2 M_2$$

Neutron attenuation

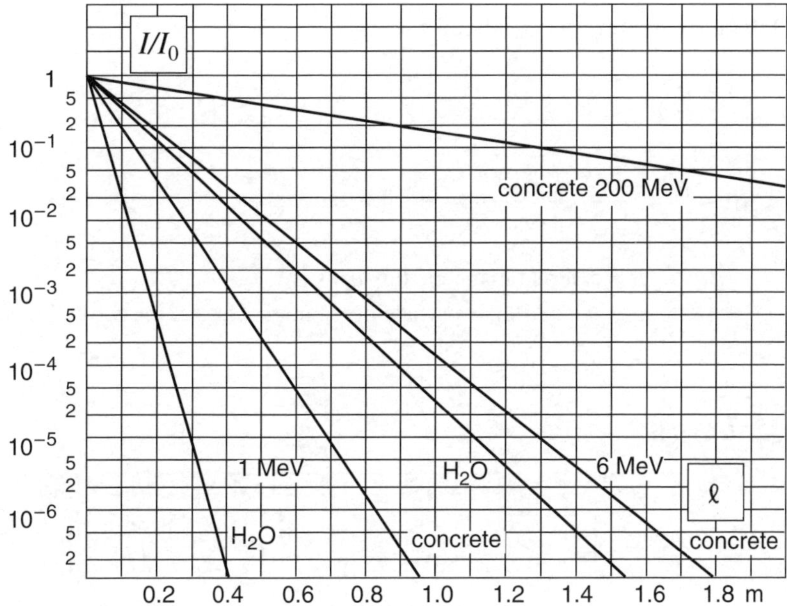

Attenuation of parallel beams in water and concrete as a function of shield thickness ℓ for different neutron energies. The relative intensity I/I_0 refers to the reduction of dose rate along the beam. This is approximately equal to the reduction of neutron flux, and includes the slow neutrons formed when the fast neutrons are moderated. The energy 6 MeV *corresponds to fission neutrons.*

Density of concrete = 2.3 g cm^{-3}.

Nuclear Fusion

Lawson Criterion

$n\tau \geqslant 10^{20}$ m^{-3} s (D-T reaction) n = plasma density

$n\tau \geqslant 10^{22}$ m^{-3} s (D-D reaction) τ = confinement time

Requisite temperature

$T \geqslant 50$ MK

8.9 Particle Physics

Cyclotron frequency and cyclotron radius

$$\omega_c = B\,\frac{q}{m}$$

$$Br = \frac{p}{q} = \frac{\sqrt{E_k^2 + 2E_k\, m_0\, c_0^2}}{c_0\, q} = \frac{mv}{q}$$

B = external magnetic flux density
p, q, m, and E_k = particle momentum, charge, mass, and kinetic energy

Magnetic rigidity

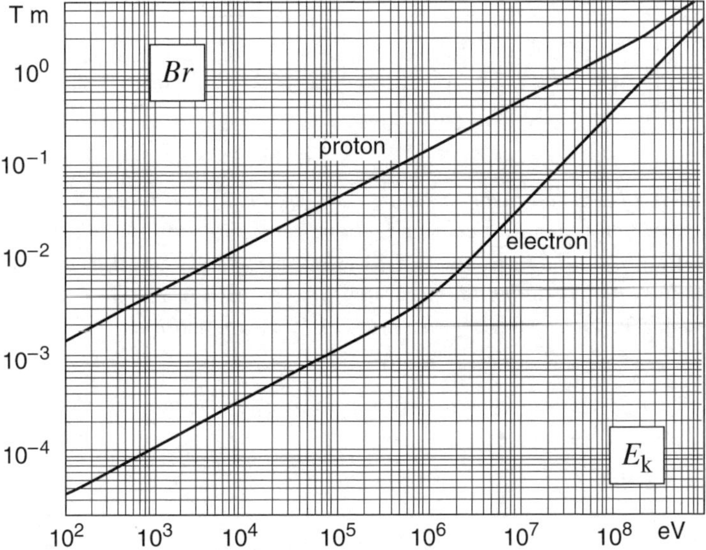

Magnetic rigidity Br as a function of kinetic energy E_k of electrons and protons.

Linear proton accelerator

$$L_n = v_n\,\frac{T}{2}$$

L_n = length of n'th tube section
T = period of alternating current
v_n = speed in n'th tube section

Betratron

$\phi_2 - \phi_1 = 2\pi r^2 B$ $\quad$ r = radius of betatron

B = flux density of the guide field

$p = Bqr$ $\quad$ ϕ = magnetic flux through the orbit

p and q = particle momentum and charge

Synchrotron radiation losses

$$\Delta E = \frac{4\pi}{3\varepsilon_0} \frac{q^2 \beta^3 \gamma^4}{\rho}$$

$\beta = \frac{v}{c_0}$

$\gamma = (1-\beta^2)^{-\frac{1}{2}}$

ρ = radius of electron orbit

ΔE = energy radiated per turn by a particle of charge q and velocity v.

Energy losses due to ionization for singly charged spin-0 bosons (Bethe-Bloch formula)

$$-\frac{\mathrm{d}E}{\mathrm{d}x} = \frac{D\,Z^2 n_\mathrm{e}}{\beta^2}\left(\ln\left(\frac{2\,m_\mathrm{e}\,c_0^2\,\beta^2\,\gamma^2}{I}\right) - \beta^2 - \frac{\delta(\gamma)}{2}\right) = \frac{Z^2}{v^2} f(v, I)$$

x = distance travelled through the medium

Z = atomic number

n_e = number density of electrons (see T – 7.5) $= Z \cdot \frac{\rho}{M}$

where ρ and M = density and atomic mass of absorber

I = ionization energy of the atoms averaged over all electrons ($\approx 10Z$ eV for $Z > 20$, see Sec. T – 7.5)

$\delta(\gamma)$ = dielectric screening correction (important only for highly relativistic particles)

$$D = \frac{4\pi\,\alpha^2\,\hbar^2}{m_\mathrm{e}} = 5.1 \cdot 10^{-25}\ \mathrm{MeV\ cm^2}$$

The differences for spin-$\frac{1}{2}$ particles are small and may be neglected in discussing the main features of energy losses due to ionization.

Range of non-relativistic particles

$$R \approx \frac{1}{m\,f(v_0, I)} \frac{E_0^2}{Z^2}$$

$E_0 = \frac{1}{2}\,m\,v_0^2$ = initial particle energy

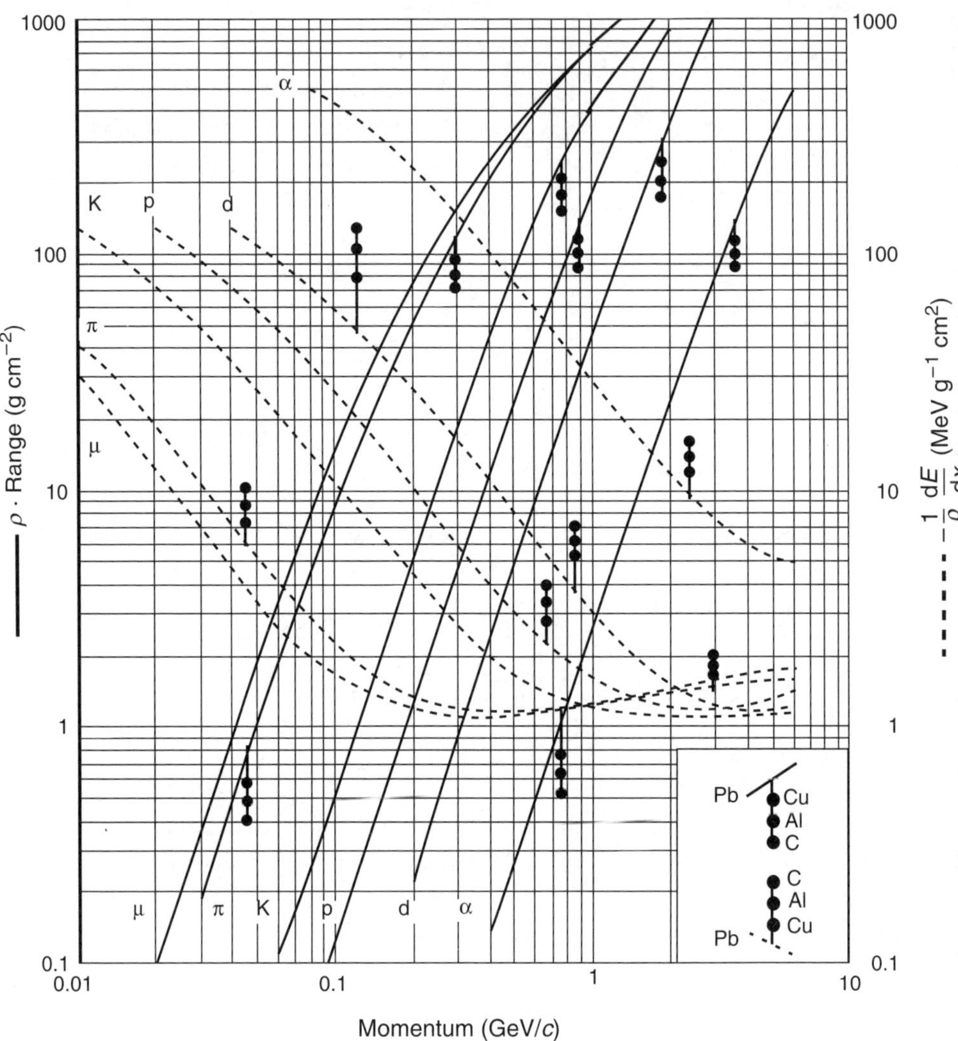

Mean range and energy losses for some charged particles in lead, copper, aluminum, and carbon according to the Bethe-Bloch formula with density effect corrections. (μ = muons, π = pions, K = kaons, p = protons, d = deuterons, α = α particles.)

Energy losses due to bremsstrahlung for relativistic particles (usually electrons)

$$-\frac{dE}{dx} = \frac{E}{L_R} \qquad\qquad E \gg \frac{m_e c_0^2}{\alpha Z^3}$$

Radiation length L_R

$$\frac{1}{L_R} = 4\left(\frac{\hbar}{m c_0^2}\right)^2 Z(Z-1)\alpha^3 n_a \ln\left(\frac{183}{Z^{1/3}}\right)$$ See also Sec. T–7.5

n_a = number density of atoms = $\frac{\rho}{M}$ where M = atomic mass

m = particle mass

Absorption of photons in matter

$$I(x) = I_0 e^{-ax} \qquad\qquad a = n_a \sigma_\gamma = \text{absorption coefficient}$$

Photon interaction cross-section σ_γ for $E_\gamma \gg m_e c_0^2/\alpha Z^{1/3}$

$$\sigma_\gamma \approx \sigma_{pair} \approx \frac{7}{9 n_a L_R}$$ σ_{pair} = cross-section for pair production

Čerenkov detector

$$\cos\theta = \frac{c_0}{n v_p}$$

θ = direction of radiation

n = material refractive index

v_p = particle velocity

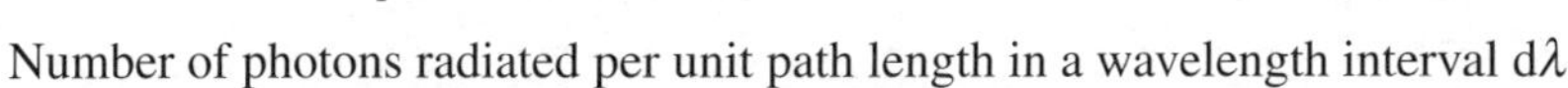

Number of photons radiated per unit path length in a wavelength interval $d\lambda$

$$N(\lambda)\,d\lambda = 2\pi\alpha\left(1 - \frac{1}{\beta^2 n^2}\right)\frac{d\lambda}{\lambda^2}$$

Electron number, muon number, and tauon number (lepton numbers)

$$L_e = N(e^-) - N(e^+) + N(\nu_e) - N(\bar{\nu}_e)$$
$$L_\mu = N(\mu^-) - N(\mu^+) + N(\nu_\mu) - N(\bar{\nu}_\mu)$$
$$L_\tau = N(\tau^-) - N(\tau^+) + N(\nu_\tau) - N(\bar{\nu}_\tau)$$

$N(\ell)$ = number of ℓ particles

Decay rate of leptons

$$\Gamma(\ell^- \to e^- + \bar{\nu}_e + \nu_\ell) = \Gamma(\ell^+ \to e^+ + \nu_e + \bar{\nu}_\ell) = K\, G_F^2\, m_\ell^{\,5}$$

ℓ = lepton (τ or μ)

K = constant

$\frac{G_F}{(\hbar c_0)^3} = 1.1664 \cdot 10^{-5}\ \text{GeV}^{-2}$ = Fermi coupling constant

Hypercharge

$$Y = b + s + c + \tilde{b} + t$$

b = baryon number
s = strangeness
c = charm
$\tilde{b}$ = beauty
t = truth
See table T – 7.2 for numerical values.

Third component of isospin

$$\tau_z = q - \frac{Y}{2}$$

q = charge in units of the fundamental charge

Conservation laws for various kinds of interaction

Conservation of		Strong inter-action	Electro-magnetic interaction	Weak interaction
energy	E	yes	yes	yes
momentum	$\boldsymbol{p}$	yes	yes	yes
angular momentum	$\boldsymbol{L}$	yes	yes	yes
electric charge	q	yes	yes	yes
lepton number	L	–	yes	yes
baryon number	b	yes	yes	yes
strangeness	s	yes	yes	no ($\Delta s = 0, \pm 1$)
parity	π	yes	yes	no
isospin	τ	yes	no	no
z component of isospin	τ_z	yes	yes	no

Kinetic energy in decay $X \to x_1 + x_2$ at rest

$$E_1 = \frac{\frac{1}{2}Q^2 + Q\, m_2\, c_0^2}{(m_1 + m_2)c_0^2 + Q} = \frac{q + 2m_2\, c_0^2}{2m_X\, c_0^2}\, Q$$

Q = Q value from Sec. 8.2

Energy distribution in decay $X \rightarrow \gamma_1 + \gamma_2$

$$\Delta E_\gamma = E_{\gamma 1} - E_{\gamma 2} = p_X c_0$$

Photon absorption

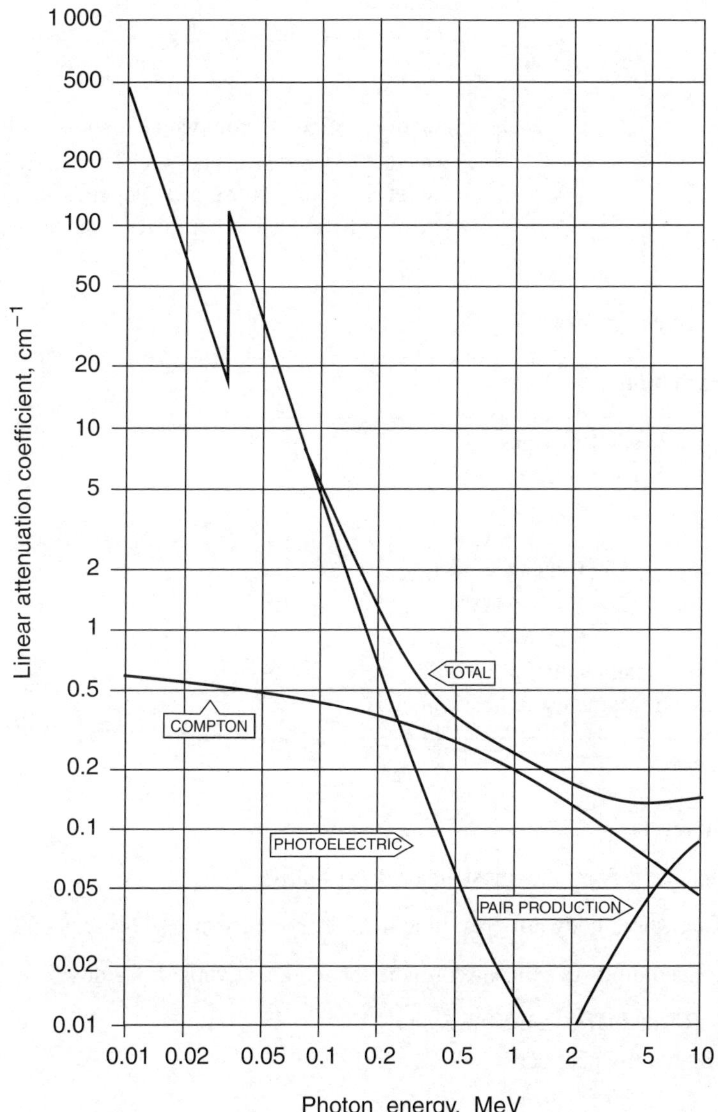

Macroscopic cross sections for absorption of photons in sodium iodide.

8.10 Radiophysics, Ionizing Radiation

Equivalent dose (for biological damage) in Sv (sievert)

$H = D\,\omega_R$

$D =$ absorbed dose, energy deposited per mass at specified location, unit: gray (Gy)
1 Gy = 1 J/kg = 6.24 EeV/kg

$\omega_R =$ radiation weight factor, which expresses long-term risk (primarily cancer and leukemia) from low-level cronic exposure. It depends upon the type of radiation and other factors, see table.

Radiation weight factors

Type of radiation		ω_R
X-rays and γ-rays, all energies		1
Electrons and myons, all energies		1
Neutrons,	< 10 keV	5
	10 – 100 keV	10
	100 keV – 2 MeV	20
	2 – 20 MeV	10
	> 20 MeV	5
Protons (other than recoils) > 2 MeV		5
Alphas, fission fragments, & heavy nuclei		20

Radiation levels

Natural annual background, most areas: 0.4 – 4 mSv

Lethal dose, whole-body dose resulting in 50% mortality in 30 days: 2.5 – 3.0 Gy

Recommended limits to exposure of radiation workers (whole-body dose):

CERN and UK: 15 mSv per year

US: 50 mSv per year

The International Nuclear Event Scale (INES)

More information: http://www.npp.hu/biztonsag/INESskala-e.htm

Level	Descriptor	Criteria
7	Major accident	External release of a large fraction of the radioactive material in a large facility (e.g. the core of a power reactor) in quantities radiologically equivalent to more than tens of thousands of terabecquerel of iodine-131. *Example:* Chernobyl, USSR, 1986.
6	Serious accident	External release of radioactive material in quantities radiologically equivalent to the order of thousands to tens of thousands of terabecquerel of iodine-131.
5	Accident with off-site risk	External release of radioactive material in quantities radiologically equivalent to the order of hundreds to thousands of terabecquerel of iodine-131. *Examples:* Three Mile Island, USA, 1979 and Windscale Pile, UK, 1957.
4	Accident without significant off-site risk	External release of radioactivity resulting in a dose to the most exposed individual off-site of the order of a few millisievert. *Examples:* Tukai Mura, Japan, 1999 and Saint Laurent, France, 1980.
3	Serious incident	External release of radioactivity above authorised limits, resulting in a dose to the most exposed individual off-site of the order of tenths of millisievert.
2	Incident	An event resulting in a dose to a worker exceeding a statutory annual dose limit and/or an event which leads to the presence of signifcant quantities of radioactivity in the installation in areas not expected by design and which require corrective action.
1	Anomaly	Anomaly beyond the authorised operating regime.

9 Statistical Physics

Quantity	Symbol	SI Unit
Number of particles	N	
Temperature	T, Θ	K
Mass	m	kg
Volume	V	m^3
Momentum	p	Ns = kg m/s
Energy	E	J = Nm
Internal energy	U	J
Chemical potential (one particle)	μ	J
Entropy	S	J/K
Pressure	p	Pa = N/m^2
Heat capacity at constant volume	C_v	J/K

9.1 The Microcanonical Ensemble

(Completely isolated system)

Entropy

$$S = k \ln W$$

W = number of microscopic states

Probability of the microscopic state *i*

$$f_i = \frac{1}{W}$$

9.2 The Canonical Ensemble

(Closed system at a given temperature)

Partition function

$$Z = \Sigma_\nu \, e^{-E_\nu/kT} = \Sigma_i \, g_i \, e^{-E_i/kT}$$

$\nu =$ state specified by quantum numbers $\nu_1, \nu_2, \ldots, \nu_\varphi$ where $\varphi =$ degrees of freedom

$g_i =$ degeneracy (statistical weight) of level i, i.e. number of states for which $E_\nu = E_i$

Probability of a system occupying state ν

$$f_\nu = \frac{1}{Z} \, e^{-E_\nu/kT}$$

Probability of a system's energy being E_i

$$f_i = \frac{g_i}{Z} \, e^{-E_i/kT}$$

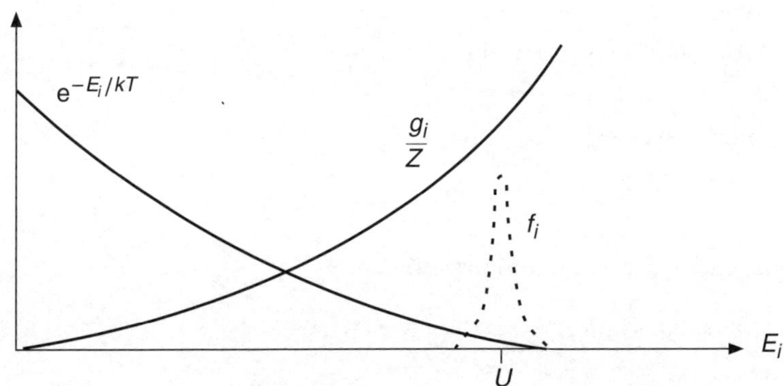

Mean value according to canonical ensemble

$$\langle A\rangle = \frac{1}{Z}\,\Sigma_\nu A_\nu\,\mathrm{e}^{-E_\nu/kT}$$

ν = eigenstates of A

In particular, for thermodynamical quantities

$$\langle A\rangle = \frac{1}{Z}\,\Sigma_i\,g_i A_i\,\mathrm{e}^{-E_i/kT}$$

Classical statistics

$$Z = \frac{1}{h^\varphi}\int \mathrm{e}^{-E(q,p)/kT}\,\mathrm{d}q\,\mathrm{d}p$$

$q = q_1, q_2, \ldots, q_\varphi$ = generalized coordinates, $\varphi = 3N$
$p = p_1, p_2, \ldots, p_\varphi$ = generalized momenta, $\varphi = 3N$
$E(q,p)$ = total energy of the state specified by p and q
h = Planck's constant

Probability of the system occupying the state specified by the coordinates p and q

$$f(q,p) = \frac{1}{h^\varphi Z}\,\mathrm{e}^{-E(q,p)/kT} \qquad 1 = \int f(q,p)\,\mathrm{d}q\,\mathrm{d}p$$

Mean value

$$\langle A\rangle = \int A(q,p)\,f(q,p)\,\mathrm{d}q\,\mathrm{d}p$$

Helmholtz' free energy

$$F = -\,kT\ln Z$$

Internal energy, entropy, and pressure

$$U = kT^2\left(\frac{\partial}{\partial T}\ln Z\right)_V$$

$$S = -\left(\frac{\partial F}{\partial T}\right)_V \qquad p = -\left(\frac{\partial F}{\partial V}\right)_T$$

Energy fluctuation

$$(\Delta E)^2 = \langle E^2\rangle - \langle E\rangle^2 = kT^2\left(\frac{\partial\langle E\rangle}{\partial T}\right)_V$$

9.3 The Grand Canonical Ensemble

(Open system at a given temperature)

Partition function

$$\mathbf{Z} = \Sigma_{\nu,\,N}\, \mathrm{e}^{(\mu N - E_{N\nu})/kT} = \Sigma_{i,N}\, g_i\, \mathrm{e}^{(\mu N - E_{N,i})/kT}$$

$\nu =$ state specified by the quantum numbers $\nu_1, \nu_2, \ldots, \nu_\varphi$ where $\varphi =$ degrees of freedom

$g_i =$ degeneracy of energy level i, i.e. number of states ν for which $E_{N\nu} = E_{N,i}$

$$\mathbf{Z} = \sum_{N=0}^{\infty} Z(N)\, \mathrm{e}^{\mu N/kT}$$

$Z(N)$ = canonical partition function of N particles as in Sec. 9.2

Probability of the system occupying the state specified by ν and N

$$f_{N\nu} = \frac{1}{\mathbf{Z}}\, \mathrm{e}^{(\mu N - E_{N\nu})/kT}$$

Probability of the system having N particles and the energy $E_{N,i}$

$$f_{N,i} = \frac{g_i}{\mathbf{Z}}\, \mathrm{e}^{(\mu N - E_{N,i})/kT}$$

Mean values of the grand canonical ensemble (negligible interaction)

$$\langle A \rangle = \frac{1}{\overline{N}}\, \Sigma_j A_j\, \overline{n}_j$$

$$\langle A \rangle \approx \frac{g}{h^3\, \overline{N}} \int A\, \overline{n}\, \mathrm{d}q\, \mathrm{d}p \qquad A = A(q,p)$$

$$\overline{N} = \Sigma_j\, \overline{n}_j \approx \frac{g}{h^3} \int \overline{n}\, \mathrm{d}q\, \mathrm{d}p$$

$\overline{n}_j$ = mean occupation number

q and p are generalized one-particle coordinates and momenta

g = spin degeneracy as in Sec. 9.5

Grand canonical potential

$$\Omega = -kT\ln \mathbf{Z} = -pV$$

$$\Omega = U - TS - \mu\overline{N}$$

Entropy and mean number of particles

$$S = -\left(\frac{\partial\Omega}{\partial T}\right)_{\mu V} \qquad \overline{N} = -\left(\frac{\partial\Omega}{\partial\mu}\right)_{TV}$$

9.4 Maxwell-Boltzmann Statistics

(Ideal gas of point particles)

$f(A)\,\mathrm{d}A$ =probability of a molecule having a value of A measured inside the element $\mathrm{d}A$

The Boltzmann distribution

$$f(\boldsymbol{r})\,\mathrm{d}x\,\mathrm{d}y\,\mathrm{d}z = \frac{\mathrm{e}^{-E_\mathrm{p}(\boldsymbol{r})/kT}\mathrm{d}x\,\mathrm{d}y\,\mathrm{d}z}{\int \mathrm{e}^{-E_\mathrm{p}(\boldsymbol{r})/kT}\mathrm{d}x\,\mathrm{d}y\,\mathrm{d}z}$$

$E_\mathrm{p}(\boldsymbol{r})$ = potential energy at $\boldsymbol{r} = (x,y,z)$

The Maxwell distribution in momentum space

$$f(\boldsymbol{p}) = \frac{\mathrm{e}^{-p^2/2mkT}}{(2\pi\, mkT)^{3/2}}$$

Distribution of velocities and energies

$$f(v) = \frac{4\pi\, m^3\, v^2}{(2\pi\, mkT)^{3/2}}\exp\left(-\frac{mv^2}{2kT}\right)$$

$$f(E) = \frac{2\sqrt{E}}{\sqrt{\pi}(kT)^{3/2}}\exp\left(-\frac{E}{kT}\right) \qquad E = \frac{1}{2}mv^2 = \frac{p^2}{2m}$$

Monoatomic Maxwell-Boltzmann gas

Degeneration parameter

$$\eta = \frac{N}{V} \frac{h^3}{(2\pi mkT)^{3/2}} = e^{\mu/kT} \ll 1$$

Mean occupation number

$$\bar{n}_j = e^{(\mu - \varepsilon_j)/kT}$$

ε_j = energy of one-particle state j

Partition functions

$$Z = \frac{1}{N!} \left(\frac{V}{h^3}\right)^N (2\pi mkT)^{3N/2}$$

$$\mathbf{Z} = \exp\left(e^{\mu/kT} \frac{V}{h^3} (2\pi mkT)^{3/2}\right)$$

Sackur-Tetrode equation for a monoatomic gas

$$S = Nk\left(\ln \frac{V}{N} + \frac{3}{2} \ln \frac{2\pi mkT}{h^2} + \frac{5}{2}\right)$$

Rotation of diatomic gas particles

Partition function

$$Z_{rot} \approx \left(\frac{T}{\Theta_r}\right)^N \qquad \text{when } T \gg \Theta_r$$

$$Z_{rot} \approx \left(1 + 3e^{-\frac{2\Theta_r}{T}}\right)^N \qquad \text{when } T \ll \Theta_r$$

(When the two atoms are identical, the results should be divided by two)

Characteristic temperature of rotation

$$\Theta_r = \frac{\hbar^2}{2I\,k}$$

Moment of inertia

$$I = \frac{m_1\, m_2}{m_1 + m_2}\, R_0^2$$ R_0 = bond length

Vibration of diatomic gas particles

Partition function

$$Z_{\text{vib}} = \frac{e^{-N\Theta_v/2T}}{(1 - e^{-\Theta_v/T})^N}$$

Characteristic temperature of vibration

$$\Theta_v = \frac{\hbar\, \omega_0}{k}$$ ω_0 = characteristic frequency as in Sec. 7.6

9.5 Ideal Bose-Einstein Gases

(Monatomic boson gas)

$$f_m(x) = \frac{1}{\Gamma(m+1)} \int_0^\infty \frac{z^m\, dz}{e^{z+x} - 1}$$

$$f_m(x) = \sum_{n=1}^{\infty} \frac{e^{-nx}}{n^{m+1}}$$ $x \geq 0$

$$f_m(0) = \zeta(m + 1)$$ ζ = Riemann's zeta function

Γ = Gamma function

$\Gamma(m + 1) = m\Gamma(m)$, $\Gamma(\tfrac{1}{2}) = \sqrt{\pi}$, $\Gamma(1) = 1$

Average occupation number

$$\bar{n}_j = \frac{1}{e^{(\varepsilon_j - \mu)/kT} - 1}$$

ε_j = energy of one-particle state j

Degeneration parameter

$$\eta = \frac{\overline{N}\, h^3}{gV(2\pi\, mkT)^{3/2}} = f_{1/2}\left(-\frac{\mu}{kT}\right)$$

Spin degeneracy

$$g = 2S + 1 \qquad S = \text{spin}$$

Internal energy

$$U = \frac{3}{2}\,\overline{N}\,k\,T\,\frac{f_{3/2}(x)}{f_{1/2}(x)} \qquad x = -\frac{\mu}{kT}$$

$$U = \frac{3}{2}\,p\,V \quad \text{(nonrelativistic)}$$

Weak degeneration

$$U = \frac{3}{2}\,\overline{N}\,k\,T\left(1 - \frac{\eta}{2^{5/2}} + \ldots\right)$$

$$C_v = \frac{3}{2}\,\overline{N}\,k\left(1 + \frac{\eta}{2^{7/2}} + \ldots\right)$$

Strong degeneration

$$U = \frac{3}{2}\,\overline{N}\,k\,T\left(\frac{T}{T_{\mathrm{B}}}\right)^{3/2}\frac{\zeta\left(\frac{5}{2}\right)}{\zeta\left(\frac{3}{2}\right)} \qquad \mu = 0$$

Condensation temperature (Bose temperature)

$$T_{\mathrm{B}} = \frac{h^2}{2\pi\, mk}\left(\frac{\overline{N}}{gV\,\zeta\left(\frac{3}{2}\right)}\right)^{2/3}$$

Partition function

$$\mathbf{Z} = \prod_j \frac{1}{(1 - \mathrm{e}^{(\mu - \varepsilon_j)/kT})}$$

Approximate values of $\zeta(x)$

x	1.5	2	2.5	3	3.5	4	4.5	5
$\zeta(x)$	2.612	1.645	1.341	1.202	1.127	1.082	1.055	1.037

9.6 Ideal Fermi-Dirac Gases

(Monatomic fermion gas)

$$F_m(x) = \frac{1}{\Gamma(m+1)} \int_0^\infty \frac{z^m \, dz}{e^{z+x}+1}$$

$$F_m(x) = \sum_{n=1}^{\infty} (-1)^n \frac{e^{-nx}}{n^{m+1}} \qquad x \geq 0$$

$$F_m(x) \approx \frac{(-x)^{m+1}}{\Gamma(m+2)} \left(1 + \frac{\pi^2}{6} m(m+1) \frac{1}{x^2}\right) \qquad \text{if } -x \gg 1$$

Γ = Gamma function. See also Sec. 9.5

Average occupation number

$$\bar{n}_j = \frac{1}{e^{(\varepsilon_j - \mu)/kT} + 1}$$

ε_j = energy of one-particle state j

Degeneration parameter

$$\eta = \frac{\bar{N} h^3}{gV(2\pi \, mkT)^{3/2}} = F_{1/2}\left(-\frac{\mu}{kT}\right)$$

g = spin degeneracy as in Sec. 9.5

Internal energy

$$U = \frac{3}{2}\,\overline{N}\,k\,T\,\frac{F_{3/2}(x)}{F_{1/2}(x)} \qquad\qquad x = -\frac{\mu}{kT}$$

$$U = \frac{3}{2}\,pV \qquad \text{(nonrelativistic)}$$

Weak degeneracy

$$U = \frac{3}{2}\,\overline{N}\,k\,T\left(1 + \frac{\eta}{2^{5/2}} + \ldots\right)$$

$$C_v = \frac{3}{2}\,\overline{N}\,k\left(1 - \frac{\eta}{2^{7/2}} + \ldots\right)$$

Strong degeneracy

$$U = \frac{3}{5}\,\overline{N}\,k\,T_{\mathrm{F}}\left(1 + \frac{5\pi^2}{12}\left(\frac{T}{T_{\mathrm{F}}}\right)^2 + \ldots\right)$$

$$\mu = k\,T_{\mathrm{F}}\left(1 - \frac{\pi^2}{12}\left(\frac{T}{T_{\mathrm{F}}}\right)^2 + \ldots\right)$$

$$C_v = \frac{1}{2}\,\pi^2\,\overline{N}\,k\,\frac{T}{T_{\mathrm{F}}}$$

Fermi temperature

$$T_{\mathrm{F}} = \frac{h^2}{2mk}\left(\frac{3\,\overline{N}}{4\pi\,gV}\right)^{2/3}$$

Partition function

$$\mathbf{Z} = \prod_j \left(1 + \mathrm{e}^{(\mu - \varepsilon_j)/kT}\right)$$

9.7 Blackbody Radiation

(Cavity radiation)

Photon gas

$$\mu = 0 \qquad\qquad E = h\nu = c_0 p$$

Average number of photons of frequency in the inverval $(\nu,\ \nu + \mathrm{d}\nu)$ inside the volume V

$$\mathrm{d}N = \frac{8\pi V}{c_0^3} \cdot \frac{\nu^2 \, \mathrm{d}\nu}{\exp\left(\dfrac{h\nu}{kT}\right) - 1}$$

Planck's radiation law, energy per volume

In frequency interval $\mathrm{d}\nu$ at frequency ν

$$\mathrm{d}e = \frac{8\pi h}{c_0^3} \cdot \frac{\nu^3 \, \mathrm{d}\nu}{\exp\left(\dfrac{h\nu}{kT}\right) - 1}$$

In wavelength interval $\mathrm{d}\lambda$ at wavelength λ

$$\mathrm{d}e = \frac{8\pi h c_0}{\lambda^5} \cdot \frac{\mathrm{d}\lambda}{\exp\left(\dfrac{h\, c_0}{kT\lambda}\right) - 1}$$

Spectral radiant exitance

$$M_\lambda = \frac{c_0}{4} \frac{\mathrm{d}e}{\mathrm{d}\lambda}$$

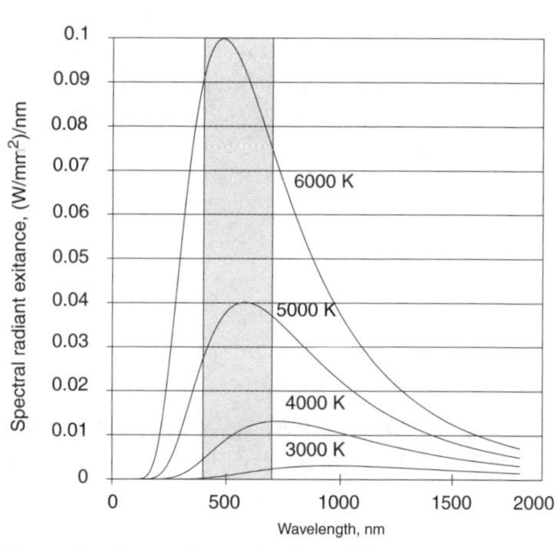

Spectral radiant exitance for four different temperatures. 6000 K corresponds to radiation from the sun. Wavelengths sensible for the human eye are shaded.

Rayleigh-Jeans formula, applicable when $h\nu \ll kT$

$$\mathrm{d}e = \frac{8\pi \nu^2}{c_0^3} kT \, \mathrm{d}\nu$$

Wien's radiation law, applicable when $h\nu \gg kT$

$$de = \frac{8\pi h \nu^3}{c_0^3} \exp\left(-\frac{h\nu}{kT}\right) d\nu$$

Total energy density

$$e = \frac{U}{V} = a\,T^4$$

$$a = \frac{8\pi^5 k^4}{15 h^3 c_0^3} = 7.5659 \cdot 10^{-16}\ \mathrm{Jm^{-3}K^{-4}}$$

Entropy

$$S = \frac{4}{3}\, a\, V\, T^3$$

Maximum of energy density; Wien's displacement law

$$\nu_m = b_\nu T \qquad b_\nu = 2.82144\, \frac{k}{h}$$

$$\lambda_m T = b_\lambda \qquad b_\lambda = \frac{h\, c_0}{4.96511\, k}$$

$$b_\nu = 5.8786 \cdot 10^{10}\ \mathrm{s^{-1}\,K^{-1}}$$

$$b_\lambda = 2.8978 \cdot 10^{-3}\ \mathrm{m\,K}$$

Radiated power per area through a small opening in wall of cavity; Stefan-Boltzmann law (Radiant exitance from blackbody)

$$M = \frac{\partial P}{\partial A} = \sigma T^4 \qquad \sigma = \frac{c_0\, a}{4}$$

$$\sigma = 5.6705 \cdot 10^{-8}\ \mathrm{Wm^{-2}\,K^{-4}}$$

10 Solid State Physics

Quantity	Symbol	SI Unit
Temperature	T, Θ	K
Mass	M, m	kg
Volume	V	m^3
Angular frequency	ω	rad/ s
Wave vector	$\boldsymbol{K}, \boldsymbol{k}$	rad/ m
Density	ρ	kg/m^3
Electron energy	ε	$J = kg\ m^2/s^2$
Electric conductivity	σ	A/ Vm
Magnetic flux density	$\boldsymbol{B}$ (B)	$T = Vs/m^2$

10.1 Crystal Structure

Volume of the primitive cell

$$V = |\boldsymbol{a} \times \boldsymbol{b} \cdot \boldsymbol{c}|$$

a, b, and ***c*** are primitive basis vectors

Primitive basis vectors *A, B,* and *C* in a reciprocal lattice

$$\boldsymbol{A} = \frac{2\pi}{V}\,\boldsymbol{b} \times \boldsymbol{c}$$

cyclic permutation

Volume of unit cell in reciprocal lattice

$$\Omega = |\boldsymbol{A} \times \boldsymbol{B} \cdot \boldsymbol{C}| = \frac{(2\pi)^3}{V}$$

Reciprocal lattices for cubic systems

Real lattice	Reciprocal lattice	Edge of cube in reciprocal lattice	Reciprocal lattice vectors $\mathbf{G}_{hk\ell} = h\mathbf{A} + k\mathbf{B} + \ell C$
sc	sc	$\frac{2\pi}{a}$	$\frac{2\pi}{a}(h\hat{x} + k\hat{y} + \ell\hat{z})$
bcc	fcc	$\frac{4\pi}{a}$	$\frac{2\pi}{a}((h+\ell)\hat{x} + (h+k)\hat{y} + (k+\ell)\hat{z})$
fcc	bcc	$\frac{4\pi}{a}$	$\frac{2\pi}{a}((h-k+\ell)\hat{x} + (h+k-\ell)\hat{y} + + (-h+k+\ell)\hat{z})$

a = cube edge in real lattice

$\hat{x}$, $\hat{y}$, $\hat{z}$, and $(h\ k\ \ell)$ refer to the cube axes in the real lattice

Distance between adjacent planes of Miller indices $(h\ k\ \ell)$

$$d_{hk\ell} = \frac{2\pi}{|\boldsymbol{G}_{hk\ell}|}$$

Example 1. Cubic systems

$$d_{hk\ell} = \frac{a}{\sqrt{h^2+k^2+\ell^2}}$$

Example 2. Tetragonal crystals

$$d_{hk\ell} = \frac{1}{\sqrt{\frac{h^2+k^2}{a^2} + \frac{\ell^2}{c^2}}} \qquad c = |\mathbf{c}|$$

Number of molecules in a primitive cell

$$n = \frac{\rho V}{M} \qquad M = \text{molecular mass}$$

10.2 Crystal Diffraction

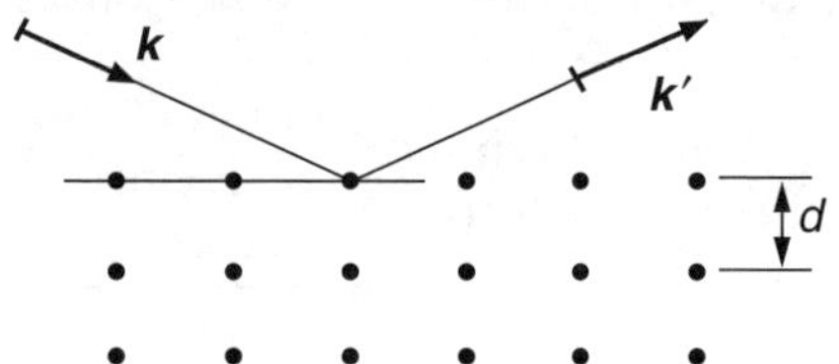

Bragg's equation

$$2\, d_{hk\ell} \sin\theta = n\lambda \quad n = 1, 2, 3, \ldots$$

$$\boldsymbol{k}' = \boldsymbol{k} + \boldsymbol{G}_{hk\ell} \qquad |\boldsymbol{k}'| = |\boldsymbol{k}|$$

λ = de Broglie wavelength of incident wave

$\boldsymbol{G}_{hk\ell}$ and $d_{hk\ell}$ are given in Sec. 10.1

Geometrical structure factor

$$S_{hk\ell} = \sum_j f_j \exp(-2\pi \mathrm{i}(hx_j + ky_j + \ell z_j))$$

The sum is to be taken over all atoms in basis.

x_j, y_j, and z_j are coordinates of atom j

Atomic form factor

$$f_j(\boldsymbol{G}) = \int n_j(\boldsymbol{r}) \mathrm{e}^{-\mathrm{i}\boldsymbol{G}\cdot\boldsymbol{r}}\, \mathrm{d}V \approx \text{constant} \cdot Z_j$$

$n_j(\boldsymbol{r})$ = electronic concentration at $\boldsymbol{r}$ of atom j

Z_j = number of electrons belonging to atom j

Intensity of the reflected wave

$$I = \text{constant} \cdot N^2\, |S_{hk\ell}|^2 \qquad N = \text{number of lattice points}$$

Temperature dependence for the intensity (Debye-Waller factor)

$$I = I_0 \exp\left(-\frac{1}{3}\, \langle u^2 \rangle\, |\boldsymbol{G}_{hk\ell}|^2\right)$$

$\langle u^2 \rangle$ is given as a function of temperature in Sec. 10.3

10.3 Lattice Vibrations

Dispersion relations of a linear, harmonic model

Example 1. Primitive structure

$$\omega^2 = \frac{2}{M} \sum_{p>0} C_p(1 - \cos pKa)$$

a

K = wavevector of phonons

C_p = force from atoms at the distance pa

Nearest-neighbour approximation

$$C_1 = C, \; C_p = 0 \text{ for } p \neq 1$$

Example 2. Two atoms in each cell

b

$$\omega^2 = \frac{C}{\mu} \pm C \sqrt{\frac{1}{\mu^2} - \frac{4 \sin^2 \frac{1}{2}Kb}{M_1 M_2}}$$

$$\mu = \frac{M_1 M_2}{M_1 + M_2} = \text{reduced mass}$$

In particular,

$$v = \frac{d\omega}{dK} = \left(\frac{C b^2}{2(M_1 + M_2)}\right)^{1/2} \quad \text{when } K \to 0+$$

v = velocity of sound

Example 3. Different force constants; two atoms per cell

b

$$\omega^2 = \frac{C_A + C_B}{2\mu} \pm \sqrt{\frac{(C_A + C_B)^2}{4\mu^2} - \frac{4C_A C_B \sin^2 \frac{1}{2}Kb}{M_1 M_2}}$$

Relations among phonon momentum $\mathbf{p}$, wavevector $\mathbf{K}$, wavelength λ, angular frequency ω, and energy E

$$\mathbf{p} = \hbar \mathbf{K} \qquad |\mathbf{K}| = \frac{2\pi}{\lambda} \qquad E = \hbar \omega$$

Phonon amplitude u_0

$$u_0^2 = 2\,\langle u^2 \rangle = \begin{cases} \dfrac{3\,\hbar}{M\,\omega} & ,\, T = 0 \\[2ex] \dfrac{6\,k_B\,T}{M\,\omega^2} & ,\, T \gg 0 \end{cases}$$

k_B = Boltzmann constant
M = atomic mass
u = displacement from equilibrium

10.4 Thermal Properties of Solids

Debye temperature Θ_D

$$k_B\,\Theta_D = \hbar\,\omega_{max} = \hbar\,v\left(\frac{6\pi^2\,N}{V}\right)^{1/3}$$

$$\frac{3}{v^3} = \frac{1}{v_\ell^3} + \frac{2}{v_t^3}$$

v_ℓ and v_t are phase velocities of longitudinal and transverse elastic waves through the medium

N = number of primitive cells

k_B = Boltzmann constant

Internal energy

$$U = \frac{9}{8}\,N\,k_B\,\Theta_D + 3N\,k_B\,T\,D\left(\frac{\Theta_D}{T}\right)$$

The Debye function

$$D(x) = \frac{3}{x^3}\int_0^x \frac{y^3\,dy}{e^y - 1}$$

Heat capacity according to Debye

$$C_v = 9N\,k_B\left(\frac{T}{\Theta_D}\right)^3 \int_0^{\Theta_D/T} \frac{x^4\,e^x\,dx}{(e^x - 1)^2}$$

Dulong and Petit law (when $T > \Theta_D$)

$$C_v = 3N\,k_B$$

Debye T^3 law (when $T \ll \Theta_D$)

$$C_v = \frac{12\pi^4}{5}\,N\,k_B\left(\frac{T}{\Theta_D}\right)^3 = \alpha T^3$$

$$\frac{12\pi^4}{5} \approx 234$$

Low-temperature metal heat capacity

$$C_v = \alpha T^3 + \gamma T$$

$$\gamma = \frac{\pi^2\,N_e\,k_B^2}{2\,\varepsilon_F}$$

N_e = number of free electrons

ε_F = Fermi energy as in Sec. 10.5

Phonon contribution to insulator heat conductivity

$$\lambda_p = \frac{1}{3}\,c\,v\,\ell\,\rho$$

c = specific heat per kg

Mean free path

$$\ell = v\,\tau$$

τ = mean interval between collisions

Wiedemann-Franz law of metals

$$\frac{\lambda}{\sigma} = L\,T \qquad \text{when } T \gg 0\text{ K}$$

λ = thermal conductivity

The Lorenz number

$$L = \frac{\pi^2}{3}\left(\frac{k_B}{e}\right)^2 = 2.44 \cdot 10^{-8}\ \text{V}^2\,\text{K}^{-2}$$

10.5 Electron Energies

Free electron model (FEM)

Radius of Fermi sphere

$$k_F = (3\pi^2 n)^{1/3}$$

$n = \dfrac{N_e}{V}$ = number of free electrons per volume

Fermi energy (Fermi level)

$$\varepsilon_F = \frac{\hbar^2 k_F^2}{2\,m_e} = k_B\,T_F$$

k_B = Boltzmann constant
T_F = Fermi temperature. See Sec. 9.6

Density of states

$$D(\varepsilon) = \frac{V}{2\pi^2}\left(\frac{2\,m_e}{\hbar^2}\right)^{3/2}\sqrt{\varepsilon}$$

V = crystal volume

In particular, at $\varepsilon = \varepsilon_F$

$$D(\varepsilon_F) = \frac{3\,N_e}{2\,\varepsilon_F}$$

Electron energies

$$\varepsilon_i(\boldsymbol{k}) = \frac{\hbar^2}{2\,m_e}\,(\boldsymbol{k} - \boldsymbol{G}_i)^2$$

$\boldsymbol{G}_i$ = reciprocal lattice vector as in Sec. 10.1

Nearly free electron model (NFEM)

Crystal potential

$$E_p\,(\boldsymbol{r}) = \sum_{\boldsymbol{G}} u_{\boldsymbol{G}}\, e^{i\,\boldsymbol{G}\cdot\boldsymbol{r}}$$

$u_{\boldsymbol{G}}$ = Fourier coefficients
$\boldsymbol{G}$ = reciprocal lattice vector

Electron wavefunction (one-dimensional case)

$$\psi_k(x) = \Sigma_G\, C_{k-G}\, \mathrm{e}^{\mathrm{i}(k-G)x}$$

Relation between coefficients C_{k-G}

$$(\lambda_k - \varepsilon_k)\, C_k + \sum_G u_G\, C_{k-G} = 0$$

$$\lambda_k = \frac{\hbar^2\, k^2}{2\, m_\mathrm{e}} = \text{energy in FEM}$$

ε_k = energy eigenvalue in NFEM

The Bloch function

$$\psi_{\boldsymbol{k}}(\boldsymbol{r}) = U_{\boldsymbol{k}}(\boldsymbol{r})\, \mathrm{e}^{\mathrm{i}\,\boldsymbol{k}\cdot\boldsymbol{r}}$$

$$U_{\boldsymbol{k}}(\boldsymbol{r}) = \Sigma_{\boldsymbol{G}}\, C_{\boldsymbol{k}-\boldsymbol{G}}\, \mathrm{e}^{-\mathrm{i}\,\boldsymbol{G}\cdot\boldsymbol{r}}$$

Bloch condition

$$U_{\boldsymbol{k}}(\boldsymbol{r}) = U_{\boldsymbol{k}}(\boldsymbol{r} + \boldsymbol{T})$$

$\boldsymbol{T}$ = translation vector in real lattice

Tight binding model

Electron energies (non-degenerate case)

sc $\quad \varepsilon(\boldsymbol{k}) = -\,\alpha - 2\gamma(\cos k_x a + \cos k_y a + \cos k_z a)$

bcc $\quad \varepsilon(\boldsymbol{k}) = -\,\alpha - 8\gamma \cos(\tfrac{1}{2}\, k_x\, a)\cos(\tfrac{1}{2}\, k_y\, a)\cos(\tfrac{1}{2}\, k_z\, a)$

fcc $\quad \varepsilon(\boldsymbol{k}) = -\,\alpha - 4\gamma(\cos(\tfrac{1}{2}\, k_x\, a)\cos(\tfrac{1}{2}\, k_y\, a) + \cos(\tfrac{1}{2}\, k_y\, a)\cos(\tfrac{1}{2}\, k_z\, a) +$

$\quad + \cos(\tfrac{1}{2}\, k_z\, a)\cos(\tfrac{1}{2}\, k_x\, a))$

α and γ are constants

10.6 Dynamics of Electron Motion

Equation of motion

$$\dot{\boldsymbol{k}} = \frac{1}{\hbar}\,\boldsymbol{F}$$

$\boldsymbol{F}$ = external force

In a magnetic field

$$\dot{\boldsymbol{k}} = -\frac{e}{\hbar^2}\,\nabla_{\boldsymbol{k}}\,\varepsilon \times \boldsymbol{B}$$

Group velocity

$$v_{\mathrm{g}} = \frac{1}{\hbar}\,\nabla_{\boldsymbol{k}}\,\varepsilon$$

Effective mass (a tensor)

$$\left(\frac{1}{m^*}\right)_{\mu\nu} = \frac{1}{\hbar^2}\,\frac{\partial^2 \varepsilon}{\partial k_\mu\,\partial k_\nu}$$

One-dimensional case

$$\frac{1}{m^*} = \frac{1}{\hbar^2}\,\frac{\partial^2 \varepsilon}{\partial\, k^2}$$

Cyclotron resonance

Cyclotron frequency

$$\omega_{\mathrm{c}} = \frac{e\,B}{m_{\mathrm{c}}} = 2\pi\,\frac{e\,B}{\hbar^2\,\dfrac{\partial S}{\partial \varepsilon}}$$

m_{c} = effective cyclotron mass

S = area of orbit in momentum space

Resonance condition in metals

$$n\,\omega_{\mathrm{c}} = \omega = 2\pi\nu$$

$n = 1, 2, 3, \ldots$

ν and ω = frequency and angular frequency of electric field

Cyclotron resonance in semiconductors when conducting band edge is a rotational ellipsoid

$$\varepsilon = \hbar^2 \left(\frac{k_x^2 + k_y^2}{2m_t} + \frac{k_z^2}{2m_\ell} \right)$$

m_t and m_ℓ = transverse and longitudinal effective mass

$$\frac{1}{m_c^2} = \frac{\cos^2 \theta}{m_t^2} + \frac{\sin^2 \theta}{m_\ell \, m_t}$$

θ = angle between ellipsoidal axis and $\boldsymbol{B}$-field

Resonance condition in semiconductors

$$\omega_c = \omega = 2\pi\nu$$

Quantization condition in deHaas-van Alphen effect

$$\frac{1}{B} = \frac{2\pi e}{\hbar S} \left(n + \frac{1}{2} \right)$$

$n = 0, 1, 2, \ldots$

Energy level in plane perpendicular to magnetic field

$$\varepsilon = \hbar \, \omega_c \left(n + \frac{1}{2} \right)$$

Relation between area S in momentum space and real area A

$$A = \left(\frac{\hbar}{e\,B} \right)^2 S$$

Electrons in an electromagnetic field

$$\begin{pmatrix} j_x \\ j_y \\ j_z \end{pmatrix} = \frac{\sigma_0}{1 + (\omega_c \tau)^2} \begin{pmatrix} 1 & -\omega_c \tau & 0 \\ \omega_c \tau & 1 & 0 \\ 0 & 0 & 1 + (\omega_c \tau)^2 \end{pmatrix} \begin{pmatrix} E_x \\ E_y \\ E_z \end{pmatrix}$$

$$\sigma_0 = \frac{n\, e^2\, \tau}{m_e} \qquad \boldsymbol{B} = B\,\hat{z}$$

$\boldsymbol{j} = (j_x, j_y, j_z)$ = current density

τ = relaxation time

$\boldsymbol{E} = (E_x, E_y, E_z)$ = electric field

10.7 Electric Properties of Solids, Semiconductor Physics

Mobility (of electrons and holes)

$$\mu_e = \frac{\langle u_e \rangle}{E} = \frac{e\,\tau_e}{m_e^*} \qquad \mu_h = \frac{\langle u_h \rangle}{E} = \frac{e\,\tau_h}{m_h^*}$$

$\langle u \rangle$ = carrier drift velocity

E = electric field strength

τ = relaxation time

m^* = effective mass

Electric conductivity

$$\sigma = ne\,\mu_e + pe\,\mu_h$$

n = number of electrons per volume

p = number of holes per volume

In particular, in a metal

$$\sigma = \frac{n\,e^2\,\tau}{m_e^*}$$

Mean free path

$$\ell = v\,\tau$$

v = carrier velocity

Current density

$$j = n\,e\,\langle u_e \rangle + p\,e\,\langle u_h \rangle = \sigma E$$

Concentration of negative (n) and positive (p) carriers

E_g = band gap

E_d = donor level

E_a = acceptor level

μ = Fermi level

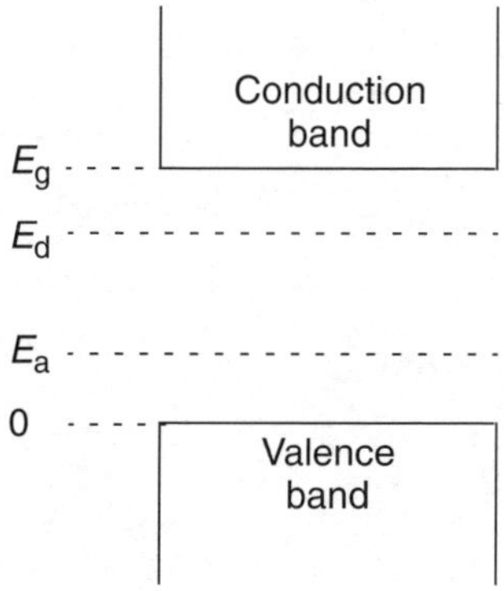

$$n = n_0 \exp\left(-\frac{E_g - \mu}{k_B T}\right) \qquad \text{when } E_g - \mu \gg k_B T$$

$$p = p_0 \exp\left(-\frac{\mu}{k_B T}\right) \qquad \text{when } \mu \gg k_B T$$

$$\left.\begin{matrix} n_0 \\ p_0 \end{matrix}\right\} = 2\left(\frac{m^*_{e,h}\, k_B T}{2\pi \hbar^2}\right)^{3/2} \approx 4.83 \cdot 10^{21}\, T^{3/2} \left(\frac{m^*_{e,h}}{m_e}\right)^{3/2}$$

(when n_0 and p_0 are measured in number per m^3 and T in K)

k_B = Boltzmann constant

Law of Mass Action

$$n\,p = n_0\, p_0 \exp\left(-\frac{E_g}{k_B T}\right) \quad \text{when} \quad \begin{cases} E_g - \mu \gg k_B T \\ \mu \gg k_B T \end{cases}$$

Temperature dependence of Fermi level in intrinsic semiconductors

$$\mu = \frac{1}{2} E_g + \frac{3}{4} k_B T \ln \frac{m^*_h}{m^*_e}$$

Neutrality condition in doped semiconductors

$$p + N_d^+ = n + N_a^-$$

N_d^+ = ionized donor concentration

N_a^- = ionized acceptor concentration

Hall effect

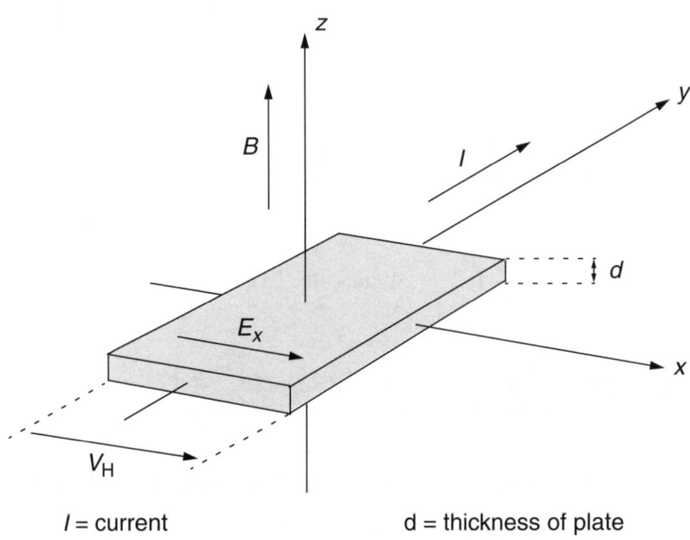

Hall voltage

$$V_H = R_H \frac{I\,B}{d}$$

Hall coefficient

$$R_H = \frac{-E_x}{j_y\,B_z} = \frac{p\mu_h^2 - n\mu_e^2}{e(p\mu_h + n\mu_e)^2}$$

10.8 Magnetism

Fundamental formulae

$$\boldsymbol{B} = \mu_0 (\boldsymbol{H} + \boldsymbol{M}) = \mu_r \mu_0 \boldsymbol{H}$$

$\boldsymbol{M}$ = magnetization
$\boldsymbol{H}$ = magnetizing field
μ_r = relative permeability

Magnetic susceptibility

$$\chi = \mu_r - 1 = \frac{M}{H}$$

Molar magnetic susceptibility

$$\chi_{mol} = \chi V_m$$

V_m = molar volume

Mass magnetic susceptibility

$$\chi_{mass} = \chi v$$

v = specific volume (= volume per mass)

Diamagnetic susceptibility

$$\chi = \frac{n \mu}{H} = -\frac{\mu_0 Z e^2 n}{6 m_e} \langle r^2 \rangle$$

n = atoms per m^3
μ = magnetic dipole moment
Z = number of electrons
$\langle r^2 \rangle$ = rms electron-nucleus distance

Paramagnetic susceptibility (Curie law)

$$\chi = \frac{n g_J J \mu_B B_J(x)}{H} = \frac{\mu_0 n p_{eff}^2 \mu_B^2}{3 k_B T} = \frac{C}{T}$$

g_J = Landé factor as in Sec. 7.4 of quantum number J
C = Curie constant
k_B = Boltzmann constant
$\mu_B = \frac{e \hbar}{2 m_e}$ = Bohr magneton

Effective number of Bohr magnetons

$$p_{eff} = g_J \sqrt{J(J+1)}$$

Brillouin function

$$B_J(x) = \frac{2J+1}{2J} \coth\left(\frac{2J+1}{2J} x\right) - \frac{1}{2J} \coth\left(\frac{x}{2J}\right)$$

$$x = \frac{g_J \, J \, \mu_B \, B}{k_B T}$$

Susceptibility of electron gas

$$\chi = \chi_{\text{Pauli}} - \chi_{\text{Landau}} = \frac{\mu_0 \, N_e \, \mu_B^2}{\varepsilon_F \, V}$$

N_e = number of free electrons within volume V
ε_F = Fermi energy as in Sec. 10.5

Ferromagnetism (Curie-Weiss law)

$$\chi = \frac{C}{T - T_c} \quad \text{for} \quad T > T_c = C\,\lambda$$

T_c = Curie temperature

$$\lambda = \frac{3 \, k_B \, T_c}{\mu_0 \, n \, g^2 \, J(J+1) \mu_B^2}$$

J = effective atomic spin
g = effective Landé factor

Weiss' effective field

$$\boldsymbol{H}_E = \lambda \boldsymbol{M}$$

Saturated magnetization M_s at spin 1/2 (for $T < T_c$)

$$M_s = n \, \mu_B \tanh \frac{\mu_0 \, \mu_B \, \lambda M_s}{k_B T}$$

Ferrimagnetism

$$\chi = \frac{(C_A + C_B) T - 2 \, \nu \, C_A \, C_B}{T^2 - T_c^2} \quad \text{when} \quad T > T_c = \nu \sqrt{C_A \, C_B}$$

C_A and C_B = Curie constants of sublattices

ν = constant depending on interacting sublattices

Antiferromagnetism

$$\chi = \frac{2\,C}{T + \Theta} \quad \text{when} \quad T > T_N = \nu\, C \qquad C = C_A = C_B$$

T_N = Néel temperature
Θ = Curie-Weiss constant

11 Astrophysics and Geophysics

Quantity	Symbol	SI Unit
Position vector	$\boldsymbol{r}$	m
Density	ρ	$\mathrm{kg/m^3}$
Force	$\boldsymbol{F}$ (F)	$\mathrm{N = kg\ m/s^2}$
Potential energy	E_p	J = Nm
Gravitational potential	V	J/kg
Gravity at sea level	$\boldsymbol{g}$ (g)	$\mathrm{m/s^2}$

11.1 Gravitation

Gravitational potential and strength of gravitational field from a homogeneous spherical body

$$V = -G\frac{M}{r}$$

$$\Phi = -\nabla V = -G\frac{M}{r^2}\hat{\boldsymbol{r}}$$

when $r > R$

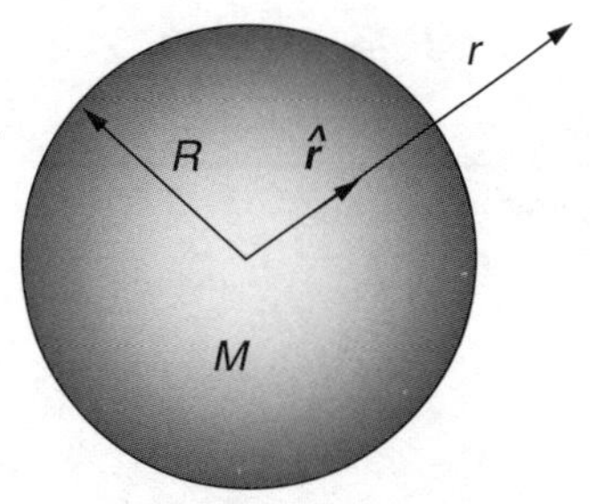

$$V = \frac{GM}{2R^3}(r^2 - 3R^2)$$

$$\Phi = -\nabla V = -G\frac{Mr}{R^3}\hat{\boldsymbol{r}}$$

when $r < R$

Φ = gravitational field strength

G = gravitational constant

M = mass of spherical body

Force on and potential energy of a mass *m* in a gravitational field

$$\boldsymbol{F} = m\Phi \qquad E_\mathrm{p} = mV$$

Potential and strength of gravitational field inside a homogeneous spherical shell

$$V = -G\frac{M}{R} \qquad \boldsymbol{\Phi} = \mathbf{0}$$

Effective acceleration of gravity

(For Newton's law of gravitations and definition of g, see section F–1.2.)

$$\boldsymbol{g}_{\mathrm{e}} = \boldsymbol{g} - \Omega \times (\Omega \times \boldsymbol{r})$$

$\boldsymbol{g} = -g\,\hat{\boldsymbol{r}}$

Ω = angular velocity

Field equations

$$\nabla\Phi = -4\pi G\rho$$

$$\nabla^2 V = 4\pi G\rho$$

Escape velocity from the earth

$$v = \sqrt{2\,g\,R}$$

R = radius of the earth

Velocity in circular orbit of radius r

$$v = R\sqrt{\frac{g}{r}}$$

Rocket equation

$$\boldsymbol{v} - \boldsymbol{v}_0 = -\boldsymbol{u}\ln\frac{m_0}{m}$$

$\boldsymbol{u}$ = velocity of ejected matter relative to rocket (= constant)

$\boldsymbol{v}_0$ = rocket velocity when mass = m_0

Kepler's planet laws

1. Each planet moves in an elliptical orbit with the sun in one of the focus points
2. The areal velocity is constant
3. $\left(\frac{T_1}{T_2}\right)^2 = \left(\frac{d_1}{d_2}\right)^3$

T = period of revolution
d = distance from sun

11.2 Astrophysics and Cosmology

Apparent and absolute magnitude

$m = \text{constant} - 2.5 \lg I$ — I = intensity (at sea level)

$M = m + 5 - 5 \lg r$ — r = distance in parsec

M = the (apparent) magnitude a star would have at a distance of 10 pc from the earth.

Hubble's law

$v = Hr$ — v = velocity

H = Hubble's constant, see Sec. T–9.4

Age of the universe

$$T_0 < \frac{1}{H}$$

Hubble's generalized law

$$c_0\, z = H\, r + \frac{H^2 r^2}{2c_0}\,(q - 1)$$

Redshift $z = \dfrac{\Delta\lambda}{\lambda} = \sqrt{\dfrac{1+v/c_0}{1-v/c_0}} - 1$

Deceleration parameter (Friedmann models with cosmological constant = 0)

$$q = -\frac{R_0 \ddot{R}_0}{\dot{R}_0^2} = \frac{4\pi\, G\, \rho_0}{3\, H^2}$$

ρ_0 = density of the universe when its scale factor (or "radius") = R_0

The universe is closed if q is $> \frac{1}{2}$

Schwarzschild radius or event horizon of black holes

$$r_s = \frac{2\, G\, M}{c_0^2} \approx 3 \text{ km}\, \frac{M}{M_\odot}$$

M = mass of black hole
$M_\odot$ = mass of the Sun

Bekenstein-Hawking formula for the entropy of a black hole

$$S = \frac{Akc_0^3}{4G\hbar}$$

A = surface area of the event horizon

For a Schwarzschild black hole (no rotational momentum)

$$S = \frac{4\pi k G M^2}{\hbar c_0} \qquad A = \frac{16\pi M^2 G^2}{c_0^4}$$

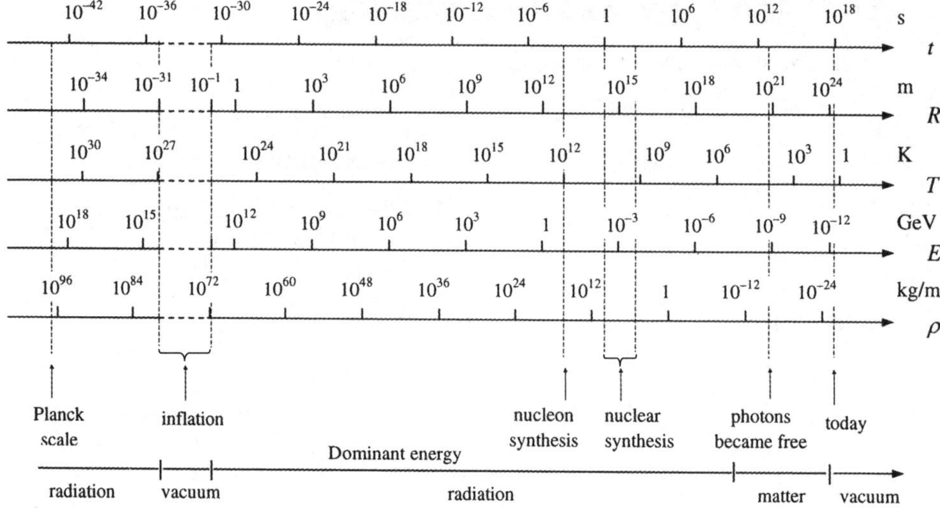

Development of the universe. t = time since Big Bang. R = expansion of an area which at the Planck time had the size of the Planck length. T = temperature. E = particle energy corresponding to the temperature T. ρ = density of the universe. During the era when radiation dominated the universe, the expansion was proportional to $t^{1/2}$, and during the era dominated by matter, the expansion was proportional to $t^{2/3}$. Since 4 Ga (giga years) it is believed that vacuum energy is dominant and that the expansion is proportional to e^{Ht}, where H is the Hubble constant. The age of the universe is 13.7 ± 0.2 Ga.

11.3 Meteorology

The barometer formula

$$\mathrm{d}p = -\rho\, g\, \mathrm{d}z$$

p = pressure

$$p = p_0 \exp\left(-\frac{M\, g\, z}{R\, T}\right)$$

Height

$$z = \frac{R\, T}{M\, g} \ln \frac{p_0}{p}$$

M = molar weight of air

R = universal gas constant

T = absolute temperature

Centrifugal force and Coriolis force

$$F_k = m\,\Omega^2 R \cos\varphi$$

$$F_{cx} = 2m\,v_y\,\Omega \sin\varphi$$

$$F_{cy} = -\,2m\,v_x\,\Omega \sin\varphi$$

Ω = angular velocity of the earth

m = particle mass

v = particle velocity relative to the earth's surface ($v_z = 0$)

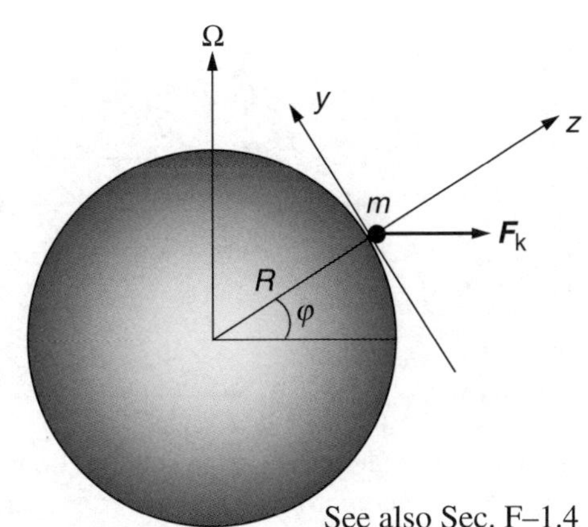

See also Sec. F–1.4

Earth's radiation balance

Global radiation

$$E_g = I \sin h + E_d$$

I = intensity of direct radiation from the sun

h = sun's altitude

E_d = diffuse solar radiation from sky

Net outgoing terrestrial radiation

$$E_{eff} = \sigma T_0^4 - E_A$$

σ = Stefan-Boltzmann constant

T_0 = temperature of earth's surface

E_A = terrestrial radiation flux from atmosphere

Net radiation (balance of radiation)

$$E_B = (1 - A)E_g - E_{eff}$$

A = fraction of global radiation reflected by the earth's surface, albedo

12 Solid Mechanics

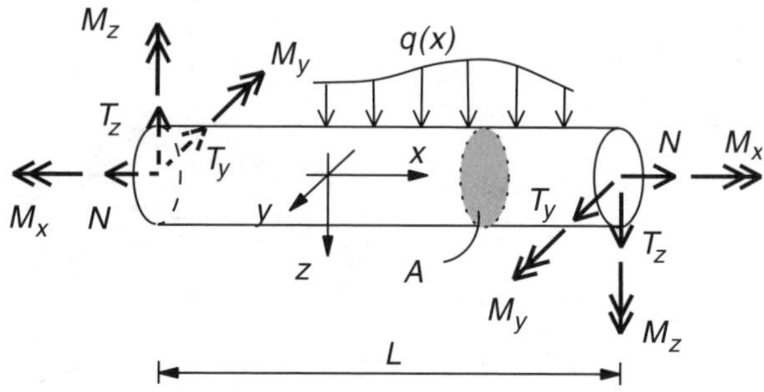

Loaded beam, length L, cross section A, and load $q(x)$, with coordinate system (origin at the geometric centre of cross section) and positive section forces and moments: normal force N, shear forces T_y and T_z, torque M_x and bending moments M_y, M_z.

Quantity	Symbol	SI Unit
Coordinate directions, with origin at geometric centre of cross-sectional area A	x, y, z	m
Normal stress in direction i (= x, y, z)	σ_i	N/m^2
Shear stress in direction j on surface with normal direction i	τ_{ij}	N/m^2
Normal strain in direction i	ε_i	–
Shear strain (corresponding to shear stress τ_{ij})	γ_{ij}	rad
Moment with respect to axis i	M, M_i	Nm
Normal force	N, P	N (= kg m/s^2)
Shear force in direction i (= y, z)	T, T_i	N
Load	$q(x)$	N/m
Cross-sectional area	A	m^2
Length	L, L_0	m
Change of length	δ	m
Displacement in direction x	u, $u(x)$, $u(x,y)$	m
Displacement in direction y	v, $v(x)$, $v(x,y)$	m
Beam deflection	$w(x)$	m
Second moment of area ($i = y, z$)	I, I_i	m^4
Modulus of elasticity (Young's modulus)	E	N/m^2
Poisson's ratio	ν	–
Shear modulus	G	N/m^2
Bulk modulus	K	N/m^2
Temperature coefficient	α	K^{-1}

12.1 Stress, Strain, and Material Relations

Normal stress σ_x

$$\sigma_x = \frac{N}{A} \text{ or } \sigma_x = \lim_{\Delta A \to 0}\left(\frac{\Delta N}{\Delta A}\right)$$

ΔN = fraction of normal force N
ΔA = cross-sectional area element

Shear stress τ_{xy} (mean value over area *A* in the *y* direction)

$$\tau_{xy} = \frac{T_y}{A} \ (= \tau_{\text{mean}})$$

Normal strain ε_x

Linear, at small deformations ($\delta << L_0$)

$$\varepsilon_x = \frac{\delta}{L_0} \text{ or } \varepsilon_x = \frac{\mathrm{d}u(x)}{\mathrm{d}x}$$

δ = change of length
L_0 = original length
$u(x)$ = displacement

Non-linear, at large deformations

$$\varepsilon_x = \ln\left(\frac{L}{L_0}\right)$$

L = actual length ($L = L_0 + \delta$)

Shear strain γ_{xy}

$$\gamma_{xy} = \frac{\partial u(x,y)}{\partial y} + \frac{\partial v(x,y)}{\partial x}$$

Linear elastic material (Hooke's law)

Tension/compression

$$\varepsilon_x = \frac{\sigma_x}{E} + \alpha \Delta T$$

ΔT = *change* of temperatur

Lateral strain

$$\varepsilon_y = -\nu\, \varepsilon_x$$

Shear strain

$$\gamma_{xy} = \frac{\tau_{xy}}{G}$$

Relationships between *G, K, E,* and ν

$$G = \frac{E}{2(1+\nu)} \qquad K = \frac{E}{3(1-2\nu)}$$

12.2 Geometric Properties of Cross-Sectional Area

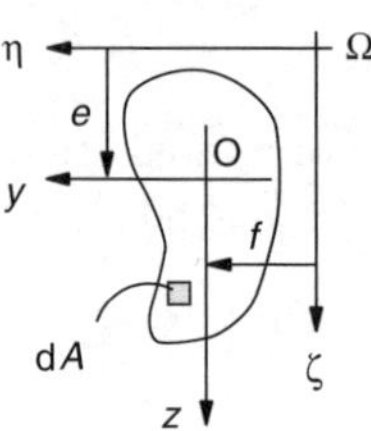

The origin of the coordinate system Oyz must be at the geometric centre of the cross section

Cross-sectional area A

$$A = \int_A \mathrm{d}A$$

dA = area element

Geometric centre (centroid)

$$e \cdot A = \int_A \zeta \, \mathrm{d}A$$

$$f \cdot A = \int_A \eta \, \mathrm{d}A$$

$e = \zeta_{gc}$ = distance from η axis to geometric centre

$f = \eta_{gc}$ = distance from ζ axis to geometric centre

First moment of area

$$S_y = \int_{A'} z\mathrm{d}A \quad \text{and} \quad S_z = \int_{A'} y\mathrm{d}A$$

A' = the "sheared" area (part of area A)

Second moment of area

$$I_y = \int_A z^2 \mathrm{d}A$$

$$I_z = \int_A y^2 \mathrm{d}A$$

$$I_{yz} = \int_A yz\mathrm{d}A$$

I_y = second moment of area with respect to the y axis

I_z = second moment of area with respect to the z axis

I_{yz} = second moment of area with respect to the y and z axes

Parallel-axis theorems

First moment of area

$$S_\eta = \int_A (z+e) \, \mathrm{d}A = eA \quad \text{and} \quad S_\zeta = \int_A (y+f) \, \mathrm{d}A = fA$$

Second moment of area

$$I_\eta = \int_A (z+e)^2 \mathrm{d}A = I_y + e^2 A, \quad I_\zeta = \int_A (y+f)^2 \mathrm{d}A = I_z + f^2 A,$$

$$I_{\eta\zeta} = \int_A (z+e)(y+f)\mathrm{d}A = I_{yz} + efA$$

Rotation of axes

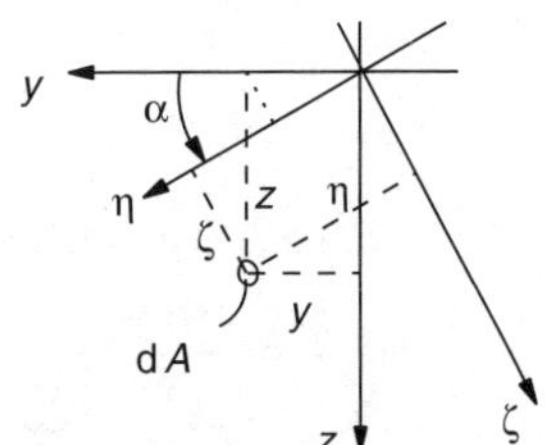

Coordinate system $\Omega\eta\zeta$ has been rotated the angle α with respect to the coordinate system Oyz

$$I_\eta = \int_A \zeta^2 \, \mathrm{d}A = I_y \cos^2\alpha + I_z \sin^2\alpha - 2I_{yz} \sin\alpha \cos\alpha$$

$$I_\zeta = \int_A \eta^2 \, \mathrm{d}A = I_y \sin^2\alpha + I_z \cos^2\alpha + 2I_{yz} \sin\alpha \cos\alpha$$

$$I_{\eta\zeta} = \int_A \zeta\eta \, \mathrm{d}A = (I_y - I_z) \sin\alpha \cos\alpha + I_{yz}(\cos^2\alpha - \sin^2\alpha) = \frac{I_y - I_z}{2} \sin 2\alpha + I_{yz} \cos 2\alpha$$

Principal moments of area

$$I_{1,2} = \frac{I_y + I_z}{2} \pm R \quad \text{where} \quad R = \sqrt{\left(\frac{I_y - I_z}{2}\right)^2 + I_{yz}^2}$$

$$I_1 + I_2 = I_y + I_z$$

Principal axes

$$\sin 2\alpha = \frac{-I_{yz}}{R} \quad \text{or} \quad \cos 2\alpha = \frac{I_y - I_z}{2R}$$

A line of symmetry is always a principal axis

Second moment of area with respect to axes through geometric centre for some symmetric areas (beam cross sections)

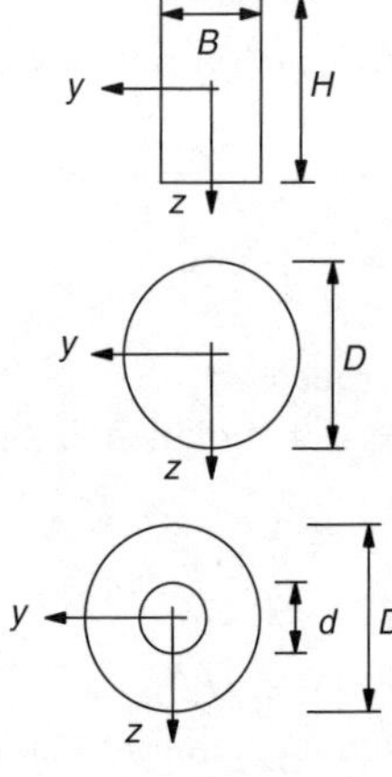

Rectangular area, base B, height H

$$I_y = \frac{BH^3}{12} \quad \text{and} \quad I_z = \frac{HB^3}{12}$$

Solid circular area, diameter D

$$I_y = I_z = \frac{\pi D^4}{64}$$

Thick-walled circular tube, diameters D and d

$$I_y = I_z = \frac{\pi}{64}(D^4 - d^4)$$

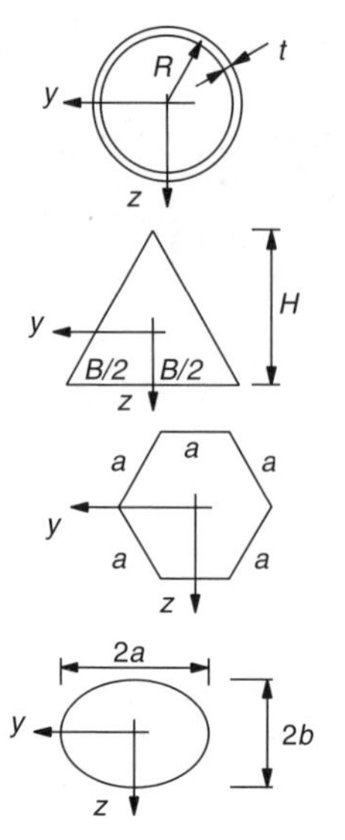

Thin-walled circular tube, radius R and wall thickness t ($t << R$)

$$I_y = I_z = \pi R^3 t$$

Triangular area, base B and height H

$$I_y = \frac{BH^3}{36} \quad \text{and} \quad I_z = \frac{HB^3}{48}$$

Hexagonal area, side length a

$$I_y = I_z = \frac{5\sqrt{3}}{16} a^4$$

Elliptical area, major axis $2a$ and minor axis $2b$

$$I_y = \frac{\pi a b^3}{4} \quad \text{and} \quad I_z = \frac{\pi b a^3}{4}$$

Semicircle, radius a (geometric centre at e)

$$I_y = \left(\frac{\pi}{8} - \frac{8}{9\pi}\right) a^4 \cong 0,110\, a^4 \quad \text{and} \quad e = \frac{4a}{3\pi}$$

12.3 One-Dimensional Bodies (bars, axles, beams)

Tension/compression of bar

Change of length

$$\delta = \frac{NL}{EA} \quad \text{or}$$

N, E, and A are constant along the bar
L = length of bar

$$\delta = \int_0^L \varepsilon(x)\mathrm{d}x = \int_0^L \frac{N(x)}{E(x)A(x)}\mathrm{d}x$$

$N(x)$, $E(x)$, and $A(x)$ may vary along the bar

Torsion of axle

Maximum shear stress

$$\tau_{\max} = \frac{M_v}{W_v}$$

M_v = torque = M_x
W_v = section modulus in torsion (given below)

Torsion (deformation) angle

$$\Theta = \frac{M_v L}{G K_v}$$

M_v = torque = M_x
K_v = section factor of torsional stiffness (given below)

Section modulus W_v and section factor K_v for some cross sections (at torsion)

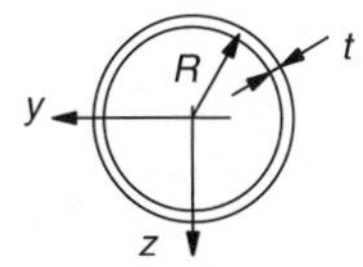

Torsion of thin-walled circular tube, radius R, thickness t, where $t << R$,

$$W_v = 2\pi R^2 t \qquad K_v = 2\pi R^3 t$$

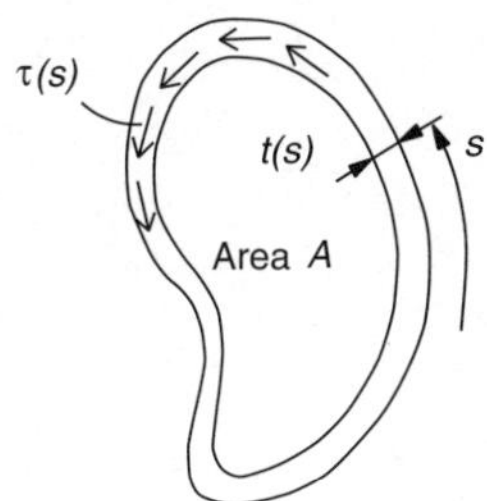

Thin-walled tube of arbitrary cross section
A = area enclosed by the tube
$t(s)$ = wall thickness
s = coordinate around the tube

$$W_v = 2At_{min} \qquad K_v = \frac{4A^2}{\oint_s [t(s)]^{-1} ds}$$

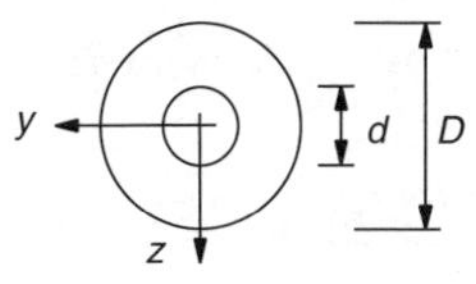

Thick-walled circular tube, diameters D and d,

$$W_v = \frac{\pi}{16}\frac{D^4 - d^4}{D} \qquad K_v = \frac{\pi}{32}(D^4 - d^4)$$

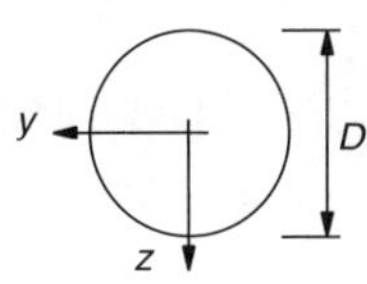

Solid axle with circular cross section, diameter D,

$$W_v = \frac{\pi D^3}{16} \qquad K_v = \frac{\pi D^4}{32}$$

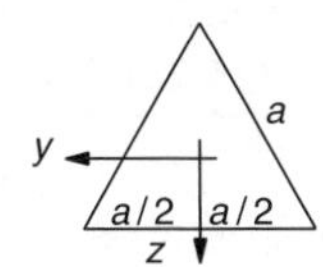

Solid axle with triangular cross section, side lengh a

$$W_v = \frac{a^3}{20} \qquad K_v = \frac{a^4\sqrt{3}}{80}$$

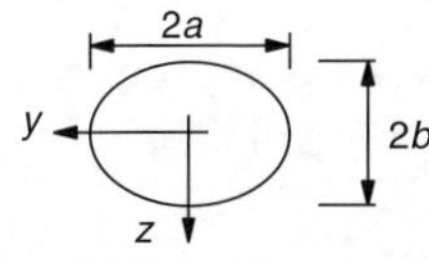

Solid axle with elliptical cross section, major axis $2a$ and minor axis $2b$

$$W_v = \frac{\pi}{2} a\, b^2 \qquad K_v = \frac{\pi a^3 b^3}{a^2 + b^2}$$

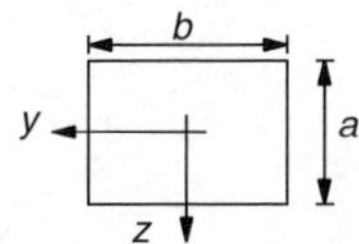

Solid axle with rectangular cross section b by a, where $b \geq a$

$$W_v = k_{Wv}\, a^2 b \qquad K_v = k_{Kv}\, a^3 b$$

for k_{Wv} and k_{Kv}, see table below

Factors k_{Wv} and k_{Kv} for some values of ratio b/a (solid rectangular cross section)

b/a	k_{Wv}	k_{Kv}
1.0	0.208	0.1406
1.2	0.219	0.1661
1.5	0.231	0.1958
2.0	0.246	0.229
2.5	0.258	0.249
3.0	0.267	0.263
4.0	0.282	0.281
5.0	0.291	0.291
10.0	0.312	0.312
∞	0.333	0.333

Bending of beam

Relationships between bending moment $M_y = M(x)$, shear force $T_z = T(x)$, and load $q(x)$ on beam

$$\frac{\mathrm{d}T(x)}{\mathrm{d}x} = -q(x)\,, \quad \frac{\mathrm{d}M(x)}{\mathrm{d}x} = T(x)\,, \quad \text{and} \quad \frac{\mathrm{d}^2M(x)}{\mathrm{d}x^2} = -q(x)$$

Normal stress

$$\sigma = \frac{N}{A} + \frac{Mz}{I}$$

I (here I_y) = second moment of area (see Section 12.2)

Maximum bending stress

$$|\sigma|_{max} = \frac{|M|}{W_b} \quad \text{where } W_b = \frac{I}{|z|_{max}}$$

W_b = section modulus (in bending)

Shear stress

$$\tau = \frac{TS_{A'}}{Ib}$$

$S_{A'}$ = first moment of "sheared" area A' (see Section 12.2)
b = length of line limiting area A'

$$\tau_{gc} = \mu\,\frac{T}{A}$$

τ_{gc} = shear stress at geometric centre
μ = the Jouravski factor

The Jouravski factor μ for some cross sections

rectangular	1.5
triangular	1.33
circular	1.33
thin-walled circular	2.0
elliptical	1.33
ideal I profile	A/A_{web}

Skew bending

Axes y and z are not principal axes:

$$\sigma = \frac{N}{A} + \frac{M_y(zI_z - yI_{yz}) - M_z(yI_y - zI_{yz})}{I_y I_z - I_{yz}^2}$$

I_y, I_z, I_{yz} = second moment of area

Axes y' and z' are principal axes:

$$\sigma = \frac{N}{A} + \frac{M_1 z'}{I_1} - \frac{M_2 y'}{I_2}$$

I_1, I_2 = principal second moment of area
M_1, M_2 = bending moment with respect to principal axis y' and z', respectively

Beam deflection $w(x)$

Differential equations

$$\frac{d^2}{dx^2}\left\{EI(x)\frac{d^2}{dx^2}w(x)\right\} = q(x)$$

when $EI(x)$ is function of x

$$EI\,\frac{d^4}{dx^4}\,w(x) = q(x)$$

when EI is constant

Homogeneous boundary conditions

Clamped beam end

$$w(*) = 0 \quad \text{and} \quad \frac{d}{dx}w(*) = 0$$

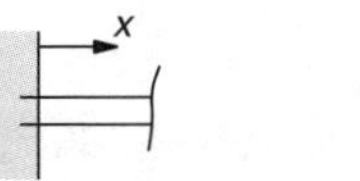

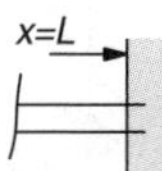

where * is the coordinate of beam end (to be entered after differentiation)

Simply supported beam end

$$w(*) = 0 \text{ and } -EI\frac{d^2}{dx^2}w(*) = 0$$

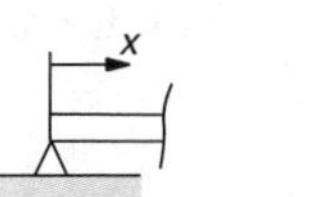

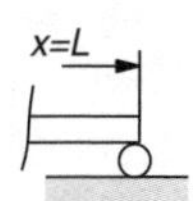

Sliding beam end

$$\frac{d}{dx}w(*) = 0 \text{ and } -EI\frac{d^3}{dx^3}w(*) = 0$$

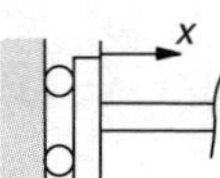

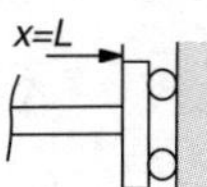

Free beam end

$$-EI\frac{d^2}{dx^2}w(*) = 0 \text{ and } -EI\frac{d^3}{dx^3}w(*) = 0$$

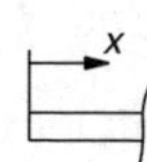

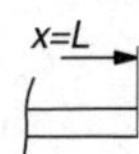

Non-homogeneous boundary conditions

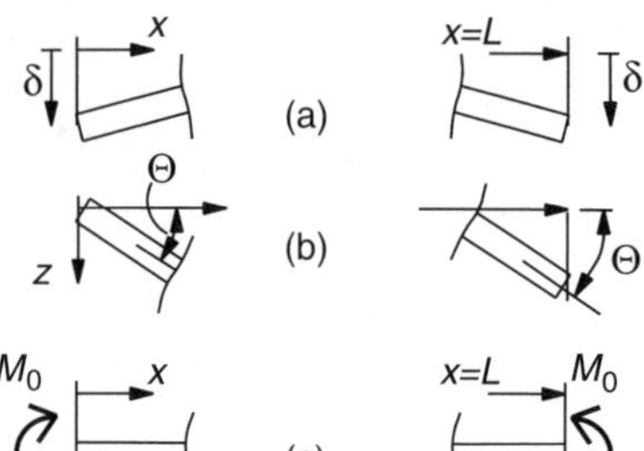

(a) Displacement δ prescribed

$w(*) = \delta$

(b) Slope Θ prescribed

$\frac{\mathrm{d}}{\mathrm{d}x}w(*) = \Theta$

(c) Moment M_0 prescribed

$-EI\frac{\mathrm{d}^2}{\mathrm{d}x^2}w(*) = M_0$

(d) Force P prescribed

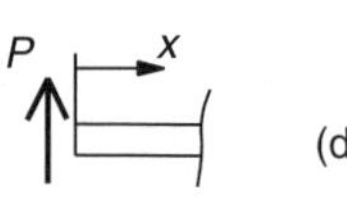

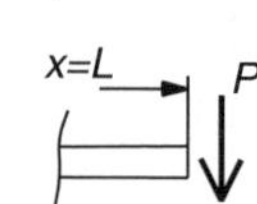

$-EI\frac{\mathrm{d}^3}{\mathrm{d}x^3}w(*) = P$

Beam on elastic bed

Differential equation

$$EI\frac{\mathrm{d}^4}{\mathrm{d}x^4}w(x) + kw(x) = q(x)$$

EI = constant bending stiffness
k = bed modulus (N/m^2)

Solution

$w(x) = w_{\mathrm{part}}(x) + w_{\mathrm{hom}}(x)$ where

$$w_{\mathrm{hom}}(x) = \{C_1\cos(\lambda x) + C_2\sin(\lambda x)\}\mathrm{e}^{\lambda x} + \{C_3\cos(\lambda x) + C_2\sin(\lambda x)\}\mathrm{e}^{-\lambda x};\ \lambda^4 = \frac{k}{4EI}$$

Boundary conditions as given above

Beam vibration

(For vibration of discrete systems, see Sections F-1.11 and F-1.12)

Differential equation

$$EI\frac{\partial^4}{\partial x^4}w(x,t) + m\frac{\partial^2}{\partial t^2}w(x,t) = q(x,t)$$

EI = constant bending stiffness
m = beam mass per metre (kg/m)
t = time

Assume solution $w(x,t) = X(x) \cdot T(t)$. Then the standing wave solution is

$T(t) = \mathrm{e}^{\mathrm{i}\omega t}$ and $X(x) = C_1\cosh(\mu x) + C_2\cos(\mu x) + C_3\sinh(\mu x) + C_4\sin(\mu x)$

where $\mu^4 = \omega^2 m/EI$

Boundary conditions (as given above) give an eigenvalue problem that provides the eigenfrequencies and eigenmodes (eigenforms) of the vibrating beam.

Axially loaded beam, stability, the Euler cases

Beam axially loaded in tension

Differential equation

$$EI\frac{d^4}{dx^4}w(x) - N\frac{d^2}{dx^2}w(x) = q(x) \qquad N = \text{normal force in tension } (N > 0)$$

Solution

$w(x) = w_{part}(x) + w_{hom}(x)$ where

$$w_{hom}(x) = C_1 + C_2\sqrt{\frac{N}{EI}}\,x + C_3\sinh\left(\sqrt{\frac{N}{EI}}\,x\right) + C_4\cosh\left(\sqrt{\frac{N}{EI}}\,x\right)$$

New boundary condition on shear force (the other boundary conditions as given above)

$$T(*) = -EI\frac{d^3}{dx^3}w(*) + N\frac{d}{dx}w(*)$$

Beam axially loaded in compression

Differential equation

$$EI\frac{d^4}{dx^4}w(x) + P\frac{d^2}{dx^2}w(x) = q(x) \qquad P = \text{normal force in compression } (P > 0)$$

Solution

$w(x) = w_{part}(x) + w_{hom}(x)$ where

$$w_{hom}(x) = C_1 + C_2\sqrt{\frac{P}{EI}}\,x + C_3\sin\left(\sqrt{\frac{P}{EI}}\,x\right) + C_4\cos\left(\sqrt{\frac{P}{EI}}\,x\right)$$

New boundary condition on shear force (the other boundary conditions as given above)

$$T(*) = -EI\frac{d^3}{dx^3}w(*) - P\frac{d}{dx}w(*)$$

Elementary cases: the Euler cases (P_c is critical load)

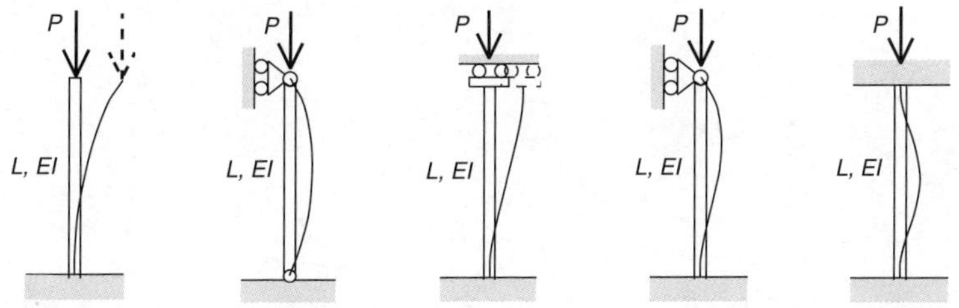

Case 1	Case 2a	Case 2b	Case 3	Case 4
P; L, EI	P; L, EI	P; L, EI	P; L, EI	P; L, EI
$P_c = \frac{\pi^2 EI}{4L^2}$	$P_c = \frac{\pi^2 EI}{L^2}$	$P_c = \frac{\pi^2 EI}{L^2}$	$P_c = \frac{2.05\pi^2 EI}{L^2}$	$P_c = \frac{4\pi^2 EI}{L^2}$

12.4 Bending of Beam – Elementary Cases

Cantilever beam

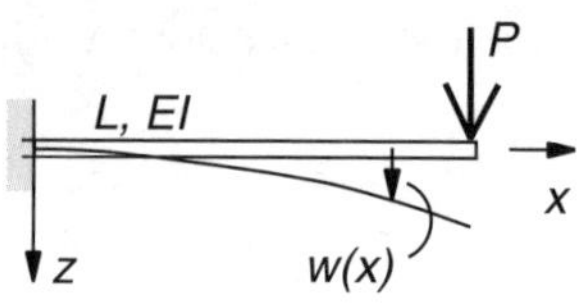

$$w(x)=\frac{PL^3}{6EI}\left(3\frac{x^2}{L^2}-\frac{x^3}{L^3}\right)$$

$$w(L)=\frac{PL^3}{3EI}\qquad w'(L)=\frac{PL^2}{2EI}$$

$$w(x)=\frac{ML^2}{2EI}\left(\frac{x^2}{L^2}\right)$$

$$w(L)=\frac{ML^2}{2EI}\qquad w'(L)=\frac{ML}{EI}$$

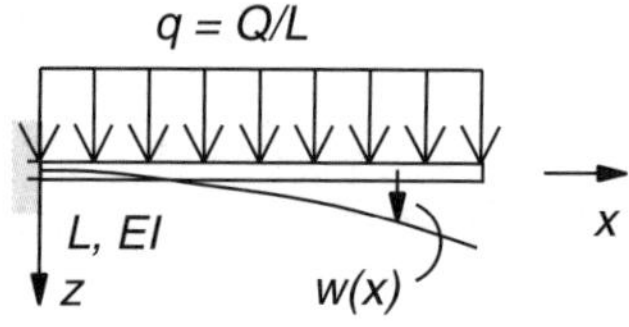

$$w(x)=\frac{qL^4}{24EI}\left(\frac{x^4}{L^4}-4\frac{x^3}{L^3}+6\frac{x^2}{L^2}\right)$$

$$w(L)=\frac{qL^4}{8EI}\qquad w'(L)=\frac{qL^3}{6EI}$$

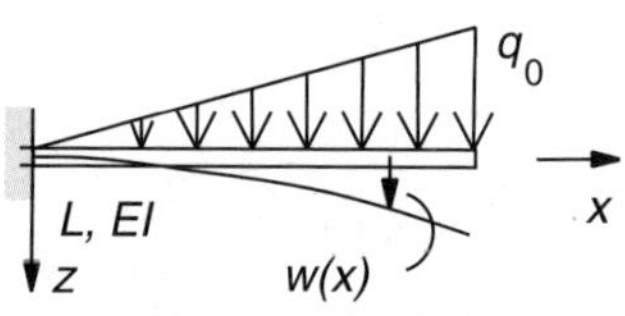

$$w(x)=\frac{q_0L^4}{120EI}\left(\frac{x^5}{L^5}-10\frac{x^3}{L^3}+20\frac{x^2}{L^2}\right)$$

$$w(L)=\frac{11q_0L^4}{120EI}\qquad w'(L)=\frac{q_0L^3}{8EI}$$

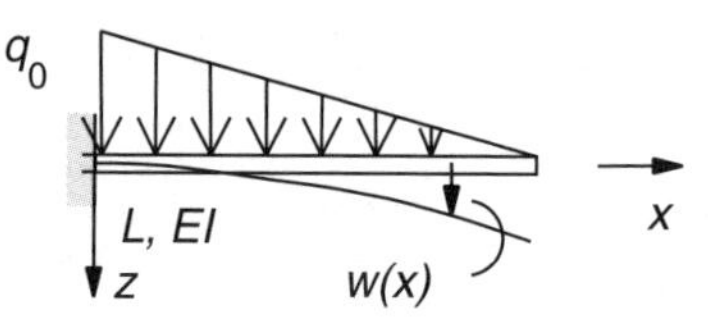

$$w(x)=\frac{q_0L^4}{120EI}\left(-\frac{x^5}{L^5}+5\frac{x^4}{L^4}-10\frac{x^3}{L^3}+10\frac{x^2}{L^2}\right)$$

$$w(L)=\frac{q_0L^4}{30EI}\qquad w'(L)=\frac{q_0L^3}{24EI}$$

Simply supported beam

Load applied at $x = \alpha L$ where $\alpha < 1$ and $\beta = 1 - \alpha$

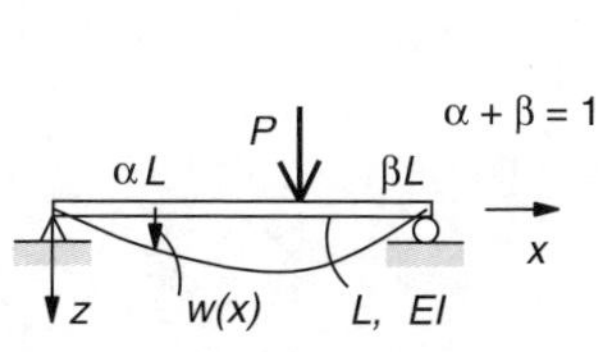

$$w(x) = \frac{PL^3}{6EI}\beta\left((1-\beta^2)\frac{x}{L} - \frac{x^3}{L^3}\right) \quad \text{for} \quad \frac{x}{L} \leq \alpha$$

$$w(\alpha L) = \frac{PL^3}{3EI}\alpha^2\beta^2. \quad \text{When } \alpha > \beta \text{ one obtains}$$

$$w_{\text{max}} = w\left(L\sqrt{\frac{1-\beta^2}{3}}\right) = w(\alpha L)\frac{1+\beta}{3\beta}\sqrt{\frac{1+\beta}{3\alpha}}$$

$$w'(0) = \frac{PL^2}{6EI}\alpha\beta(1+\beta) \quad w'(L) = -\frac{PL^2}{6EI}\alpha\beta(1+\alpha)$$

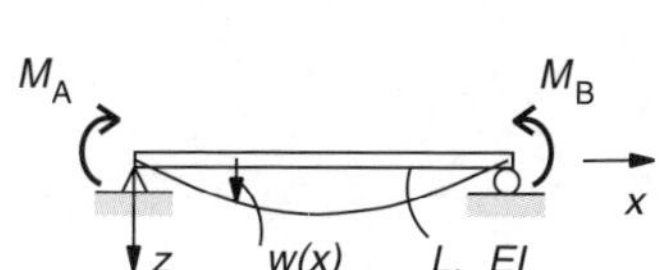

$$w(x) = \frac{L^2}{6EI}\left\{M_{\text{A}}\left(2\frac{x}{L} - 3\frac{x^2}{L^2} + \frac{x^3}{L^3}\right) + M_{\text{B}}\left(\frac{x}{L} - \frac{x^3}{L^3}\right)\right\}$$

$$w'(0) = \frac{M_{\text{A}}L}{3EI} + \frac{M_{\text{B}}L}{6EI} \qquad w'(L) = -\frac{M_{\text{A}}L}{6EI} - \frac{M_{\text{B}}L}{3EI}$$

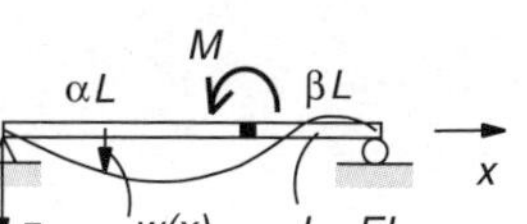

$$w(x) = \frac{ML^2}{6EI}\left((1-3\beta^2)\frac{x}{L} - \frac{x^3}{L^3}\right) \quad \text{for} \quad \frac{x}{L} \leq \alpha$$

$$w'(0) = \frac{ML}{6EI}(1-3\beta^2) \qquad w'(L) = \frac{ML}{6EI}(1-3\alpha^2)$$

$$w(x) = \frac{QL^3}{24EI}\left(\frac{x^4}{L^4} - 2\frac{x^3}{L^3} + \frac{x}{L}\right)$$

$$w(L/2) = \frac{5QL^3}{384EI} \qquad w'(0) = -w'(L) = \frac{QL^2}{24EI}$$

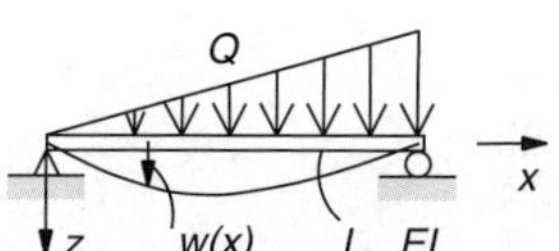

$$w(x) = \frac{QL^3}{180EI}\left(3\frac{x^5}{L^5} - 10\frac{x^3}{L^3} + 7\frac{x}{L}\right)$$

$$w'(0) = \frac{7QL^2}{180EI} \qquad w'(L) = -\frac{8QL^2}{180EI}$$

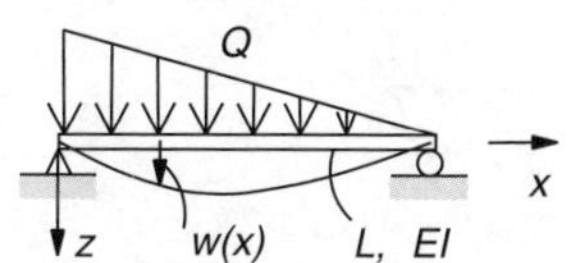

$$w(x) = \frac{QL^3}{180EI}\left(-3\frac{x^5}{L^5} + 15\frac{x^4}{L^4} - 20\frac{x^3}{L^3} + 8\frac{x}{L}\right)$$

$$w'(0) = \frac{8QL^2}{180EI} \qquad w'(L) = -\frac{7QL^2}{180EI}$$

Clamped – simply supported beam and clamped – clamped beam

Load applied at $x = \alpha L$ where $\alpha < 1$ and $\beta = 1 - \alpha$

Only redundant reactions are given. For deflections, use superposition of solutions for simply supported beams.

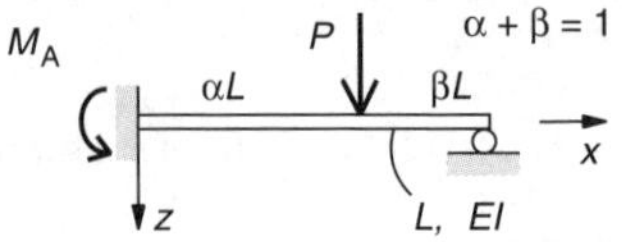

$$M_A = \frac{PL}{2}\beta(1-\beta^2)$$

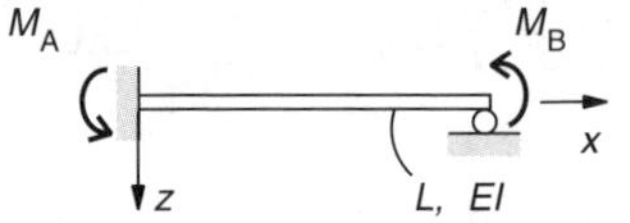

$$M_A = \frac{M_B}{2}$$

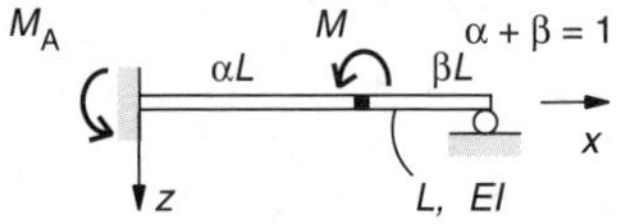

$$M_A = \frac{M}{2}(1-3\beta^2)$$

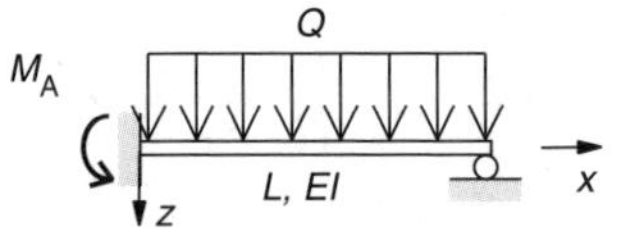

$$M_A = \frac{QL}{8}$$

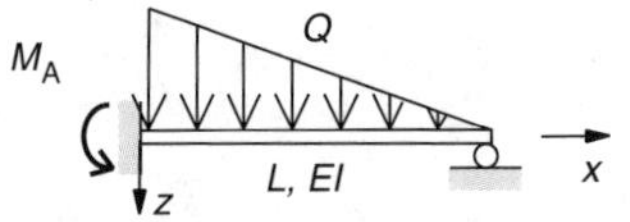

$$M_A = \frac{2QL}{15}$$

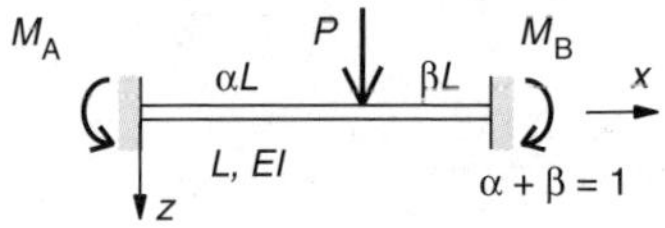

$$M_A = PL\alpha\beta^2 \qquad M_B = PL\alpha^2\beta$$

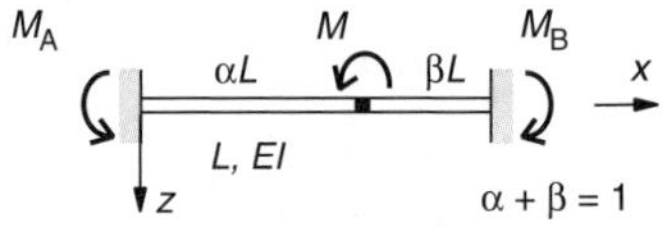

$$M_A = -M\beta(1-3\alpha) \qquad M_B = M\alpha(1-3\beta)$$

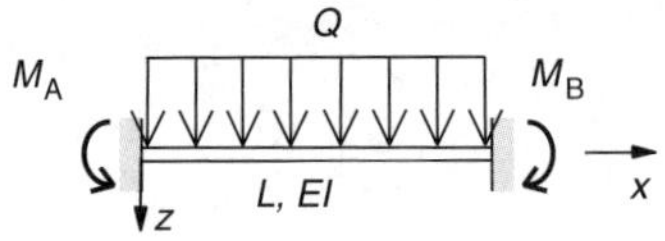

$$M_A = M_B = \frac{QL}{12}$$

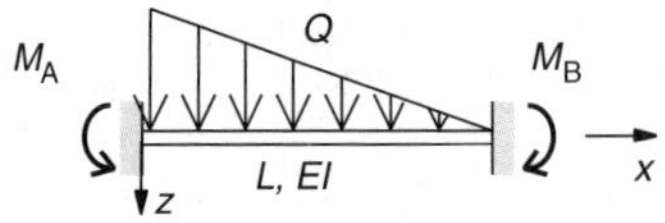

$$M_A = \frac{QL}{10} \qquad M_B = \frac{QL}{15}$$

12.5 Material Fatigue

Fatigue limits (notations)

Load	Alternating	Pulsating
Tension/compression	$\pm\sigma_u$	$\sigma_{up} \pm \sigma_{up}$
Bending	$\pm\sigma_{ub}$	$\sigma_{ubp} \pm \sigma_{ubp}$
Torsion	$\pm\tau_{uv}$	$\tau_{uvp} \pm \tau_{uvp}$

The Haigh diagram

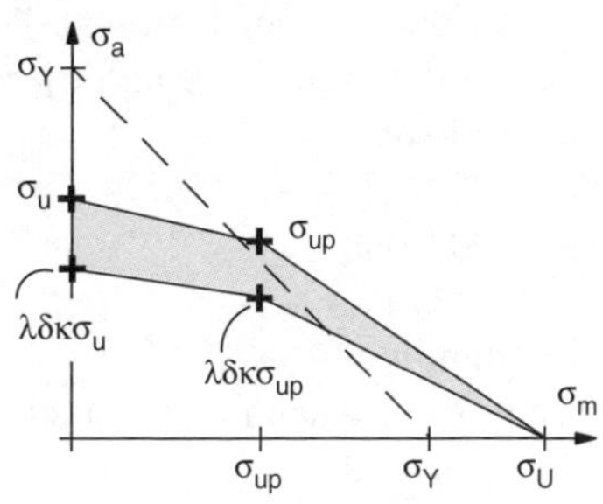

σ_a = stress amplitude
σ_m = mean stress
σ_Y = yield limit
σ_U = ultimate strength
σ_u, σ_{up} = fatigue limits
λ, δ, κ = factors reducing fatigue limits
(similar diagrams for $\sigma_{ub}, \sigma_{ubp}$ and τ_{uv}, τ_{uvp})

Factors reducing fatigue limits

Surface finish κ

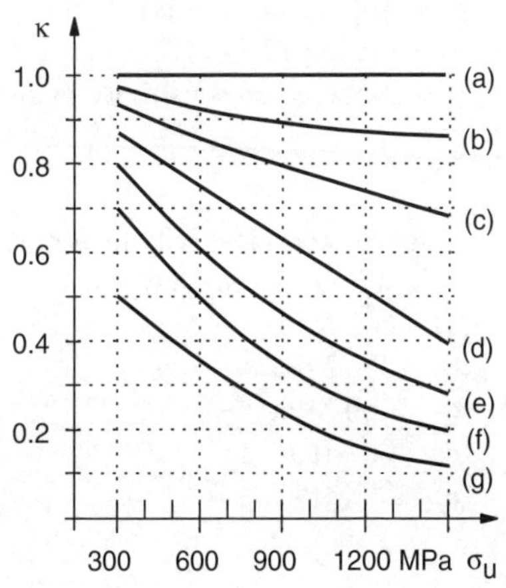

Factor κ reducing the fatigue limit due to surface irregularities

(a) polished surface ($\kappa = 1$)
(b) ground
(c) machined
(d) standard notch
(e) rolling skin
(f) corrosion in sweet water
(g) corrosion in salt water

Volume factor λ (due to process)

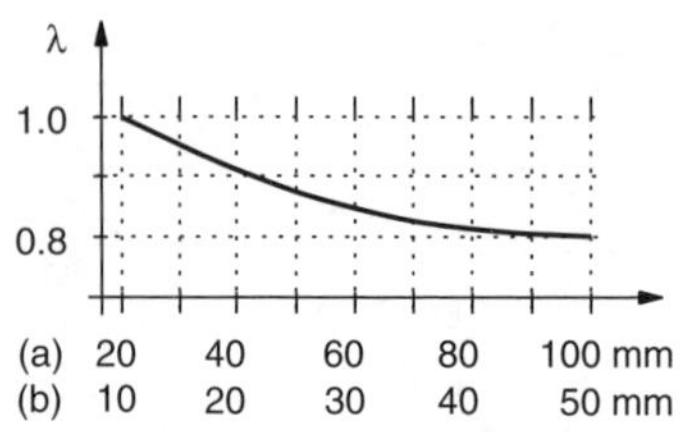

Factor λ reducing the fatigue limit due to size of raw material

(a) diameter at circular cross section
(b) thickness at rectangular cross section

Volume factor δ (due to geometry)

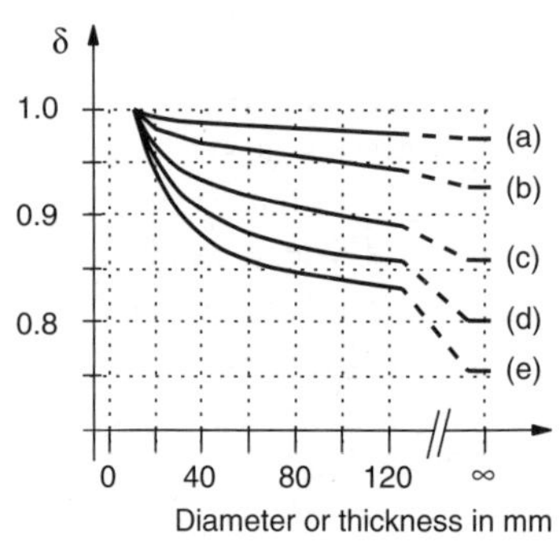

Factor δ reducing the fatigue limit σ_{ub} and τ_{uv} due to loaded volume.
Steel with ultimate strength σ_U =
(a) 1500 MPa
(b) 1000 MPa
(c) 600 MPa
(d) 400 MPa
(e) aluminium
Factor δ = 1 when fatigue notch factor $K_f > 1$ is used

Fatigue notch factor K_f (at stress concentration)

$$K_f = 1 + q(K_t - 1)$$

K_t = stress concentration factor (see Section 12.8)
q = fatigue notch sensitivity factor

Fatigue notch sensitivity factor *q*

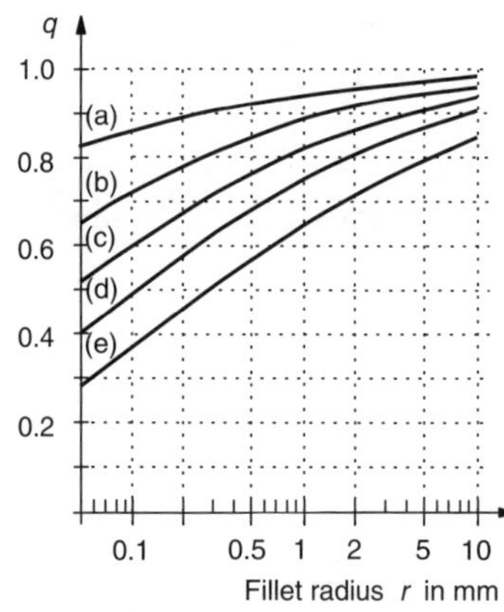

Fatigue notch sensitivity factor q for steel with ultimate strength σ_U =
(a) 1600 MPa
(b) 1300 MPa
(c) 1000 MPa
(d) 700 MPa
(e) 400 MPa

Wöhler diagram

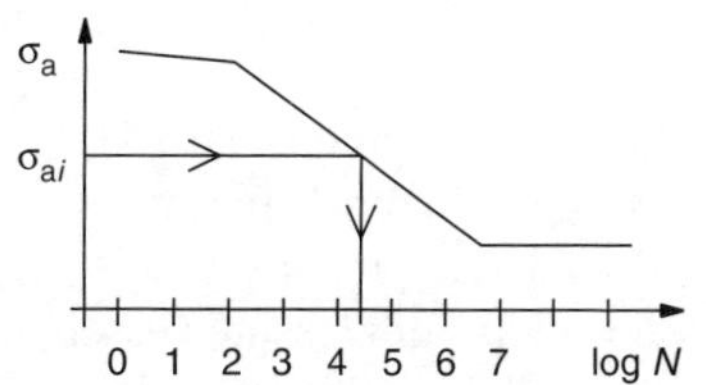

σ_{ai} = stress amplitude
N_i = fatigue life (in cycles) at stress amplitude σ_{ai}

Damage accumulation *D*

$$D = \frac{n_i}{N_i}$$

n_i = number of loading cycles at stress amplitude σ_{ai}
N_i = fatigue life at stress amplitude σ_{ai}

Palmgren-Miner's rule

Failure when

$$\sum_{i=1}^{I} \frac{n_i}{N_i} = 1$$

n_i = number of loading cycles at stress amplitude σ_{ai}
N_i = fatigue life at stress amplitude σ_{ai}
I = number of loading stress levels

Fatigue data (cyclic, constant-amplitude loading)

The following fatigue limits may be used *only* when solving exercise problems. For a real design, data should be taken from latest official standard and *not* from this table.[1]

Material	Tension		Bending		Torsion	
	alternating MPa	pulsating MPa	alternating MPa	pulsating MPa	alternating MPa	pulsating MPa
Carbon steel						
141312-00	± 110	110 ± 110	± 170	150 ± 150	± 100	100 ± 100
141450-1	± 140	130 ± 130	± 190	170 ± 170	± 120	120 ± 120
141510-00	± 230					
141550-01	± 180	160 ± 160	± 240	210 ± 210	± 140	140 ± 140
141650-01	± 200	180 ± 180	± 270	240 ± 240	± 150	150 ± 150
141650			± 460			
Stainless steel 2337-02, $\sigma_u = \pm 270$ MPa						
Aluminium SS 4120-02, $\sigma_{ub} = \pm 110$ MPa; SS 4425-06, $\sigma_u = \pm 120$ MPa						

[1] Data in this table has been collected from B Sundström (editor): Handbok och Formelsamling i Hållfasthetslära, Institutionen för hållfasthetslära, KTH, Stockholm, 1998.

12.6 Multi-Axial Stress States

Stresses in thin-walled circular pressure vessel

$$\sigma_t = p\frac{R}{t} \text{ and } \sigma_x = p\frac{R}{2t} \quad (\sigma_z \approx 0)$$

σ_t = circumferential stress
σ_x = longitudinal stress
p = internal pressure
R = radius of pressure vessel
t = wall thickness ($t << R$)

Rotational symmetry in structure and load (plane stress, i.e. $\sigma_z = 0$)

Differential equation for rotating circular plate

$$\frac{d^2u}{dr^2} + \frac{1}{r}\frac{du}{dr} - \frac{u}{r^2} = -\frac{1-\nu^2}{E}\rho\omega^2 r$$

$u = u(r)$ = radial displacement
ρ = density
ω = angular rotation (rad/s)

Solution

$$u(r) = u_{\text{hom}} + u_{\text{part}} = A_0 r + \frac{B_0}{r} - \frac{1-\nu^2}{8E}\rho\omega^2 r^3$$

Stresses

$$\sigma_r(r) = A - \frac{B}{r^2} - \frac{3+\nu}{8}\rho\omega^2 r^2 \quad \text{and} \quad \sigma_\phi(r) = A + \frac{B}{r^2} - \frac{1+3\nu}{8}\rho\omega^2 r^2$$

where

$$A = \frac{E\,A_0}{1-\nu} \quad \text{and} \quad B = \frac{E\,B_0}{1+\nu}$$

Boundary conditions
σ_r or u must be known on inner and outer boundary of the circular plate

Shrink fit

$$\delta = u_{\text{outer}}(p) - u_{\text{inner}}(p)$$

δ = difference of radii
p = contact pressure
u = radial displacement as function of p

Plane stress and plane strain (plane state)

Plane stress (in *xy*-plane) when $\sigma_z = 0$, $\tau_{xz} = 0$, and $\tau_{yz} = 0$
Plane strain (in *xy*-plane) when $\tau_{xz} = 0$, $\tau_{yz} = 0$, and $\varepsilon_z = 0$ or constant

Stresses in direction α (plane state)

$$\sigma(\alpha) = \sigma_x \cos^2\alpha + \sigma_y \sin^2\alpha + 2\tau_{xy}\cos\alpha\,\sin\alpha$$

$$\tau(\alpha) = -(\sigma_x - \sigma_y)\sin\alpha\,\cos\alpha + \tau_{xy}(\cos^2\alpha - \sin^2\alpha)$$

$\sigma(\alpha)$ = normal stress in direction α
$\tau(\alpha)$ = shear stress on surface with normal in direction α

Principal stresses $\sigma_{1,2}$ and principal directions at plane stress state

$$\sigma_{1,2} = \sigma_c \pm R = \frac{\sigma_x + \sigma_y}{2} \pm \sqrt{\left(\frac{\sigma_x - \sigma_y}{2}\right)^2 + \tau_{xy}^2}$$

$$\sin(2\psi_1) = \frac{\tau_{xy}}{R} \quad \text{or} \quad \cos(2\psi_1) = \frac{\sigma_x - \sigma_y}{2R}$$

ψ_1 = angle from x axis (in xy plane) to direction of principal stress σ_1

Strain in direction α (plane state)

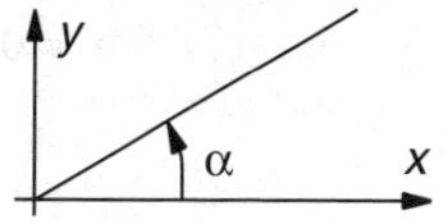

$$\varepsilon(\alpha) = \varepsilon_x \cos^2\alpha + \varepsilon_y \sin^2\alpha + \gamma_{xy} \sin\alpha \cos\alpha$$

$$\gamma(\alpha) = (\varepsilon_y - \varepsilon_x)\sin(2\alpha) + \gamma_{xy}\cos(2\alpha)$$

$\varepsilon(\alpha)$ = normal strain in direction α

$\gamma(\alpha)$ = shear strain of element with normal in direction α

Principal strains and principal directions (plane state)

$$\varepsilon_{1,2} = \varepsilon_c \pm R = \frac{\varepsilon_x + \varepsilon_y}{2} \pm \sqrt{\left(\frac{\varepsilon_x - \varepsilon_y}{2}\right)^2 + \left(\frac{\gamma_{xy}}{2}\right)^2}$$

$$\sin(2\psi_1) = \frac{\gamma_{xy}}{2R} \quad \text{or} \quad \cos(2\psi_1) = \frac{\varepsilon_x - \varepsilon_y}{2R}$$

ψ_1 = angle from x axis (in xy plane) to direction of principal strain ε_1

Principal stresses and principal directions at three-dimensional stress state

The determinant

$$|\mathbf{S} - \sigma \mathbf{I}| = 0$$

gives three roots (the principal stresses)

$$\text{Stress matrix } \mathbf{S} = \begin{bmatrix} \sigma_x & \tau_{xy} & \tau_{xz} \\ \tau_{yx} & \sigma_y & \tau_{yz} \\ \tau_{zx} & \tau_{zy} & \sigma_z \end{bmatrix}$$

(contains the nine stress components σ_{ij})

$$\text{Unit matrix } \mathbf{I} = \begin{bmatrix} 1 & 0 & 0 \\ 0 & 1 & 0 \\ 0 & 0 & 1 \end{bmatrix}$$

Direction of principal stress σ_i (i = 1, 2, 3) is given by

$$(\mathbf{S} - \sigma_i \mathbf{I}) \cdot \mathbf{n}_i = \mathbf{0}$$

and

$$\mathbf{n}_i^{\mathrm{T}} \cdot \mathbf{n}_i = 1$$

n_{ix}, n_{iy} and n_{iz} are the elements of the unit vector $\mathbf{n}_i$ in the direction of principal stress σ_i

(^{T} means transpose)

Principal strains and principal directions at three-dimensional stress state

Use shear strain $\varepsilon_{ij} = \gamma_{ij}/2$ for $i \neq j$

The determinant

$|\mathbf{E} - \varepsilon\,\mathbf{I}| = 0$

gives three roots (the principal strains)

Strain matrix $\mathbf{E} = \begin{bmatrix} \varepsilon_x & \varepsilon_{xy} & \varepsilon_{xz} \\ \varepsilon_{yx} & \varepsilon_y & \varepsilon_{yz} \\ \varepsilon_{zx} & \varepsilon_{zy} & \varepsilon_z \end{bmatrix}$

$\mathbf{I}$ = unit matrix

Direction of principal strain ε_i (i = 1, 2, 3) is given by

$(\mathbf{E} - \varepsilon_i\,\mathbf{I}) \cdot \mathbf{n}_i = \mathbf{0}$

and

$\mathbf{n}_i^{\mathrm{T}} \cdot \mathbf{n} = 1$

n_{ix}, n_{iy} and n_{iz} are the elements of the unit vector $\mathbf{n}_i$ in the direction of principal strain ε_i

(^{T} means transpose)

Hooke's law, including temperature term (three-dimensional stress state)

$$\varepsilon_x = \frac{1}{E}[\sigma_x - \nu(\sigma_y + \sigma_z)] + \alpha\Delta T$$

$$\varepsilon_y = \frac{1}{E}[\sigma_y - \nu(\sigma_z + \sigma_x)] + \alpha\Delta T$$

$$\varepsilon_z = \frac{1}{E}[\sigma_z - \nu(\sigma_x + \sigma_y)] + \alpha\Delta T$$

$$\gamma_{xy} = \frac{\tau_{xy}}{G} \quad \gamma_{yz} = \frac{\tau_{yz}}{G} \quad \gamma_{zx} = \frac{\tau_{zx}}{G}$$

α = temperature coefficient

ΔT = change of temperature (relative to the temperature not giving any stress)

Effective stress

The Huber-von Mises effective stress (the deviatoric stress hypothesis)

$$\sigma_e^{\mathrm{vM}} = \sqrt{\sigma_x^2 + \sigma_y^2 + \sigma_z^2 - \sigma_x\sigma_y - \sigma_y\sigma_z - \sigma_z\sigma_x + 3\tau_{xy}^2 + 3\tau_{yz}^2 + 3\tau_{zx}^2}$$

$$= \sqrt{\frac{1}{2}\{(\sigma_1 - \sigma_2)^2 + (\sigma_2 - \sigma_3)^2 + (\sigma_3 - \sigma_1)^2\}}$$

The Tresca effective stress (the shear stress hypothesis)

$$\sigma_e^{\mathrm{T}} = \max\,[|\sigma_1 - \sigma_2|, |\sigma_2 - \sigma_3|, |\sigma_3 - \sigma_1|] = \sigma_{\max}^{\mathrm{pr}} - \sigma_{\min}^{\mathrm{pr}}$$

(pr = principal stress)

12.7 Energy Methods – the Castigliano Theorem

Strain energy u per unit of volume

Linear elastic material and uni-axial stress

$$u = \frac{\sigma\varepsilon}{2}$$

Total strain energy U in beam loaded in tension/compression, torsion, bending, and shear

$$U_{\text{tot}} = \int_0^L \left\{ \frac{N(x)^2}{2EA(x)} + \frac{M_t(x)^2}{2GK_v(x)} + \frac{M_{\text{bend}}(x)^2}{2EI(x)} + \beta \frac{T(x)^2}{2GA(x)} \right\} \mathrm{d}x$$

M_t = torque = M_x

M_{bend} = bending moment = M_y

K_v = section factor of torsional stiffness

β = shear factor, see below

Cross section	β	μ
	6/5	3/2
	10/9	4/3
	2	2
	A/A_{web}	A/A_{web}

Shear factor β

$$\beta = \frac{A}{I^2} \int_A \left(\frac{S_{A'}}{b} \right)^2 \mathrm{d}A$$

β is given for some cross sections in the table (μ is the Jouravski factor, see Section 12.3 One-Dimensional Bodies)

Elementary case: Pure bending

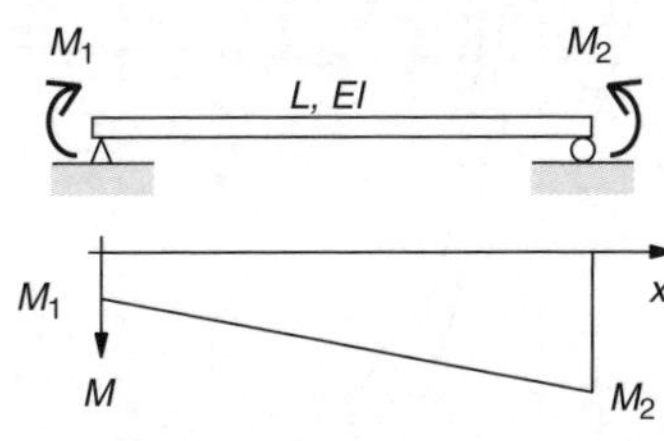

Only bending moment M_{bend} is present. The moment varies linearly along the beam with moments M_1 and M_2 at the beam ends. One has

$$M_{\text{bend}}(x) = M_1 + (M_2 - M_1)x/L,$$

which gives

$$U_{\text{tot}} = \frac{L}{6EI}\{M_1^2 + M_1M_2 + M_2^2\}$$

The second term is negative if M_1 and M_2 have different signs

The Castigliano theorem

$$\delta = \frac{\partial U}{\partial P} \quad \text{and} \quad \Theta = \frac{\partial U}{\partial M}$$

δ = displacement in the direction of force P of the point where force P is applied

Θ = rotation (change of angle) at moment M

12.8 Stress Concentration

Tension/compression

Maximum normal stress at a stress concentration is

$\sigma_{max} = K_t \sigma_{nom}$, where K_t and σ_{nom} are given in the diagrams

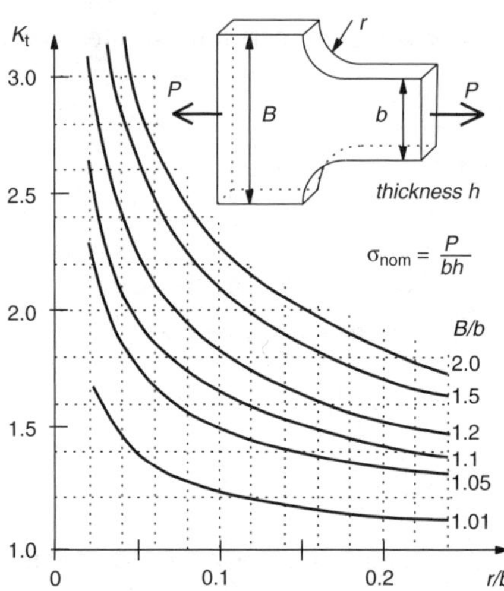

Tension of flat bar with shoulder fillet

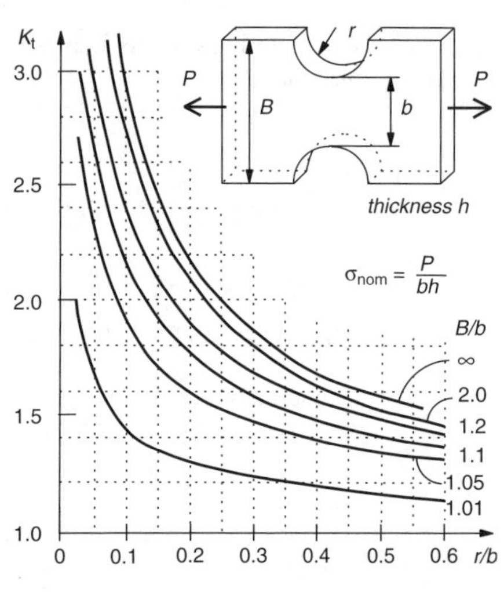

Tension of flat bar with notch

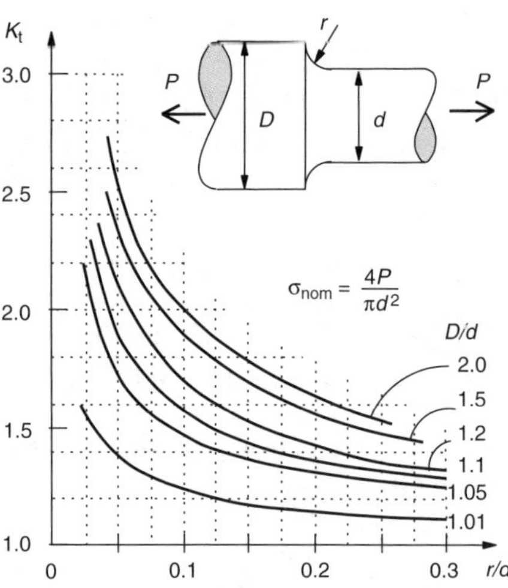

Tension of circular bar with shoulder fillet

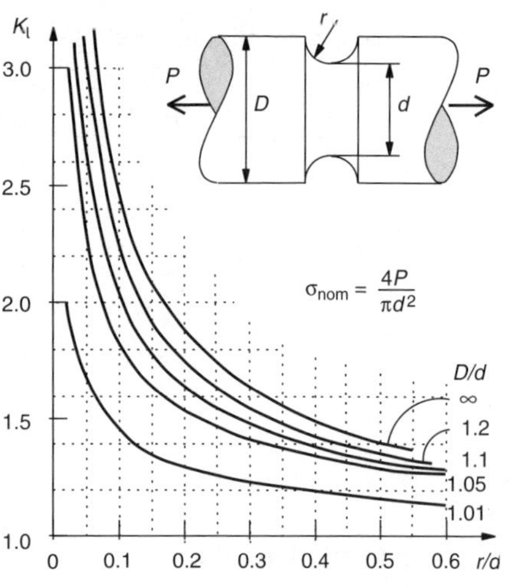

Tension of circular bar with U-shaped groove

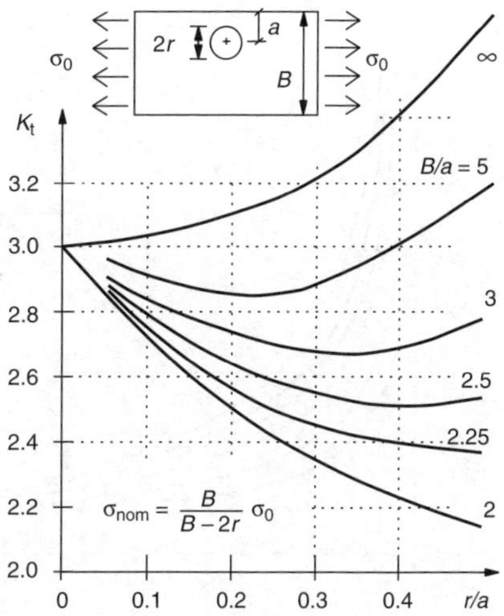

Tension of flat bar with hole

Bending

Maximum normal stress at a stress concentration is

$\sigma_{max} = K_t\sigma_{nom}$, where K_t and σ_{nom} are given in the diagrams

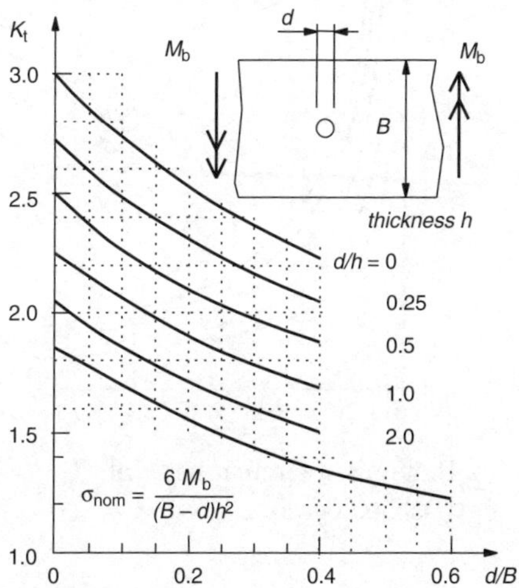

Bending of flat bar with hole

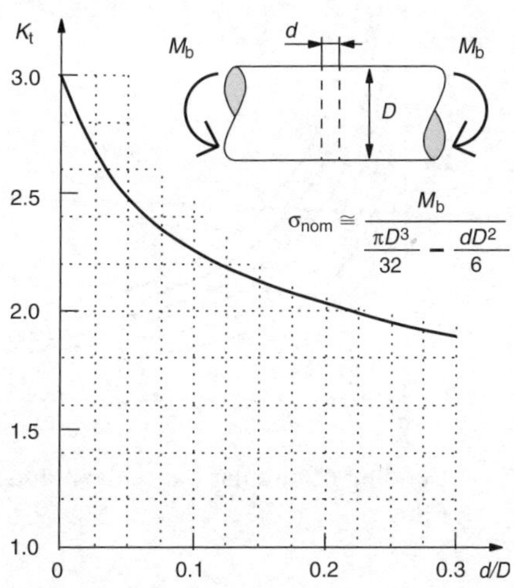

Bending of circular bar with hole

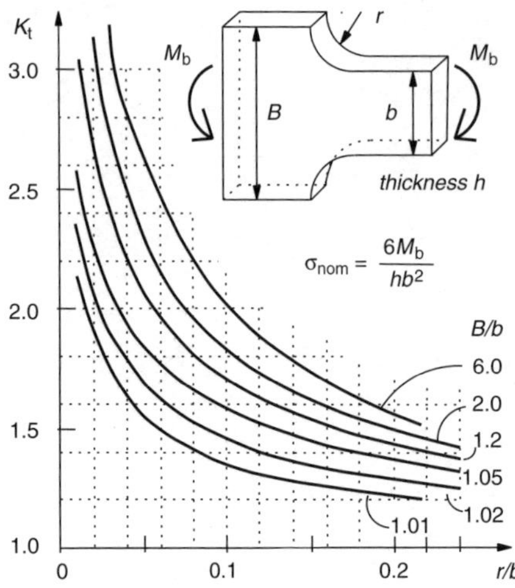

Bending of flat bar with shoulder fillet

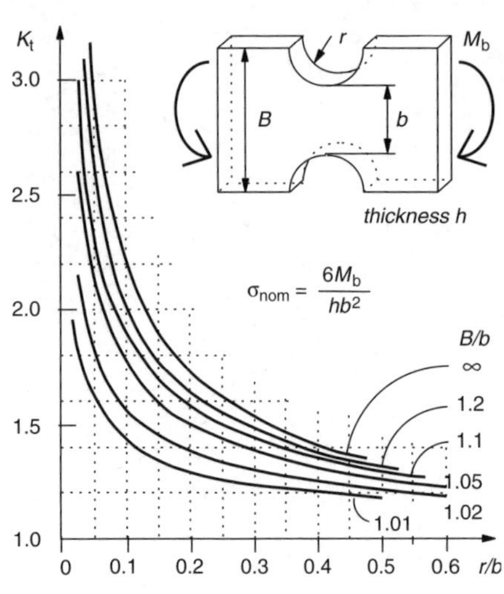

Bending of flat bar with notch

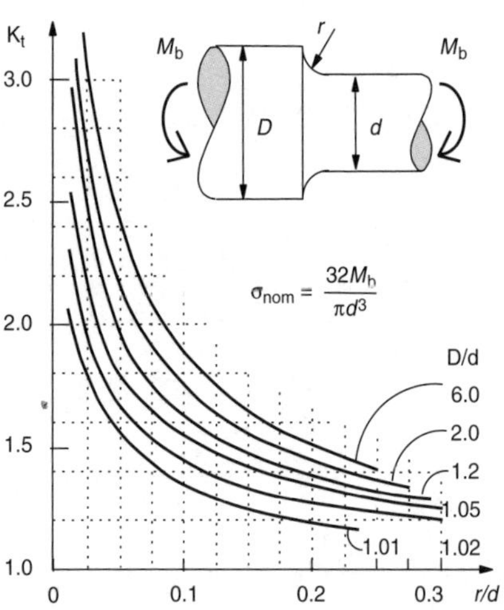

Bending of circular bar with shoulder fillet

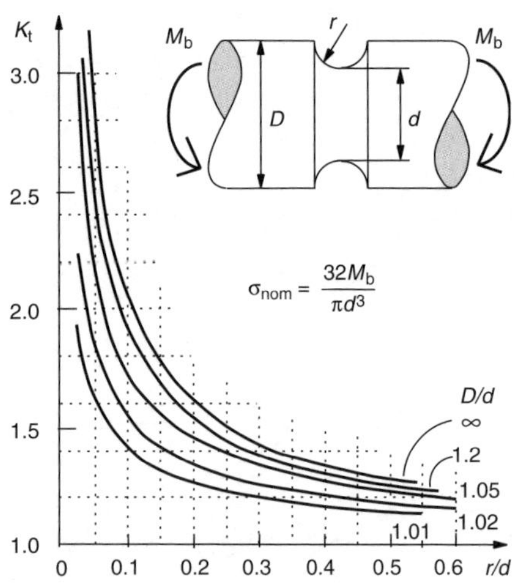

Bending of circular bar with U-shaped groove

Torsion

Maximum shear stress at a stress concentration is

$\tau_{max} = K_t \tau_{nom}$, where K_t and τ_{nom} are given in the diagrams

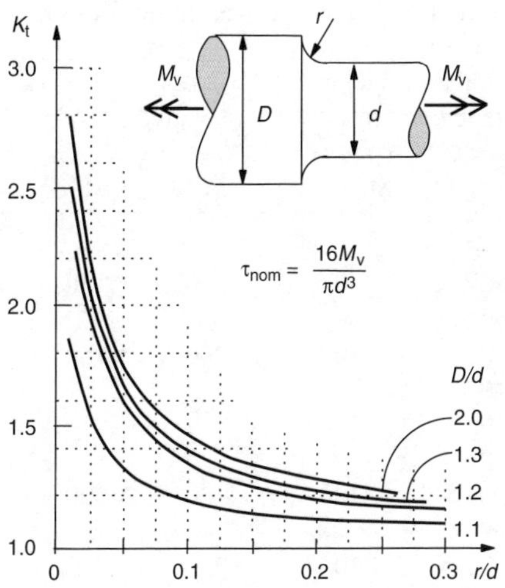

K_t
r
M_v
D
d
$\tau_{nom} = \frac{16M_v}{\pi d^3}$
D/d
2.0
1.3
1.2
1.1
3.0
2.5
2.0
1.5
1.0
0
0.1
0.2
0.3
r/d

Torsion of circular bar with shoulder fillet

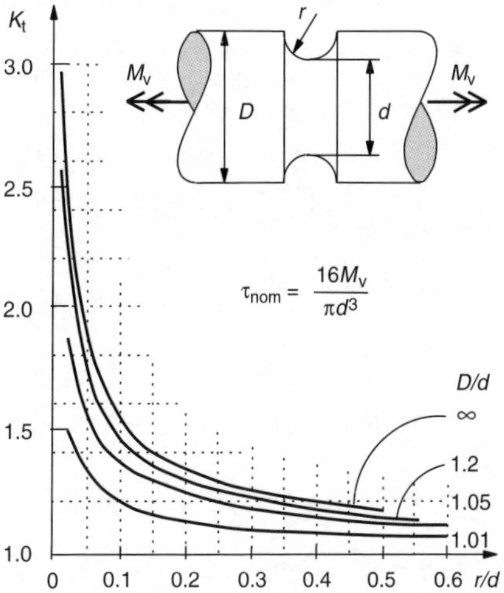

K_t
r
M_v
D
d
$\tau_{nom} = \frac{16M_v}{\pi d^3}$
D/d
∞
1.2
1.05
1.01
3.0
2.5
2.0
1.5
1.0
0
0.1
0.2
0.3
0.4
0.5
0.6
r/d

Torsion of circular bar with notch

K_t
r
d/4
$\frac{7d}{8}$
d
$\tau_{nom} = \frac{16M_v}{\pi d^3}$
4.0
3.5
3.0
2.5
2.0
0
0.05
0.10
r/d

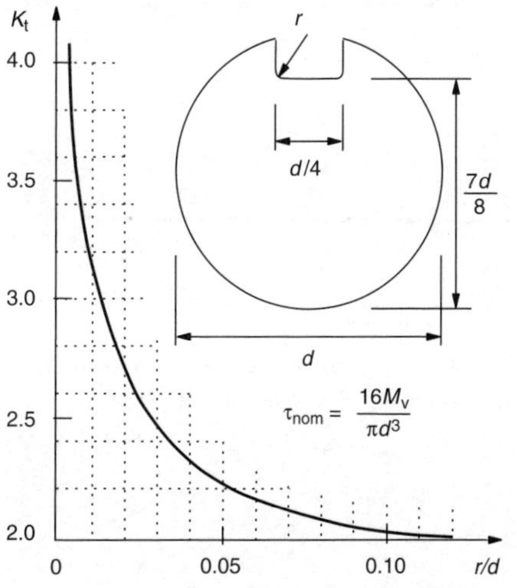

Torsion of bar with longitudinal keyway

K_t
d
M_v
D
$\tau_{nom} \cong \frac{M_v}{\frac{\pi D^3}{16} - \frac{dD^2}{6}}$
2.0
1.5
1.0
0
0.1
0.2
0.3
d/D

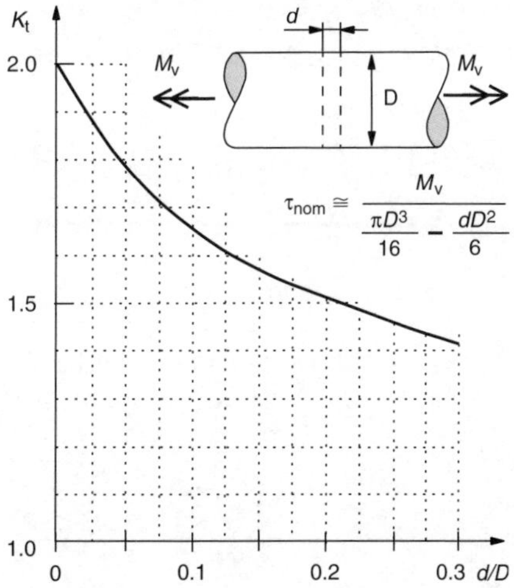

Torsion of circular bar with hole

12.9 Material data

The following material properties may be used *only* when solving exercise problems. For a real design, data should be taken from latest official standard and *not* from this table (two values for the same material means different qualities).[1]

Material	Young's modulus E GPa	ν –	$\alpha 10^6$ K^{-1}	Ultimate strength MPa	Yield limit tension/ compression MPa	bending MPa	torsion MPa
Carbon steel							
141312-00	206	0.3	12	360 460	>240	260	140
141450-1	205	0.3		430 510	>250	290	160
141510-00	205	0.3		510 640	>320		
141550-01	205	0.3		490 590	>270	360	190
141650-01	206	0.3	11	590 690	>310	390	220
141650	206	0.3		860	>550	610	
					Offset yield strength $R_{p0.2}$ ($\sigma_{0.2}$)		
Stainless steel							
2337-02	196	0.29	16.8	>490	>200		
Aluminium							
SS 4120-02	70		23	170 215	>65		
SS 4120-24	70		23	220 270	>170		
SS 4425-06	70		23	>340	>270		

[1] Data in this table has been collected from B Sundström (editor): Handbok och Formelsamling i Hållfasthetslära, Institutionen för hållfasthetslära, KTH, Stockholm, 1998.

M

Mathematical Formulae

1 Mathematical Constants
2 Algebra
3 Geometric Formulae
4 Trigonometric Identities
5 Derivatives
6 Integrals
7 Taylor Series
8 Special Polynomials and Associated Functions
9 Vector Analysis
10 Special Coordinate Systems
11 Laplace Transforms
12 Fourier Series
13 Fourier Transforms
14 Differential Equations
15 Numerical Methods
16 Error Estimation

1 Mathematical Constants

π = 3.14159 26535 89793 23846 26433 83279 50288 41971 69 …

e = 2.71828 18284 59045 23536 02874 71352 66249 77572 47 …

$$= \lim_{n\to\infty}\left(1+\frac{1}{n}\right)^n$$

γ = 0.57721 56649 … = Euler's constant =

$$= \lim_{n\to\infty}\left(1+\frac{1}{2}+\frac{1}{3}+\ldots+\frac{1}{n}-\ln n\right)$$

ϕ = 1.61803 39887 49894 8482… = Golden ratio = $(1+\sqrt{5})/2$

Values for some commonly used expressions

$\sqrt{2}$ = 1.4142	lg 2 = 0.3010	ln 2 = 0.6931
$\sqrt{3}$ = 1.7320	lg 3 = 0.4771	ln 3 = 1.0986
$\sqrt{5}$ = 2.2361	lg 5 = 0.6990	ln 5 = 1.6094
$\sqrt{10}$ = 3.1623	lg 10 = 1	ln 10 = 2.3026
$\sqrt{\pi}$ = 1.7725	lg π = 0.4971	ln π = 1.1447
$\sqrt{e}$ = 1.6487	lg e = 0.4343	ln e = 1
$\sin\frac{\pi}{6} = \cos\frac{\pi}{3} = \frac{1}{2} = 0.5$		$\sqrt[3]{2}$ = 1.2599
$\sin\frac{\pi}{4} = \cos\frac{\pi}{4} = \frac{1}{\sqrt{2}} = 0.7071$		$\sqrt[3]{3}$ = 1.4422
$\sin\frac{\pi}{3} = \cos\frac{\pi}{6} = \frac{\sqrt{3}}{2} = 0.8660$		$\sqrt[3]{10}$ = 2.1544

Feigenbaum numbers for the onset of chaos

α = 2.50290 7875 …

δ = 4.66920 1609 …

More information:
http://dgleahy.com/dgl/p15.html

Some exact values of trigonometric functions

degrees	radians	sin	cos	tan	cot
0	0	0	1	0	$\mp\infty$
15	$\frac{\pi}{12} \approx 0.2618$	$\frac{1}{4}(\sqrt{6}-\sqrt{2})$	$\frac{1}{4}(\sqrt{6}+\sqrt{2})$	$2-\sqrt{3}$	$2+\sqrt{3}$
30	$\frac{\pi}{6} \approx 0.5236$	$\frac{1}{2}$	$\frac{\sqrt{3}}{2}$	$\frac{1}{\sqrt{3}}$	$\sqrt{3}$
45	$\frac{\pi}{4} \approx 0.7854$	$\frac{1}{\sqrt{2}}$	$\frac{1}{\sqrt{2}}$	1	1
60	$\frac{\pi}{3} \approx 1.0472$	$\frac{\sqrt{3}}{2}$	$\frac{1}{2}$	$\sqrt{3}$	$\frac{1}{\sqrt{3}}$
75	$\frac{5\pi}{12} \approx 1.3090$	$\frac{1}{4}(\sqrt{6}+\sqrt{2})$	$\frac{1}{4}(\sqrt{6}-\sqrt{2})$	$2+\sqrt{3}$	$2-\sqrt{3}$
90	$\frac{\pi}{2} \approx 1.5708$	1	0	$\pm\infty$	0
105	$\frac{7\pi}{12} \approx 1.8326$	$\frac{1}{4}(\sqrt{6}+\sqrt{2})$	$-\frac{1}{4}(\sqrt{6}-\sqrt{2})$	$-(2+\sqrt{3})$	$\sqrt{3}-2$
120	$\frac{2\pi}{3} \approx 2.0944$	$\frac{\sqrt{3}}{2}$	$-\frac{1}{2}$	$-\sqrt{3}$	$-\frac{1}{\sqrt{3}}$
135	$\frac{3\pi}{4} \approx 2.3562$	$\frac{1}{\sqrt{2}}$	$-\frac{1}{\sqrt{2}}$	-1	-1
150	$\frac{5\pi}{6} \approx 2.6180$	$\frac{1}{2}$	$-\frac{\sqrt{3}}{2}$	$-\frac{1}{\sqrt{3}}$	$-\sqrt{3}$
165	$\frac{11\pi}{12} \approx 2.8798$	$\frac{1}{4}(\sqrt{6}-\sqrt{2})$	$-\frac{1}{4}(\sqrt{6}+\sqrt{2})$	$\sqrt{3}-2$	$-(2+\sqrt{3})$
180	$\pi \approx 3.1416$	0	-1	0	$\mp\infty$
195	$\frac{13\pi}{12} \approx 3.4034$	$-\frac{1}{4}(\sqrt{6}-\sqrt{2})$	$-\frac{1}{4}(\sqrt{6}+\sqrt{2})$	$2-\sqrt{3}$	$2+\sqrt{3}$
210	$\frac{7\pi}{6} \approx 3.6652$	$-\frac{1}{2}$	$-\frac{\sqrt{3}}{2}$	$\frac{1}{\sqrt{3}}$	$\sqrt{3}$
225	$\frac{5\pi}{4} \approx 3.9270$	$-\frac{1}{\sqrt{2}}$	$-\frac{1}{\sqrt{2}}$	1	1
240	$\frac{4\pi}{3} \approx 4.1888$	$-\frac{\sqrt{3}}{2}$	$-\frac{1}{2}$	$\sqrt{3}$	$\frac{1}{\sqrt{3}}$
255	$\frac{17\pi}{12} \approx 4.4506$	$-\frac{1}{4}(\sqrt{6}+\sqrt{2})$	$-\frac{1}{4}(\sqrt{6}-\sqrt{2})$	$2+\sqrt{3}$	$2-\sqrt{3}$
270	$\frac{3\pi}{2} \approx 4.7124$	-1	0	$\pm\infty$	0
285	$\frac{19\pi}{12} \approx 4.9742$	$-\frac{1}{4}(\sqrt{6}+\sqrt{2})$	$\frac{1}{4}(\sqrt{6}-\sqrt{2})$	$-(2+\sqrt{3})$	$\sqrt{3}-2$
300	$\frac{5\pi}{3} \approx 5.2360$	$-\frac{\sqrt{3}}{2}$	$\frac{1}{2}$	$-\sqrt{3}$	$-\frac{1}{\sqrt{3}}$
315	$\frac{7\pi}{4} \approx 5.4978$	$-\frac{1}{\sqrt{2}}$	$\frac{1}{\sqrt{2}}$	-1	-1
330	$\frac{11\pi}{6} \approx 5.7596$	$-\frac{1}{2}$	$\frac{\sqrt{3}}{2}$	$-\frac{1}{\sqrt{3}}$	$-\sqrt{3}$
345	$\frac{23\pi}{12} \approx 6.0214$	$-\frac{1}{4}(\sqrt{6}-\sqrt{2})$	$\frac{1}{4}(\sqrt{6}+\sqrt{2})$	$\sqrt{3}-2$	$-(2+\sqrt{3})$
360	$2\pi \approx 6.2832$	0	1	0	$\mp\infty$

2 Algebra

Some identities

$$(a+b)^n = \sum_{k=0}^{n} \binom{n}{k} a^{n-k} b^k \quad \text{where } \binom{n}{k} = \frac{n!}{k!(n-k)!} \text{ (binomial coefficients)}$$

$$a^2 - b^2 = (a-b)(a+b)$$

$$a^3 + b^3 = (a+b)(a^2 - ab + b^2)$$

$$a^3 - b^3 = (a-b)(a^2 + ab + b^2)$$

Second degree equation

$$ax^2 + bx + c = 0 \qquad x^2 + px + q = 0$$

$$x = \frac{-b \pm \sqrt{b^2 - 4ac}}{2a} \qquad x = -\frac{p}{2} \pm \sqrt{\left(\frac{p}{2}\right)^2 - q}$$

$$x_1 + x_2 = -p \qquad x_1 \cdot x_2 = q$$

Arithmetic series

$$t_i = t_1 + (i-1)\,d \qquad s_n = \sum_{i=1}^{n} t_i = n\,\frac{t_1 + t_n}{2}$$

Geometric series

$$t_i = t_1\,k^{i-1} \qquad s_n = t_1 \sum_{i=0}^{n-1} k^i = \frac{t_1(1-k^n)}{1-k}$$

Logarithms

$$\log_a x = \frac{\log_c x}{\log_c a} = \frac{\ln x}{\ln a}$$

$$\log xy = \log x + \log y$$

$$\log (x/y) = \log x - \log y$$

$$\log x^n = n \log x$$

Factorial and semifactorials

$$n! = 1 \cdot 2 \cdot 3 \cdot \ldots \cdot n \qquad 0! = 1$$

$$(2n-1)!! = 1 \cdot 3 \cdot 5 \cdot \ldots \cdot (2n-1)$$

$$(2n)!! = 2 \cdot 4 \cdot 6 \cdot \ldots \cdot 2n$$

Stirling's formula for $n \gg 1$

$$n! = n\,(n-1)(n-2)\ldots 3\cdot 2\cdot 1 \approx \sqrt{2\pi n}\left(\frac{n}{\mathrm{e}}\right)^n$$

$$\ln n! \approx n \ln n - n + \frac{1}{2}\ln 2\pi n$$

Compound interest

Compound amount at the end of n years when a principal P is deposited

$$A = P(1+r)^n \qquad r = \text{interest rate (in decimals) compounded annually}$$

Accumulated amount at the end of n years when a principal P is deposited at the end of each year (amount of an annuity)

$$A = P\frac{(1+r)^n - 1}{r}$$

Present amount of an annuity in which the yearly payment at the end of each of n years is P

$$A = P\frac{1-(1+r)^{-n}}{r}$$

Complex numbers

$$z = x + \mathrm{i}y = r(\cos\varphi + \mathrm{i}\sin\varphi) = r\,\mathrm{e}^{\mathrm{i}\varphi}$$

$$\arg z = \varphi + n\,2\pi = \arctan\frac{y}{x}$$

y imaginary axis

z

$r = |z|$

φ

x

real axis

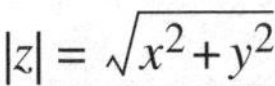

$$|z| = \sqrt{x^2+y^2}$$

$$|z\,w| = |z|\cdot|w| \qquad \arg(z\,w) = \arg z + \arg w \qquad \text{See also Sec. M–4}$$

$$|z/w| = |z|/|w| \qquad \arg(z/w) = \arg z - \arg w$$

Complex conjugate

$$z^* = x - \mathrm{i}y$$

de Moivre formula

$$z^n = r^n(\cos n\varphi + \mathrm{i}\sin n\varphi)$$

Cauchy-Schwarz inequality

$$|a_1b_1 + a_2b_2 + \ldots + a_nb_n|^2 \le (|a_1|^2 + |a_2|^2 + \ldots + |a_n|^2)(|b_1|^2 + |b_2|^2 + \ldots + |b_n|^2)$$

The equality holds if and only if $a_1/b_1 = a_2/b_2 = \ldots = a_n/b_n$

Prime number factorization of odd numbers

	0	**100**	**200**	**300**	**400**	**500**	**600**	**700**	**800**	**900**
1	–	–	3·67	7·43	–	3·167	–	–	3^2·89	17·53
3	–	–	7·29	3·101	13·31	–	32·67	19·37	11·73	3·7·43
5	–	3·5·7	5·41	5·61	3^4·5	5·101	5·11^2	3·5·47	5·7·23	5·181
7	–	–	3^2·23	–	11·37	3·13^2	–	7·101	3·269	–
9	3^2	–	11·19	3·103	–	–	3·7·29	–	–	3^2·101
11	–	3·37	–	–	3·137	7·73	13·47	3^2·79	–	–
13	–	–	3·71	–	7·59	3^3·19	–	23·31	3·271	11·83
15	3·5	5·23	5·43	3^2·5·7	5·83	5·103	3·5·41	5·11·13	5·163	3·5·61
17	–	3^2·13	7·31	–	3·139	11·47	–	3·239	19·43	7·131
19	–	7·17	3·73	11·29	–	3·173	–	–	3^2·7·13	–
21	3·7	11^2	13·17	3·107	–	–	3^3·23	7·103	–	3·307
23	–	3·41	–	17·19	3^2·47	–	7·89	3·241	–	13·71
25	5^2	5^3	3^2·5^2	5^2·13	5^2·17	3·5^2·7	5^4	5^2·29	3·5^2·11	5^2·37
27	3^3	–	–	3·109	7·61	17·31	3·11·19	–	–	3^2·103
29	–	3 · 43	–	7·47	3·11·13	23^2	17·37	3^6	–	–
31	–	–	3·7·11	–	–	3^2·59	–	17·43	3·277	7^2·19
33	3·11	7·19	–	3^2·37	–	13·41	3·211	–	7^2·17	3·311
35	5·7	3^3·5	5·47	5·67	3·5·29	5·107	5·127	3·5·7^2	5·167	5·11·17
37	–	–	3·79	–	19·23	3·179	7^2·13	11·67	3^3·31	–
39	3·13	–	–	3·113	–	7^2·11	3^2·71	–	–	3·313
41	–	3·47	–	11·31	3^2·7^2	–	–	3·13·19	29^2	–
43	–	11·13	3^5	7^3	–	3·181	–	–	3·281	23·41
45	3^2·5	5 · 29	5·7^2	3·5·23	5·89	5·109	3·5·43	5·149	5·13^2	3^3·5·7
47	–	3·7^2	13·19	–	3·149	–	–	3^2·83	7·11^2	–
49	7^2	–	3·83	–	–	3^2·61	11·59	7·107	3·283	13·73
51	3·17	–	–	3^3·13	11·41	19·29	3·7·31	–	23·37	3·317
53	–	3^2·17	11·23	–	3·151	7·79	–	3·251	–	–
55	5·11	5·31	3·5·17	5·71	5·7·13	3·5·37	5·131	5·151	5·3^2·19	5·191
57	3·19	–	–	3·7·17	–	–	3^2·73	–	–	3·11·29
59	–	3·53	7·37	–	3^3·17	13·43	–	3·11·23	–	7·137
61	–	7·23	3^2·29	19^2	–	3·11·17	–	–	3·7·41	31^2
63	3^2·7	–	–	3·121	–	–	3·13·17	7·109	–	3^2·107
65	5·13	3·5·11	5·53	5·73	3·5·31	5·113	5·7·19	3^2·5·17	5·173	5·193
67	–	–	3·89	–	–	3^4·7	23·29	13·59	3·17^2	–
69	3·23	13^2	–	3^2·41	7·67	–	3·223	–	11·79	3·17·19
71	–	3^2·19	–	7·53	3·157	–	11·61	3·257	13·67	–
73	–	–	3·7·13	–	11·43	3·191	–	–	3^2·97	7·139
75	3·5^2	5^2·7	5^2·11	3·5^3	5^2·19	5^2·23	3^3·5^2	5^2·31	53·7	3·5^2·13
77	7·11	3·59	–	13·29	3^2·53	–	–	3·7·37	–	–
79	–	–	3^2·31	–	–	3·193	7·97	19·41	3·293	11·89
81	3^4	–	–	3·127	13·37	7·83	3·227	11·71	–	3^2·109
83	–	3·61	–	–	3·7·23	11·53	–	3^3·29	–	–
85	5·17	5·37	3·5·19	5·7 · 11	5·97	3^2·5·13	5·137	5·157	3·5·59	5·197
87	3·29	11·71	7·41	3^2·43	–	–	3·229	–	–	3·7·47
89	–	3^3·7	17^2	–	3·163	19·31	13·53	3·263	7·127	23·43
91	7·13	–	3·97	17·23	–	3·197	–	7·113	3^4·11	–
93	3·31	–	–	3·131	17·29	–	3^2·7·11	13·61	19·47	3·331
95	5·19	3·5·13	5·59	5·79	3^2·5·11	5·7·17	5·139	3·5·53	5·179	5·199
97	–	–	3^3·11	–	7·71	3·199	17·41	–	3·13·23	–
99	3^2·11	–	13·23	3·7·19	–	–	3·233	17·47	29·31	3^3·37
	0	**100**	**200**	**300**	**400**	**500**	**600**	**700**	**800**	**900**

3 Geometric Formulae

A = area
P = perimeter
V = volume
S = surface area

Triangle

$$A + B + C = 180°$$

$$A = \frac{1}{2}\, bh = \frac{1}{2}\, ab \sin C = \sqrt{s(s-a)(s-b)(s-c)}$$

$$\text{where } s = \frac{1}{2}\,(a + b + c) = \frac{1}{2}\, P$$

$a^2 = b^2 + c^2 - 2\, bc \cos A$ (law of cosines)

$\dfrac{a}{\sin A} = \dfrac{b}{\sin B} = \dfrac{c}{\sin C}$ (law of sines)

Right triangle, Theorem of Pythagoras

$$c^2 = a^2 + b^2$$

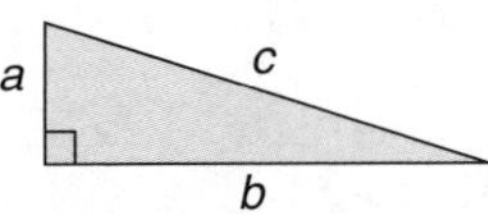

Parallelogram

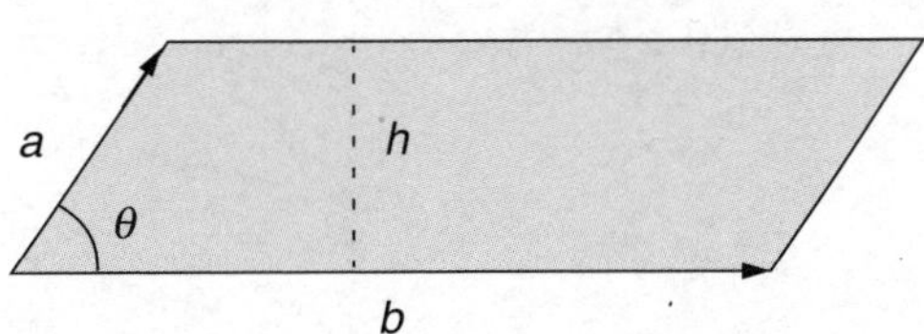

$$A = bh = ab \sin \theta = |\boldsymbol{a} \times \boldsymbol{b}|$$

Trapezoid

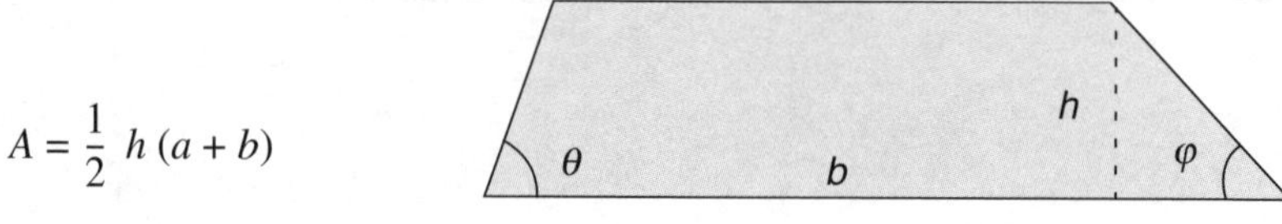

$$A = \frac{1}{2}\, h\,(a + b)$$

$$P = a + b + h\left(\frac{1}{\sin\theta} + \frac{1}{\sin\varphi}\right)$$

Circle

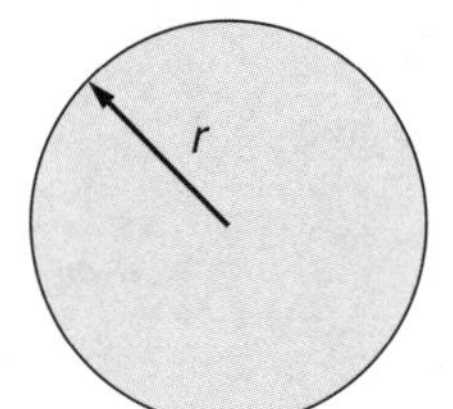

$$A = \pi r^2$$

$$P = 2\pi r$$

Equation in rectangular coordinates

$$(x - x_0)^2 + (y - y_0)^2 = r^2$$

Sector of circle

$$A = \frac{1}{2}\, sr = \frac{1}{2}\, r^2\,\theta$$

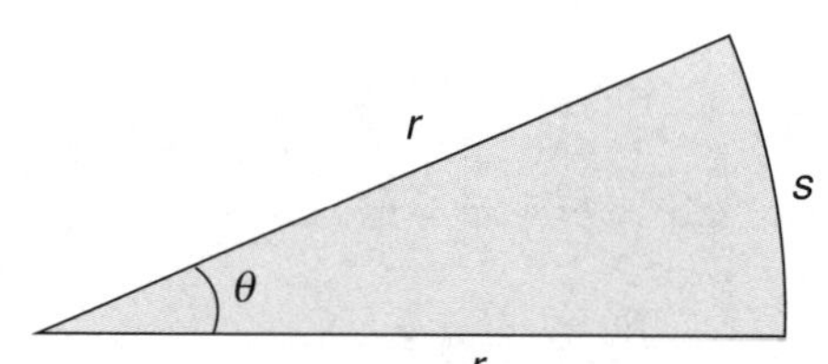

Segment of circle

$$A = \frac{1}{2}\, r^2\,(\theta - \sin\theta)$$

$$a^2 = d(2r - d)$$

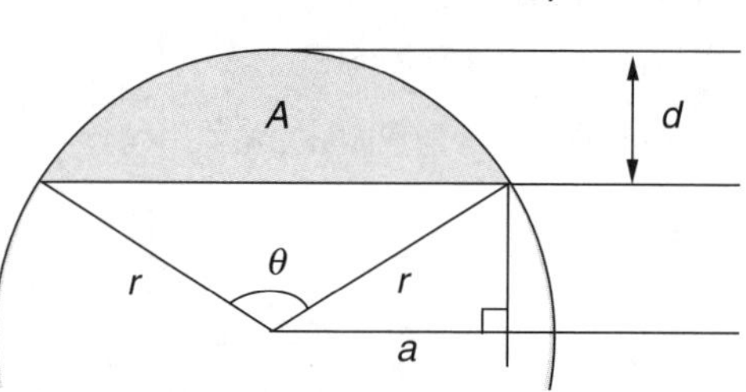

Theorem of chords

$$XA \cdot XB = XC \cdot XD$$

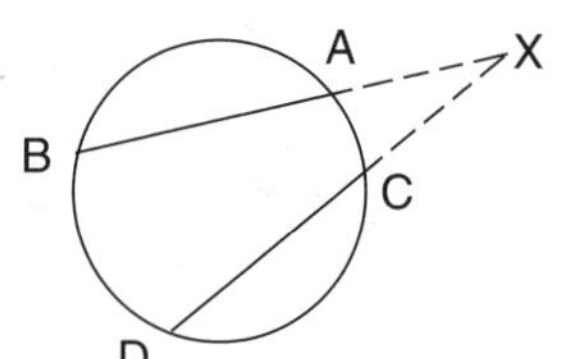

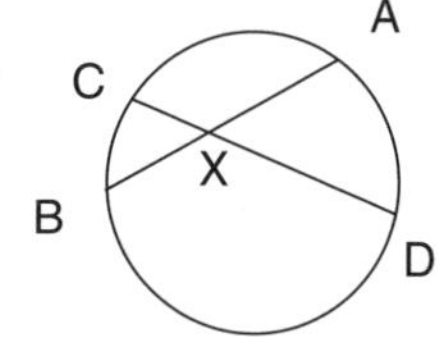

Ellipse

$$A = \pi ab$$

$$P \approx 2\pi \sqrt{\frac{1}{2}(a^2+b^2)}$$

$$e = d/a$$

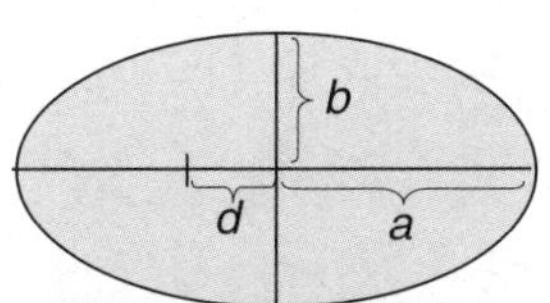

e = excentricity
d = distance from centre to foci

Equation in rectangular coordinates

$$\frac{(x-x_0)^2}{a^2}+\frac{(y-y_0)^2}{b^2}=1$$

Segment of a parabola

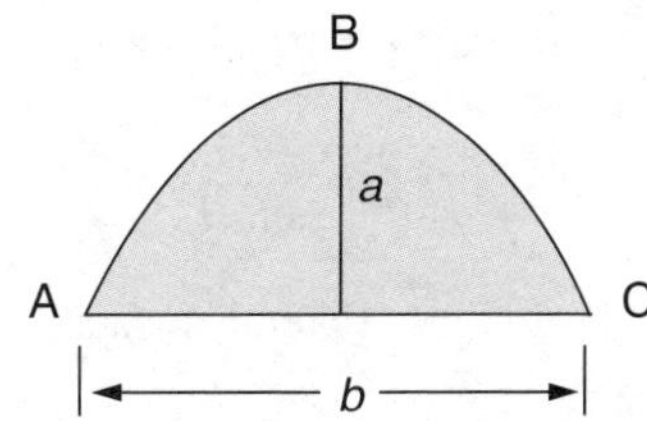

$$A = \frac{2}{3}\, ab$$

Length of arc ABC =

$$= \frac{1}{2}\sqrt{b^2+16a^2}+\frac{b^2}{8a}\ln\left(\frac{4a+\sqrt{b^2+16a^2}}{b}\right)$$

Equation in rectangular coordinates

$$(y-y_0) = \frac{1}{4d}\,(x-x_0)^2$$

vertex at (x_0, y_0) and
focus at $(x_0, y_0 + d)$

Parallelepiped

$$V = Ah = abc \sin\theta$$

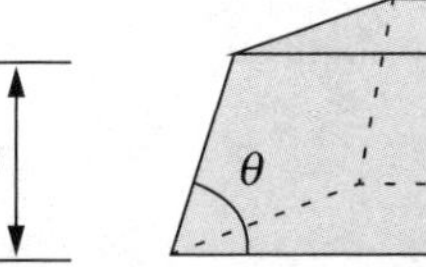

Sphere

$$V = \frac{4}{3}\,\pi r^3$$

$$S = 4\pi r^2$$

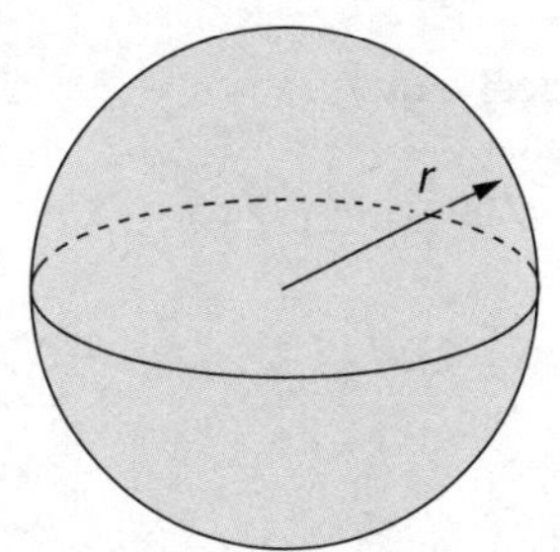

Equation in rectangular coordinates

$$(x-x_0)^2+(y-y_0)^2+(z-z_0)^2 = r^2$$

Shortest distance on the Earth's surface between two points P_1 *and* P_2 *(Great circle distance)*

$$d = \frac{2\pi\Omega}{360} R \qquad \cos\Omega = \sin v_1 \sin v_2 + \cos v_1 \cos v_2 \cos(u_1 - u_2)$$

Ω = angle (in degrees) between points P_1 and P_2 measured at the Earth's centre

u_i = degree of longitude (E – W) of point P_i

v_i = degree of latitude (N – S) of point P_i

R = radius of the Earth

Circular cylinder

$$V = \pi r^2 h = \pi r^2 \ell \sin\theta$$

$$S = 2\pi r \ell = \frac{2\pi r h}{\sin\theta}$$

(lateral surface)

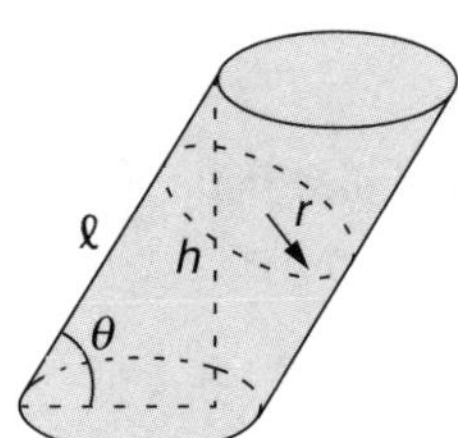

Circular cone

$$V = \frac{1}{3} \pi r^2 h$$

$$S = \pi r \sqrt{r^2 + h^2} = \pi r \ell$$

(lateral surface)

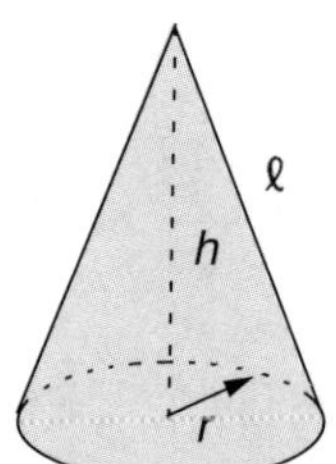

Pyramid

$$V = \frac{1}{3} Ah$$

Height of a regular tetrahedron

$$h = a\sqrt{\frac{2}{3}}$$

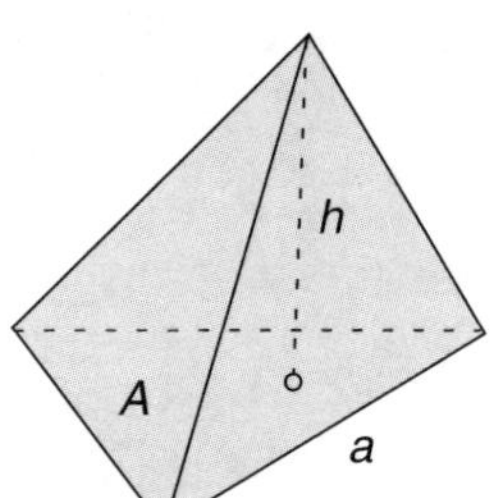

Spherical cap

$$V = \frac{1}{3} \pi h^2 (3r - h)$$

$$S = 2\pi r h$$

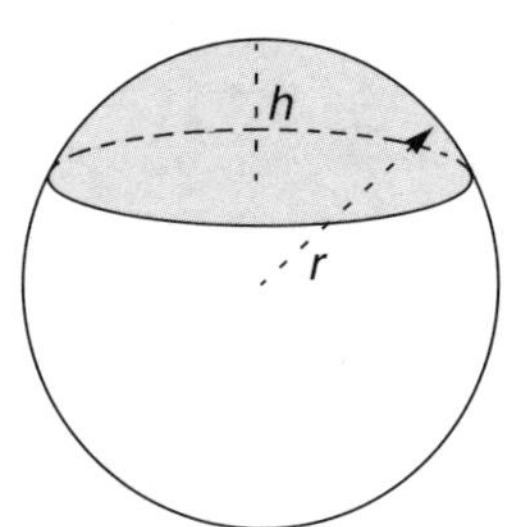

Ellipsoid

$$V = \frac{4}{3}\pi a\, b\, c$$

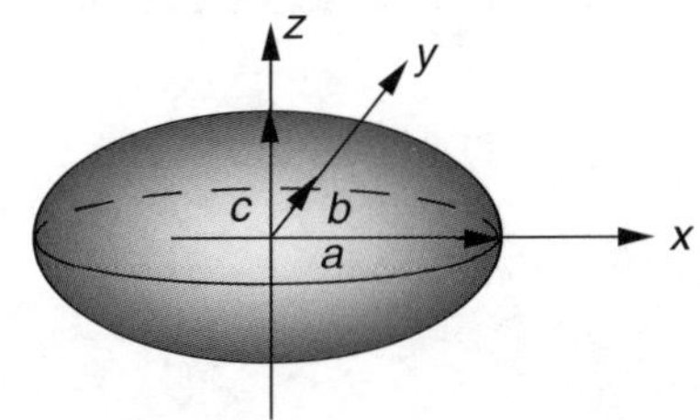

Equation in rectangular coordinates

$$\frac{(x-x_0)^2}{a^2}+\frac{(y-y_0)^2}{b^2}+\frac{(z-z_0)^2}{c^2}=1$$

Elliptic paraboloid

$$V = \frac{1}{2}\pi a\, b\, h$$

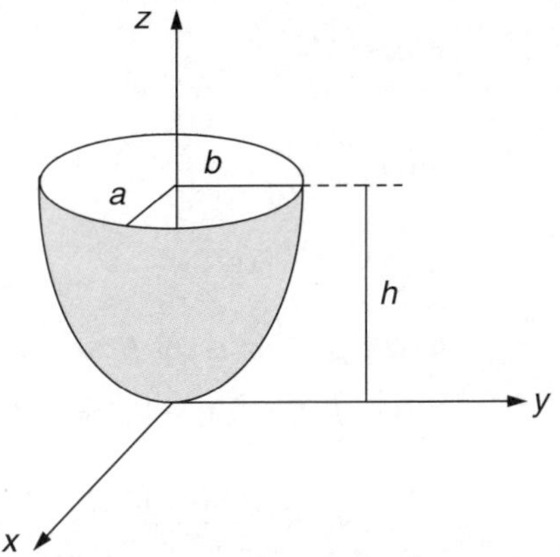

Equation in rectangular coordinates

$$\frac{x^2}{a^2}+\frac{y^2}{b^2}=\frac{z}{h}$$

Torus

$$V = \frac{1}{4}\pi^2 (r_1 + r_2)(r_1 - r_2)^2$$

$$S = \pi^2 (r_1^{\,2} - r_2^{\,2})$$

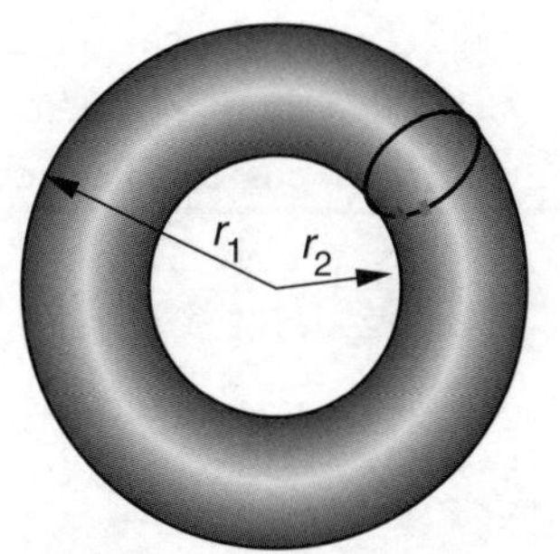

Pappus's centroid theorem (Guldin's rules)

Area of rotational surface= (length of generating curve) ·
· (distance traversed by its CM)

Volume of rotational body= (area of generating surface) ·
· (distance traversed by its CM).

4 Trigonometric Identities

Basic identities

$$\tan\alpha = \frac{\sin\alpha}{\cos\alpha} \qquad \sec\alpha = \frac{1}{\cos\alpha}$$

$$\cot\alpha = \frac{\cos\alpha}{\sin\alpha} = \frac{1}{\tan\alpha} \qquad \csc\alpha = \frac{1}{\sin\alpha}$$

$$\sin^2\alpha + \cos^2\alpha = 1$$

$$\sin(\alpha \pm \beta) = \sin\alpha\cos\beta \pm \cos\alpha\sin\beta$$

$$\cos(\alpha \pm \beta) = \cos\alpha\cos\beta \mp \sin\alpha\sin\beta$$

$$\tan(\alpha \pm \beta) = \frac{\tan\alpha \pm \tan\beta}{1 \mp \tan\alpha\tan\beta}$$

$$\sin 2\alpha = 2\sin\alpha\cos\alpha$$

$$\cos 2\alpha = \cos^2\alpha - \sin^2\alpha = 2\cos^2\alpha - 1 = 1 - 2\sin^2\alpha$$

$$\sin 3\alpha = 3\sin\alpha - 4\sin^3\alpha$$

$$\cos 3\alpha = 4\cos^3\alpha - 3\cos\alpha$$

$$\sin^2\frac{\alpha}{2} = \frac{1}{2}(1 - \cos\alpha) \qquad \tan\frac{\alpha}{2} = \frac{\sin\alpha}{1+\cos\alpha} = \pm\sqrt{\frac{1-\cos\alpha}{1+\cos\alpha}}$$

$$\cos^2\frac{\alpha}{2} = \frac{1}{2}(1 + \cos\alpha) \qquad \tan 2\alpha = \frac{2\tan\alpha}{1-\tan^2\alpha}$$

$$a\sin\alpha + b\cos\alpha = m\sin(\alpha + \varphi) \qquad m = \sqrt{a^2+b^2}$$

$$\tan\varphi = \frac{b}{a}$$

$$\sin\alpha + \sin\beta = 2\sin\frac{1}{2}(\alpha+\beta)\cos\frac{1}{2}(\alpha-\beta)$$

$$\sin\alpha - \sin\beta = 2\cos\frac{1}{2}(\alpha+\beta)\sin\frac{1}{2}(\alpha-\beta)$$

$$\cos\alpha + \cos\beta = 2\cos\frac{1}{2}(\alpha+\beta)\cos\frac{1}{2}(\alpha-\beta)$$

$$\cos\alpha - \cos\beta = -2\sin\frac{1}{2}(\alpha+\beta)\sin\frac{1}{2}(\alpha-\beta)$$

Euler's identities

$$\sin\alpha = \frac{1}{2\mathrm{i}}(\mathrm{e}^{\mathrm{i}\alpha} - \mathrm{e}^{-\mathrm{i}\alpha}) \qquad \mathrm{i} = \sqrt{-1}$$

$$\cos\alpha = \frac{1}{2}(\mathrm{e}^{\mathrm{i}\alpha} + \mathrm{e}^{-\mathrm{i}\alpha}) \qquad \mathrm{e}^{\mathrm{i}\alpha} = \cos\alpha + \mathrm{i}\sin\alpha$$

Complex numbers

$$x + \mathrm{i}\,y = r(\cos\alpha + \mathrm{i}\sin\alpha) \qquad r = \sqrt{x^2+y^2}$$

$$\tan\alpha = \frac{y}{x}$$

See also Sec. M – 2

Hyperbolic functions

$$\sinh x = \frac{1}{2}(\mathrm{e}^{x} - \mathrm{e}^{-x}) \qquad \tanh x = \frac{\sinh x}{\cosh x}$$

$$\cosh x = \frac{1}{2}(\mathrm{e}^{x} + \mathrm{e}^{-x}) \qquad \coth x = \frac{\cosh x}{\sinh x}$$

5 Derivatives

General rules

$$f = u + v \Rightarrow f' = u' + v'$$
$$f = uv \Rightarrow f' = u'v + uv'$$
$$f = \frac{u}{v} \Rightarrow f' = \frac{u'v - uv'}{v^2}$$

$$\frac{dy}{dx} = \frac{dy}{du}\frac{du}{dx} \quad \text{where } y = F(u),\ u = f(x)$$

$$df = \frac{\partial f}{\partial x}\,dx + \frac{\partial f}{\partial y}\,dy + \ldots \quad \text{where } f = f(x, y, \ldots)$$

Logarithmic differentiation

$$f = u^a v^b w^c \ldots$$
$$\ln f = a \ln u + b \ln v + c \ln w + \ldots$$
$$\frac{df}{f} = a\frac{du}{u} + b\frac{dv}{v} + c\frac{dw}{w} + \ldots$$

Lagrange's method

Find maximum and minimum of the function $f(x_1 \ldots x_n)$ when

$$g_1(x_1 \ldots x_n) = 0$$
$$g_2(x_1 \ldots x_n) = 0$$
$$\vdots$$
$$g_m(x_1 \ldots x_n) = 0$$

Solutions are found by solving a system of $n + m$ equations

$$\frac{\partial F}{\partial x_i} = 0 \text{ for } i = 1, 2, \ldots, n \text{ and } g_j = 0 \text{ for } j = 1, 2, \ldots, m$$

where

$$F = f + \lambda_1 g_1 + \lambda_2 g_2 + \ldots + \lambda_m g_m$$

Special derivatives

function $f(x)$		derivative $\dfrac{df}{dx}$
x^n		$n\,x^{n-1}$
$\dfrac{1}{x}$		$-\dfrac{1}{x^2}$
$\sqrt{x}$		$\dfrac{1}{2\sqrt{x}}$
$\ln x$	$(x > 0)$	$\dfrac{1}{x}$
e^x		e^x
$\sin x$		$\cos x$
$\cos x$		$-\sin x$
$\tan x$		$1 + \tan^2 x = \dfrac{1}{\cos^2 x}$
$\cot x$		$-(1 + \cot^2 x) = -\dfrac{1}{\sin^2 x}$
a^x	$(a > 0)$	$a^x \ln a$
$\log_a x$	$(a, x > 0, a \neq 1)$	$\dfrac{1}{x}\log_a e = \dfrac{1}{x \ln a}$
$\arcsin x$		$\dfrac{1}{\sqrt{1-x^2}}$
$\arccos x$		$-\dfrac{1}{\sqrt{1-x^2}}$
$\arctan x$		$\dfrac{1}{1+x^2}$

6 Integrals

General rules

$$\int_a^b (u + v)\,dx = \int_a^b u\,dx + \int_a^b v\,dx$$

$$\int_a^b u\,dx = \int_a^c u\,dx + \int_c^b u\,dx$$

$$\int_a^b u(x)\,dx = \int_{v(a)}^{v(b)} u(v)\,\frac{dx}{dv}\,dv = \int_{v(a)}^{v(b)} u(v)\,\frac{1}{\frac{dv}{dx}}\,dv \qquad v = v(x)$$

$$\int_a^b f(g(x))\,g'(x)\,dx = \int_{g(a)}^{g(b)} f(t)\,dt$$

$$\int_a^b u(x)\,v(x)\,dx = U(b)v(b) - U(a)v(a) - \int_a^b U(x)\,v'(x)\,dx$$

$U(x)$ = primitive function to $u(x)$

$$\int_a^b u\,v^{(n)}\,dx = [uv^{(n-1)} - u'v^{(n-2)} + u''v^{(n-3)} - \ldots]_a^b + (-1)^n \int_a^b v\,u^{(n)}\,dx$$

$$\frac{d}{dx}\int_{g(x)}^{h(x)} f(x, y)\,dy = f(x, h(x))h'(x) - f(x, g(x))g'(x) + \int_{g(x)}^{h(x)} \frac{\partial}{\partial x} f(x, y)\,dy$$

Cauchy-Schwarz inequality

$$\left|\int_a^b f(x)\,g(x)\,dx\right|^2 \leq \left\{\int_a^b |f(x)|^2\,dx\right\}\left\{\int_a^b |g(x)|^2\,dx\right\}$$

Indefinite integrals (constants are omitted)

function	primitive function	
x^n	$\frac{1}{n+1}\,x^{n+1}$	$n \neq -1$

$\dfrac{1}{x-a}$	$\ln\lvert x-a\rvert$	$x \neq a$
$\dfrac{f'(x)}{f(x)}$	$\ln f(x)$	
a^x	$\dfrac{a^x}{\ln a}$	$(a>0,\ a\neq 1)$
$\dfrac{1}{x^2-a^2}$	$\dfrac{1}{2a}\ln\dfrac{\lvert x-a\rvert}{\lvert x+a\rvert}$	$\lvert x\rvert \neq a$
$\dfrac{1}{(ax+b)(px+q)}$	$\dfrac{1}{bp-aq}\ln\dfrac{\lvert px+q\rvert}{\lvert ax+b\rvert}$	
$\dfrac{x}{(ax+b)(px+q)}$	$\dfrac{1}{bp-aq}\left(\dfrac{b}{a}\ln\lvert ax+b\rvert-\dfrac{q}{p}\ln\lvert px+q\rvert\right)$	
$\dfrac{1}{x^2+a^2}$	$\dfrac{1}{a}\arctan\dfrac{x}{a}$	
$\dfrac{x}{x^2+a^2}$	$\dfrac{1}{2}\ln(x^2+a^2)$	
$\sqrt{x^2+a^2}$	$\dfrac{x}{2}\sqrt{x^2+a^2}+\dfrac{a^2}{2}\ln(x+\sqrt{x^2+a^2})$	
$\sqrt{x^2-a^2}$	$\dfrac{x}{2}\sqrt{x^2-a^2}-\dfrac{a^2}{2}\ln\lvert x+\sqrt{x^2-a^2}\rvert$	
$\sqrt{a^2-x^2}$	$\dfrac{x}{2}\sqrt{a^2-x^2}+\dfrac{a^2}{2}\arcsin\dfrac{x}{a}$	
$\dfrac{1}{\sqrt{x^2+a^2}}$	$\ln(x+\sqrt{x^2+a^2})$	
$\dfrac{1}{\sqrt{x^2-a^2}}$	$\ln\lvert x+\sqrt{x^2-a^2}\rvert$	
$\dfrac{1}{\sqrt{a^2-x^2}}$	$\arcsin\dfrac{x}{a}$	
$\dfrac{1}{(x^2+a^2)^{3/2}}$	$\dfrac{x}{a^2\sqrt{x^2+a^2}}$	
$\sin ax$	$-\dfrac{1}{a}\cos ax$	

$\sin^2 ax$	$\frac{1}{2a}\ (ax - \sin ax \cos ax)$
$\frac{1}{\sin ax}$	$\frac{1}{a}\ \ln \lvert\tan \frac{ax}{2}\rvert$
$x \sin ax$	$\frac{\sin ax}{a^2} - \frac{x \cos ax}{a}$
$x^2 \sin ax$	$\frac{2x}{a^2}\ \sin ax + \left(\frac{2}{a^3} - \frac{x^2}{a}\right)\cos ax$
$\cos ax$	$\frac{1}{a}\ \sin ax$
$\cos^2 ax$	$\frac{1}{2a}\ (ax + \sin ax \cos ax)$
$\frac{1}{\cos ax}$	$\frac{1}{a}\ \ln \left\lvert\tan\left(\frac{\pi}{4} + \frac{ax}{2}\right)\right\rvert$
$x \cos ax$	$\frac{\cos ax}{a^2} + \frac{x \sin ax}{a}$
$x^2 \cos ax$	$\frac{2x}{a^2}\ \cos ax + \left(\frac{x^2}{a} - \frac{2}{a^3}\right)\sin ax$
$\frac{1}{\cos^2 ax}$	$\frac{1}{a}\ \tan ax$
$\frac{1}{\sin^2 ax}$	$-\frac{1}{a}\ \cot ax$
$\tan ax$	$-\frac{1}{a}\ \ln \lvert\cos ax\rvert$
$\ln ax$	$x \ln ax - x$
e^{ax}	$\frac{e^{ax}}{a}$
xe^{ax}	$\frac{e^{ax}}{a}\left(x - \frac{1}{a}\right)$
x^2e^{ax}	$\frac{e^{ax}}{a}\left(x^2 - \frac{2x}{a} + \frac{2}{a^2}\right)$
x^ne^{ax}	$\frac{x^ne^{ax}}{a} - \frac{n}{a}\int x^{n-1}e^{ax}\,dx$
	$= \frac{e^{ax}}{a}\left(x^n - \frac{nx^{n-1}}{a} + \frac{n(n-1)x^{n-2}}{a^2} - \ldots\ \frac{(-1)^n n!}{a^n}\right)$ if n = positive integer

Definite integrals ($a > 0$)

$$\int_0^a \sqrt{a^2 - x^2}\,dx = \frac{a^2\pi}{4}$$

$$\int_0^\infty e^{-ax^2}\,dx = \frac{1}{2}\sqrt{\frac{\pi}{a}}$$

$$\int_0^\infty x^n e^{-ax}\,dx = \frac{1}{a^{n+1}}\,\Gamma(n+1)$$

where Γ is the gamma-function $\Gamma(m+1) = m\,\Gamma(m)$, $\Gamma(1) = 1$

$$\Gamma\left(\frac{1}{2}\right) = \sqrt{\pi}$$

$$\int_0^\infty x^m e^{-ax^2}\,dx = I_m = \frac{1}{2a^{(m+1)/2}}\,\Gamma[(m+1)/2] \qquad I_m = \left(\frac{m-1}{2a}\right)I_{m-2}$$

$$\int_0^\infty \frac{x}{e^x - 1}\,dx = 2\int_0^\infty \frac{x}{e^x + 1}\,dx = \frac{\pi^2}{6}$$

$$\int_0^\infty \frac{x^3}{e^x - 1}\,dx = \frac{\pi^4}{15}$$

$$\int_0^\infty e^{-ax}\cos bx\,dx = \frac{a}{a^2 + b^2}$$

$$\int_0^\infty e^{-ax}\sin bx\,dx = \frac{b}{a^2 + b^2}$$

$$\int_0^\pi \sin mx \sin nx\,dx = \begin{cases} 0 & m, n \text{ integers; } m \neq n \\ \pi/2 & m, n \text{ integers; } m = n \end{cases}$$

$$\int_0^\pi \cos mx \cos nx\,dx = \begin{cases} 0 & m, n \text{ integers; } m \neq n \\ \pi/2 & m, n \text{ integers; } m = n \end{cases}$$

$$\int_0^\pi \sin mx \cos nx\,dx = \begin{cases} 0 & m, n \text{ integers; } m + n \text{ even} \\ 2m/(m^2 - n^2) & m, n \text{ integers; } m + n \text{ odd} \end{cases}$$

$$\int_0^{\pi/2} \sin^n x\,dx = \int_0^{\pi/2} \cos^n x\,dx = \begin{cases} \dfrac{(n-1)!!}{n!!} & n \text{ odd} \\[2ex] \dfrac{(n-1)!!}{n!} \cdot \dfrac{\pi}{2} & n \text{ even} \end{cases}$$

$$\int_0^\infty \frac{\sin^2 ax}{x^2}\,\mathrm{d}x = \frac{\pi a}{2}$$

($n!! = n\,(n-2)\,(n-4)\cdot\ldots\cdot n_1$ where $n_1 = 1$ or 2)

$$\int_0^\infty \frac{\sin ax}{x}\,\mathrm{d}x = \frac{\pi}{2}$$

Divergence theorem (Gauss' theorem, Green's theorem)

$$\int_V \nabla\cdot\boldsymbol{A}\,\mathrm{d}V = \int_S \boldsymbol{A}\cdot\mathrm{d}\boldsymbol{S}$$

where S is a closed surface bounding a region of volume V and $\mathrm{d}\boldsymbol{S} = \boldsymbol{e}_\mathrm{N}\,\mathrm{d}S$ where $\boldsymbol{e}_\mathrm{N}$ is the positive (outward) normal.

Stokes' theorem

$$\oint_C \boldsymbol{A}\cdot\mathrm{d}\boldsymbol{r} = \int_S (\nabla\times\boldsymbol{A})\cdot\mathrm{d}\boldsymbol{S}$$

where S is a two-sided surface bounded by a closed non-intersecting curve C (simple closed curve).

Green's theorem in the plane

$$\oint_C (P\,\mathrm{d}x + Q\,\mathrm{d}y) = \iint_R \left(\frac{\partial Q}{\partial x} - \frac{\partial P}{\partial y}\right)\mathrm{d}x\,\mathrm{d}y$$

where R is the area bounded by the simple closed curve C.

Miscellaneous theorems

$$\int_V \nabla\times\boldsymbol{A}\,\mathrm{d}V = \int_S \mathrm{d}\boldsymbol{S}\times\boldsymbol{A}$$

$$\int_C f\,\mathrm{d}r = \int_S \mathrm{d}\boldsymbol{S}\times\nabla f$$

7 Taylor Series

Taylor formula

$$f(x) = f(a) + \frac{1}{1!} f'(a)(x-a) + \frac{1}{2!} f''(a)(x-a)^2 + \frac{1}{3!} f'''(a)(x-a)^3 +$$

$$+ \ldots + \frac{1}{n!} f^{(n)}(\xi)(x-a)^n$$

where $\xi \in [a, x]$

Special cases for $a = 0$ (Maclaurin series)

$$e^x = 1 + \frac{1}{1!} x + \frac{1}{2!} x^2 + \frac{1}{3!} x^3 + \ldots$$

$$\sin x = \frac{1}{1!} x - \frac{1}{3!} x^3 + \frac{1}{5!} x^5 - \ldots$$

$$\cos x = 1 - \frac{1}{2!} x^2 + \frac{1}{4!} x^4 - \ldots$$

$$\tan x = x + \frac{1}{3} x^3 + \frac{2}{15} x^5 + \ldots \qquad |x| < \frac{\pi}{2}$$

$$\cot x = \frac{1}{x} - \frac{1}{3} x - \frac{1}{45} x^3 + \ldots \qquad 0 < |x| < \pi$$

$$\ln(1+x) = x - \frac{1}{2} x^2 + \frac{1}{3} x^3 - \frac{1}{4} x^4 + \ldots \qquad |x| < 1$$

$$(1+x)^a = 1 + \frac{a}{1!} x + \frac{a(a-1)}{2!} x^2 + \ldots \qquad |x| < 1$$

$$\sqrt{1+x} = 1 + \frac{1}{2} x - \frac{1}{8} x^2 + \frac{1}{16} x^3 + \ldots$$

$$\frac{1}{\sqrt{1+x}} = 1 - \frac{1}{2} x + \frac{3}{8} x^2 - \frac{5}{16} x^3 + \ldots$$

$$\arctan x = x - \frac{1}{3}\, x^3 + \frac{1}{5}\, x^5 - \ldots \qquad |x| < 1$$

$$\arcsin x = x + \frac{x^3}{6} + \frac{3}{40}\, x^5 + \ldots \qquad |x| < 1$$

$$\cosh x = 1 + \frac{1}{2!}\, x^2 + \frac{1}{4!}\, x^4 + \ldots$$

$$\sinh x = x + \frac{1}{3!}\, x^3 + \frac{1}{5!}\, x^5 + \ldots$$

8 Special Polynomials and Associated Functions

(See also Sec. M – 12)

Hermite polynomials

$H_0(x) = 1$

$H_1(x) = 2x$

$H_2(x) = 4x^2 - 2$

$H_3(x) = 8x^3 - 12x$

$H_4(x) = 16x^4 - 48x^2 + 12$

...

Legendre polynomials

$P_0(x) = 1$

$P_1(x) = x$

$P_2(x) = \frac{1}{2}(3x^2 - 1)$

$P_3(x) = \frac{1}{2}(5x^3 - 3x)$

...

$P_0(\cos\theta) = 1$

$P_1(\cos\theta) = \cos\theta$

$P_2(\cos\theta) = \frac{1}{4}(1 + 3\cos 2\theta)$

$P_3(\cos\theta) = \frac{1}{8}(3\cos\theta + 5\cos 3\theta)$

...

Associated Legendre functions

$P_1^1(x) = (1 - x^2)^{1/2}$

$P_2^1(x) = 3x(1 - x^2)^{1/2}$

$P_2^2(x) = 3(1 - x^2)$

...

$P_3^1(x) = \frac{3}{2}(5x^2 - 1)(1 - x^2)^{1/2}$

$P_3^2(x) = 15x(1 - x^2)$

$P_3^3(x) = 15(1 - x^2)^{3/2}$

...

Laguerre polynomials

$L_0(x) = 1$

$L_1(x) = -x + 1$

$L_2(x) = x^2 - 4x + 2$

Associated Laguerre polynomials

$L_1^1(x) = -1$

$L_2^1(x) = 2x - 4$

$L_2^2(x) = 2$

...

$L_3^1(x) = -3x^2 + 18x - 18$

$L_3^2(x) = -6x + 18$

$L_3^3(x) = -6$

...

Chebyshev polynomials

$T_0(x) = 1$

$T_1(x) = x$

$T_2(x) = 2x^2 - 1$

$T_3(x) = 4x^3 - 3x$

$T_4(x) = 8x^4 - 8x^2 + 1$

$T_5(x) = 16x^5 - 20x^3 + 5x$

...

$1 = T_0$

$x = T_1$

$x^2 = \frac{1}{2}(T_0 + T_2)$

$x^3 = \frac{1}{4}(3T_1 + T_3)$

$x^4 = \frac{1}{8}(3T_0 + 4T_2 + T_4)$

$x^5 = \frac{1}{16}(10T_1 + 5T_3 + T_5)$

...

Bessel functions of the first kind of order 0 **and** 1

$$J_0(x) = 1 - \frac{x^2}{2^2} + \frac{x^4}{2^2 \cdot 4^2} - \frac{x^6}{2^2 \cdot 4^2 \cdot 6^2} + \dots$$

$$J_1(x) = \frac{x}{2} - \frac{x^3}{2^2 \cdot 4} + \frac{x^5}{2^2 \cdot 4^2 \cdot 6} - \frac{x^7}{2^2 \cdot 4^2 \cdot 6^2 \cdot 8} + \dots$$

9 Vector Analysis

Scalar product

$$\boldsymbol{A}\cdot\boldsymbol{B} = |\boldsymbol{A}|\cdot|\boldsymbol{B}|\cos(\boldsymbol{A},\boldsymbol{B}) = A_x B_x + A_y B_y + A_z B_z$$

$$|\boldsymbol{A}| = \sqrt{A_x^2 + A_y^2 + A_z^2}$$

Vector product

$$|\boldsymbol{A}\times\boldsymbol{B}| = |\boldsymbol{A}|\,|\boldsymbol{B}|\sin(\boldsymbol{A},\boldsymbol{B})$$

$$\boldsymbol{A}\times\boldsymbol{B} = -\boldsymbol{B}\times\boldsymbol{A} = \begin{vmatrix} \hat{\boldsymbol{x}} & \hat{\boldsymbol{y}} & \hat{\boldsymbol{z}} \\ A_x & A_y & A_z \\ B_x & B_y & B_z \end{vmatrix} =$$

$$= (A_y B_z - A_z B_y)\,\hat{\boldsymbol{x}} + (A_z B_x - A_x B_z)\,\hat{\boldsymbol{y}} + (A_x B_y - A_y B_x)\,\hat{\boldsymbol{z}}$$

$$\boldsymbol{A}\cdot(\boldsymbol{B}\times\boldsymbol{C}) = \boldsymbol{B}\cdot(\boldsymbol{C}\times\boldsymbol{A}) = \boldsymbol{C}\cdot(\boldsymbol{A}\times\boldsymbol{B})$$
$$(\boldsymbol{A}\times\boldsymbol{B})\cdot(\boldsymbol{C}\times\boldsymbol{D}) = (\boldsymbol{A}\cdot\boldsymbol{C})(\boldsymbol{B}\cdot\boldsymbol{D}) - (\boldsymbol{A}\cdot\boldsymbol{D})(\boldsymbol{B}\cdot\boldsymbol{C})$$
$$(\boldsymbol{A}\times\boldsymbol{B})\times\boldsymbol{C} = \boldsymbol{B}(\boldsymbol{A}\cdot\boldsymbol{C}) - \boldsymbol{A}(\boldsymbol{B}\cdot\boldsymbol{C})$$
$$\boldsymbol{A}\times(\boldsymbol{B}\times\boldsymbol{C}) = \boldsymbol{B}(\boldsymbol{A}\cdot\boldsymbol{C}) - \boldsymbol{C}(\boldsymbol{A}\cdot\boldsymbol{B})$$

Miscellaneous formulae involving $\nabla = \hat{\boldsymbol{x}}\frac{\partial}{\partial x} + \hat{\boldsymbol{y}}\frac{\partial}{\partial y} + \hat{\boldsymbol{z}}\frac{\partial}{\partial z}$

$$\Delta f = \nabla^2 f = \nabla\cdot(\nabla f) = \frac{\partial^2 f}{\partial x^2} + \frac{\partial^2 f}{\partial y^2} + \frac{\partial^2 f}{\partial z^2} \qquad \text{(Laplace operator)}$$
$$\nabla\times(\nabla f) = \boldsymbol{0}$$
$$\nabla\cdot(\nabla\times\boldsymbol{A}) = 0$$
$$\nabla\times(\nabla\times\boldsymbol{A}) = \nabla(\nabla\cdot\boldsymbol{A}) - \nabla^2\boldsymbol{A}$$
$$\nabla\cdot(f\boldsymbol{A}) = (\nabla f)\cdot\boldsymbol{A} + f(\nabla\cdot\boldsymbol{A})$$
$$\nabla\times(f\boldsymbol{A}) = f(\nabla\times\boldsymbol{A}) + (\nabla f)\times\boldsymbol{A}$$
$$\nabla\cdot(\boldsymbol{A}\times\boldsymbol{B}) = \boldsymbol{B}\cdot(\nabla\times\boldsymbol{A}) - \boldsymbol{A}\cdot(\nabla\times\boldsymbol{B})$$
$$\nabla\times(\boldsymbol{A}\times\boldsymbol{B}) = (\boldsymbol{B}\cdot\nabla)\boldsymbol{A} - (\boldsymbol{A}\cdot\nabla)\boldsymbol{B} + \boldsymbol{A}(\nabla\cdot\boldsymbol{B}) - \boldsymbol{B}(\nabla\cdot\boldsymbol{A})$$
$$\nabla(\boldsymbol{A}\cdot\boldsymbol{B}) = (\boldsymbol{A}\cdot\nabla)\boldsymbol{B} + (\boldsymbol{B}\cdot\nabla)\boldsymbol{A} + \boldsymbol{A}\times(\nabla\times\boldsymbol{B}) + \boldsymbol{B}\times(\nabla\times\boldsymbol{A})$$

$$\operatorname{grad} f = \nabla f$$
$$\operatorname{div}\boldsymbol{A} = \nabla\cdot\boldsymbol{A}$$
$$\operatorname{curl}\boldsymbol{A} = \operatorname{rot}\boldsymbol{A} = \nabla\times\boldsymbol{A}$$

10 Special Coordinate Systems

Cylindrical coordinates (ρ, φ, z)

$$x = \rho \cos \varphi$$

$$y = \rho \sin \varphi$$

$$z = z$$

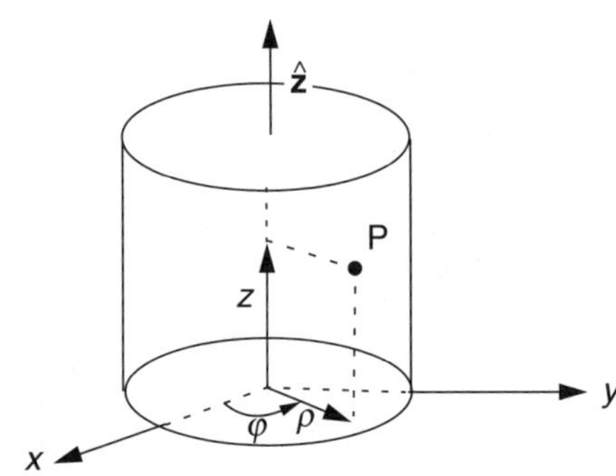

$$\mathrm{d}V = \rho \, \mathrm{d}\rho \, \mathrm{d}\varphi \, \mathrm{d}z$$

$$\nabla \cdot \boldsymbol{A} = \frac{1}{\rho} \frac{\partial}{\partial \rho} (\rho A_\rho) + \frac{1}{\rho} \frac{\partial}{\partial \varphi} A_\varphi + \frac{\partial}{\partial z} A_z$$

$$\nabla f = \frac{\partial f}{\partial \rho} \hat{\boldsymbol{\rho}} + \frac{1}{\rho} \frac{\partial f}{\partial \varphi} \hat{\boldsymbol{\varphi}} + \frac{\partial f}{\partial z} \hat{\boldsymbol{z}}$$

$$\nabla^2 f = \frac{\partial^2 f}{\partial \rho^2} + \frac{1}{\rho} \frac{\partial f}{\partial \rho} + \frac{1}{\rho^2} \frac{\partial^2 f}{\partial \varphi^2} + \frac{\partial^2 f}{\partial z^2} = \frac{1}{\rho} \frac{\partial}{\partial \rho} \left(\rho \frac{\partial f}{\partial \rho} \right) + \frac{1}{\rho^2} \frac{\partial^2 f}{\partial \varphi^2} + \frac{\partial^2 f}{\partial z^2}$$

(See also Section F–1.5)

$$\nabla \times \boldsymbol{A} = \frac{1}{\rho} \left(\frac{\partial}{\partial \varphi} A_z - \frac{\partial}{\partial z} (\rho A_\varphi) \right) \hat{\boldsymbol{\rho}} + \left(\frac{\partial}{\partial z} A_\rho - \frac{\partial A_z}{\partial \rho} \right) \hat{\boldsymbol{\varphi}} + \frac{1}{\rho} \left(\frac{\partial}{\partial \rho} (\rho A_\varphi) - \frac{\partial A_\rho}{\partial \varphi} \right) \hat{\boldsymbol{z}}$$

$$\int_{\boldsymbol{r}_0}^{\boldsymbol{r}_1} \boldsymbol{A} \cdot \mathrm{d}\boldsymbol{r} = \int_{\mathrm{P}_0}^{\mathrm{P}_1} (A_\rho \, \mathrm{d}\rho + A_\varphi \rho \, \mathrm{d}\varphi + A_z \, \mathrm{d}z)$$

Spherical coordinates (r, θ, ϕ)

$$x = r \sin\theta \cos\phi$$

$$y = r \sin\theta \sin\phi$$

$$z = r \cos\theta$$

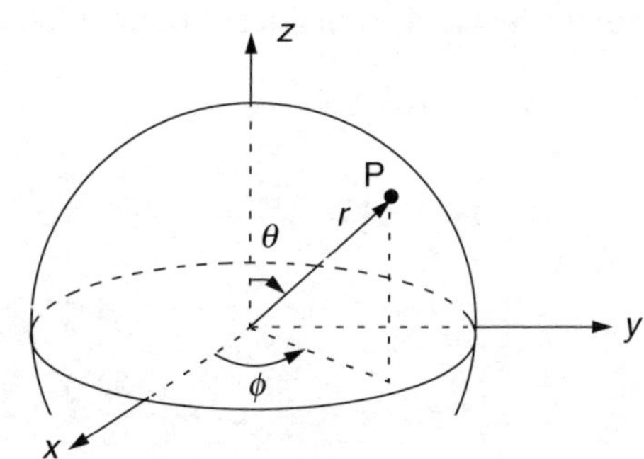

$$\mathrm{d}V = r^2 \sin\theta \,\mathrm{d}r\,\mathrm{d}\theta\,\mathrm{d}\phi$$

$$\nabla \cdot \boldsymbol{A} = \frac{1}{r^2}\frac{\partial}{\partial r}(r^2 A_r) + \frac{1}{r\sin\theta}\frac{\partial}{\partial\theta}(\sin\theta\, A_\theta) + \frac{1}{r\sin\theta}\frac{\partial}{\partial\phi}A_\phi$$

$$\nabla f = \frac{\partial f}{\partial r}\hat{\boldsymbol{r}} + \frac{1}{r}\frac{\partial f}{\partial\theta}\hat{\boldsymbol{\theta}} + \frac{1}{r\sin\theta}\frac{\partial f}{\partial\phi}\hat{\boldsymbol{\phi}}$$

$$\nabla^2 f = \frac{1}{r^2}\frac{\partial}{\partial r}\left(r^2\frac{\partial f}{\partial r}\right) + \frac{1}{r^2\sin\theta}\frac{\partial}{\partial\theta}\left(\sin\theta\frac{\partial f}{\partial\theta}\right) + \frac{1}{r^2\sin^2\theta}\frac{\partial^2 f}{\partial\phi^2} =$$

$$= \frac{1}{r}\frac{\partial^2}{\partial r^2}(rf) + \frac{1}{r^2\sin\theta}\frac{\partial}{\partial\theta}\left(\sin\theta\frac{\partial f}{\partial\theta}\right) + \frac{1}{r^2\sin^2\theta}\frac{\partial^2 f}{\partial\phi^2}$$

(See also Section F–1.5)

$$\nabla\times\boldsymbol{A} = \frac{1}{r^2\sin\theta}\left(\left(\frac{\partial}{\partial\theta}(r\sin\theta\, A_\phi)\right) - \frac{\partial}{\partial\phi}(rA_\theta)\right)\hat{\boldsymbol{r}} +$$

$$+ \frac{1}{r\sin\theta}\left(\frac{\partial}{\partial\phi}A_r - \frac{\partial}{\partial r}(r\sin\theta\, A_\phi)\right)\hat{\boldsymbol{\theta}} + \frac{1}{r}\left(\frac{\partial}{\partial r}(rA_\theta) - \frac{\partial}{\partial\theta}A_r\right)\hat{\boldsymbol{\phi}}$$

$$\int_{\boldsymbol{r}_0}^{\boldsymbol{r}_1} \boldsymbol{A}\cdot\mathrm{d}\boldsymbol{r} = \int_{P_0}^{P_1}(A_r\,\mathrm{d}r + A_\theta r\,\mathrm{d}\theta + A_\phi r\sin\theta\,\mathrm{d}\phi)$$

Any orthogonal curvelinear coordinates u_1, u_2, u_3

$$\boldsymbol{A} = A_1\hat{\boldsymbol{u}}_1 + A_2\hat{\boldsymbol{u}}_2 + A_3\hat{\boldsymbol{u}}_3$$

$$\operatorname{div}\boldsymbol{A} = \nabla\cdot\boldsymbol{A} = \frac{1}{h_1h_2h_3}\left[\frac{\partial}{\partial u_1}(h_2h_3A_1) + \frac{\partial}{\partial u_2}(h_3h_1A_2) + \frac{\partial}{\partial u_3}(h_1h_2A_3)\right]$$

$$\operatorname{grad} f = \nabla f = \frac{1}{h_1}\frac{\partial f}{\partial u_1}\hat{\boldsymbol{u}}_1 + \frac{1}{h_2}\frac{\partial f}{\partial u_2}\hat{\boldsymbol{u}}_2 + \frac{1}{h_3}\frac{\partial f}{\partial u_3}\hat{\boldsymbol{u}}_3$$

$$\Delta f = \nabla^2 f = \frac{1}{h_1h_2h_3}\left[\frac{\partial}{\partial u_1}\left(\frac{h_2h_3}{h_1}\frac{\partial f}{\partial u_1}\right) + \frac{\partial}{\partial u_2}\left(\frac{h_3h_1}{h_2}\frac{\partial f}{\partial u_2}\right) + \frac{\partial}{\partial u_3}\left(\frac{h_1h_2}{h_3}\frac{\partial f}{\partial u_3}\right)\right]$$

$$\operatorname{curl}\boldsymbol{A} = \nabla\times\boldsymbol{A} = \frac{1}{h_1h_2h_3}\begin{vmatrix} h_1\hat{\boldsymbol{u}}_1 & h_2\hat{\boldsymbol{u}}_2 & h_3\hat{\boldsymbol{u}}_3 \\ \dfrac{\partial}{\partial u_1} & \dfrac{\partial}{\partial u_2} & \dfrac{\partial}{\partial u_3} \\ h_1A_1 & h_2A_2 & h_3A_3 \end{vmatrix} = \frac{1}{h_2h_3}\left[\frac{\partial}{\partial u_2}(h_3A_3) - \frac{\partial}{\partial u_3}(h_2A_2)\right]\hat{\boldsymbol{u}}_1$$

$$+ \frac{1}{h_1h_3}\left[\frac{\partial}{\partial u_3}(h_1A_1) - \frac{\partial}{\partial u_1}(h_3A_3)\right]\hat{\boldsymbol{u}}_2 + \frac{1}{h_1h_2}\left[\frac{\partial}{\partial u_1}(h_2A_2) - \frac{\partial}{\partial u_2}(h_1A_1)\right]\hat{\boldsymbol{u}}_3$$

For cylindrical coordinates

$$h_1^2 = 1, h_2^2 = \rho^2, h_3^2 = 1$$

For spherical coordinates

$$h_1^2 = 1, h_2^2 = r^2, h_3^2 = r^2\sin^2\theta$$

11 Laplace Transforms

Definition

$$F(s) = \int_0^\infty e^{-st} f(t)\, dt,\ s > \alpha \text{ where } \alpha \text{ is some constant}$$

Transform	**Original function**
$F(s)$	$f(t)$
$F(s+a)$	$e^{-at} f(t)$
$e^{-as} F(s)$	$\begin{cases} f(t-a); & t-a>0 \\ 0 & ; t-a<0 \end{cases}$
$\frac{1}{a} F\left(\frac{s}{a}\right)$	$f(at)$
$F(as) \quad (a>0)$	$\frac{1}{a} f\left(\frac{t}{a}\right)$
$\frac{d^n F(s)}{ds^n}$	$(-t)^n f(t)$
$\int_s^\infty F(\sigma)\, d\sigma$	$\frac{f(t)}{t}$
$\frac{1}{2\pi i} \int_{c-i\infty}^{c+i\infty} F_1(\sigma) F_2(s-\sigma)\, d\sigma$	$f_1(t)\, f_2(t)$
$F_1(s)\, F_2(s)$	$\int_0^t f_1(\tau) f_2(t-\tau)\, d\tau = \int_0^t f_1(t-\tau) f_2(\tau)\, d\tau$
$s\,F(s) - f(0)$	$f'(t)$
$s^2 F(s) - [s\,f(0) + f'(0)]$	$f''(t)$
$s^n F(s) - [s^{n-1} f(0) + \ldots + f^{(n-1)}(0)]$	$f^{(n)}(t)$
$\frac{1}{s} F(s) + \frac{1}{s} \left[\int_0^t f(\tau)\, d\tau\right]_{t=+0}$	$\int_0^t f(\tau)\, d\tau$

Transform	**Original function**	
$\lim\limits_{s\to 0} s\,F(s)$	$\lim\limits_{t\to\infty} f(t)$	
$\lim\limits_{s\to\infty} s\,F(s)$	$\lim\limits_{t\to 0} f(t)$	

Special transforms

Transform	Original function	
1	$\delta(t) = \begin{cases} +\infty & \text{if } t = 0 \\ 0 & \text{if } t \neq 0 \end{cases}$	Dirac function
$\dfrac{1}{s}$	1	Step function
$\dfrac{1}{s^2}$	t	
$\dfrac{1}{s^3}$	$\dfrac{1}{2}\,t^2$	
$\dfrac{1}{s^{n+1}}$	$\dfrac{t^n}{\Gamma(n+1)}$	$n>0$, $\Gamma(n+1)=n!$ if $n=1,2,3,\ldots$
$\dfrac{1}{s+a}$	e^{-at}	
$\dfrac{1}{(s+a)^2}$	te^{-at}	
$\dfrac{s}{(s+a)^2}$	$(1-at)\,e^{-at}$	
$\dfrac{1}{1+as}$	$\dfrac{1}{a}\,e^{-\frac{t}{a}}$	
$\dfrac{a}{s^2+a^2}$	$\sin at$	
$\dfrac{a}{s^2-a^2}$	$\sinh at$	
$\dfrac{s}{s^2+a^2}$	$\cos at$	
$\dfrac{s}{s^2-a^2}$	$\cosh at$	
$\dfrac{1}{s(s+a)}$	$\dfrac{1}{a}\,(1-e^{-at})$	

Transform		**Original function**
$\dfrac{1}{s(1+as)}$		$1-\mathrm{e}^{-\frac{t}{a}}$
$\dfrac{1}{(s+a)(s+b)}$		$\dfrac{\mathrm{e}^{-bt}-\mathrm{e}^{-at}}{a-b}$
$\dfrac{s}{(s+a)(s+b)}$		$\dfrac{a\,\mathrm{e}^{-at}-b\,\mathrm{e}^{-bt}}{a-b}$
$\dfrac{a}{(s+b)^2+a^2}$		$\mathrm{e}^{-bt}\sin at$
$\dfrac{s+b}{(s+b)^2+a^2}$		$\mathrm{e}^{-bt}\cos at$
$\dfrac{1}{s^2+2\zeta\omega s+\omega^2}$		
	$\zeta=0$	$\dfrac{1}{\omega}\sin\omega t$
	$\zeta<1$	$\dfrac{1}{\omega\sqrt{1-\zeta^2}}\,\mathrm{e}^{-\zeta\omega t}\sin(\omega\sqrt{1-\zeta^2}\;t)$
	$\zeta=1$	$t\,\mathrm{e}^{-\omega t}$
	$\zeta>1$	$\dfrac{1}{\omega\sqrt{\zeta^2-1}}\,\mathrm{e}^{-\zeta\omega t}\sinh(\omega\sqrt{\zeta^2-1}\;t)$
$\dfrac{s}{s^2+2\zeta\omega s+\omega^2}$		
	$\zeta<1$	$\dfrac{1}{\sqrt{1-\zeta^2}}\,\mathrm{e}^{-\zeta\omega t}\sin(\omega\sqrt{1-\zeta^2}\;t+\tau)$
		$\tau=\arctan\dfrac{\omega\sqrt{1-\zeta^2}}{-\zeta\omega}$
	$\zeta=1$	$(1-\omega t)\,\mathrm{e}^{-\omega t}$
	$\zeta>1$	$\dfrac{1}{\sqrt{\zeta^2-1}}\,\mathrm{e}^{-\zeta\omega t}\sinh(\omega\sqrt{\zeta^2-1}\;t+\tau)$
		$\tau=\arctan\dfrac{\omega\sqrt{\zeta^2-1}}{-\zeta\omega}$
	$\zeta=0$	$\cos\omega t$

Transform	**Original function**
$\dfrac{a}{(s^2+a^2)(s+b)}$	$\dfrac{1}{\sqrt{a^2+b^2}}[\sin(at-\phi)+\sin\phi\, e^{-bt}]$ $\phi=\arctan\dfrac{a}{b}$
$\dfrac{s}{(s^2+a^2)(s+b)}$	$\dfrac{1}{\sqrt{a^2+b^2}}[\cos(at-\phi)-\cos\phi\, e^{-bt}]$ $\phi=\arctan\dfrac{a}{b}$
$\dfrac{1}{s(s+a)(s+b)}$	$\dfrac{1}{ab}+\dfrac{ae^{-bt}-be^{-at}}{ab(b-a)}$
$\dfrac{1}{(s+a)(s+b)(s+c)}$	$\dfrac{(b-c)e^{-at}+(c-a)e^{-bt}+(a-b)e^{-ct}}{(b-a)(c-a)(b-c)}$
$\dfrac{1}{(s+a)^3}$	$\dfrac{1}{2}\, t^2\, e^{-at}$
$\dfrac{s}{(s+a)(s+b)(s+c)}$	$\dfrac{a(b-c)e^{-at}+b(c-a)e^{-bt}+c(a-b)e^{-ct}}{(b-a)(b-c)(a-c)}$
$\dfrac{as}{(s^2+a^2)^2}$	$\dfrac{t}{2}\sin at$
$\dfrac{1}{\sqrt{s}}$	$\dfrac{1}{\sqrt{\pi t}}$
$\dfrac{1}{\sqrt{s}}F(\sqrt{s})$	$\dfrac{1}{\sqrt{\pi t}}\int_0^\infty e^{-\sigma^2/4t}f(\sigma)\,d\sigma$

12 Fourier Series

Fourier series of a function $f(x)$

$$\frac{a_0}{2} + \sum_{k=1}^{\infty} \left(a_k \cos \frac{k\pi x}{L} + b_k \sin \frac{k\pi x}{L} \right)$$

$f(x)$ integrable and periodic function with period $2L$.

Fourier coefficients

$$a_k = \frac{1}{L} \int_{-L}^{L} f(x) \cos \frac{k\pi x}{L} \, dx \qquad k = 0, 1, 2, \ldots$$

$$b_k = \frac{1}{L} \int_{-L}^{L} f(x) \sin \frac{k\pi x}{L} \, dx \qquad k = 1, 2, 3, \ldots$$

The integration can be done for any interval of $2L$.

If $f(x) = f(-x)$ then $b_k = 0$ and $a_k = \frac{2}{L} \int_0^L f(x) \cos \frac{k\pi x}{L} \, dx$

If $f(x) = -f(-x)$ then $a_k = 0$ and $b_k = \frac{2}{L} \int_0^L f(x) \sin \frac{k\pi x}{L} \, dx$

Complex Fourier series of a function $f(x)$

$$\sum_{k=-\infty}^{\infty} c_k e^{ik\pi x/L} \qquad c_k = \frac{1}{2L} \int_{-L}^{L} f(x) e^{-ik\pi x/L} \, dx$$

$$\begin{cases} c_k = \frac{1}{2}(a_k - ib_k), & k \geq 0 \\ c_{-k} = \frac{1}{2}(a_k + ib_k), & k \geq 0 \end{cases} \qquad \begin{cases} a_k = (c_k - c_{-k}), & k \geq 0 \\ b_k = i(c_k - c_{-k}), & k \geq 1 \end{cases}$$

Bessel's identity

$$\frac{1}{2} a_0^2 + \sum_{k=1}^{n} (a_k^2 + b_k^2) = \frac{1}{L} \int_{-L}^{L} f^2(x) \, dx - \frac{1}{L} \int_{-L}^{L} (f(x) - s_n(x))^2 \, dx$$

where $s_n(x) = \frac{a_0}{2} + \sum_{k=1}^{n} (a_k \cos \frac{k\pi x}{L} + b_k \sin \frac{k\pi x}{L})$ and $f^2(x)$ is integrable

Bessel's inequality

$$\frac{1}{2}a_0^2 + \sum_{k=1}^{\infty}(a_k^2 + b_k^2) \le \frac{1}{L}\int_{-L}^{L} f^2(x)\,dx$$

Parseval's formula

$$\frac{1}{2}a_0^2 + \sum_{k=1}^{\infty}(a_k^2 + b_k^2) = \frac{1}{L}\int_{-L}^{L} f^2(x)\,dx$$

if and only if $\lim\limits_{n\to\infty}\int_{-L}^{L}(f(x) - s_n(x))^2\,dx = 0$

Dirichlet's integral for a function $f(x)$ **with period** $2L = 2\pi$

$$s_n(x) = \frac{1}{2\pi}\int_{-\pi}^{\pi} f(x+u)\frac{\sin(n+\frac{1}{2})u}{\sin\frac{1}{2}u}\,du$$

Some special Fourier series

$$f(x) = \begin{cases} 1 & 0 < x < \pi \\ -1 & -\pi < x < 0 \end{cases}$$

$$\frac{4}{\pi}\left(\frac{\sin x}{1} + \frac{\sin 3x}{3} + \frac{\sin 5x}{5} + \ldots\right)$$

$$f(x) = |x| = \begin{cases} x & 0 < x < \pi \\ -x & -\pi < x < 0 \end{cases}$$

$$\frac{\pi}{2} - \frac{4}{\pi}\left(\frac{\cos x}{1^2} + \frac{\cos 3x}{3^2} + \frac{\cos 5x}{5^2} + \ldots\right)$$

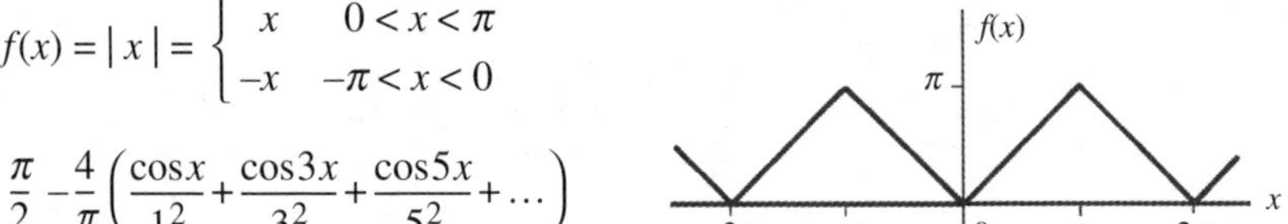

$$f(x) = x,\ -\pi < x < \pi$$

$$2\left(\frac{\sin x}{1} - \frac{\sin 2x}{2} + \frac{\sin 3x}{3} - \ldots\right)$$

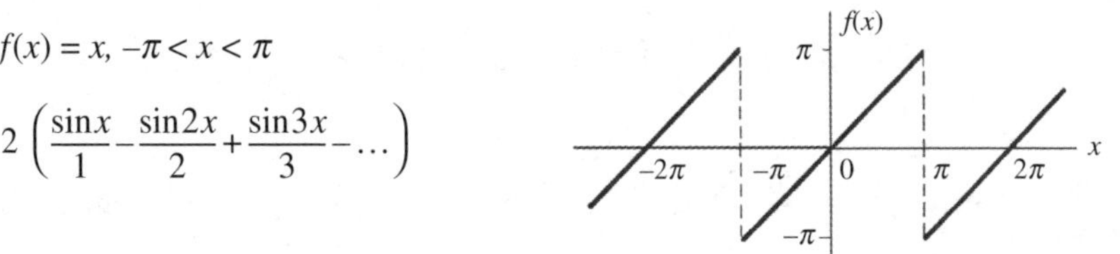

$f(x) = x,\ 0 < x < 2\pi$

$$\pi - 2\left(\frac{\sin x}{1} + \frac{\sin 2x}{2} + \frac{\sin 3x}{3} + \dots\right)$$

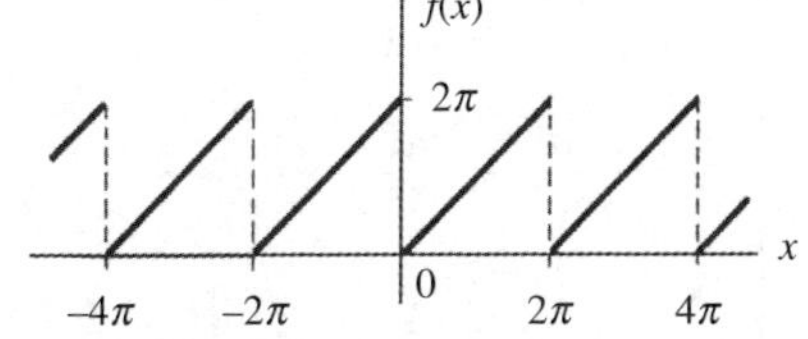

$f(x) = |\sin x|,\ -\pi < x < \pi$

$$\frac{2}{\pi} - \frac{4}{\pi}\left(\frac{\cos 2x}{1\cdot 3} + \frac{\cos 4x}{3\cdot 5} + \frac{\cos 6x}{5\cdot 7} + \dots\right)$$

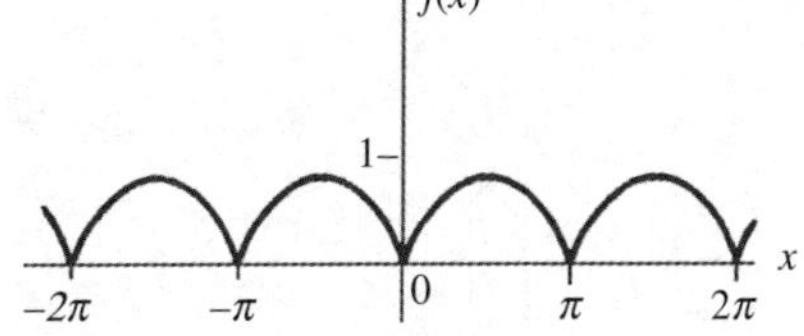

$$f(x) = \begin{cases} \sin x & 0 < x < \pi \\ 0 & \pi < x < 2\pi \end{cases}$$

$$\frac{1}{\pi} + \frac{1}{2}\sin x - \frac{2}{\pi}\left(\frac{\cos 2x}{1\cdot 3} + \frac{\cos 4x}{3\cdot 5} + \frac{\cos 6x}{5\cdot 7} + \dots\right)$$

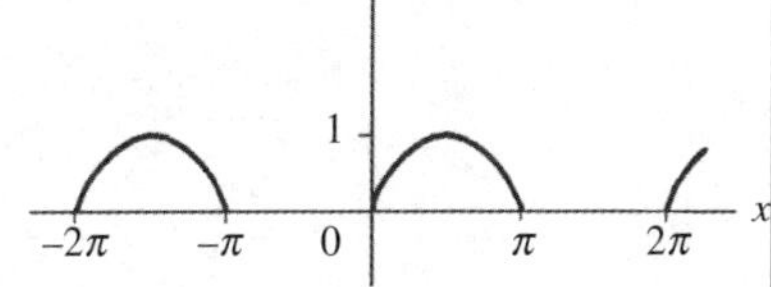

$$f(x) = \begin{cases} \cos x & 0 < x < \pi \\ -\cos x & -\pi < x < 0 \end{cases}$$

$$\frac{8}{\pi}\left(\frac{\sin 2x}{1\cdot 3} + \frac{2\sin 4x}{3\cdot 5} + \frac{3\sin 6x}{5\cdot 7} + \dots\right)$$

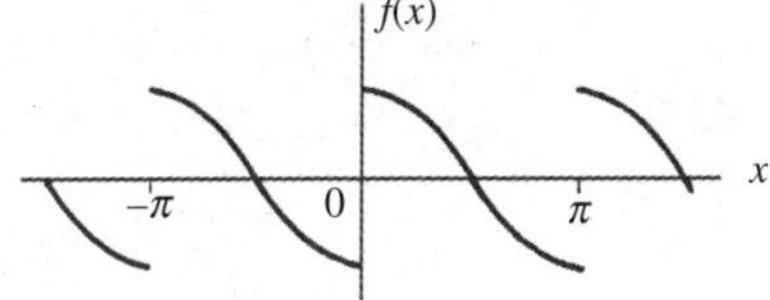

$f(x) = x^2,\ -\pi < x < \pi$

$$\frac{\pi^2}{3} - 4\left(\frac{\cos x}{1^2} - \frac{\cos 2x}{2^2} + \frac{\cos 3x}{3^2} - \dots\right)$$

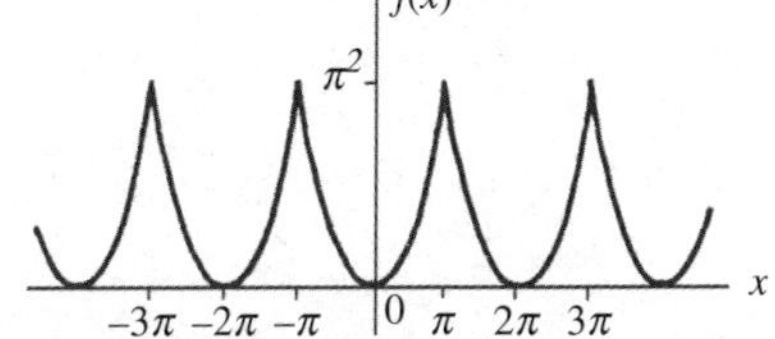

$f(x) = x(\pi - x),\ 0 < x < \pi$

$$\frac{\pi^2}{6} - \left(\frac{\cos 2x}{1^2} + \frac{\cos 4x}{2^2} + \frac{\cos 6x}{3^2} + \dots\right)$$

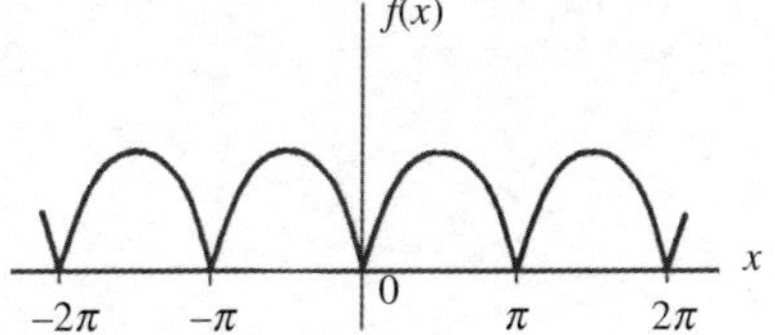

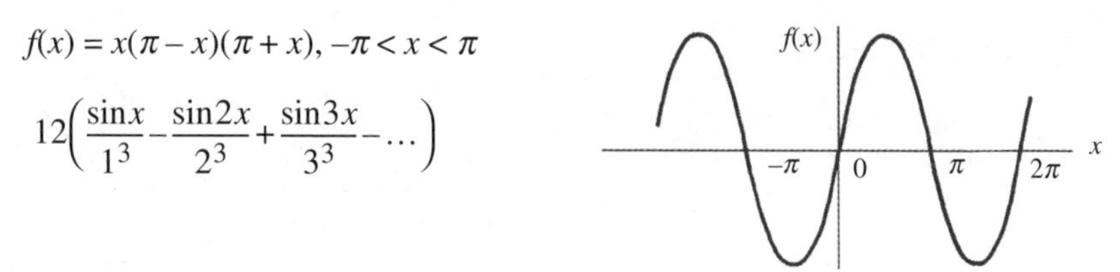

$$f(x) = x(\pi - x)(\pi + x), \ -\pi < x < \pi$$

$$12\left(\frac{\sin x}{1^3} - \frac{\sin 2x}{2^3} + \frac{\sin 3x}{3^3} - \dots\right)$$

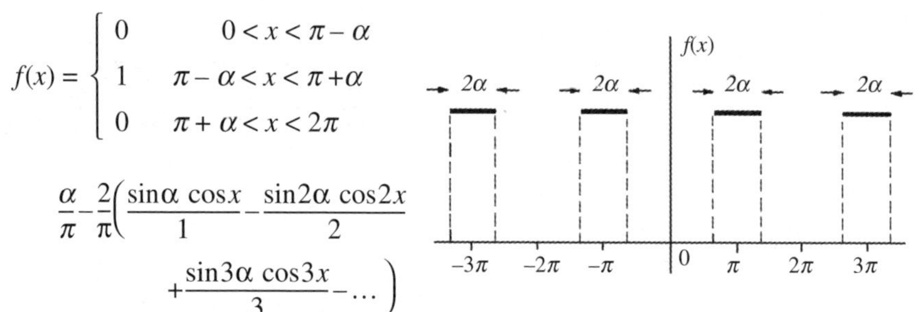

$$f(x) = \begin{cases} 0 & 0 < x < \pi - \alpha \\ 1 & \pi - \alpha < x < \pi + \alpha \\ 0 & \pi + \alpha < x < 2\pi \end{cases}$$

$$\frac{\alpha}{\pi} - \frac{2}{\pi}\left(\frac{\sin\alpha \cos x}{1} - \frac{\sin 2\alpha \cos 2x}{2} + \frac{\sin 3\alpha \cos 3x}{3} - \dots\right)$$

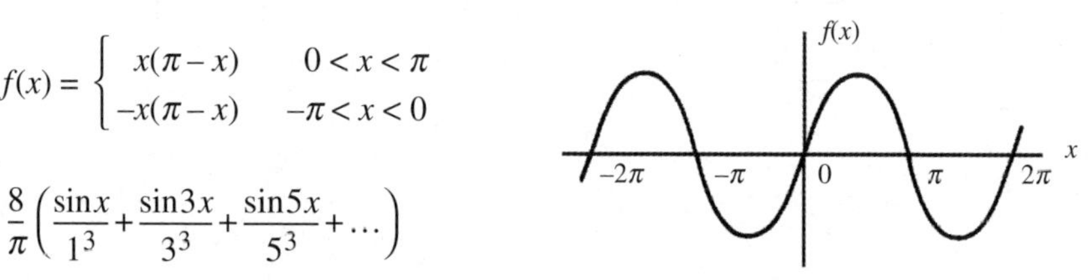

$$f(x) = \begin{cases} x(\pi - x) & 0 < x < \pi \\ -x(\pi - x) & -\pi < x < 0 \end{cases}$$

$$\frac{8}{\pi}\left(\frac{\sin x}{1^3} + \frac{\sin 3x}{3^3} + \frac{\sin 5x}{5^3} + \dots\right)$$

13 Fourier Transforms

Definition of the Fourier transform of $f(t)$ and formula of the inverse Fourier transform

$$F(\omega) = \int_{-\infty}^{\infty} f(t)e^{-i\omega t}dt \qquad f(t) = \frac{1}{2\pi}\int_{-\infty}^{\infty} F(\omega)e^{i\omega t}d\omega$$

Plancherel's formula (Parseval's identity)

$$\int_{-\infty}^{\infty} |f(t)|^2 dt = \frac{1}{2\pi}\int_{-\infty}^{\infty} |F(\omega)|^2 d\omega$$

F and G = Fourier transforms of f and g

$$\int_{-\infty}^{\infty} f(t)\overline{g(t)}\,dt = \frac{1}{2\pi}\int_{-\infty}^{\infty} F(\omega)\overline{G(\omega)}d\omega$$

The bar denotes complex conjugate

The Fourier sine transform and its inverse

$$F_S(\omega) = \int_0^{\infty} f(x)\sin\omega x\,dx \qquad f(x) = \frac{2}{\pi}\int_0^{\infty} F_S(\omega)\sin\omega x\,d\omega$$

The Fourier cosine transform and its inverse

$$F_C(\omega) = \int_0^{\infty} f(x)\cos\omega x\,dx \qquad f(x) = \frac{2}{\pi}\int_0^{\infty} F_C(\omega)\cos\omega x\,d\omega$$

General properties of Fourier transforms (a and c are constants)

$f(t)$	$F(\omega)$	
$f(t-a)$	$e^{-ia\omega} F(\omega)$	(a real)
$e^{iat} f(t)$	$F(\omega - a)$	(a real)
$f(ct)$	$\frac{1}{\lvert c \rvert} F\left(\frac{\omega}{c}\right)$	(c real, $c \neq 0$)
$f(c(t-a))$	$\frac{1}{\lvert c \rvert} e^{-1a\omega} F\left(\frac{\omega}{c}\right)$	(a and c real, $c \neq 0$)
$e^{iat} f(ct)$	$\frac{1}{\lvert c \rvert} F\left(\frac{\omega - a}{c}\right)$	(a and c real, $c \neq 0$)
$t^n f(t)$	$i^n F^{(n)}(\omega)$	($F^{(n)}(\omega) = n^{\text{th}}$ derivative)
$f^{(n)}(t)$	$(i\omega)^n F(\omega)$	($f^{(n)}(t) = n^{\text{th}}$ derivative)
$F(t)$	$2\pi f(-\omega)$	
$\overline{f(t)}$	$\overline{F(-\omega)}$	(complex conjugate)
$f(t)\, g(t)$	$\frac{1}{2\pi} F * G(\omega) = \frac{1}{2\pi} \int_{-\infty}^{\infty} F(\omega - \tau) G(\tau)\, d\tau$	
$f * g(t) = \int_{-\infty}^{\infty} f(t-\tau) g(\tau)\, d\tau$	$F(\omega)\, G(\omega)$	($*$ is called convolution)

Special Fourier transforms (constant $c > 0$)

$f(t)$	$F(\omega)$	
$\begin{cases} 1, & \lvert t \rvert < c \\ 0, & \lvert t \rvert > c \end{cases}$	$\frac{2 \sin c\omega}{\omega}$	
$\frac{1}{t^2 + c^2}$	$\frac{\pi}{c} e^{-c\lvert\omega\rvert}$	
$\frac{t}{t^2 + c^2}$	$-\pi i\, e^{-c\lvert\omega\rvert} \operatorname{sgn} \omega$	$\operatorname{sgn} \omega = \begin{cases} -1, & \omega < 0 \\ +1, & \omega > 0 \end{cases}$
$\frac{1}{2c} e^{-c\lvert t \rvert}$	$\frac{1}{\omega^2 + c^2}$	
$\frac{i}{2} e^{-c\lvert t \rvert} \operatorname{sgn} t$	$\frac{\omega}{\omega^2 + c^2}$	

Special Fourier sine transforms (constant $c > 0$)

$f(x),\ x > 0$	$F_S(\omega),\ \omega > 0$
$\begin{cases} 1, & 0 < x < c \\ 0, & x > c \end{cases}$	$\dfrac{1-\cos c\omega}{\omega}$
x^{-1}	$\dfrac{\pi}{2}$
$x^{-1/2}$	$\sqrt{\dfrac{\pi}{2\omega}}$
$\dfrac{x}{x^2+c^2}$	$\dfrac{\pi}{2}\,e^{-c\omega}$
e^{-cx}	$\dfrac{\omega}{\omega^2+c^2}$
$x\,e^{-cx^2}$	$\sqrt{\dfrac{\pi}{c}}\,\dfrac{\omega}{4c}\,e^{-\omega^2/4c}$
$\dfrac{\sin cx}{x}$	$\dfrac{1}{2}\ln\left\lvert\dfrac{\omega+c}{\omega-c}\right\rvert$
$\dfrac{\cos cx}{x}$	$\begin{cases} 0 & \omega < c \\ \pi/4 & \omega = c \\ \pi/2 & \omega > c \end{cases}$
$\dfrac{\sin cx}{x^2}$	$\begin{cases} \pi\omega/2 & \omega < c \\ \pi c/2 & \omega \geq c \end{cases}$

Special Fourier cosine transforms (constant $c > 0$)

$f(x),\ x > 0$	$F_C(\omega),\ \omega > 0$	
$\begin{cases} 1, & 0 < x < c \\ 0, & x > c \end{cases}$	$\dfrac{\sin c\omega}{\omega}$	
$\dfrac{1}{x^2+c^2}$	$\dfrac{\pi}{2c}\,\mathrm{e}^{-c\omega}$	
e^{-cx}	$\dfrac{c}{\omega^2+c^2}$	
e^{-cx^2}	$\dfrac{1}{2}\sqrt{\dfrac{\pi}{c}}\,\mathrm{e}^{-\omega^2/4c}$	
$x^{-1/2}$	$\sqrt{\dfrac{\pi}{2\omega}}$	
$\ln\left(\dfrac{x^2+a^2}{x^2+c^2}\right)$	$\dfrac{\pi}{\omega}(\mathrm{e}^{-c\omega}-\mathrm{e}^{-a\omega})$	$(a > 0)$
$\dfrac{\sin cx}{x}$	$\begin{cases} \pi/2 & \omega < c \\ \pi/4 & \omega = c \\ 0 & \omega > c \end{cases}$	
$\sin cx^2$	$\sqrt{\dfrac{\pi}{8c}}\left(\cos\dfrac{\omega^2}{4c} - \sin\dfrac{\omega^2}{4c}\right)$	
$\cos cx^2$	$\sqrt{\dfrac{\pi}{8c}}\left(\cos\dfrac{\omega^2}{4c} + \sin\dfrac{\omega^2}{4c}\right)$	

14 Differential Equations

Differential equation with separable variables

Equation

$$F_1(x)G_1(y)\mathrm{d}x + F_2(x)G_2(y)\mathrm{d}y = 0$$

Solution

$$\int \frac{F_1(x)}{F_2(x)}\,\mathrm{d}x + \int \frac{G_2(y)}{G_1(y)}\,\mathrm{d}y = C$$

"Homogeneous" differential equation

Equation

$$y' - P(v) = 0 \quad v = y/x$$

Solution

$$\int \frac{\mathrm{d}v}{P(v) - v} + c = \ln x \quad \text{if } P(v) \neq v$$

$$y = cx \quad \text{if } P(v) = v$$

Linear equation of first order

Equation

$$y' + f(x) \cdot y = g(x)$$

Solution

$$y(x) = (Q(x) + C)\,\mathrm{e}^{-F(x)}$$

F is a primitive function to f

Q is a primitive function to $g(x)\mathrm{e}^{F(x)}$

See Sec. M–13 for numerical solutions

Linear equation of second order with constant coefficients

Equation

$$y'' + 2ay' + by = F(x)$$

Solution

y = solution of homogeneous equation + particular integral

See Sec. M – 13 for numerical solution

Homogeneous equation

$$y'' + 2ay' + by = 0$$

Auxiliary equation

$$f(\lambda) \equiv \lambda^2 + 2a\lambda + b = 0$$

Solution of homogeneous equation

1. Roots of auxiliary equation are real and unequal, $\lambda_1 \neq \lambda_2$

$$y = A_1\,e^{\lambda_1 x} + A_2\,e^{\lambda_2 x}$$

2. Roots of auxiliary equation are real and equal, $\lambda_1 = \lambda_2 = \lambda$

$$y = A_1\,e^{\lambda x} + A_2\,x\,e^{\lambda x}$$

3. Roots of auxiliary equation are complex, $\lambda_{1,2} = -a \pm i\sqrt{b-a^2} = -a \pm i\omega$

$$y = e^{-ax}(A_1\,e^{i\omega x} + A_2\,e^{-i\omega x}) = e^{-ax}(B_1 \cos \omega x + B_2 \sin \omega x) =$$
$$= Ae^{-ax} \cos(\omega x + \varphi)$$

Particular integral for different F(x)

$F(x)$ is polynomial

$$F(x) = k_0 + k_1 x + k_2 x^2 + \ldots$$
$$y = A_0 + A_1 x + A_2 x^2 + \ldots$$

$F(x)$ is trigonometric function

$$F(x) = K_1 \cos \omega x + K_2 \sin \omega x$$

$y = A_1 \cos \omega x + A_2 \sin \omega x$ if $a \neq 0$ or $b \neq \omega^2$

$y = \dfrac{K_1 x}{2\omega} \sin \omega x - \dfrac{K_2 x}{2\omega} \cos \omega x$ if $a = 0$ and $b = \omega^2$

$F(x)$ is exponential function

$$F(x) = Ke^{kx}$$

1. k is not root of auxiliary equation

$$y = \frac{K}{f(k)}\,e^{kx}$$

2. k is single root of auxiliary equation

$$y = \frac{K}{f'(k)}\,x\,e^{kx}$$

3. k is double root of auxiliary equation

$$y = \frac{K}{f''(k)}\,x^2\,e^{kx}$$

Some named differential equations

Hermite equation

$$y'' - 2xy' + 2ny = 0$$

Solution for $n = 0, 1, 2, \ldots$

$$y = H_n(x) = (-1)^n\,e^{x^2}\,\frac{d^{(n)}}{dx^n}\,(e^{-x^2})$$

$H_n(x)$ = Hermite polynomials (see Sec. M–8)

Legendre equation

$$(1 - x^2)y'' - 2xy' + \ell(\ell+1)y = 0$$

Solution for $\ell = 0, 1, 2, \ldots$

$$y = P_\ell(x) = \frac{1}{2^\ell \ell!}\,\frac{d^{(\ell)}}{dx^\ell}\,(x^2 - 1)^\ell$$

$P_\ell(x)$ = Legendre polynomials (see Sec. M–8)

Associated Legendre equation

$$(1 - x^2)y'' - 2xy' + [\,\ell(\ell+1) - \frac{m^2}{1-x^2}\,]\,y = 0$$

Solution for $\ell, m = 0, 1, 2, \ldots \quad (m \leq \ell)$

$$y = P_\ell^m(x) = (1 - x^2)^{m/2} \frac{\mathrm{d}^{(m)}}{\mathrm{d}x^m} P_\ell(x)$$

$P_\ell^m(x)$ = associated Legendre functions (see Sec. M–8)

Laguerre equation

$$xy'' + (1 - x)y' + n\,y = 0$$

Solution for $\ell = 0, 1, 2, \ldots$

$$y = L_\ell(x) = \mathrm{e}^x \frac{\mathrm{d}^{(\ell)}}{\mathrm{d}x} (x^\ell\, \mathrm{e}^{-x})$$

$L_\ell(x)$ = Laguerre polynomials (see Sec. M–8)

Associated Laguerre equation

$$xy'' + (m + 1 - x)y' + (n - m)y = 0$$

Solution for $\ell, m = 0, 1, 2, \ldots (m \leq \ell)$

$$y = L_\ell^m(x) = \frac{\mathrm{d}^{(m)}}{\mathrm{d}x^m} L_\ell(x)$$

$L_\ell^m(x)$ = Associated Laguerre polynomials (see Sec. M–8)

Bessel's differential equation

$$x^2y'' + xy' + (x^2 - n^2)y = 0 \qquad n \geq 0$$

Solutions are called *Bessel functions of order n.*

$$J_n(x) = \sum_{k=0}^{\infty} \frac{(-1)^k (x/2)^{n+2k}}{k!\,\Gamma(n+k+1)}$$

$J_n(x)$ = Bessel functions of the first kind of order n

$$J_{-n}(x) = \sum_{k=0}^{\infty} \frac{(-1)^k (x/2)^{2k-n}}{k!\,\Gamma(k+1-n)}$$

15 Numerical Methods

Newton-Raphson method of solving $f(x) = 0$

$$x_{n+1} = x_n - f(x_n)/f'(x_n)$$

Crout's method to solve a system of linear equations

$$\begin{pmatrix} a_{11} & a_{12} & \cdots & a_{1n} \\ \vdots & \vdots & & \vdots \\ a_{n1} & a_{n2} & \cdots & a_{nn} \end{pmatrix} \begin{pmatrix} x_1 \\ \vdots \\ x_n \end{pmatrix} = \begin{pmatrix} y_1 \\ \vdots \\ y_n \end{pmatrix}$$

Unknown: $x_1, x_2, \dots x_n$

I. Calculate $b_{11}, b_{12}, \dots b_{1n}, b_{21}, b_{22}, \dots b_{nn}$

$$\begin{cases} b_{ij} = a_{ij} - \sum_{k=1}^{j-1} b_{ik} b_{kj} & \text{for } i \geq j \\ b_{ij} = \dfrac{1}{b_{ii}} (a_{ij} - \sum_{k=1}^{i-1} b_{ik} b_{kj}) & \text{for } i < j \end{cases}$$

II. Calculate $z_1, z_2, \dots z_n$

$$z_i = \frac{1}{b_{ii}} (y_i - \sum_{k=1}^{i-1} b_{ik} b_{kj})$$

III. Calculate $x_n, x_{n-1}, \dots x_1$

$$x_i = z_i - \sum_{k=i+1}^{n} b_{ik} x_k$$

Simpson's method of solving $\int_a^b f(x)\,\mathrm{d}x$

$$\int_a^b f(x)\,\mathrm{d}x = \frac{h}{3} [f(x_0) + 4 f(x_1) + 2 f(x_2) + \dots$$
$$+ 2 f(x_{n-2}) + 4 f(x_{n-1}) + f(x_n)] + R$$

where $x_0 = a, x_n = b, x_i = x_0 + ih$
(n must be even)

$$R = -\frac{(b-a)}{180} h^4 f^{(4)}(\xi), \quad a \leq \xi \leq b$$

Numerical methods to solve $y' = f(x, y)$

Polygon method

Given

$$y' = f(x, y)$$
$$y(x_0) = y_0$$

$(n + 1)$:st point

$$x_{n+1} = x_n + h$$
$$y_{n+1} = y_n + h \cdot f(x_n, y_n)$$

Runga-Kutta method

Given

$$y' = f(x, y)$$
$$y(x_0) = y_0$$

Auxiliary quantities

$$k_1 = hf(x_n, y_n)$$
$$k_2 = hf(x_n + \tfrac{1}{2}h, y_n + \tfrac{1}{2}k_1)$$
$$k_3 = hf(x_n + \tfrac{1}{2}h, y_n + \tfrac{1}{2}k_2)$$
$$k_4 = hf(x_n + h, y_n + k_3)$$

$(n + 1)$:st point

$$x_{n+1} = x_n + h$$
$$y_{n+1} = y_n + \tfrac{1}{6}(k_1 + 2k_2 + 2k_3 + k_4)$$

The global truncation error is of the order h^4.

Runge-Kutta method of solving $y'' = f(x, y, y'), y(x_0) = y_0, y'(x_0) = y'_0$

Iteration: $x_{n+1} = x_n + h$

$$y_{n+1} = y_n + hy'_n + \frac{h}{6}(k_1 + k_2 + k_3)$$

$$y'_{n+1} = y'_n + \frac{1}{6}(k_1 + 2k_2 + 2k_3 + k_4)$$

$$k_1 = hf(x_n, y_n, y'_n)$$

$$k_2 = hf(x_n + \frac{h}{2}, y_n + \frac{h}{2}y'_n, y'_n + \frac{k_1}{2})$$

$$k_3 = hf(x_n + \frac{h}{2}, y_n + \frac{h}{2}y'_n + \frac{k_1 h}{4}, y'_n + \frac{k_2}{2})$$

$$k_4 = hf(x_n + h, y_n + hy'_n + \frac{k_2 h}{2}, y'_n + k_3)$$

16 Error Estimation

(For logarithmic differentiation see Sec. M–5)

Average (arithmetic mean)

$$\langle x \rangle = \frac{\sum_{i=1}^{n} x_i}{n}$$

Weighted average

$$\langle x \rangle = \sum_{i=1}^{n} \omega_i x_i \,/ \sum_{i=1}^{n} \omega_i$$

Standard deviation (mean error) of an individual measurement

$$s = \sqrt{\frac{\sum_{i=1}^{n} (x_i - \langle x \rangle)^2}{n-1}} = \sqrt{\frac{\sum x_i^2 - n\langle x \rangle^2}{n-1}} \approx \frac{1.25}{n} \sum |x_i - \langle x \rangle|$$

Standard deviation (mean error) of average

$$s_{\text{ave}} = \frac{s}{\sqrt{n}}$$

Standard deviation of n counted, Poisson distributed pulses

$$\sigma = \sqrt{n}$$

Probability that number of pulses at time $t = n$

$$P(t,n) = e^{-\lambda t}(\lambda t)^n / n! \qquad \lambda = \text{intensity}$$

Repeated intervals

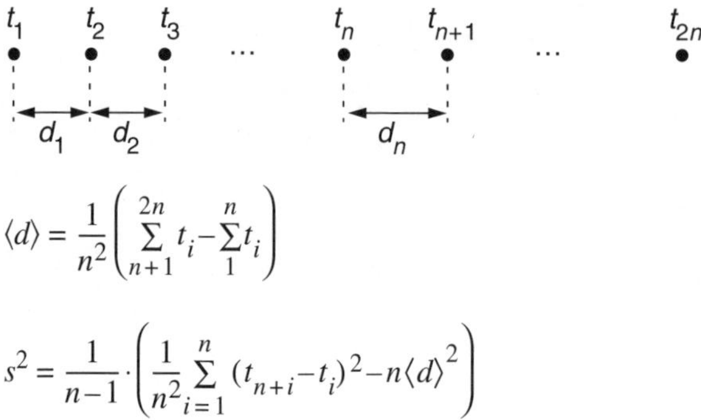

$$\langle d\rangle = \frac{1}{n^2}\left(\sum_{n+1}^{2n} t_i - \sum_{1}^{n} t_i\right)$$

$$s^2 = \frac{1}{n-1}\cdot\left(\frac{1}{n^2}\sum_{i=1}^{n}(t_{n+i}-t_i)^2 - n\langle d\rangle^2\right)$$

Propagation of errors (mean error and maximum error)

$$\Delta Z = \sqrt{\left(\frac{\partial Z}{\partial x}\right)^2(\Delta x)^2 + \left(\frac{\partial Z}{\partial y}\right)^2(\Delta y)^2 + \ldots}$$

$$|\Delta Z_{\max}| = \left|\frac{\partial Z}{\partial x}\Delta x_{\max}\right| + \left|\frac{\partial Z}{\partial y}\Delta y_{\max}\right| + \ldots$$

Weighted average of measurements with different errors (confidence intervals)

$$\langle z\rangle = \frac{\Sigma(1/\Delta z_i)^2\langle z_i\rangle}{\Sigma(1/\Delta z_i)^2}$$

Error of weighted average

$$\Delta z = [\Sigma(1/\Delta z_i)^2]^{-1/2}$$

Confidence interval for $n = f + 1$ measurements of quantity x

$$\langle x\rangle - t_{\mathrm{p}}(f)s_{\mathrm{ave}} < X < \langle x\rangle + t_{\mathrm{p}}(f)s_{\mathrm{ave}}$$

Confidence interval for n_1 measurements of quantity u_1, n_2 measurements of quantity u_2, etc. (total of N quantities), with standard deviation s of total set of measurements

$$\langle u_j \rangle - t_p(f)\frac{s}{\sqrt{n_j}} < U_j < \langle u_j \rangle + t_p(f)\frac{s}{\sqrt{n_j}}$$

$$f = \sum_j n_j - N$$

Lengths of confidence intervals (t_p-values) for four confidence levels P

f	$P = 70\,\%$	$P = 95\,\%$	$P = 99\,\%$	$P = 99.9\,\%$
1	1.96	12.7	63.7	636.6
2	1.39	4.3	9.9	31.6
3	1.25	3.2	5.8	12.9
4	1.19	2.8	4.6	8.6
5	1.16	2.6	4.0	6.9
7	1.12	2.4	3.5	5.4
10	1.09	2.2	3.2	4.6
25	1.06	2.1	2.8	3.7
100	1.04	2.0	2.6	3.4
∞	1.04	1.96	2.58	3.29

Least squares fit of straight line $y = kx + \ell$ to a set of measured points x_i, y_i (linear regression)

$$k = \frac{\Sigma(x_i - \langle x \rangle)\, y_i}{\Sigma(x_i - \langle x \rangle)^2} = \frac{n\,\Sigma x_i y_i - \Sigma x_i\, \Sigma y_i}{n\,\Sigma x_i^2 - (\Sigma x_i)^2}$$

$$\ell = \langle y \rangle - k\langle x \rangle = \frac{1}{n}\,(\Sigma y_i - k\,\Sigma x_i)$$

$$(\Delta k)^2 = \frac{1}{\Sigma(x_i - \langle x \rangle)^2}\;\frac{\Sigma(y_i - kx_i - \ell)^2}{n-2}$$

$$(\Delta \ell)^2 = \left(\frac{1}{n} + \frac{\langle x \rangle^2}{\Sigma(x_i - \langle x \rangle)^2}\right)\frac{\Sigma(y_i - kx_i - \ell)^2}{n-2}$$

Correlation coefficient

$$\rho = \frac{n\,\Sigma x_i y_i - \Sigma x_i\, \Sigma y_i}{(n\,\Sigma x_i^2 - (\Sigma x_i)^2)\,(n\,\Sigma y_i^2 - (\Sigma y_i)^2)} \qquad -1 \le \rho \le 1$$

Equations for least squares fit of a polynomial $y = a_n x^n + a_{n-1} x^{n-1} + \ldots + a_0$ **to a set of measured points** x_i, y_i

$$\begin{pmatrix} n & \Sigma x_i & \Sigma x_i^2 & \Sigma x_i^3 & \ldots & \Sigma x_i^n \\ \Sigma x_i & \Sigma x_i^2 & \Sigma x_i^3 & \Sigma x_i^4 & \ldots & \Sigma x_i^{n+1} \\ \vdots & \vdots & \vdots & \vdots & & \vdots \\ \Sigma x_i^n & \Sigma x_i^{n+1} & \Sigma x_i^{n+2} & \Sigma x_i^{n+3} & \ldots & \Sigma x_i^{2n} \end{pmatrix} \begin{pmatrix} a_0 \\ a_1 \\ \vdots \\ a_n \end{pmatrix} = \begin{pmatrix} \Sigma y_i \\ \Sigma x_i y_i \\ \vdots \\ \Sigma x_i^n y_i \end{pmatrix}$$

$a_0, a_1, a_2, \ldots$ can be found with Crout's method, see Sec. M – 15.

Frequency and integral functions of the normal distribution

$$f(u) = \frac{1}{\sqrt{2\pi}} \mathrm{e}^{-\frac{1}{2}u^2}$$

$$\Phi(x) = \int_{-\infty}^{x} f(u)\,\mathrm{d}u$$

The normal distribution

$$\Phi(x) = \frac{1}{\sqrt{2\pi}} \int_{-\infty}^{x} e^{-t^2/2}\, dt$$

x	.00	.01	.02	.03	.04	.05	.06	.07	.08	.09
0.0	.5000	.5040	.5080	.5120	.5160	.5199	.5239	.5279	.5319	.5359
0.1	.5398	.5438	.5478	.5517	.5557	.5596	.5636	.5675	.5714	.5753
0.2	.5793	.5832	.5871	.5910	.5948	.5987	.6026	.6064	.6103	.6141
0.3	.6179	.6217	.6255	.6293	.6331	.6368	.6406	.6443	.6480	.6517
0.4	.6554	.6591	.6628	.6664	.6700	.6736	.6772	.6808	.6844	.6879
0.5	.6915	.6950	.6985	.7019	.7054	.7088	.7123	.7157	.7190	.7224
0.6	.7257	.7291	.7324	.7357	.7389	.7422	.7454	.7486	.7517	.7549
0.7	.7580	.7611	.7642	.7673	.7704	.7734	.7764	.7794	.7823	.7852
0.8	.7881	.7910	.7939	.7967	.7995	.8023	.8051	.8078	.8106	.8133
0.9	.8159	.8186	.8212	.8238	.8264	.8289	.8315	.8340	.8365	.8389
1.0	.8413	.8438	.8461	.8485	.8508	.8531	.8554	.8577	.8599	.8621
1.1	.8643	.8665	.8686	.8708	.8729	.8749	.8770	.8790	.8810	.8830
1.2	.8849	.8869	.8888	.8907	.8925	.8944	.8962	.8980	.8997	.9015
1.3	.9032	.9049	.9066	.9082	.9099	.9115	.9131	.9147	.9162	.9177
1.4	.9192	.9207	.9222	.9236	.9251	.9265	.9279	.9292	.9306	.9319
1.5	.9332	.9345	.9357	.9370	.9382	.9394	.9406	.9418	.9429	.9441
1.6	.9452	.9463	.9474	.9484	.9495	.9505	.9515	.9525	.9535	.9545
1.7	.9554	.9564	.9573	.9582	.9591	.9599	.9608	.9616	.9625	.9633
1.8	.9641	.9649	.9656	.9664	.9671	.9678	.9686	.9693	.9699	.9706
1.9	.9713	.9719	.9726	.9732	.9738	.9744	.9750	.9756	.9761	.9767
2.0	.97725	.97778	.97831	.97882	.97932	.97982	.98030	.98077	.98124	.98169
2.1	.98214	.98257	.98300	.98341	.98382	.98422	.98461	.98500	.98537	.98574
2.2	.98610	.98645	.98679	.98713	.98745	.98778	.98809	.98840	.98870	.98899
2.3	.98928	.98956	.98983	.99010	.99036	.99061	.99086	.99111	.99134	.99158
2.4	.99180	.99202	.99224	.99245	.99266	.99286	.99305	.99324	.99343	.99361
2.5	.99379	.99396	.99413	.99430	.99446	.99461	.99477	.99492	.99506	.99520
2.6	.99534	.99547	.99560	.99573	.99585	.99598	.99609	.99621	.99632	.99643
2.7	.99653	.99664	.99674	.99683	.99693	.99702	.99711	.99720	.99728	.99736
2.8	.99744	.99752	.99760	.99767	.99774	.99781	.99788	.99795	.99801	.99807
2.9	.99813	.99819	.99825	.99831	.99836	.99841	.99846	.99851	.99856	.99861
3.0	.99865									
3.1	.99903									
3.2	.99931									
3.3	.99952									
3.4	.99966									
3.5	.99977									
3.6	.99984									
3.7	.99989									
3.8	.99993									
3.9	.99995									
4.0	.99997									

$\alpha = 1 - \Phi(\lambda_\alpha)$

α	λ_α	α	λ_α
0.10	1.2816	0.001	3.0902
0.05	1.6449	0.0005	3.2905
0.025	1.9600	0.0001	3.7190
0.010	2.3263	0.00005	3.8906
0.005	2.5758	0.00001	4.2649

Appendices

A. The Greek Alphabet

(*Italic letters*)

α	A	alpha	ι	I	iota	ρ	P	rho
β	B	beta	κ	K	kappa	σ, ς	Σ	sigma
γ	Γ	gamma	λ	Λ	lambda	τ	T	tau
δ	Δ	delta	μ	M	mu	υ	Υ	upsilon
ε, ϵ	E	epsilon	ν	N	nu	φ, ϕ	Φ	phi
ζ	Z	zeta	ξ	Ξ	xi	χ	X	chi
η	H	eta	o	O	omikron	ψ	Ψ	psi
θ, ϑ	Θ	theta	π	Π	pi	ω	Ω	omega

B. Notation and Writing Rules

Values of physical quantities, units, and prefixes

A physical quantity is expressed as the product of a numerical value and a unit. For example, the equatorial radius of the earth $R = 6.38 \cdot 10^6$ m, or with a prefix $R = 6.38$ Mm. Do not put a period after the unit unless it indicates the end of a sentence.

Do not use multiple prefixes. For example, the mass turned into energy by the sun $= 4 \cdot 10^9$ kg/s. Do not write 4 Gkg/s. Use GWh instead of MkWh and mg instead of µkg.

There must not be a space between the prefix and the unit. If the notation of a unit includes several letters, the leading letter should be interpreted as a prefix if this is possible. Example: mN means millinewton and Nm means newtonmetre. If necessary, a dot can be placed between the units, for example T · m = teslametre (not terametre).

When to use italic and bold fonts

Symbols representing physical quantities are written in italics with Latin or Greek letters. For vector quantities a medium bold font should be used.

If an index is attached to a symbol, the index should be in italics if it represents a physical quantity. A letter index representing a series of numbers should also be printed in italics. For all other indices a regular non-italic font should be used.

Digits, units, and prefixes before units are always printed with a regular font. Also use a regular font for mathematical functions and constants such as sin, cos, ln, and e.

When to use capital letters

Names of units are written with a small initial letter, for example second, metre, joule, watt, tesla, and kelvin. Notice: use a capital C in degree Celsius.

Abbreviations for units have an upper case initial letter if the unit has been named after a person, for example J (named after James Prescott Joule), Sv (named after Rolf Sievert), and W (named after James Watt), but m for metre, and s for second. Exception: AU = astronomical unit.

Names of the elements are written with a lower case initial letter.

Dimensions

Most physical constants and variables have a dimension. We often refer to mass, length, and time as the fundamental dimensions. The dimensions of a quantity are the same in every frame of reference. Two values can never be added unless they have the same dimensions.

It is possible for a quantity to have units but no dimensions. For example, a watch that runs slow and loses 3 seconds a day has a rate of loss = 3 s/day. This can also be expressed as 0.05 s/h or in the form $3.5 \cdot 10^{-5}$, which gives the fractional loss of time in any interval.

The following characters are used to designate dimensions:
dim(*length*) = L   dim(*mass*) = M   dim(*time*) = T
dim(*electric current*) = I   dim(*temperature*) = Θ   dim(*amount of substance*) = N
dim(*luminous intensity*) = J

Indices to designators for the elements

Nuclear reactions: ${}^{A}_{Z}\mathrm{X}^{N}$ A = mass number (= $Z + N$), N = number of neutrons
Z = atomic number (= number of protons)

Chemical reactions: X^{q}_{n} q = ionic charge (including sign)
n = number of atoms

C. History of the Elements

… after a name means several discoverers, Eng = English, Ger = German, Gr = Greek, Lat = Latin, Sw = Swedish. Names of Swedish persons have a bold font. Ytterby, a mine in the Stockholm archipelago, has named four elements.

	Element	Discovery		Etymology of the name
1	hydrogen	1766	Cavendish	Gr hydro = water
2	helium	1868	Janssen	Gr helios = sun
3	lithium	1817	**J. A. Arfwedson**	Gr lithos = stone
4	beryllium	1798	Vauquelin	Gr beryllos = beryl (a mineral)
5	boron	1807	Davy	Arabic buraq and persian burah
6	carbon	Prehistoric element		Lat carbo = charcoal
7	nitrogen	1772	D. Rutherford[1]	Gr nitron = soda
8	oxygen	1774	Priestley[1]	Gr oxys = acid
9	fluorine	1886	Moissan	Lat fluere = flow or flux
10	neon	1898	Ramsay …	Gr neos = new
11	sodium	1807	Davy	Eng soda. Symbol Na from the Natron valley, near Cairo, Egypt.
12	magnesium	1808	Davy	Gr Magnesia = district in Thessaly
13	aluminium	1827	Woehler	Lat alumen = alum (a salt)
14	silicon	1824	**J. J. Berzelius**	Lat silex = flint
15	phosphorus	1669	Brand	Gr phosphoros = light-bringing
16	sulphur	1777	Lavoisier[2]	Known since ancient times
17	chlorine	1810	Davy	Gr kloros = greenish yellow
18	argon	1894	Rayleigh …	Gr argos = inactive
19	potassium	1807	Davy	Lat kallium and Eng potash
20	calcium	1808	**J. J. Berzelius …**	Lat caix = lime
21	scandium	1879	**Lars F. Nilson**	Lat Scandia = Scandinavia
22	titanium	1791	Gregor	Titans = giants in Gr mythology
23	vanadium	1801	del Rio[3]	Vanadis = name of a Scandinavian goddess, also called Freya
24	chromium	1797	Vauquelin	Gr kroma = colour
25	manganese	1774	**C. W. Scheele**	Lat magnes = magnet
26	iron	Used since 1200 BC		Symbol Fe from Lat ferrum = iron
27	cobalt	1735	**Georg Brandt**	Ger Kobalt = an evil spirit
28	nickel	1751	**Axel Cronstedt**	Ger Nickel = an evil and deceptive spirit
29	copper	Used since 5000 BC		Probably from Lat name of Cyprus
30	zinc	1746	Marggraf[4]	Old Ger word for zinc
31	gallium	1875	Boisbaudran	Lat Gallia = France
32	germanium	1886	Winkler	Lat Germania = Germany
33	arsenic	1250	A. Magnus	Lat arsenicum. Gr arsenikos = male
34	selenium	1817	**J. J. Berzelius**	Gr selene = old word for moon
35	bromine	1826	Balard[5]	Gr bromos = stench, awful odour
36	krypton	1898	Ramsay …	Gr kryptos = hidden

	Element	Discovery		Etymology of the name
37	rubidium	1861	Bunsen …	Lat rubidius = deepest red
38	strontium	1808	Davy	Strontian = a town in Scotland
39	yttrium	1794	Gadolin	Ytterby, see comment at top
40	zirconium	1789	Klaproth[6]	Arabic zargun = gold colour
41	niobium	1801	Hatchett	Niobe was daughter of Gr god Tantalus
42	molybdenum	1778	**C. W. Scheele**[7]	Gr molybdos = lead
43	technetium	1937	Perrier, Segre	Gr teknetos = artificial
44	ruthenium	1828	Osann[8]	Lat Ruthenia = Russia
45	rhodium	1803	Wollaston	Gr rhodon = rosy or pinkish
46	palladium	1803	Wollaston	Asteroid Pallas, discovered 1802[9]
47	silver	Known since 3500 BC		Symbol Ag from Lat argentum = silver
48	cadmium	1817	Strohmeyer	Gr cadmia fornakum ≈ "furnace zinc"
49	indium	1863	Reich, Richter	From the intense indigo blue colour that indium salts impart to flames
50	tin	Known since 2500 BC		Symbol Sn from Lat stannum = tin
51	antimony	Prehistoric element		Lat antimonium = not found alone, symbol Sb from Lat stibium = stibnite, a mineral
52	tellurium	1782	Reichenstein	Lat tellus = earth
53	iodine	1811	Courtois	Gr iodes = violet
54	xenon	1898	Ramsay …	Gr xenon = stranger
55	cesium	1860	Bunsen …	Lat caesius = blue sky
56	barium	1779	**C. W. Scheele**[10]	Gr barus = heavy
57	lanthanum	1839	**Carl Mosander**	Gr lantanein = to lie hidden
58	cerium	1803	Klaproth	Asteroid Ceres, discovered 1801
59	praseodymium	1885	Welsbach	Gr prasinos = leekgreen, referring to the green salts, Gr didymos = twin
60	neodymium	1885	Welsbach	Gr neos = new, Gr didymos = twin
61	promethium	1945	Marinsky …	In Gr mythology Prometheus stole fire from the gods and gave it to humans
62	samarium	1879	Boisbaudran	Named after Russian mining engineer Samarskij
63	europium	1896	Demarcay	Europe
64	gadolinium	1880	de Marignac	Johan Gadolin = Finnish chemist
65	terbium	1843	**Carl Mosander**	Ytterby, see comment at top
66	dysprosium	1886	Boisbaudran	Gr dysprositos = hard to get at
67	holmium	1879	**Per T. Cleve**[11]	Lat Holmia = Stockholm
68	erbium	1842	**Carl Mosander**	Ytterby, see comment at top
69	thulium	1879	**Per T. Cleve**	Thule = early name of Scandinavia
70	ytterbium	1878	Marignac	Ytterby, see comment at top
71	lutetium	1907	Urbain	Lat Lutetia = ancient Roman name for Paris
72	hafnium	1923	Coster, Hevesy	Lat Hafnia = Copenhagen
73	tantalum	1802	**Gustav Ekeberg**	In Gr mythology Tantalus was a son of Zeus

	Element	Discovery		Etymology of the name
74	tungsten	1779	Woulfe[12]	Sw "tung sten" = heavy stone
75	rhenium	1925	Noddack …	Lat Rhenus = the Rhine
76	osmium	1803	Tennant	Gr osmos = a smell
77	iridium	1804	Tennant	Lat iris = rainbow
78	platinum	1735	Ulloa	Spanish platina ≈ "small silver"
79	gold	Used before 9000 BC		Symbol Au from Lat aurum = beautiful dawn
80	mercury	Used in Egypt 1500 BC		Mercury was a god in Roman mythology
81	thallium	1861	Crookes	Gr thallos = green or young twig
82	lead	Prehistoric element		Symbol Pb from Lat plumbum = water-works
83	bismuth	1753	Geoffroy	Ger "weisse Masse" = white mass
84	polonium	1898	P and M Curie	Poland = home country of Marie Curie
85	astatine	1940	Corson …	Gr astatos = not stable
86	radon	1900	Dorn	Radon ≈ decendent from radium[13]
87	francium	1939	Perrey	France
88	radium	1898	P and M Curie	Lat radius = ray
89	actinium	1899	Debierne	Gr acktis or actinos = bream or ray
90	thorium	1828	**J. J. Berzelius**	Thor = Nordic god of war and thunder
91	protactinium	1913	Fajans, …	Gr protos = first, Gr actinos = beam or ray
92	uranium	1789	Klaproth	The planet Uranus, discovered 1781[14]
93	neptunium	1940	McMillan …	The planet Neptun[15]
94	plutonium	1941	Seaborg …	The planet Pluto[16]
95	americium	1944	Seaborg …	America
96	curium	1944	Seaborg …	Pierre and Marie Curie
97	berkelium	1949	Seaborg …	Berkeley = city in California
98	californium	1950	Seaborg …	California and University of California
99	einsteinium	1952	[17]	Albert Einstein
100	fermium	1952	Seaborg …	Enrico Fermi
101	mende-levium	1955	Seaborg …	Dmitrij Mendeleyev, Russian chemist[18]
102	nobelium	1957	[19]	**Alfred Nobel**
103	lawrencium	1961	Ghiorso …	Ernest Lawrence, inventor of the cyclotron
104	rutherfor-dium	1964	[20]	Ernest Rutherford
105	dubnium	1968	[20]	Institute for Nuclear Research in Dubna, Russia
106	seaborgium	1974	[20]	Glenn Seaborg, chemist and physicist from USA
107	bohrium	1981	Münzenberg …	Niels Bohr, Danish physicist
108	hassium	1984	Münzenberg …	Lat Hassia = the province Hasse in Germany
109	meitnerium	1982	Münzenberg …	Lise Meitner, nuclear physicist from Austria

	Element	Discovery		Etymology of the name
110	darmstadtium	1994	[21]	Reserch center GSI in Darmstadt, Germany
111	roentgenium	1994	[21]	Wilhelm Röntgen
112	–	1996	[21]	
113	–	2003	[22]	
114	–	1999	[22]	
115	–	2003	[22]	

[1] Nitrogen and oxygen were discovered independently by **Carl Wilhelm Scheele.**

[2] Sulphur is known since ancient times. Lavoisier classified it as an element.

[3] Vanadium was rediscovered and named in 1830 by **Nils G. Sefström.**

[4] Zinc was produced in India and China during the Middle Ages.

[5] Balard was only 23 years old when he published a paper in which he announced the discovery of a new element that he called muride. The Academie Francaise did not accept the proposed name.

[6] Impure zirconium was first isolated by **Jöns Jakob Berzelius.**

[7] **Carl Wilhelm Scheele** recognized molybdenite as a distinct ore of a new element, and **Peter Jacob Hjelm** prepared molybdenum in an impure form in 1782.

[8] The discovery of ruthenium is often credited Karl Klaus, who obtained pure metal in 1844.

[9] In Gr mythology Palladium was a wodden statue of goddess Pallas Athene (or Athena).

[10] Barium was first isolated in 1808 by Humphry Davy.

[11] Holmium was also discovered by Soret and Delafontaine but was named by **Per Teodor Cleve.**

[12] Tungsten was also discovered in 1781 by **Carl Wilhelm Scheele,** who gave the element its English name.

[13] Radon was once called niton from Lat nitens = shining.

[14] In Greek mythology Uranus represented the sky or heaven.

[15] In Roman mythology Neptune was the chief god of the sea.

[16] In Greek mythology Pluto was a god both of death and of fertility or abundance.

[17] Einsteinium was first discovered in the debris from the "Mike" thermonuclear explosion in the South Pacific in 1952. Shortly thereafter it was produced by Seaborg and collaborators.

[18] In 1869 Mendeleyev devised the periodic table of the elements.

[19] A group of scientists from Great Britain, Sweden, and the USA claimed to have produced nobelium in 1957. The experiment could not be reproduced. Seaborg and collaborators produced nobelium in 1958.

[20] Both Lawrence Berkeley National Laboratory in California and the Institute for Nuclear Research in Dubna near Moscow claim to have been first to create elements 104, 105, and 106.

[21] Elements 110, 111, and 112 were first produced by a team led by Sigurd Hofmann at the research center Gesellschaft für Schwerionenforschung GmbH (or GSI for short) in Darmstadt, Germany.

[22] Elements 113, 114, and 115 were first produced by scientists from the Joint Institute for Nueclear Research, Dubna, Russia and the Lawrence Livermore National Laboratory, Berkeley, California.

D. Famous Physicists

More information on famous scientists can be found at the following internet sites.

WWW

http://scienceworld.wolfram.com/biography/
http://www-history.mcs.st-andrews.ac.uk/history/ (Mostly mathematicians)
http://www.physics.ucla.edu/~cwp/ (Famous women)
http://dbhs.wvusd.k12.ca.us/Gallery/GalleryMenu.html (A photo gallery)

Ampère, André M. (1775–1836)

French physicist who applied mathematical formulae to the relation between electric currents and magnetism. His first paper was presented one week after he heard of the Danish physicist Hans Christian Ørsted's observations that electric currents produce magnetic fields. He later invented the solenoid and he showed that parallel wires with current in the same direction attract and wires with current in opposite directions repel each other. In 1814 he independently derived Avogadro's law and later also developed a wave theory of heat. He was greatly affected by his father's execution during the French Revolution and by his first wife's untimely death. He personally formulated the words on his tombstone: "Tandem felix" (happy at last).

Ångström, Anders Jonas (1814–1874)

Swedish physicist who was one of the founders of spectroscopy. He proposed in 1853 the relationship between emission and absorption spectra of chemical elements, which was more clearly articulated by Gustav Kirchhoff in 1859. In 1861, Ångström discovered hydrogen lines in the solar spectrum and subsequently confirmed the likely existence of other elements in the sun. In 1867 he initiated spectral studies of auroras and, a year later, published an authoritative map of the sun's spectral lines with wavelengths expressed in the unit today known as the ångström unit.

Archimedes of Syracuse (ca. 287–212 BC)

Archimedes was the greatest scientist and mathematician of ancient times. He calculated the area of a sphere and an ellipse and discovered that the volume of a sphere is two thirds of the volume of the smallest cylinder that can contain it. He showed that π has a value between 3 10/71 and 3 1/7. Archimedes was also an outstanding engineer who formulated the principle of buoyancy (today called Archimedes' principle), invented the catapult, the Archimedes' screw, a device still used, and constructed lenses which could focus the sun's light. It is said that he was killed by a Roman soldier when he had snapped at him "Don't perturb my circles", referring to a geometric figure he had outlined in the sand.

Becquerel, A. Henri (1852–1908)

French physicist who in 1896 accidentally discovered radioactivity while investigating fluorescence in uranium salts. For this he shared the 1903 Nobel Prize in physics with Marie and Pierre Curie. He discovered that the radiation from uranium and radium comprised electrons. The only place these electrons could be coming from was from within the atoms. At a stroke this observation destroyed the nineteenth century conception of atomic structure.

Bohr, Niels (1885-1962)

Danish physicist who was the first to apply quantum theory to atomic structure and thus founding the modern quantum theory of matter. His atom model from 1913 assumed that electrons orbit the nucleus at precise distances from the nucleus without loosing energy, and that the angular momentum associated with an allowed motion is an integral multiple of $\hbar$. He proposed that radiation is emitted when an electron jumps from one orbit (one quantum number) to another. In 1922 Bohr was awarded the Nobel Prize for physics. In 1943 Bohr and his family moved to the United States, where he participated in the atomic bomb project at Los Alamos. After the war he returned to Denmark.

Boltzmann, Ludwig (1844–1906)

Austrian physicist who (with J. W. Gibbs) developed the branch of physics called statistical mechanics. In the 1870s he obtained the Maxwell-Boltzmann distribution and explained the second law of thermodynamics by applying laws of probability to atomic motion. His work was opposed by many European physicists, and depressed and in poor physical health he committed suicide in 1906. Shortly thereafter the French scientist Jean Perrin verified much of his work.

de Broglie, Louis (1892–1987)

French physicist who in 1923 proposed the wave nature of material particles, which was experimentally confirmed for the electron in 1927. The wave concept of the electron was included in Schrödinger's wave mechanical picture of the atom. De Broglie has also written numerous popular works, including *New Perspectives in Physics* (1962). In 1929 he received the Nobel Prize for physics.

Celsius, Anders (1701–1744)

Swedish astronomer who in 1742 published the first precise determination of two fixed points on a temperature scale and thus formed a true international scale. He suggested zero degrees to be the boiling point of water and 100 degrees to be the freezing point. The opposite appeared some years later with no specific originator. Celsius also studied the Earth's magnetic field and published descriptions of the northern light. As professor of astronomy at Uppsala University he built an observatory, but only three years after it was built he died in tuberculosis.

Compton, Arthur Holly (1892–1962)

American physicist who described the behaviour of X-rays when they interact with electrons. In 1923 he found that when X-rays strike graphite they are scattered and their wavelengths are increased. This discovery, known as the "Compton effect", was the first proof that X-rays can behave like particles. For this discovery he shared the 1927 Nobel Prize in physics with C.T.R. Wilson, a Scottish physicist who received his share for inventing the cloud chamber, which is used to view tracks of charged nuclear particles. Compton suggested the name "photon" for the light quantum.

Copernicus, Nicholaus (1473–1543)

Polish astronomer and mathematician who is often considered the founder of modern astronomy. Polish name: Mikolaj Kopernik. By postulating (1) the rotation of the Earth, (2) revolution of the planets around the sun, and (3) the tilt of the Earth's rotational axis, Copernicus worked out a heliocentric model of the solar system in full mathematical detail. The sun was in this model slightly offset from the centre of the solar system. Out of fear that his ideas would get him into trouble with the church, his paper *De Revolutionibus Orbium Coelestium*, which was outlined in 1514, was delayed and published just before his death in 1543.

Curie, Marie, born Sklodowska (1867–1934)

Polish-French physicist and chemist who gave the name "radioactivity" to the emission of radiation from atoms. Together with her husband Pierre Curie she studied radioactivity which in 1898 led to the discovery of the elements radium and polonium. In 1903 they shared the Nobel Prize in physics with Henri Becquerel. In 1911 Marie Curie received a second Nobel Prize, this time in chemistry for the isolation of pure radium. The Curies refined eight tons of raw ore to produce one gram of radium. In 1935 Marie and Pierre's daughter, Irène Joliot-Curie, and son-in-law Frédéric Joliot-Curie received the Nobel Prize for chemistry. Marie Curie died of leukemia caused by overexposure to radioactivity.

Curie, Pierre (1859–1906)

French physicist and chemist, married to Marie Curie. He discovered the phenomenon of piezoelectricity and constructed a torsion balance with a tolerance of 0.01 mg. He also discovered that the magnetic susceptibility of paramagnetic materials is inversely proportional to the absolute temperature (Weiss-Curie law) and that there exists a critical temperature above which the magnetic properties disappear (the Curie temperature). Pierre Curie joined his wife's work with researching the nature of radioactivity, and for this he shared the 1903 Nobel Prize for physics with Henri Becquerel and his wife. In 1906 he died in a traffic accident in Paris, at the age of 46.

Dirac, Paul A. M. (1902–1984)

English mathematician and theoretical physicist who in 1930 predicted the existence of antimatter. The positron (anti-electron) was subsequently discovered in 1932 by Carl Anderson. Dirac also developed a relativistic version of the Schrödinger equation, known as the Dirac equations, and provided the first rigorous description of the spin of elementary particles. In 1933 he shared the Nobel Prize for physics with Erwin Schrödinger.

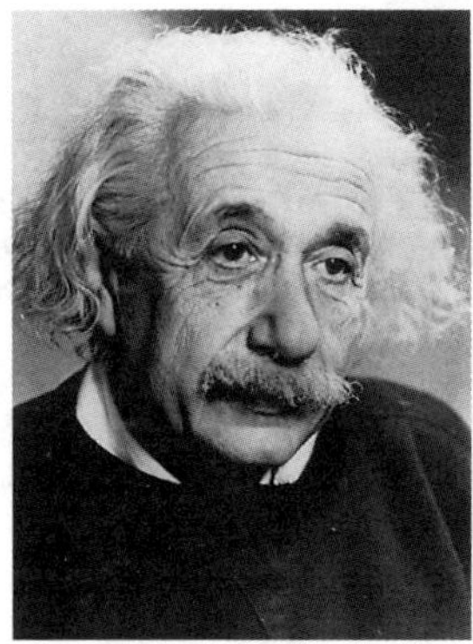

Einstein, Albert (1879–1955)

Einstein is the most well-known scientist of the 20th century. In 1905 he published three papers in *Annalen der Physik*, the leading German physics journal. In the first paper he proposed that light could behave as packets of energy and that the energy of these light quanta, as he called them, were proportional to the frequency. He went far beyond Planck's work and used this interpretation to explain the photoelectric effect. The second paper concerned statistical mechanics and explained the Brownian motion. In the third paper "*On the Electrodynamics of Moving Bodies*" he proposed the theory of special relativity, which was based on two postulates: (1) the physical laws are the same in all inertial reference systems, and (2) the speed of light is the same in every inertial reference system. Later in 1905 Einstein derived the equivalence between mass and energy ($E = mc^2$). Late 1915 Einstein published his General Theory of Relativity, which postulated that uniform acceleration and a gravitational field are equivalent. It predicted that the light from distant stars would bend slightly when passing near the sun, which was confirmed during an eclipse of the sun in 1919. Einstein was born in Germany, at the age of 22 he became a Swiss citizen, and in 1933 he moved to the United States and became a US citizen in 1940. When he in 1921 received the Nobel Prize for physics he was awarded for his explanation of the photoelectric effect, which shows that the world was still sceptical to his ideas on relativity.

Picture: Pressens Bild AB

More about Einstein: http://www.westegg.com/einstein/
and http://www.aip.org/history/einstein/

Faraday, Michael (1791–1867)

English chemist and physicist who is known for his pioneering experiments in electricity and magnetism and who is one of the greatest experimentalists of all time. He found the principle of induction and invented the dynamo, which produces electricity by mechanical means. His research of electrolysis led to the laws known today as Faraday's laws of electrolysis. Faraday discovered that an intense magnetic field can rotate the plane of polarized light, which today is known as the Faraday effect. He also introduced the concept of fields to describe magnetic and electric forces and advocated the law of energy conservation.

Fermi, Enrico (1901–1954)

Italian physicist who in 1938 escaped the fascism in Italy and emigrated to the United States, where he became a citizen in 1944. In 1938 he was awarded the Nobel Prize for physics for the discovery of artificially radioactive elements, produced by neutron irradiation, and for the discovery of nuclear reactions induced by slow neutrons. He divided elementary particles into two groups, today known as fermions and bosons, depending of their spin characteristics. Fermi is well-known for leading a group of scientists who on 2 December 1942 achieved the first man-made and self-sustaining nuclear chain reaction, which led to the con- struction of the atomic bomb and nuclear power plants.

Feynman, Richard P. (1918–1988)

American physicist, raconteur, and musician, who invented diagrams, now called Feynman diagrams, that describe the behaviour of systems of interacting particles. For this and for works within quantum field theories and quantum electrodynamics Feynman shared the 1965 Nobel Prize in physics with Tomonaga and Schwinger. Other works on particle spin and what he called "patrons" (quarks) contributed to the current theory of quarks. He is well-known for his ability to clearly explain many esoteric physical ideas in words that are easy to understand. His lectures have been published and widely read. In the USA he is known from his TV broadcasted demonstration of the reason to the accident with the Challenger space shuttle in 1986.

Read more and watch his lectures in New Zealand:
http://www.feynman.com/
http://www.vega.org.uk/series/lectures/feynman/index.php

Fresnel, Augustin (1788–1827)

French physicist who made fundamental contributions to theoretical and applied optics. Fresnel rejected the view derived from Newton that light consists of particles (corpuscles) and established a wave theory on a firm mathematical and experimental basis. He developed the Fresnel equations of reflection and refraction and developed a mathematical theory of refraction and polarization. In 1819 Poission found that as a consequence of Fresnel's theory, the centre of the shadow of a diffracting disk should be illuminated. This unexpected effect was subsequently observed, verifying Fresnel's theory.

Galilei, Galileo (1564–1642)

Italian scientist who showed that all bodies fall at the same rate. This opposed Aristotelian physics, which assumed that speed of fall is proportional to weight. Galilei described dynamics and statics and emphasized the use of mathematics. He said that motion is continuous and can only be altered by the application of a force. In 1604 he observed a supernova and when he failed to determine its parallax, he concluded that the star must be very distant. In 1609 Galilei built a 20-power telescope and found sunspots, craters on the moon, the four largest satellites of Jupiter, and the phases of Venus. He found the Copernican theory to be correct, but the church forced him to recant his conviction. In 1633 the church condemned him to life imprisonment for "vehement suspicion of heresy".

Gamow, George (1904–1968)

Russian-born American physicist who contributed significantly to our knowledge of nuclear reactions within stars, in particular he found that stars tend to become hotter when their hydrogen is depleted. Gamow also worked out a theory of alpha decay in terms of tunnelling through the nuclear potential barrier. He strongly supported the Big Bang theory of Georges Lemaitre. Gamow is well-known as an author of, for example, "Mr Tompkins Explores the Atom", "Atomic Energy in Cosmic and Human Life", "The Creation of the Universe", "Matter, Earth, and Sky", "A Planet Called Earth", "Thirty Years that Shook Physics", and "A Star Called the Sun".

http://www.colorado.edu/physics/Web/Gamow/life.html

Contributed by *R. Igor Gamow.*

Hawking, Stephen W. (1942–)

British theoretical physicist who in 1973 combined general relativity with quantum theory to predict that black holes radiate as if they have a temperature proportional to their surface gravity and therefore evaporate over time. When Hawking was 21 years old it was discovered that he has Lou Gehrig's disease (amyotrophic lateral scleroris, ALS) and it was predicted that he would die young. He is since around 1970 confined to a wheel chair, and after a tracheostomy operation in 1985 he is unable to speak without the aid of a computer voice synthesizer. Hawking, who today is one of the world's most well-known physicists, has published several books, among them *A Brief History of Time*, which has been translated to 33 languages and has sold 10 million copies.

http://www.hawking.org.uk

Heisenberg, Werner (1901–1976)

German theoretical physicist who is best know for his uncertainty principle. In 1925 Heisenberg invented matrix mechanics, the first version of quantum mechanics. For this work and for the extension of it into a complete mathematical theory of the behaviour of atoms and their components, he was in 1932 awarded the Nobel Prize for physics. During World War II he headed the unsuccessful German nuclear weapon project.

Hubble, Edwin (1889–1953)

American astronomer, the founder of extragalactic astronomy, who in 1925 composed the classification scheme for the structure of galaxies which is still in use. He demonstrated that the Andromeda nebula was far outside the Milky Way and found observational evidence for an expanding universe. Hubble used cepheid variables to determine the distance to other galaxies and proposed, in 1929, the Hubble law for distances to other galaxies, but his original value of the Hubble constant (526 km s^{-1} Mpc^{-1}) was not very accurate.

Huygens, Christiaan (1629–1695)

Dutch astronomer, mathematician and physicist who proposed a wave theory of light. In 1656 he invented the pendulum clock, which greatly increased the accuracy of time measurements. In the early 1650s he and his elder brother discovered a new method of grinding and polishing lenses, and with one of his lenses he 1655 discovered a Saturn satellite and the rings of Saturn. Huygens also published work on probability theory, stated that the centre of gravity moves uniformly in a straight line, and found the mathematical expression for the centrifugal force. But he disagreed with Newton's theory of gravity.

Joule, James P. (1818–1889)

English physicist who found that the power lost in a resistor is given by $P = R\,I^2$, a law known as Joule's law. He found that the heat produced by different kinds of energy was proportional to the energy expended, and thus established the equivalence between heat and mechanical energy. In 1853 Joule and William Thomson (Lord Kelvin) observed that when an ideal gas expands at a low temperature without performing work (isenthalpic expansion), its temperature falls. This phenomenon, which today is known as the Joule-Thomson effect, is applied in cryogenetic work, ie work done in the temperature range 0–100 K.

Kelvin, William Thomson (1824–1907)

William Thomson, who became Lord Kelvin of Largs (Scotland) in 1892, was one of Great Britain's foremost scientists and inventors. He published more than 650 scientific papers and patented some 70 inventions. Thomson was born in Belfast, Ireland, and in 1832 the family moved to Glasgow, Scotland. Young Thomson entered the university when he was ten. He is best known for developing the Kelvin temperature scale and for observing the Joule-Thomson effect. His participation in the transatlantic submarine cable project in 1866 formed the basis of a large personal fortune.

Kepler, Johannes (1571–1630)

German astronomer and mathematician who used the astronomical observations of Tycho Brahe to formulate the three laws of planetary motion that are named after him. Kepler studied optics and developed the concept of rays. He also improved the early telescopes, invented a convex eyepiece and discovered a means of determining the magnifying power of lenses. In his work Kepler was side-tracked by his interest in mystic notions. For example, he believed in "the music of spheres" and like many astronomers in those days he threw himself into casting horoscopes.

Kirchhoff, Gustav (1824–1887)

German physicist who, in collaboration with Robert William Bunsen, developed the science of spectrum analysis. Kirchhoff showed that each element, when heated to incan descence, produced a characteristic pattern of emission lines. This led to the discovery of cesium (in 1860) and rubidium (in 1861). His investigations of the emissive power of the radiation of a black body led Planck to a quantum hypothesis in 1900. He also formulated Kirchhoff's laws for electric circuits.

Maxwell, James Clerk (1831–1879)

Scottish physicist who made numerous contributions to the advancement of science. He is best known for the four Maxwell equations, which he published in 1873. This mathematical formulation of Faraday's theories of electricity and lines of force is one of the great achievements of 19th century physics. Independently of Boltzmann he formulated a kinetic theory of gases where the behaviour was explained with probability analysis for a large number of particles. He also calculated that the rings of Saturn must consist of a large number of small particles and proposed that light is an electromagnetic phenomenon.

Meitner, Lise (1878–1968)

Austrian-born nuclear physicist who was the theoretical physics collaborator of Otto Hahn in the research that led to the discovery of nuclear fission. She named the process fission and in 1939, with her nephew Otto R. Frisch, she published a theoretical interpretation of Hahn's experiment. With Hahn she, in 1917, discovered the most stable isotope of protactinium, element 91. She also described internal conversion and (in 1923) the Auger effect. (Auger independently discovered the effect in 1925.) She also worked with the relation between beta and gamma rays. In 1907 she moved to Berlin but, because she was Jewish, she fled to Stockholm in 1938. In 1960 Meitner retired to live in England.

Newton, Isaac (1642–1727 according to the old style calendar, 1643–1727 according to the Gregorian time)

Sir Isaac Newton is by many considered to be the single most important contributor to modern science. Several years before Leibniz he invented the differential and integral calculus, but he did not publish his results until after Leibniz had published his. Newton constructed the first reflecting telescope and worked out the mathematics of planetary motions. Halley persuaded Newton to publish his calculations, and in 1687 he published *Philosophiae Naturalis Principia Mathematica*, which probably is the most important and influential work on physics of all times. The first book of *Principia* explained Newton's three laws of motion. In the second book he stated explicit principles of scientific methods which applied universally to all branches of science, and in book three he proposed a universal gravitational force with which he could explain the causes of the tides and their major variations, the precession of the Earth's axis, and the motion of the moon and the planets. He also observed that white light from the sun changed into a spectrum of colours when the beam passed a glass prism. He believed that light consisted of small particles, or corpuscles, and that white light really was a mixture of different types of corpuscles. During his studies of optics he also observed Newton's rings. Newton retired 1693 after suffering a nervous breakdown. *Opticks* was published 1704, but the major research was done long before. It has been discovered that Newton had large amounts of mercury in his body, which is probably due to his alchemist experiments which from 1679 took a great part of his free time.

http://www.newton.cam.ac.uk/newton.html
http.//www.bbc.co.uk/history/historic_figures/newton_isaac.shtml

Pauli, Wolfgang (1900–1958)

Austrian theoretical physicist who was awarded the 1945 Nobel Prize for physics for his exclusion principle from 1925. This principle was a breakthrough in physics and chemistry that explained the structure of the periodic table of the elements in terms of the quantum theory. In 1931 he proposed the existence of a particle with no electrical charge and little or no mass. This particle, called a neutrino by Enrico Fermi, was finally observed in 1956. Following the outbreak of World War II he moved to the United States and became a US citizen in 1946. He later returned to Zürich. By many of his colleagues Pauli was looked upon as the archetype of a theorist, ignorant of practical matters. According to the "Pauli effect", stated by Gamow, all experiments self-destructed when Pauli was present in the laboratory.

Planck, Max (1858–1947)

German physicist who made significant contributions to optics, thermodynamics, statistical mechanics, physical chemistry, and other fields. In 1900 he published an equation which describes the blackbody spectrum. This equation, now called the Planck distribution, indicated that energy exists in discrete packages or quanta ($E = h\nu$), which makes Planck the originator of quantum physics. The full consequences of this revolutionary discovery was recognized many years later, and in 1918 he was awarded the Nobel Prize for physics. The fundamental constant h is now called the Planck constant. He lost his eldest son during World War I, and in 1945 his other son, Erwin, was executed for plotting to assassinate Hitler.

Röntgen, Wilhelm Conrad (1845–1923)

German physicist who in 1895 discovered X-rays, for which he received the first Nobel Prize for physics in 1901. He demonstrated the medical and scientific use of X-rays, and this knowledge quickly spread throughout Europe and the United States.

Rutherford, Ernest (1871–1937)

New Zealand-born British physicist who established the nuclear model of the atom. He distinguished two kinds of radioactivity, which he called alpha and beta, and showed that alpha particles were doubly charged helium ions. He also coined the term "half- life" of radioactive elements. In 1908 he received the Nobel Prize for chemistry (although he considered himself to be a physicist) for his investigations into the disintegration of the elements and the chemistry of radioactive substances. By bombarding metal foils with alpha particles Rutherford could in 1911 announce his version of the structure of the atom with a very small tightly packed and charged nucleus. In 1919 he was the first to artificially transmute one element into another. He bombarded nitrogen with alpha particles and found traces of hydrogen and oxygen.

Schrödinger, Erwin (1887–1961)

Austrian theoretical physicist who in 1926 invented wave mechanics, which is an alternative way, to Heisenberg's matrix mechanics, to formulate quantum mechanics. The central equation is now called the Schrödinger equation, which, in most cases, has turned out to be simpler to solve than Heisenberg's equations. In 1933 he shared the Nobel Prize for physics with Paul Dirac. From 1940 he was a professor at the Dublin Institute for Advanced Studies in Ireland. In 1956 he retired and returned to Vienna. Schrödinger published a collection of poems and the popular science book *What is life*?

Tesla, Nikola (1856–1943)

Inventor and electrical engineer who was granted more than 100 US patents. He was born i Croatia, he studied at the University of Prague and worked as an engineer in Budapest before he in 1884 moved to the United States to become a US citizen in 1889. He invented an AC power system with an efficient polyphase induction motor, far better than Edison's DC generators. He sold the patent to George Westinghouse for 1 million dollars + $1 per horsepower as royalty. With this money he founded the Tesla Electric Company and built his own laboratory. But his lab burnt to the ground in 1895 and everything was lost, because he had no insurance. He invented a high-frequency transformer, called the Tesla coil, which made AC power transmission practical. Tesla was highly eccentric and absolutely impractical with money. In the 1880s he quit working for Edison and when he found no other job he had to dig ditches for two years.

Volta, Alessandro (1745–1827)

Italian physicist who is best known for his invention from 1800 of the voltaic pile, the first electric battery. He used plates of copper and zinc separated by disks of cardboard moistened with a salt solution. He demonstrated his invention to Napoleon, who became so impressed that he appointed Volta a Count and Senator of Lombardy (a region in northern Italy). In 1775 he invented the electrophorus, a device that, once electrically charged by being rubbed, could transfer charge to other objects. He also discovered and isolated methane gas.

van der Waals, Johannes Diderik (1837–1923)

Dutch physicist who is best known for his work in physical chemistry. In 1910 he was awarded the Nobel Prize for physics for his research on gases and liquids which in 1873 led to the equation of state that has his name. With this law the existence of condensation and a critical temperature of gases could be predicted. In 1880 he formulated his "law of corresponding states" and in 1893 a theory for capillary phenomena. His equation of state assumes weak forces between molecules, now called "van der Waals forces".

Watt, James (1736–1819)

Scottish engineer and inventor who made several major improvements on the inefficient steam engine of his time. His first patent in 1779 was a chamber for condensing the steam. Although Watt did not invent the steam engine, his improved engine was really the first practical device for efficiently converting heat into useful work. This was a key stimulus to the Industrial Revolution. In 1774 Watt emigrated to England and by 1790 he had earned enough money to retire to his estate near Birmingham.

E. Events in the History of Physics and Engineering

4000 BC The wheel is invented.

2500 BC Wagons are used in China and Egypt.

440 BC Greek philosophers Leucippus and Democritus maintain that all matter consists of almost infinitely tiny and indivisible particles which they call atoms (= "indivisible").

250 BC Archimedes describes the principle of the lever in exact mathematical terms.

1200 The optics of (convex) lenses is developed.

1512 Nicolaus Copernicus discretely circulates *Commentariolus* where he states that the earth orbits the sun.

1543 Nicolaus Copernicus publishes *De revolutionibus*, his complete theory of a heliocentric solar system.

1585 Galileo Galilei makes experimental tests of Aristotelian mechanics, resulting in total refutation of ancient conceptions.

1590 A two-lens microscope is developed and manufactured by Dutch lensmakers Hans and Zacharias Jansen.

1592 Galileo Galilei invents the first thermometer.

1609 Galieo Galilei improves the telescope invented by Hans Lippershey in Holland 1608 and makes epoch-making astronomical observations.

1609 Johannes Kepler publishes his first law of planetary motion.

1656 Christiaan Huygens invents the pendulum clock.

1666 Calculus is invented by Isaac Newton. In 1673 Gottfried Wilhelm Leibniz does the same, but with a superior notation which is still used.

1676 Christiaan Huygens proposes a wave theory of light.

1687 Isaac Newton publishes *"Principia"* (*Mathematical Principles of Natural Philosophy*), which includes Newton's laws of motion and his theory of gravitation.

1704 Isaac Newton publishes his theory of colour and light in *Opticks*.

1705 Edmund Halley predicts that the comet of 1682 will return in 1758.

1738 Daniel Bernoulli publishes his law concerning the pressure of flowing fluids.

1747 Benjamin Franklin in America and William Watson in England independently conclude that all materials possess an electrical "fluid" which neither can be created nor destroyed. Soon after they both originate the principle of conservation of charge.

1765 James Watt's first modification of Thomas Newcomen's steam engine.

1785 French physicist Charles Coulomb formulates the law of forces between electric charges, now known as Coulomb's law.

1798 The metre and the kilogram are defined by the French Academy of Sciences.

1800 The electric battery is invented by Alessandro Volta.

1801 Thomas Young conceives the principle of optical interference.

1805 John Dalton states that matter is made up of very small particles or "atoms".

1814 Augustin Fresnel establishes a wave theory of light.

1820 Danish physicist Hans Christian Ørsted discovers electromagnetic induction.

1821 Michael Faraday constructs the first electric motor.

1825 The first railroad locomotive, the *Active*, constructed by George Stephenson from England, pulls railroad cars carrying a total of 450 people.

1825 English chemist Humphry Davy invents a safety lamp for miners.

1826 German physicist Georg Simon Ohm discovers his law of electric currents.

1839 The French Academy of Sciences announces the photographic process of Louis J. M. Daguerre.

1842 Austrian physicist Christian Doppler observes the phenomenon today known as the Doppler effect.

1848 William Thomson (Lord Kelvin) proposes an absolute temperature scale.

1862 French physicist Jean Foucault successfully measures the speed of light.

1866 Swedish inventor Alfred Nobel invents dynamite.

1866 James Maxwell and Ludwig Boltzmann independently of each other obtain the Maxwell-Boltzmann kinetic theory of gases.

1869 Russian chemist Dmitry Mendeleyev proposes the Periodic System of the elements.

1873 James Clerk Maxwell publishes *A Treatise on Electricity and Magnetism* which includes his four basic equations for electric and magnetic fields.

1873 Johannes Diderik van der Waals derives an equation of state for non-ideal gases.

1876 Alexander Graham Bell invents the telephone.

1877 Thomas Alva Edison demonstrates the first "talking machine" (a phonograph). Subsequently the gramophone is patented in 1887 by Emile Berliner.

1879 Thomas Alva Edison invents the first commercially practical incandescent light bulb.

1887 Albert Michelson and Edward Morley fail to detect the ether, which is probably the most significant negative experiment in the history of science.

1895 X-rays are discovered by Wilhelm Conrad Röntgen.

1896 Henri Becquerel observes radioactivity.

1897 British physicist Joseph John Thomson discovers the electron while experimenting with radiation from cathode ray tubes.

1898 Marie and Pierre Curie discover radium.

1900 Max Planck announces his theoretical research on blackbody radiation.

1901 Radio signals transmitted from Cornwall, England, are successfully received at Newfoundland by Italian physicist Guglielmo Marconi. This is the breakthrough for radio communication.

1903 Orville and Wilbur Wright become the first to fly a machine heavier than air.

1905 Albert Einstein explains the photoelectric effect as an effect of light behaving as light quanta with energy dependent on the frequency.

1905 Albert Einstein provides convincing evidence that the Brownian motion is a proof of the existance of atoms and molecules.

1905 Albert Einstein publishes his theory of special relativity.

1905 Albert Einstein proposes the equivalence between mass and energy.

1908 Dutch physicist Heike Kamerlingh-Onnes produces liquid helium.

1911 Heike Kamerlingh-Onnes disovers superconductivity.

1911 Ernest Rutherford discovers the atomic nucleus.

1913 Niels Bohr publishes his theory of atomic structure and emission of light.

1915 Albert Einstein publishes his general theory of relativity.

1923 Arthur Compton discovers that X-rays can behave as particles.

1923 Louis de Broglie proposes a wave theory of material particles.

1923 Edwin Hubble shows that the Andromeda nebula is far outside our galaxy.

1925 Wolfgang Pauli proposes his exclusion principle.

1926 Erwin Schrödinger introduces wave mechanics (quantum mechanics).

1926 Scottish engineer John Baird transmits television pictures.

1927 Werner Heisenberg develops his principle of uncertainty.

1927 Belgian priest (!) and astronomer Georges Lemaitre proposes the big bang theory for the origin of the universe.

1928 Paul A. M. Dirac proposes a relativistic quantum theory.

1929 Edwin Hubble finds evidence for an expanding universe.

1930 Paul A. M. Dirac predicts the existence of antimatter.

1930 Pluto, the nineth planet, is discovered by Clyde Tombaugh.

1931 Wolfgang Pauli proposes the existence of a neutral particle (the neutrino) emitted in beta decay.

1931 American physicist Ernest Lawrence builds the first working cyclotron.

1932 The positron is discovered by Carl Anderson.

1932 James Chadwick discovers the neutron.

1932 The German engineer Ernst Ruska constructs the first electron microscope.

1934 Irène and Frédéric Joliot-Curie discover artificially induced radioactivity.

1935 Scottish physicist Robert Alexander Watson-Watt makes the first workable radar device.

1939 The first flight with a jet engine, constructred by German engineer Hans Pabst von Ohain.

1942 Enrico Fermi and colleagues build the first nuclear fission reactor.

1947 Englishman Dennis Gabor invents holography.

1948 John Bardeen, Walter Brattain, and William Shockley invent the first transistor.

1954 Charles Townes constructs the first operational maser.

1956 The neutrino is observed by Clyde Cowan and Frederick Reines.

1957 The Soviet Union launches the first earth-orbiting satellite – Sputnik.

1959 American physicists Jack Kilby and Robert Noyce independently of each other devise the first integrated circuits.

1960 Theodore Maiman constructs the first ruby laser, and Ali Javan constructs the first helium-neon laser.

1964 American physicists Murray Gell-Mann and, independently, George Zweig propose the quark theory of elementary particles.

1965 Arno Penzias and Robert Wilson discover cosmic microwave background radiation.

1967 A pulsar is observed for the first time by radio astronomers Jocelyn Bell Burnell and Antony Hewish at Cambridge, England.

1969 United States' astronaut Neil Armstrong becomes the first man to set foot on the moon.

1975 The first personal computer, the Altair 8800, goes on the market.

1979 The compact disk is invented in a joint project by Philips and Sony.

1983 Carlo Rubbia from Italy and collaborators observe the W and Z particles using the method of stochastic cooling invented by Simon van der Meer from Holland.

1986 Johannes George Bednorz from Germany and Karl A. Müller from Switzerland produce the first high-temperature superconductors.

1994 Evidence of the top quark is discovered by investigators at Fermilab.

1995 Bose-Einstein condensation of an atomic gas is produced by Eric Cornell and Carl Wieman at Boulder, Colorado, and, independently, by Wolfgang Ketterle at MIT, Cambridge, USA.

1998 Researchers at the Super Kamiokande neutrino detector in Japan report evidence that neutrinos have mass.
http://www.phys.hawaii.edu/~superk/ WWW

1998 By studies of distant Ia supernovae, two teams of astronomers discover that the expansion of the universe is accelerating.
http://www.noao.edu/outreach/press/9810.html WWW

F. Conspicuous Physical Properties

The following paragraphs show some examples of the remarkable fine-tuning of many physical properties.

1. **Half-life of the ^{8}Be nuclide**
   Carbon can only be produced inside stars, at a temperature of 10^8 K, through a process where three helium nuclei combine to produce one carbon nucleus. In the first step of this process two helium nuclei combine to form the beryllium nuclide ^{8}Be, which then combines with another helium nucleus to form carbon.

   In isolation ^{8}Be would decay in about 10^{-16} seconds. If the half-life of ^{8}Be had been much shorter, carbon could not have been formed and life, as we know it, would not exist.

2. **The value of the gravitational constant**
   The luminosity of stars is proportional to the seventh power of the gravitational constant G. This means that if G were twice its actual value, the stars would be 128 times brighter, and therefore their lives would be 128 times shorter, and thus life highly improbable. On the other hand, if G were smaller than it is, the formation of stars and planets would be a lot more difficult or even impossible.

3. **The slow creation of deuterium**
   The existence of long-lived sources of energy (the stars) is a consequence of the slow conversion of hydrogen into helium. Hydrogen is still the most abundant element in the universe because only some 30 % of it was converted into helium during the first three minutes of the cosmic evolution. This is due to the fact that the nuclear forces between protons are just strong enough to allow some nuclear reactions, but not too strong to make all protons combine into deuterium and helium nuclei.

   If almost all hydrogen had been turned into helium, the ordinary stars would not exist today. If helium stars had been created during the early years of the universe, they would have produced energy by "burning" helium. Such stars are, however, very short lived. In their lifetime, maybe 100 million years, it would not have been possible for life to develop.

   In the interior of stars, deuterium is produced by the conversion of a proton into a neutron by the reaction $p \rightarrow n + e^+ + \nu$, and hence $p + p \rightarrow D + e^+ + \nu$. This reaction is governed by the weak force, and is extremely slow. Other nuclear reactions are 10^{18} times faster, and if this reaction were as fast as other nuclear reactions, the sun would have exploded in a very short time, almost like a giant hydrogen bomb.

4. **The opacity of the outer layers of stars and the strength of electromagnetism**
Another factor that ensures the long lives of stars is the opacity of their outer layers. In this area the electromagnetic forces dominate the particles and ensure that the radiation coming from the nucleus is absorbed, and thus high temperatures are maintained in the interior of the stars. If it were not for this absorption, the sun would have been extinguished within a day.

   If the strength of electromagnetism had been slightly stronger, then the main sequence of stars, on which stars spend most of their lives, would consist entirely of red dwarfs, losing heat mainly by convection and life-discouragingly cold. If electromagnetism had been slightly weaker, then all main sequence stars would be blue, i.e. very hot and short-lived.

5. **The mass of electrons**
An increase of the electron's mass by a factor of 10 would have made the energies required for the various organic reactions of life ten times greater and thus much more difficult to attain. On the other hand, a reduction of the electron's mass by a factor of ten would have made organic molecules very unstable, being immediately destroyed by solar radiation.

6. **The mass of protons and neutrons**
If the mass of the proton were somehow increased by 0.2 %, the proton would, like the free neutron, become an unstable particle. Now the neutron is slightly heavier than the proton. If the mass of neutrons were slightly less, neutrons outside atoms would not decay, and then all protons would have changed irreversibly into neutrons during the Big Bang.

7. **The strength of the nuclear strong force**
If the nuclear strong force were slightly stronger (~ 2 %) the formation of protons out of quarks would be blocked, and hydrogen, the basis for production of all atoms, would not exist. On the other hand, if the nuclear force were a little weaker (~ 5 %) a neutron would not be able to bind to a proton to create deuterium, and no other elements apart from hydrogen would exist.

8. **The strength of the weak force**
If the weak force had been appreciably stronger, then the Big Bang's nuclear burning would have proceeded past helium and all the way to iron. Fusion-powered stars would then be impossible.

   Elements heavier than iron are produced in supernova explosions, as the exploding stars lose their outer layers that are rich in heavy elements. These layers are blasted off by neutrinos that interact with them via the weak force alone. Its extreme weakness allows neutrinos to escape from a supernova's collapsing core. Still, the force is just strong enough to hurl into space the outer-layer atoms needed for constructing heavier elements.

A weakening of the weak force by a factor of ten would have led to a universe consisting mainly of helium, and in such a universe supernova explosions would not occur.

9. Properties of water

Almost all liquids become heavier as the temperature decreases, but water, which is important for the form of life that we know, has its maximum density at 4 °C. This means that seas and lakes freeze from the surface and not from the bottom. Further, the fact that the solid state of water is less dense than the liquid state (so ice floats) is a unique property in nature. This property of water might not be absolutely necessary for life, but if ice were denser than water, it would sink to the bottom of lakes and oceans and remain there until most of the reservoir had been solidly frozen.

Water also has a higher specific heat than almost any organic compound. This property allows water to be a store of heat and so stabilize the environment. The thermal conductivity of water is also higher than that of most liquids, which again permits water to act as a temperature stabilizer for the environment. Water has, moreover, a higher heat of vaporization than any known substance. This makes water the best possible coolant by evaporation, and living creatures make extensive use of it for their temperature control.

Main rcferences:

http://www.leaderu.com/truth/3truth12.html

Cosmology – The Structure and Evolution of the Universe, G Contopoulos and D Kotsakis, Springer-Verlag, 1987. ISBN 3-540-16922-9.

Addenda

On the following blank pages you may write formulae which you cannot find elsewhere in the Physics Handbook and attach additional tables and diagrams that are needed.

Comprehensive physics encylopedias can (in December 2005)
be found at
http://scienceworld.wolfram.com/physics
http://hyperphysics.phy-astr.gsu.edu/hbase/hframe.html

Addenda

Addenda

Index